International Series in Operations Research & Management Science

Volume 358

The book series **International Series in Operations Research and Management Science** encompasses the various areas of operations research and management science. Both theoretical and applied books are included. It describes current advances anywhere in the world that are at the cutting edge of the field. The series is aimed especially at researchers, advanced graduate students, and sophisticated practitioners.

The series features three types of books:

- Advanced expository books that extend and unify our understanding of particular areas.
- Research monographs that make substantial contributions to knowledge.
- Handbooks that define the new state of the art in particular areas. Each handbook will be edited by a leading authority in the area who will organize a team of experts on various aspects of the topic to write individual chapters. A handbook may emphasize expository surveys or completely new advances (either research or applications) or a combination of both.

The series emphasizes the following four areas:

Mathematical Programming: Including linear programming, integer programming, nonlinear programming, interior point methods, game theory, network optimization models, combinatorics, equilibrium programming, complementarity theory, multiobjective optimization, dynamic programming, stochastic programming, complexity theory, etc.

Applied Probability: Including queuing theory, simulation, renewal theory, Brownian motion and diffusion processes, decision analysis, Markov decision processes, reliability theory, forecasting, other stochastic processes motivated by applications, etc.

Production and Operations Management: Including inventory theory, production scheduling, capacity planning, facility location, supply chain management, distribution systems, materials requirements planning, just-in-time systems, flexible manufacturing systems, design of production lines, logistical planning, strategic issues, etc.

Applications of Operations Research and Management Science: Including telecommunications, health care, capital budgeting and finance, economics, marketing, public policy, military operations research, humanitarian relief and disaster mitigation, service operations, transportation systems, etc.

This book series is indexed in Scopus.

Teodor Gabriel Crainic • Michel Gendreau •
Antonio Frangioni
Editors

Combinatorial Optimization and Applications

A Tribute to Bernard Gendron

Springer

Editors
Teodor Gabriel Crainic
CIRRELT and Département d'analytique,
opérations et technologies de l'information,
École des sciences de la gestion
Université du Québec à Montréal
Montréal, QC, Canada

Michel Gendreau
CIRRELT and Département de
mathématiques et de génie industriel
Polytechnique Montréal
Montreal, QC, Canada

Antonio Frangioni
Dipartimento di Informatica
Università di Pisa
Pisa, Italy

ISSN 0884-8289 ISSN 2214-7934 (electronic)
International Series in Operations Research & Management Science
ISBN 978-3-031-57602-7 ISBN 978-3-031-57603-4 (eBook)
https://doi.org/10.1007/978-3-031-57603-4

This Springer imprint is published by the registered company Springer Nature Switzerland AG
The registered company address is: Gewerbestrasse 11, 6330 Cham, Switzerland

Contents

Bernard Gendron and Operations Research

Teodor Gabriel Crainic, Antonio Frangioni, and Michel Gendreau

1 Introduction

In July 2022, we lost our dear friend and colleague, Professor Bernard Gendron, who made major contributions with long-lasting impacts to the science and practice of *Operations Research* (*OR*) and *Combinatorial Optimization* (*CO*). This book aims to be both a fitting way to honour the memory and scientific legacy of Professor Gendron, and a useful reference for researchers and graduate students in OR, CO, and related fields.

Inspired by the exceptional contributions made by Professor Gendron, and written by acknowledged researchers who collaborated closely with him, the book presents a state-of-the-art view of the field through a combination of surveys, expository articles, and focused methodological and applied chapters. The range of topics addressed is as broad as the interests of Professor Gendron, and includes modelling, algorithmic, and application aspects of packing, facility location, network design, scheduling, routing, and transportation planning. The content of the book is

T. G. Crainic (✉)
CIRRELT and Département d'analytique, opérations et technologies de l'information, École des sciences de la gestion, Université du Québec à Montréal, Montréal, QC, Canada
e-mail: TeodorGabriel.Crainic@cirrelt.net

A. Frangioni
Dipartimento di Informatica, Università di Pisa, Pisa, Italy
e-mail: frangio@di.unipi.it

M. Gendreau
CIRRELT and Département de mathématiques et de génie industriel, Polytechnique Montréal, Montreal, QC, Canada
e-mail: michel.gendreau@cirrelt.net

T. G. Crainic et al. (eds.), *Combinatorial Optimization and Applications*,
International Series in Operations Research & Management Science 358,
https://doi.org/10.1007/978-3-031-57603-4_1

briefly presented in Sect. 3, following a short in-memoriam text. We conclude this introductory chapter with a few words of thanks.

2 Bernard Gendron (1966–2022): Friend and Colleague[1]

On July 17, 2022, our friend and colleague Bernard Gendron passed away, after an unforgiving illness. Bernard was Professor of Operations Research in the Département d'informatique et de recherche opérationnelle (DIRO—Computer Science and Operations Research) of Université de Montréal and a senior CIRRELT member (the Interuniversity Research Centre on Enterprise Networks, Logistics and Transportation).

Bernard started displaying his innate talent for research and making important contributions to Operations Research, Combinatorial Optimization, and Transportation Science while working toward his master's degree at DIRO. His master's thesis addressed a challenging, recently-defined, logistics problem through network-design modelling and a parallel exact enumerative algorithm. In this, Bernard was an important member of the team that started the Montreal research work on parallel optimization, and one of the pioneers developing such methodology to address transportation and logistics planning challenges. The Canadian Operational Research Society (CORS) acknowledged the quality of the research, and Bernard's writing skills, by awarding him the First Prize of the 1992 student-article competition.

Bernard followed with an exceptional Ph.D. dissertation, still at DIRO, on multicommodity capacitated fixed charge network design applied to planning transportation and logistics systems. The thesis included several important contributions to the understanding, modelling, and solution of network design problems, as well as to the development of parallel optimization, for which he received the 1995 Best Dissertation Award of the Transportation Science Section of INFORMS. Impressed by the quality of his work, DIRO hired Bernard as a tenure-track faculty member, even before his dissertation defence.

In the years that followed, Bernard pursued a very active academic career combining top-level research with teaching and transfer activities. In these, he showed that he was a true great Operations Research scientist, harmoniously blending applications and methodology to push the science envelope and address relevant and challenging problems of substantial industrial and societal interest. From the very beginning of his research work, Bernard was highly motivated by such applications, in particular in transportation and logistics (system and service network design for freight carriers and distribution networks, location of facilities, and forestry problems), scheduling of tasks and personnel, and fixed and wireless telecommunication networks planning and management. As it is typical of the field, addressing practical problems led him to study and develop the relevant methodological part of Operations Research and Mathematical Programming, focusing on

[1] Earlier versions of this text have been published in a number of journals and web sites.

Combinatorial Optimization and Integer Programming, e.g., graph theory, location, network design, and routing.

To address the highly challenging problems raised by the generic and applied models he studied, Bernard left no stone unturned and explored a wide range of mathematical optimization methodologies. He cunningly employed existing techniques and proposed significant methodological advances in as diverse fields as Lagrangian relaxation, column generation, Benders' decomposition, logic-based constraint programming, cutting planes and enumerative (Branch & Bound & Cut) algorithms, and meta- and math-heuristic solution methods. Since the early days of his career, he was a pioneer and an avid proponent of parallel optimization.

In applied and methodological work alike, Bernard deployed his creativity and considerable knowledge to both modelling and algorithmic development. For the applications he addressed, Bernard searched for the mathematical formulation that would simultaneously capture the essence of the problem at hand and be amenable to an efficient solution method, in order to yield the numerical results required by the application. To reach this goal, Bernard applied the significant understanding and tools developed studying the more abstract, generic problem classes in Combinatorial Optimization, reformulation techniques and polyhedral analysis, in particular.

Bernard had both a passion and a special knack for continuously revisiting so called "old" models and problems, and finding new and original reformulations and solution approaches. He never stopped studying general models with many practical applications, and proposing novel ways of viewing, decomposing, and approaching them. The many variants of the multicommodity network flow and design problems formed one of the fields of work about which he was most passionate, yielding some of his more important and long-lasting contributions to science and Operations Research. The chapters detailing the various network design formulations (Crainic et al. 2021) and exact solution methods (Crainic & Gendron 2021), in the "Network Design with Applications to Transportation and Logistics" book he recently co-edited (Crainic et al. 2021), synthesizes his immense heritage in the field. This unabated curiosity is the hallmark of a truly exceptional researcher, such as he was.

A large part of Bernard's work was performed in the context of very lively international collaborations with colleagues from top institutions and companies from around the world. These include MIT in Boston, EPFL in Lausanne, ILOG in Paris, École Centrale in Lille, and the universities of Pisa, Lisbon, Nice-Sophia-Antipolis, Clermont-Ferrand (Blaise-Pascal), Versailles, and Valenciennes, among others. These international collaborations also gave rise to several participations in specialized doctoral schools as both organizer and lecturer.

Bernard's talent as a researcher created many interesting opportunities. Among these, he held two important chairs: an industrial one on data intelligence for logistics, supported by the company Purolator and the Natural Sciences and Engineering Research Council of Canada, and a chair on green technologies, funded by the government of Quebec, dealing with the transformations in the transportation sector.

Besides his primary interest in science and research, Bernard cared immensely for the scientific and academic community, both in Montreal and abroad. Over time, this led him to accept a number of duties and tasks in a variety of roles. In 2008, Bernard was selected to become the first director of CIRRELT, the Interuniversity Research Centre on Enterprise Networks, Logistics and Transportation. Born from the merger of various research groups in five different universities, CIRRELT flourished under Bernard's skillful stewardship of eight years and became one of the leading research centres in logistics and transportation in the world. A few years later, Bernard was asked to become Adjunct Vice-Rector of Research, Discovery, Creation, and Innovation of Université de Montréal in 2021. Unfortunately, illness prevented him from accomplishing as much as he aimed for in that position.

Bernard displayed the same energy and gift for stewardship with respect to learned societies, which brought him great appreciation and respect in the community. He held many positions within CORS, from President of the Montreal section to Council member to President (2004–2005). CORS recognized Bernard's outstanding contributions to Canadian Operations Research by bestowing him three major distinctions: the Award of Merit (2010), the Service Award (2006), and the Practice Prize (2004).

Bernard was also deeply committed to the INFORMS community, in particular those interested by transportation and logistics and the interplay between Operations Research developments and the performance of applications. He was thus first elected President of the Transportation Science Section of INFORMS (2001–2002, after serving as Vice-President, 2000–2001, and Secretary-Treasurer, 1998–2000), before becoming the first President of the Section on Transportation Science and Logistics (2002–2003).

Bernard was very active in scientific publishing. After serving as Associate Editor of *INFOR* from 2004 to 2006, he was its Editor in Chief from 2006 to 2014. He has also been Area Editor in "Combinatorial Optimization" of *RAIRO-Operations Research* since 2019, and a member of the editorial boards of *Constraint Programming Letters* and the *EURO Journal on Transportation and Logistics* since 2006 and 2012, respectively. Always at the forefront of transformations, he was also one of the founders and the Section Editor in "Computational Aspects and Applications" of the *Open Journal of Mathematical Optimization*, the first completely Gold Open Access journal in the OR area. Bernard was also involved in the organization of several conferences and workshops as member of their organizing or scientific committees. These scientific events include, among others, the *CORS/INFORMS International Meeting* of 2009, the *IFORS Triennial Meeting* in 2017, as well as several editions of the *Odysseus*, *CPAIOR*, and *Optimization Days* conferences.

Throughout his career, Bernard has been deeply committed to his students, both graduate and undergraduate. Bernard not only shared his knowledge, but he also wanted to give them the opportunity to grow as scientists and human beings. Younger colleagues also recall his reassuring presence as a mentor and an example of a true academic.

All those who have known Bernard have a deep respect for his scientific abilities and his numerous professional accomplishments, but they especially dearly remember the man who Bernard was: warm, smiling, and with an incomparable laughter. His hard-working attitude never was disjoint from his heart-felt, true respect of others; together with his renowned sense of humor, this made him a wonderful colleague to both have full-night work stints with, and share small talk about hockey or the history of Quebec over a beer. We will sorely miss him.

3 Contents of the Book

The book is subdivided into two parts. The first one comprises twelve chapters and is dedicated to methodologically-oriented developments. The chapters address some fundamental families of OR and CO problems, in particular knapsack, packing, spanning trees, facility location, and network design. They focus on models and integer-programming formulation issues, which increase the difficulty to address the corresponding problem, e.g., multiple network echelons, interwoven layers of decision, and complex capacity-utilization measures or cost structures, as well as on associated algorithmic challenges and developments, including decomposition algorithms, comprising Lagrangian (dual) and Benders' (primal) approaches.

The seven chapters of Part II of the book are dedicated to developments that are more closely tied to specific real-world applications in complex settings, including health-staff scheduling and routing, city logistics, fleet management and scheduling, pricing and location of infrastructure, and optimization of public transport systems. We now provide a more detailed introduction to each of the chapters, where we also highlight the connection of the corresponding research line with the work of Bernard Gendron. In so doing, we have greatly benefited from the material provided by the authors of each chapter. Hence, this introductory chapter owes in fact to all the authors of the book.

3.1 Part I: Methodological Developments

Chapter 2, *Variable Neighbourhood Search with Dynamic Exploration for the Set Union Knapsack Problem*, by Igor Machado Coelho, Saïd Hanafi, Raca Todosijevic, Bernard Gendron, and Mustapha Ratli. The authors address a problem setting where one is given a set of elements, each with a positive weight, and a set of items, each with a positive profit and a corresponding set of associated elements. An item may be included in the knapsack only if all its associated elements are included. The goal is to select the items to be included in the knapsack such that the total profit of the selected items is maximized and the capacity of the knapsack is not exceeded. The chapter proposes a Variable Neighbourhood Search meta-heuristic and shows, on a set of benchmark instances from the literature, that this method is competitive with the state-of-the-art heuristics for the problem. The research presented in this chapter was initiated during the period when Bernard Gendron, a friend rather than just a colleague for the authors, was co-director of the Laboratoire International Associé LIA/IRP between LAMIH CNRS 8201 and CIRRELT, and extended within the framework of the French chair in Brazil of Saïd Hanafi. Bernard Gendron was always interested in addressing advanced knapsack problems through meta-heuristics, so the authors believe themselves fortunate to have had his valuable contributions to the results presented in this chapter.

Chapter 3, *Common-Flow Formulations for the Diameter Constrained Spanning Tree Problem*, by Luis Gouveia, Markus Leitner, and Ivana Ljubic. This chapter proposes and computationally evaluates new formulations for specific variants of the Spanning and Steiner Tree Problems with diameter constraints. The research described in the chapter is closely related to important parts of the scientific work of Bernard Gendron, who has laid important foundations for many network optimization problems by making invaluable contributions in studying and improving standard multi-commodity flow formulations for various network design, location, and routing problems. In particular, Bernard Gendron has shown in some of his seminal works how disaggregation techniques can be applied to solve network flow problems with piecewise linear costs, and the concepts of common and uncommon flows studied in this chapter can be seen as an application of disaggregation techniques to multicommodity flows.

Chapter 4, *Models and Methods for Two-Level Uncapacitated Facility Location Problem*, by Bernard Gendron, Paul-Virak Khuong, and Frédéric Semet. This chapter is a synthesis of the research work carried out by Bernard Gendron with his co-authors over 15 years on two-level facility location problems. The initial motivation to explore this line of research came from an applied project performed in cooperation with a multi-channel company to design its distribution network. In addition to the scientific results obtained, the chapter highlights some of the scientific interests of Bernard Gendron. First, the initial motivation for this research is an illustration of how, as an operations researcher, he was deeply interested in applying OR techniques to address real-life problems. Second, he liked to focus on modelling and studying the properties of the models. He has never stopped questioning and proposing new modelling ideas. The chapter describes the extensive work done on the different models for two-level facility location problems and the linear relaxation bounds that can be obtained. Last, the chapter presents a Lagrangian-based approach to the problem considered, which was one of Bernard Gendron's favorite techniques, since it can lead to very efficient matheuristics with a performance guarantee. This work provides foundations on which future research can be conducted along two lines. First, regarding the problem extension to the case where the distribution network design is based on contracts with different carriers in a multi-period context. Second, in terms of methods, the study of new approaches to efficiently address the two-level facility location problem with modular costs has great practical interest.

Chapter 5, *Facility Location: A Guide to Modeling and Solving Complex Problem Variants via Lagrangian Relaxation*, by Sanjay Dominik Jena. The author wishes to dedicate this chapter to Bernard Gendron, who was not only his former Ph.D. supervisor, but especially a colleague and a friend. The topic of this chapter mirrors the focal point of their joint research: Lagrangian relaxation methods for complex facility location problems, driven by industrial requirements. The work in this chapter extends beyond the previous joint research, including some recent work inspired, in one way or another, by Bernard Gendron's enduring influence, and in particular by his passion and relentless pursuit for stronger mixed-integer formulations, facility location, and Lagrangian relaxation.

Chapter 6, *Bin Packing Problems for Capacity Planning and Last Mile Applications*, by Maria Elena Bruni, Teodor Gabriel Crainic, and Guido Perboli. This chapter is focused on the representation of capacity restrictions within consolidation-based freight transportation and logistics planning, and the important role bin packing methodologies may play to enhance the accuracy and relevance of OR models and methods for the strategic and tactical planning of such firms and systems. The research work touches directly two of the major fields of interest and contributions of Bernard Gendron: location problems and, especially, network design. It also recalls his continuous desire to explore new modelling and algorithmic avenues. The authors heartfeltly dedicate this chapter to Bernard Gendron, who was not only a colleague but a dear friend. In their own words: "While writing it, the many opportunities of discussing these scientific issues with him came to mind: seminars, conferences, but also over lunch or dinner and while we were gathered around the coffee machine or sipping a glass of wine. We think that the best way to remember Bernard and his legacy is to recall his heart; the heart that gave him the passion to move forward in research when a solution seemed out of reach; the heart with which he shared his knowledge with students; the heart with which he gave himself to friends and colleagues; the heart with which he loved his wife and family. We are certain that reading these lines now he will be laughing, with that contagious laughter we all remember so well. He is for ever in our minds and hearts."

Chapter 7, *Models for Network Flow and Network Design Problems with Piecewise Linear Costs*, by Bernard Fortz, Bernard Gendron, and Luis Gouveia, presents a review of some pioneering work of Bernard Gendron on multicommodity flow and network design problems with piecewise linear costs. The chapter emphasizes the importance of methodological advances in developing models for these problems, pointing out relevant modelling contributions by Bernard Gendron on this topic. In particular, the ideas of variable and constraint disaggregation, allowing to develop structural valid inequalities or to eliminate variables in an efficient pre-processing step, are illustrated with several examples. The chapter therefore provides an introduction to, and complements, other chapters of the book that emphasize the careful and innovative ways in which Bernard Gendron has developed and applied decomposition techniques such as Lagrangian relaxation and Benders' decomposition to these models.

Chapter 8, *A New Technique for Modeling Piecewise Linear Costs in Network Design Problems*, by Mike Hewitt and Fabien Lehuédé. The chapter is concerned with modelling piecewise linear cost functions within Scheduled Service Network Design (SSND) problems, a setting often encountered in planning consolidation-based transportation systems, in particular when the unit rates to be paid depend on the quantity of flow to be shipped. The authors show the great potential of the approach they propose compared to the classical ones, in a line of research that follows on and extends Bernard Gendron's work on network optimization with piecewise linear costs, from its pioneering results on linear relaxation, disaggregation, and cutting-plane approaches (Croxton et al. 2003a,b), to generalizations to Network Flow and Multicommodity Capacitated Network Design (MCND) problems (Croxton et al. 2007; Frangioni & Gendron 2009; Gendron & Gouveia 2017; Frangioni & Gendron 2021). Recalling that the SSND is a MCND defined on the time-space network, the authors dedicate their work "to Bernard in memory of both his kindness and research".

Chapter 9, *Multi-layer Network Design for Consolidation-based Transportation Planning*, by Teodor Gabriel Crainic. Multi-layer networks are used to model complex, multi-component systems, frequently encountered in transportation and telecommunication planning, each component being represented by a network layer and an arc in a given layer being defined with respect to a set of arcs in one or several other layers. Multi-layer network design aims to simultaneously design all layer-specific networks to satisfy at minimum overall cost a given set of origin-destination demands. The chapter recalls and enhances the basic definitions and formulations (Crainic et al. 2022), discusses the challenges arising from the connectivity relations and requirements among the design, flow, and attribute decision variables associated with the arcs of interconnected layers, and proposes new formulations for rich networks, with more than two layers and connectivity relations involving several layers simultaneously. The chapter is closely related to the work of Bernard Gendron. In the words of the author "Bernard was my first and one of the brightest PhD students I had the privilege to supervise. We became very good friends and research partners, exploring together many exciting areas of OR, including location (e.g., Gendron and Crainic 1995), parallel optimization (e.g., Gendron and Crainic 1994a,b; Bourbeau et al. 2000), and network design (e.g., Crainic et al. 2021; Crainic and Gendron 2021). In research, Bernard aimed, "locally", for a refined and relevant representation of the problem at hand supported by a powerful solution approach, and, "globally", for a better understanding and continuing development of our profession's methodological treasure chest, as illustrated in Crainic et al. (2022)".

Chapter 10, *Separable Lagrangian Decomposition for Quasi-Separable Problems*, by Antonio Frangioni, Bernard Gendron, and Enrico Gorgone, compares different forms of decomposition for problems whose Lagrangian relaxations can be efficiently computed due to partial, but not complete, separability, once the complicating constraints are relaxed; a new reformulation of the master problem, related to the Lagrangian Decomposition approach, is shown to be computationally superior to both non-separable and trivially-separable versions. The chapter builds on the very specific knack—it could perhaps even be called a passion—that Bernard Gendron had for decomposition methods and the very many different structures that they can reveal even for the same problem. This started early in his career (Crainic et al. 2001), it was a constant line of research, yielding very significant methodological results (e.g., Frangioni and Gendron 2009; Frangioni and Gendron 2013), and possibly reached a culmination in later works (e.g., Crainic et al. 2022), where new forms of decomposition are proposed for the very same "classical" problem, displaying a remarkable capability of innovating even for problems studied for more than 20 years. This has been one of the leading sources of inspirations leading to the development of the ambitious `SMS++` project (Frangioni et al. 2024), which precisely aims at making this form of sophisticated solution algorithms more accessible to researchers and practitioners. In this, as in many other cases, the legacy of Bernard Gendron's work will keep inspiring future researchers.

Chapter 11, *Decomposition-based Algorithms for Mixed-integer Linear Programs with Integer Subproblems*, by Mariá-Isabel Restrepo Ruiz, Bernard Gendron, and Louis-Martin Rousseau. This chapter studies a special class of problems whose Benders' reformulation exhibits integer variables both in the master problem and the subproblems, while a number of master-problem variables are subject to cardinality constraints. Two novel decomposition-based algorithms exploiting that structure are proposed to address these problems. Their efficiency and effectiveness are tested on two classes of problems: multi-commodity facility location and multi-activity shift scheduling problem. The work in this chapter came from the wish to extend and test some of the ideas developed in the first author's Ph.D. thesis, and thus address two of the main research interests of Bernard Gendron: the development of decomposition methods and the solution of practical large-scale problems. As often in the work of Bernard Gendron, the ideas and approaches proposed therein will be useful to address other practical problems with similar characteristics, continuing to honour the memory of Bernard Gendron and the numerous contributions he made to OR.

Chapter 12, *Perspectives on Using Benders Decomposition to Solve Two-Stage Stochastic Mixed-Integer Programs*, by Mike Hewitt and Walter Rei, discusses accelerating enhancements to the standard Benders' decomposition approaches applied to two-stage stochastic programs. Two strategies are proposed and studied related to, either transferring information from the subproblems to the master, thus strengthening the latter's formulation, or applying the reverse approach sending information from the master to the subproblems, thus improving the quality of the cuts that can be obtained from them. This is related to Bernard Gendron's research, widely recognized for numerous contributions to the field of discrete optimization, particularly for the innovative work on integer programming methods based on decomposition strategies for fixed-charge network design formulations. Although the work presented in this chapter was not done in direct collaboration with Bernard Gendron, it embodies the kind of methodological developments he championed.

Chapter 13, *Decomposition Methods for Choice-Based Optimization Models*, by Shadi Sharif Azadeh, Meritxell Pacheco and Michel Bierlaire. In Choice-Based Optimization (CBO) problems, individual's decisions are explicitly incorporated in the CO framework. The probabilistic expression of the choice model with endogenous variables, such as price and capacity, causes nonlinearity and non-convexity. A Lagrangian decomposition method, coupled with scenario decomposition and scenario grouping, is proposed to address his challenge. This work originated with a collaboration with Bernard Gendron, which started in 2015, when he was having his sabbatical at the École Polytechnique Fédérale de Lausanne (EPFL). In the words of the authors, "his vast and deep knowledge of large-scale optimization was crucial for investigating different decomposition methods for the CBO framework, and eventually settling on Lagrangian decomposition, since this approach allows to address the tractability of the CBO problems for large-scale instances. Additionally, Bernard Gendron suggested that the proposed method can potentially be combined with other decomposition techniques (e.g., Benders' decomposition), according to the model requirements, which will in the future allow to further enhance and expand the applicability of the method to the application areas discussed in the chapter introduction, such as revenue management and transportation-related problems".

3.2 Part II: Application-Oriented Developments

Chapter 14, *The Static Elevator Dispatching Problem with Destination Control*, by Camille Richer, Michel Bierlaire and Fabian Torres, deals with vehicle dispatching in vertical transportation systems, where individuals must be assigned to vehicles to be transported between different floors. The chapter focuses on the case of perfect information on passenger arrivals and their origin and destination floors. The authors present a novel network-flow model with two objectives, reducing the average passenger-journey times and cutting down on the energy use of the elevator group. The sensitivity of the problem to parameter variation is studied and possible gains in energy consumption are identified. In the words of the authors, "Bernard Gendron and Michel Bierlaire have been close friends since 1993, when they were working on their Ph.D.s. Their paths have crossed many times, mainly during conferences around the world, but it was only recently that they had the opportunity to work together on a research project, during Bernard's academic visit to EPFL in 2016. Some of this research is published and is also the subject of Chap. 13 in this book. Although Bernard was not directly involved in it, the contents of this chapter are well aligned with some of Bernard Gendron's research, in particular scheduling and transportation problems. He was interested in mixed integer programming and network-flow formulations that were primarily applied to transportation and networks. The authors would have loved to have his opinion on this complex vertical transportation system, as it exemplifies the type of problems he liked to solve."

Chapter 15, *Flow-based Robustness in Consistent Home Care Service Delivery*, by Paola Cappanera and Maria Grazia Scutellà. This chapter addresses a complex planning problem that arises in home-care service delivery. In a multi-period time horizon, closely related decisions need to be made: scheduling of patients' requests, patient-caregiver assignment, and caregiver routing. The solutions must be robust with respect to demand variability and ensure service regularity (consistency), which frail patients and their families particularly value. Robustness is addressed through a non-standard cardinality-constrained exact model, as opposed to an existing approximate model in the literature. In the proposed exact formulation, the uncertainty paradigm is modelled in terms of a constrained longest path on a suitably defined auxiliary graph, thus testifying the power and the relevance of network design in addressing applications related to the fields of logistics, transportation and, more specifically, health care. The authors acknowledge the outstanding results achieved by Bernard Gendron in network design models and methodologies, and his capabilities to cleverly apply them in many relevant application contexts, as testified, among other things, by the profound scientific discussions they had with him, which have been a guide for the results presented in this chapter.

Chapter 16, *Production Inventory Technician Routing Problem: a Bi-Objective Post-Sales Application*, by Alline Zanette, Michel Gendreau and Walter Rei. The chapter considers a complex problem inspired from a real-life application in post-sales maintenance services. The problem under study combines features of production-routing and technician routing-and-scheduling problems. A deterministic model aiming at minimizing costs and customer dissatisfaction is proposed and it is solved using Benders' decomposition. An extensive sensitivity analysis is performed to evaluate how much the customer satisfaction level could be improved. Bernard Gendron, with his keen interest in combinatorial problems arising in the context of supply chains, would have certainly been very interested by the production-routing problem addressed in this chapter. Furthermore, relying on a solution approach based on Benders' decomposition would have also felt very familiar to Bernard Gendron.

Chapter 17, *Express Package Delivery Optimization Using Walkers, Cargo Tricycles and Delivery Trucks*, by Patrick Meredith-Karam, Jia Hui Jiang, Sina Bahrami and Matthew J. Roorda. This chapter presents a real-world City Logistics application, using sophisticated modelling and algorithmic techniques for the optimization of express deliveries in an urban context by exploiting the synergies and complementarities among several different delivery systems. The authors state that "this work is one of the many outcomes from the Purolator Chair, a consortium of researchers from Université de Montréal, University of Toronto, U.Q.A.M., and McMaster University that Bernard Gendron led from 2018 with intellectual rigour, sincerity, and always with good nature and laughter. The research contributed to the business case for Purolator's real-life pilot e-cargo cycle program in Toronto that began in 2022. Bernard Gendron reviewed an early version of the model, gave helpful and encouraging feedback for the student researchers, and had plans to collaborate to integrate with his research and apply the work to Montreal and other cities in Canada; the authors deeply regret that he could not read the final version and see the outcomes of the Toronto e-cargo cycle pilot."

Chapter 18, *Integrated Location, Sizing, and Pricing for EV Charging Stations*, by Miguel F. Anjos, Ikram Bouras, Luce Brotcorne, Alemseged G. Weldeyesus, Clémence Alasseur, and Riadh Zorgati. The authors propose a bilevel optimization model to support decision-making relative to deploying electric vehicle (EV) charging stations, while taking into account the users' behaviour. The charging-service provider acts at the upper level by making decisions on the location, size, and dynamic pricing of the charging stations to maximize its profit. Users make up the second level and select their available charging stations according to their preference lists. The strategic bilevel optimization problem is addressed using an optimality-conditions-based single-level reformulation complemented by relaxing the complementary constraints. This chapter originated from a partnership between Bernard Gendron and Miguel Anjos, starting in 2016, to rise to the challenge of helping Hydro-Québec achieve the optimal placement of charging points for electric vehicles. The subsequent 5 years saw them thoroughly discuss every aspect of this topic that perfectly joined their expert knowledge of transportation and energy, respectively. This chapter presents work carried out by Miguel Anjos and co-authors in the wake of Bernard Gendron's passing. The authors contribute the chapter to the book in appreciation of Bernard's deep scientific impact and as a token of friendship.

Chapter 19, *Improving Service Quality of an Urban Rail Transit Line by Integrating Passenger and Freight Train Transportation*, by Krissada Tundulyasaree, Layla Martin, Rolf van Lieshout, and Tom van Woensel. This chapter also focuses on City Logistics-related issues, namely, integrating freight movements into public transport. The authors propose a model for the joint optimization of the train and rolling-stock schedules together with the freight allocation, to maximize the carrier profits from freight and passengers alike, under the user-behaviour hypothesis that passengers will choose rail only when the travel time is sufficiently short. Experimental results are extremely promising for this integrative City Logistics strategy, as the flexibility of freight allows the carrier to move it during low-demand periods, while the additional train departures increase supply and thus decrease passenger travel times. The Mixed-Integer Linear Programming model proposed mirrors Bernard Gendron's extensive research in CO, where he delved into polyhedral theory, relaxation, and decomposition methods for integer programming models. Furthermore, the chapter's emphasis on practical applications in transportation resonates with Bernard Gendron's application-inspired research, particularly in transportation, logistics, and telecommunications.

Chapter 20, *Two-stage Stochastic Optimization for the Extended Aircraft Arrival Management Problem Under Uncertainty*, by Fabian Bastin, Sonia Cafieri, Ahmed Khassiba, and Marcel Mongeau, addresses the management of aircraft arrivals in the context of extended arrival manager systems, for which uncertainty is significant when predicting expected times to start the approach phase and landing times. The authors propose a high-level multi-stage stochastic optimization formulation, considering several air-network points of interest and practical operational constraints, and study in some depth the two-stage case, which corresponds to recent studies in the field. Extended numerical experiments, performed on realistic instances, provide the means to analyse the impact of problem characteristics and modelling choices (e.g., inclusion of chance constraints in the first stage) and show, in particular, that the stochastic solutions are more robust than their deterministic counterparts. This chapter builds upon research collaboration with Bernard Gendron, who served as a co-supervisor for Ahmed Khassiba's Ph.D. thesis on aircraft arrival scheduling under uncertainty. A seasoned expert in Mixed-Integer Linear Programming, Bernard Gendron proposed improved solution strategies for such challenges, seamlessly applying his expertise from urban and road transportation to the realm of air transportation. He contributed to consolidating decomposition strategies in stochastic programming, paving the way for future developments in air traffic management. Given the ongoing growth in air traffic, Bernard Gendron's insights have been crucial for developing better OR algorithms and improving the efficiency and safety of air transportation.

Acknowledgments First and foremost, we wish to address our warmest thanks to the authors of the different chapters who accepted so graciously our invitation to contribute to the book. We also want to thank the reviewers, who dedicated time and effort to help the authors and us to improve the final product.

Teodor Gabriel Crainic is Professor of Operations Research, Logistics and Transportation in the Department of Analytics, Operations, and Information Technology of the School of Management, Université du Québec à Montréal, Adjunct Professor, Department of Computer Science and Operations Research, Université de Montréal, and Senior Researcher, Interuniversity Research Centre on Enterprise Networks, Logistics and Transportation (CIRRELT), Canada.

Antonio Frangioni is Professor in the Computer Science Department of the University of Pisa, Italy.

Michel Gendreau is Professor of Operations Research in the Department of Mathematics and Industrial Engineering of Polytechnique Montréal, Adjunct Professor, Department of Computer Science and Operations Research, Université de Montréal, and Senior Researcher, Interuniversity Research Centre on Enterprise Networks, Logistics and Transportation (CIRRELT), Canada.

References

Bourbeau, B., Gendron, B., & Crainic, T. G. (2000) Branch-and-bound parallelization strategies applied to a depot location and container fleet management problem. *Parallel Computing, 26*(1):27–46.

Crainic, T. G., Frangioni, A., & Gendron, B. (2001). Bundle-based relaxation methods for multicommodity capacitated network design. *Discrete Applied Mathematics, 112*(1–3), 73–99.

Crainic, T. G., Gendreau, M., & Gendron, B. (2021). Fixed-charge network design problems. In T. G. Crainic, M. Gendreau, & B. Gendron (Eds.), *Network design with applications in transportation and logistics* (chap. 2, pp. 15–28). Springer.

Crainic, T. G., & Gendron, B. (2021). Exact methods for fixed-charge network design. In T. G. Crainic, M. Gendreau, & B. Gendron (Eds.), *Network design with applications in transportation and logistics* (chap. 3, pp. 29–89). Springer.

Crainic, T. G., Gendron, B., & Kazemzadeh, M. R. A. (2022). A taxonomy of multilayer network design and a survey of transportation and telecommunication applications. *European Journal of Operational Research, 303*(1), 161–179.

Croxton, K. L., Gendron, B., & Magnanti, T. L. (2003a). A comparison of mixed-integer programming models for nonconvex piecewise linear cost minimization problems. *Management Science, 49*(9), 1268–1273.

Croxton, K. L., Gendron, B., & Magnanti, T. L. (2003b). Models and methods for merge-in-transit operations. *Transportation Science, 37*(1), 1–22.

Croxton, K. L., Gendron, B., & Magnanti, T. L. (2007). Variable disaggregation in network flow problems with picewise linear costs. *Operations Research, 55*(1), 146–157.

Frangioni, A., & Gendron, B. (2009). 0–1 reformulations of the multicommodity capacitated network design problem. *Discrete Applied Mathematics, 157*(6), 1229–1241.

Frangioni, A., & Gendron, B. (2013). A stabilized structured dantzig-wolfe decomposition method. *Mathematical Programming, 140*, 45–76.

Frangioni, A., & Gendron, B. (2021). Piecewise linear cost network design. In: T. G. Crainic, M. Gendreau, B. Gendron (Eds.), Network design with applications to transportation and logistics (chap. 6, pp. 167–185). Springer.

Frangioni, A., Durbano Lobato, R., van Ackooij, W., Iardella, N., De Pasquale, F., Bacci, T., Gorgone, E., Meoli, D., Calandrini, E., & Ghezelsoflu, A. (2024). `SMS++`, a structured modeling system for mathematical models. https://gitlab.com/smspp/smspp-project

Gendron, B., & Crainic, T. G. (1994a). Parallel implementations of bounding procedures for multicommodity capacitated network design problems. Publication CRT-94-45, Centre de recherche sur les transports, Université de Montréal, Montréal, QC, Canada.

Gendron, B., & Crainic, T. G. (1994b). Parallel branch-and-bound algorithms: Survey and synthesis. *Operations Research, 42*(6), 1042–1066.

Gendron, B., & Crainic, T. G. (1995). A branch-and-bound algorithm for depot location and container fleet management. *Location Science, 3*(1), 39–53.

Gendron, B., & Gouveia, L. (2017). Reformulations by discretization for piecewise linear integer multicommodity network flow problems. *Transportation Science, 51*(2), 629–649.

Part I
Methodological Developments

Variable Neighborhood Search with Dynamic Exploration for the Set Union Knapsack Problem

Igor Machado Coelho, Saïd Hanafi, Raca Todosijevic, Mustapha Ratli, and Bernard Gendron

1 Introduction

In the set-union knapsack problem (SUKP), introduced in Goldschmidt et al. (1994), we are given a set of elements, each having a positive weight, and a set of items, where each item has a positive profit and a corresponding set of associated elements. An item is included in a knapsack of fixed weight capacity only if all its associated elements are included in the knapsack. The goal of the SUKP is to select the items to be included in the knapsack so that the total profit of the selected items is maximized and the capacity of the knapsack is not exceeded.

Formally, the SUKP is defined as follows. We are given a set N of n elements and a set M of m items. Each item $i \in M$ is characterized by its profit $c_i > 0$ and by its set of associated elements $N_i \subseteq N$. For any subset $S \subseteq M$, the profit of S is given by $\sum_{i \in S} c_i$, while the weight of S is $\sum_{j \in \cup_{i \in S} N_i} a_j$, where $a_j > 0$ is the weight of element j. The objective of the SUKP is to select a maximum profit subset of M,

---

Bernard Gendron is deceased at time of publication.

I. M. Coelho (✉)
Programa de Pós-Graduação em Ciências Computacionais, Universidade do Estado do Rio de Janeiro, Rio de Janeiro, Brazil

Instituto de Computação, Universidade Federal Fluminense, Niterói, Brazil
e-mail: igor.machado@ime.uerj.br; imcoelho@ic.uff.br

S. Hanafi · R. Todosijevic · M. Ratli
LAMIH, Université Polytechnique de Hauts de France, CNRS UMR 8201, Valenciennes, France
e-mail: Said.Hanafi@uphf.fr; Raca.Todosijevic@uphf.fr; Mustapha.Ratli@uphf.fr

B. Gendron
CIRRELT, Université de Montréal, Montréal, QC, Canada

T. G. Crainic et al. (eds.), *Combinatorial Optimization and Applications*,
International Series in Operations Research & Management Science 358,
https://doi.org/10.1007/978-3-031-57603-4_2

while respecting the capacity $b > 0$ of the knapsack. The set formulation of the SUKP can then be stated as follows:

$$(SUKP)\begin{cases} \max & \sum_{i \in S} c_i \\ s.t. & \sum_{j \in \cup_{i \in S} N_i} a_j \leq b \\ & S \subseteq M \end{cases}$$

For any subset $Z \subseteq N$, the weight of Z is given by $a(Z) = \sum_{j \in Z} a_j$. Without loss of generality, we assume that $\max\{a(N_i) : i \in M\} \leq b < a(N)$, otherwise, either some items would never be selected, or all the items would be selected. It may be observed that the SUKP includes the knapsack problem as a special case when $N_i \cap N_{i'} = \emptyset$ for all $i, i' \in M$ and $|N_i| = 1$ for all $i \in M$. Since the knapsack problem is NP-hard, the SUKP is NP-hard as well.

We derive from this set formulation a *mixed integer programming* (MIP) model that uses two sets of binary variables, one associated with the items and the other with the elements. More precisely, we define for each item $i \in M$ and for each element $j \in N$, respectively, the binary variables:

$$x_i = \begin{cases} 1 & \text{if item } i \text{ is selected} \\ 0 & \text{otherwise} \end{cases}$$

and

$$y_j = \begin{cases} 1 & \text{if element } j \text{ is selected} \\ 0 & \text{otherwise} \end{cases}$$

Using these variables, the MIP formulation is given as follows:

$$(MIP)\begin{cases} \max x_0 = \sum_{i \in M} c_i x_i & & (1) \\ s.t. \sum_{j \in N} a_j y_j \leq b & & (2) \\ x_i \leq y_j & i \in M, j \in N_i & (3) \\ x_i, y_j \in \{0, 1\} & i \in M, j \in N_i & (4) \end{cases}$$

The objective, (1), is to maximize the total profit associated with the selected items. Constraint (2) ensures that the capacity of the knapsack is not exceeded by the selected elements. Constraints (3) imply that the selection of item i forces the selection of all elements in N_i.

The SUKP has various domain-specific applications, including (a) information security systems and financial decision making (Arulselvan, 2014; Kellerer et al.,

2004), (b) flexible manufacturing machine scheduling (Goldschmidt et al., 1994; Khuller et al., 1999; Tang & Denardo, 1988), (c) database partitioning (Goldschmidt et al., 1993), and (d) data stream compression (Yang et al., 2016).

An interesting and practical application of SUKP arises in the context of software engineering called "Dependency Hell" (Fan et al., 2020). Nowadays, every *feature* $f \in M$ in a software can have several *dependencies* $d \in N_f$ (with possibly different versions as well), thus combining all of them in a working software can be quite challenging (or even impossible) on practice (thus the term "hell"). Considering each feature f as an item (from set of features M) with profit c_f and dependency d as an element (from set of dependencies N) with weight a_d, we have a SUKP variant for this interesting software engineering problem. A dependency can be naturally reused for several features, without extra weight. Note that *weight*, in this context, may correspond to some storage space in *megabytes*, that can be limited to b for large projects. The weight can also be adapted to any other software metric that we need to constrain, such as security risk exposure, code coverage, etc. (Esteves et al., 2019).

Several algorithms have been proposed to solve the SUKP. In Goldschmidt et al. (1994), the authors developed an exact *dynamic programming* algorithm for the SUKP and derived sufficient conditions for having polynomial run time. Arulselvan (2014) showed that a greedy algorithm for the *budgeted maximum coverage* problem approximates the SUKP within a constant factor. Following the same research direction, Taylor (2016) proposed an approximation algorithm relying on an algorithm for the *densest k-subhypergraph* problem.

Heuristic solution approaches for the SUKP have been proposed recently. The first heuristic for the SUKP has been presented in He et al. (2018). The authors developed a *binary artificial bee colony algorithm* (BABC) and used greedy repairing and optimization to handle infeasibility of a solution. Feng et al. (2019) proposed several versions of *discrete moth search*, while Ozsoydan and Baykasoglu (2019) developed a hybrid approach that combines a *genetic algorithm* and *particle swarm optimization*. Wei and Hao (2019) proposed an *iterated two-phase local search* (I2PLS) approach for solving the SUKP. Their heuristic is based on exploring neighborhood structures in an *intensification* phase in order to generate local optima and a *diversification* phase based on frequency information to escape local optima traps. The intensification phase employs a hybrid local search routine that combines *variable neighborhood descent* (VND) and *tabu search* (TS) procedures. VND explores two neighborhood structures, while TS explores one. All examined neighborhood structures are based on the *swap(p, q)* operator that allows removing p selected items from a solution and adding q non-selected items. The three neighborhood structures consist in: (1) removing two selected items from a solution and adding one non-selected item; (2) removing one selected item from a solution and adding two non-selected items; (3) exchanging two selected items with two non-selected items. I2PLS may be considered as the current state-of-the-art heuristic for the SUKP. It was executed 100 times on each test instance, with a time limit of 500 s per run, and succeeded to establish best known solutions for most instances in a set of 30 benchmark instances, introduced in He et al. (2018).

In this paper, we propose a simple, but effective variable neighborhood search (VNS) heuristic for solving the SUKP. Similarly to I2PLS, our heuristic may be seen as an iterated two-phase approach, since it also includes intensification and diversification phases. However, compared to I2PLS, our heuristic does not use frequency information nor any long-term memory to diversify the search. In addition, our heuristic does not make use of a tabu list in the intensification phase. Nevertheless, our heuristic is competitive with I2PLS, sometimes providing better results in shorter time. Because of its simplicity, our heuristic may be regarded as a *less is more approach* (LIMA), see, e.g., Brimberg et al. (2017), Lu et al. (2020), Liao et al. (2020), Mikić et al. (2019), Mladenović et al. (2016, 2020). The underlying idea of LIMA approach is to make a heuristic as simple as possible, but at the same time at least competitive with the current state-of-the-art heuristic.

The paper is organized as follows. In Sect. 2, we present the main ingredients of our VNS heuristic, including constructive heuristics to generate initial solutions, the neighborhood structures, and the intensification and diversification phases. Section 3 presents our computational experiments on the same set of 30 benchmark instances tested in Wei and Hao (2019). The conclusion summarizes our results and identifies avenues for future research.

2 Main Ingredients of the Variable Neighborhood Search

The metaheuristic variable neighborhood search (VNS), introduced in Mladenović and Hansen (1997), is based upon systematic changes of neighborhoods both in an intensification phase, to find local optima, and in a diversification phase, to escape local optima traps by a shaking procedure. VNS systematically changes neighborhood structures during the search for a near-optimal solution, based on the following three observations: (1) A local optimum relative to one neighborhood structure is not necessarily a local optimum for another neighborhood structure; (2) A global optimum is a local optimum with respect to all neighborhood structures; and (3) Empirical evidence shows that, for many problems, a large majority of the local optima are relatively close to each other, see, e.g., Hansen et al. (2017) for more details on VNS.

To describe our algorithms, we need additional notation. For each element $j \in N$, $M_j = \{i \in M : j \in N_i\}$ is the set of items covered by element j, with $m_j = |M_j|$. Let x denote a vector in $\{0, 1\}^m$, and define the associated index sets $M^v(x) = \{i \in M : x_i = v\}$ for $v \in \{0, 1\}$. Similarly, let y denote a vector in $\{0, 1\}^n$, and define the associated index sets $N^v(y) = \{j \in N : y_j = v\}$ for $v \in \{0, 1\}$. Note that we have $M = M^0(x) \cup M^1(x)$ and $N = N^0(y) \cup N^1(y)$.

A solution s of the SUKP is represented by two binary vectors, i.e., $s = (x, y)$ with $x \in \{0, 1\}^m$ and $y \in \{0, 1\}^n$ where $x_i = 1$ for item $i \in M$ and $y_j = 1$ for each element $j \in N$ corresponding to the selection of i and j, respectively. We define the solution space as $S = \{(x, y) \in \{0, 1\}^m \times \{0, 1\}^n : ay \leq b, x_i \leq y_j, i \in M, j \in N_i\}$. In such solution space, once the values of x are known, the values of y may be derived directly by setting $y_j = \max\{x_i : i \in M_j\}$. Hence, we restrict our space search X to the binary variables x, i.e., $X = \{x \in \{0, 1\}^m : \exists y, ay \leq b, y_j = \max\{x_i : i \in M_j\},\ j \in N\}$.

In this section we present the main ingredients of the proposed VNS based heuristics. First, we present the procedure used to generate an initial solution and then we describe the neighborhood structures exploited by our VNS heuristic. Next, we present the local search and the shaking procedures used to intensify and to diversify the search, respectively. Finally, we summarize the overall approach.

2.1 Constructive Heuristics to Generate Initial Solutions

We propose a dynamic constructive heuristic that starts from a null solution and iteratively extends the current partial solution until a complete solution is constructed. For the SUKP, a null solution corresponds to a solution that sets all variables to 0 and a complete solution is a feasible solution such that each add move (adding an item) leads to an infeasible solution. These constructive heuristics differ in the order in which items are processed. For the knapsack problem, an effective constructive heuristic sorts items in decreasing order of their ratios profit/weight. The proposed constructive heuristics for the SUKP use the same principle based on the ratio $r_i = \frac{c_i}{w_i^*}$ where the weight w_i^* is estimated. Arulselvan (2014) proposes the estimation $w_i^1 = \sum_{j \in N_i} \frac{a_j}{m_j}$, while Wei and Hao (2019) use $w_i^2 = a(N_i)$.

These constructive heuristics can be implemented in static or dynamic ways. In a static constructive heuristic, first, all items are sorted in decreasing order of their ratios and the initial solution is the null solution (i.e., $x = y = 0$). Then, all items are processed in the given order and added one by one, while ensuring that the capacity of the knapsack is never exceeded. Each time an item is added to the knapsack, the solution, the residual capacity, the weight of the selected items and the objective function value are updated. In the pseudo-code described in Algorithm 1, denoted SH^h, we use the following notation: the ratio of each item i is denoted as $r_i^h = \frac{c_i}{w_i^h}$ for $h \in \{1, 2\}$; the variable β denotes the residual capacity of the current solution; the set J represents the set of selected elements in the current solution, thus allowing to compute the current weight $a(N_i \setminus J)$ for each item i.

Based on the observation that the ratios may change their values if they are computed after the selection of some items, we propose dynamic constructive heuristics. In a dynamic constructive heuristic, we start with a null solution and, at each iteration, we compute the estimated weight and select the item that maximizes the current ratio among non-selected items. The procedure stops when there is no item that can be added without violating the capacity of the knapsack. The pseudo-code of the proposed dynamic constructive heuristics is described in Algorithm 2,

Algorithm 1: Static constructive heuristic (SH^h)

```
Input: r, b, a(.), c, N, m
Output: x
Sort items in decreasing order according of r_i^h
x ← (0, ..., 0), J ← ∅, β ← b, z ← 0
for i=1 to m do
    if a(N_i \ J) ≤ β then
        x_i ← 1
        foreach j ∈ N_i do
            y_j ← 1
            β ← β − a(N_i \ J)
            z ← z + c_i
            J ← J ∪ N_i
        end
    end
end
return x
```

where the set I contains the items that are not selected in the current solution. The current estimated weight w_i^h, for $i \in I$ and $h \in \{1, 2\}$, is computed using the following rules:

- $w_i^1 = \sum_{j \in N_i} \frac{a_j}{m'_j}$ with $m'_j = |\{i \in M \setminus I : j \in N_i\}|$
- $w_i^2 = a(N_i \setminus J)$

2.2 Neighborhood Structures

The neighborhood structures for the SUKP are based on the standard *flip* operator used to change the value of only one variable in the current solution x. The resulting value of the variable x_i affected by the flip operator is set to its complementary value $1 - x_i$. More precisely $Flip(x, i) = x + (1 - 2x_i)e^i$ where e^i is the n-dimensional vector composed of value 1 at position i and 0 elsewhere.

Using the flip operator, we define *Add* and *Drop* neighborhoods that contain all solutions obtained from a given solution x by, respectively, adding and removing one item from the knapsack:

- $Add(x) = \{Flip(x, p) : p \in M^0(x)\}$
- $Drop(x) = \{Flip(x, q) : q \in M^1(x)\}$

Note that $M^0(x)$ contains the indices of the items not included in the knapsack, while $M^1(x)$ contains the indices of the items selected in the knapsack. Since all profits are positive, we observe that only an *Add* move may lead to an improving solution, while for all $x' \in Drop(x)$, we have $cx' < cx$.

Algorithm 2: Dynamic constructive heuristic (DH^h)

Input: $w, b, a(.), c, M, N$
Output: x
$x \leftarrow (0, \ldots, 0)$, $J \leftarrow \emptyset$, $I \leftarrow M$, $z \leftarrow 0$, $\beta \leftarrow b$
$Stop \leftarrow False$
while *not* $Stop$ **do**
  Compute estimated weight w_i^h for $i \in I$
  $i^* \leftarrow argmax \left\{ \frac{c_i}{w_i^h} : i \in I, a(N_i \setminus J) \leq \beta \right\}$
  if i^* *exists* **then**
    $x_{i^*} \leftarrow 1$
    foreach $j \in N_{i^*}$ **do**
      $y_j \leftarrow 1$
      $\beta \leftarrow \beta - a(N_{i^*} \setminus J)$, $z \leftarrow z + c_{i^*}$
      $I \leftarrow I \setminus \{i^*\}$, $J \leftarrow J \cup N_{i^*}$
      Update w_i^h, for $i \in I$
    end
  else
    $Stop \leftarrow True$
  end
end
return x

In order to characterize optimal solutions of the SUKP, we introduce the following definition.

Definition 1 A feasible solution is called a *boundary* solution if $Add(x) \cap X = \emptyset$, i.e., $Flip(x, p) \notin X$, $p \in M^0(x)$.

From this definition, it follows that a local optimum with respect to the Add neighborhood is a boundary solution. Consequently, this holds for an optimal solution, as stated in Proposition 1.

Proposition 1 *Any optimal solution of the SUKP is a boundary solution.*

Proof If x is an optimal solution, it is a local optimum with respect to the *Add* neighborhood and therefore, a boundary solution. □

In order to escape from local optima traps of the *Add* neighborhood, we introduce the following more complex neighborhood structure:

$$\mathcal{N}_k(x) = \{x' : d(x, x') = |M^1(x')| - |M^1(x)| + 2k;\ x' \ boundary\ solution\}$$

where $d(x, x')$ denotes the *Hamming* distance between solutions x and x', i.e., $d(x, x') = \sum_{i \in M} |x_i - x'_i|$.

The set $\mathcal{N}_k(x)$ contains all solutions that may be obtained from the current solution x by dropping k items and by adding several items until the obtained solution is a new boundary solution. Note that, by removing k items and inserting

k' different items from the previously removed k, one gets a Hamming distance of $d(x, x') = k + k'$. Naturally, the final number of items $|M^1(x')|$ decreases by k and increases by k', meaning that $|M^1(x')| = |M^1(x)| - k + k'$, so $d(x, x') = |M^1(x')| - |M^1(x)| + 2k$. This neighborhood structure has a huge cardinality, even for $k \leq 2$.

We propose the following nested procedure to explore this neighborhood in a dynamic manner. After dropping k items, the resulting solution is used as a starting point for a local search with respect to the *Add* neighborhood. In this way, we actually perform a nested VND (Mladenović & Hansen, 1997). Algorithm 4 presents the pseudo-code of the function **NestedVND**$(x, \mathcal{N}_2)$ for exploring neighborhood $\mathcal{N}_2$. The procedure **NestedVND**$(x, \mathcal{N}_1)$ for exploring neighborhood $\mathcal{N}_1$ may be derived from **NestedVND**$(x, \mathcal{N}_2)$ by replacing the for loop **For each** $p, q \in M^1(x)$ **do** by **For each** $p \in M^1(x)$ **do** and by deleting the line $x' = Flip(x', q)$. A first improvement strategy is used for the local search within the Add neighborhood, described in Algorithm 5, i.e., as soon as an improving solution is detected, it is set to be the new incumbent solution and the search is resumed from that solution. To satisfy the definition of $\mathcal{N}_k(x)$, we make sure that the local search procedure does not add the dropped elements p and q.

Algorithm 3: Function $NestedVND(x, \mathcal{N}_1)$

```
Stop ← False
while not Stop do
    Stop ← True
    x' ← x
    foreach p ∈ M¹(x) do
        x' ← Flip(x', p)
        x'' ← LS_Add(x', {p})
        if c(x'') > c(x) then
            Stop ← False
            x ← x''
        end
    end
end
```

2.3 *Sequential Variable Neighborhood Descent*

Calls to the procedures **NestedVND**$(x, \mathcal{N}_1)$ and **NestedVND**$(x, \mathcal{N}_2)$ alternate within the sequential VND scheme shown in Algorithm 6. More precisely, as soon as **NestedVND**$(x, \mathcal{N}_1)$ is trapped in a local optimum, **NestedVND**$(x, \mathcal{N}_2)$ is invoked. If it succeeds to escape from the trap, the search is resumed using **NestedVND**$(x, \mathcal{N}_1)$. When both procedures do not succeed in finding an improving solution, the sequential VND stops.

Algorithm 4: Function $NestedVND(x, \mathcal{N}_2)$

```
Stop ← False
while not Stop do
    Stop ← True
    x' ← x
    foreach p, q ∈ M^1(x) do
        x' ← Flip(x', p)
        x' ← Flip(x', q)
        x'' ← LS_Add(x', {p, q})
        if c(x'') > c(x) then
            Stop ← False
            x ← x''
        end
    end
end
```

Algorithm 5: Function $LS\_Add(x, \mathcal{L})$

```
Stop ← False
while not Stop do
    Select x' ∈ Add(x) : c(x') > c(x), x ∉ L
    if x' exists then
        x ← x'
    else
        Stop ← True
    end
end
```

2.4 Shaking Procedure

In order to escape from local optima traps encountered by the sequential VND, we develop a shaking procedure, **Shake**(x, k), that aims to diversify the search. As input, it requires a solution x and a parameter k. As output, it returns a solution obtained from solution x by removing $\min\left(ShakeMax, \frac{|M^1(x)|}{2}\right)$ items at random, and by adding k new items also chosen at random. During experiments, we decided to adopt parameter $ShakeMax = 20$.

2.5 General Variable Neighborhood Search

All the above procedures are used as main ingredients of a *general* VNS (GVNS) shown in Algorithm 7. The proposed GVNS alternates between sequential VND and shaking procedures until a predefined stopping condition is met. In our case, the stopping condition is a maximum CPU time t_{max}. Besides parameter t_{max}, our

Algorithm 6: Function $SeqVND(x)$

```
Stop ← False, k ← 1
while not Stop do
    x' ← x
    if k = 1 then
        NestedVND(x', N1)
    else
        NestedVND(x', N2)
    end
    if c(x') > c(x) then
        x ← x', k ← 1
    else
        if k = 2 then
            Stop ← True
        k ← k + 1
    end
end
```

GVNS has one additional parameter k_{max} used to control the level of diversification. To generate an initial solution, the proposed GVNS uses one of the procedures presented in Sect. 2.1.

3 Computational Results

This section presents the computational results obtained by the implementation of the proposed GVNS heuristic in C++. Experiments are performed on a computer Intel(R) Core(TM) i7-7820HQ CPU @ 2.90 GHz, 2901 MHz, 4 Core(s), 8 Logical Processor(s) with 32 GB (RAM). We test our algorithm on 30 benchmark instances introduced in He et al. (2018) and also used in Wei and Hao (2019) for evaluating the performance of I2PLS. We present the benchmark instances in Sect. 3.1, then the results of the constructive and GVNS heuristics in Sects. 3.2 and 3.3, respectively. Finally, Sect. 3.4 compares the performance of our GVNS with that of I2PLS.

3.1 Benchmark Instances

He et al. (2018) proposed an instance generation method that makes use of the following parameters:

- $\alpha = \frac{\sum_{i \in M} \sum_{j \in N} R_{ij}}{m \times n}$, where R is an $m \times n$ binary matrix such that $R_{ij} = 1 \iff j \in N_i$
- $\beta = \frac{\sum_{j \in N} a_j}{b}$

Algorithm 7: $GVNS(k_{max}, t_{max})$

```
x ← Initial_Solution()
k ← 1
repeat
    while k ≤ k_max do
        x' ← x
        Shaking(x', k)
        SeqVND(x')
        if c(x') > c(x) then
            x ← x', k ← 1
        else
            x' ← x, k ← k + 1
        end
    end
until CPU_Time() > t_max
```

The term α denotes a density factor, where $\alpha = 1$ means that all items are connected to every element; β denotes the weight factor, where $\beta = 1$ means that capacity if full at b.

He et al. (2018) generated three classes with 10 instances for each:

- $n > m$: $(n, m) \in \{100, 200, 300, 400, 500\} \times \{85, 185, 285, 385, 485\}$
- $n = m$: $m \in \{100, 200, 300, 400, 500\}$
- $n < m$: $(n, m) \in \{85, 185, 285, 385, 485\} \times \{100, 200, 300, 400, 500\}$

Table 1 presents these 30 benchmark instances,[1] including their dimension, i.e., the number of items m and the number of elements n, as well as the best known value (BKV) for each of them. The instances are shown in two groups: those with values of $\alpha = 0.1$ and $\beta = 0.75$ (SUKP01 to SUKP15) on the left, and those with values of $\alpha = 0.15$ and $\beta = 0.85$ on the right (SUKP16 to SUKP30).

3.2 Results with Constructive Heuristics

We performed experiments with the four constructive heuristics (SH^1, DH^1, SH^2 and DH^2) presented in Sect. 2.1. For each constructive heuristic CH, Table 2 shows the value of the best feasible solution found (VCH) and the CPU time (TCH), with best results highlighted in bold. Table 3 presents the same values, but in relative gap notation over BKV (less is better). Bold values in Tables 2 and 3 indicate the best results.

It is clear from the results shown in Table 2 that DH^1 has the best performance, finding the best value for all 30 instances, in a very short computational time (on

[1] The generated instances and source-code are publicly available at the OptFrame GitHub repository: https://github.com/optframe/instances-sukp.

Table 1 Instances with $\alpha = 0.1$ and $\beta = 0.75$ on the left (SUKP01 to SUKP15), and with $\alpha = 0.15$ and $\beta = 0.85$ on the right (SUKP16 to SUKP30)

Instance	m	n	BKV	Instance	m	n	BKV
SUKP01	85	100	12,045	SUKP16	85	100	12,369
SUKP02	100	85	13,283	SUKP17	100	85	12,479
SUKP03	100	100	14,044	SUKP18	100	100	13,508
SUKP04	185	200	13,696	SUKP19	185	200	11,298
SUKP05	200	185	13,521	SUKP20	200	185	14,215
SUKP06	200	200	12,522	SUKP21	200	200	12,317
SUKP07	285	300	11,568	SUKP22	285	300	11,802
SUKP08	300	285	11,563	SUKP23	300	285	12,607
SUKP09	300	300	12,817	SUKP24	300	300	11,858
SUKP10	385	400	10,600	SUKP25	385	400	10,506
SUKP11	400	385	11,484	SUKP26	400	385	11,209
SUKP12	400	400	11,665	SUKP27	400	400	11,325
SUKP13	485	500	11,321	SUKP28	485	500	10,220
SUKP14	500	485	11,771	SUKP29	500	485	10,238
SUKP15	500	500	11,249	SUKP30	500	500	10,381
$\alpha = 0.1$	$\beta = 0.75$			$\alpha = 0.15$	$\beta = 0.85$		

average, 0.002 s). In spite of this evident superior performance, we note that the best values found by DH^1 are quite far from the best known values (see Table 1). Specifically on Table 3, it is shown that DH^1 is still 11.6% away from BKV, on average. In addition, even if it is superior to the other alternatives as a standalone constructive heuristic, DH^1 is not necessarily the most effective when embedded in our GVNS heuristic. This is why, in the next subsection, we report the results of our GVNS heuristic when finding initial solutions with the four constructive heuristics.

3.3 Results with GVNS

Table 4 reports the objective function values, best and average over 10 runs of GVNS. We show four versions of the GVNS heuristic that differ only by the constructive heuristic used to generate the initial solution. We denote as VNS^h for $h = 1, \cdots, 4$ the variant of GVNS where the initial solution is generated by SH^1, DH^1, SH^2 and DH^2, respectively. For each instance and each run, the CPU time limit of VNS^h is set to $\max(m, n)$. Bold values indicate when the proposed methods successfully achieve the best known values.

For all proposed GVNS variants, we observe a deviation of 0.9% on average between the best values found by GVNS and the best known values (varying between 0% and 2.3%, in the worst case). Moreover, our GVNS could identify the best known solutions for all instances, except SUKP23, SUKP24 and SUKP29, with respective gaps of 0.8%, 2.3% and 0.2%. Table 5 presents the time to obtain the best solutions for each method (VNS^1 to VNS^4).

Table 2 Computational experiments with constructive heuristics

Instance	Value of best solution for each constructive heuristic				Computational time (seconds) for each constructive heuristic			
	VSH^1	VDH^1	VSH^2	VDH^2	TSH^1	TDH^1	TSH^2	VDH^2
SUKP01	9868	**10,764**	8909	8640	0.001	0.000	0.000	0.001
SUKP02	11,350	**12,379**	11,862	10,113	0.001	0.000	0.000	0.000
SUKP03	10,933	**11,924**	11,393	11,393	0.000	0.000	0.000	0.001
SUKP04	10,173	**12,428**	10,154	10,154	0.000	0.000	0.001	0.002
SUKP05	9387	**11,942**	10,371	9019	0.000	0.001	0.000	0.002
SUKP06	9833	**11,397**	8930	8541	0.000	0.000	0.001	0.002
SUKP07	9642	**10,832**	9276	8799	0.001	0.001	0.001	0.005
SUKP08	7820	**10,113**	8098	8098	0.000	0.001	0.001	0.004
SUKP09	9308	**11,955**	9155	9131	0.001	0.001	0.001	0.005
SUKP10	7362	**9587**	7707	7643	0.001	0.001	0.001	0.006
SUKP11	8616	**9968**	8225	8443	0.001	0.001	0.001	0.007
SUKP12	8647	**10,796**	9041	8641	0.003	0.005	0.006	0.029
SUKP13	8649	**10,068**	8866	8613	0.001	0.003	0.001	0.012
SUKP14	8657	**10,595**	8645	8645	0.000	0.003	0.002	0.010
SUKP15	8858	**10,552**	8858	8858	0.001	0.002	0.003	0.013
SUKP16	6640	**11,480**	7524	7524	0.000	0.000	0.000	0.001
SUKP17	8594	**10,300**	9000	9000	0.000	0.000	0.000	0.001
SUKP18	9427	**10,740**	9311	9311	0.000	0.000	0.001	0.002
SUKP19	6793	**10,547**	7297	6607	0.000	0.002	0.000	0.004
SUKP20	10,544	**11,264**	9461	9806	0.001	0.001	0.001	0.006
SUKP21	8184	**10,980**	8162	8162	0.001	0.002	0.000	0.006
SUKP22	8010	**10,217**	7770	7755	0.001	0.002	0.002	0.009
SUKP23	7517	**10,764**	7517	7517	0.001	0.001	0.000	0.005
SUKP24	7253	**9566**	7948	7666	0.000	0.001	0.001	0.005
SUKP25	6892	**9880**	6898	6461	0.001	0.002	0.001	0.007
SUKP26	7648	**9322**	7581	7140	0.000	0.002	0.001	0.008
SUKP27	6924	**10,064**	7032	6643	0.002	0.005	0.003	0.018
SUKP28	7594	**8860**	7332	7332	0.001	0.003	0.003	0.012
SUKP29	7287	**9467**	7742	7742	0.002	0.003	0.004	0.016
SUKP30	6888	**8781**	7202	7105	0.001	0.003	0.003	0.014
Best	0	30	0	0	–	–	–	–
Avg t.(s)	–	–	–	–	0.001	0.002	0.001	0.007

From this table, we see that the different methods spent on average 56.9% of the available time $\max(m, n)$ to obtain the best solution. This implies that the maximum time could be roughly divided by 2, without affecting solution quality. However, this value also increases proportionally to the size of the instance, ranging from 34%, 58%, 63%, 62% and 66%, respectively for sizes 100, 200, 300, 400 and 500. This demonstrates that, proportionally, more time is required for larger instances.

Table 3 Computational experiments with constructive heuristics: relative gaps over SUKP (less is better)

Instance	BKV	Relative gap (%)			
		VSH^1	VDH^1	VSH^2	VDH^2
SUKP01	12,045	18.1	**10.6**	26.0	28.3
SUKP02	13,283	14.6	**6.8**	10.7	23.9
SUKP03	14,044	22.2	**15.1**	18.9	18.9
SUKP04	13,696	25.7	**9.3**	25.9	25.9
SUKP05	13,521	30.6	**11.7**	23.3	33.3
SUKP06	12,522	21.5	**9.0**	28.7	31.8
SUKP07	11,568	16.6	**6.4**	19.8	23.9
SUKP08	11,563	32.4	**12.5**	30.0	30.0
SUKP09	12,817	27.4	**6.7**	28.6	28.8
SUKP10	10,600	30.5	**9.6**	27.3	27.9
SUKP11	11,484	25.0	**13.2**	28.4	26.5
SUKP12	11,665	25.9	**7.4**	22.5	25.9
SUKP13	11,321	23.6	**11.1**	21.7	23.9
SUKP14	11,771	26.5	**10.0**	26.6	26.6
SUKP15	11,249	21.3	**6.2**	21.3	21.3
SUKP16	12,369	46.3	**7.2**	39.2	39.2
SUKP17	12,479	31.1	**17.5**	27.9	27.9
SUKP18	13,508	30.2	**20.5**	31.1	31.1
SUKP19	11,298	39.9	**6.6**	35.4	41.5
SUKP20	14,215	25.8	**20.8**	33.4	31.0
SUKP21	12,317	33.6	**10.9**	33.7	33.7
SUKP22	11,802	32.1	**13.4**	34.2	34.3
SUKP23	12,607	40.4	**14.6**	40.4	40.4
SUKP24	11,858	38.8	**19.3**	33.0	35.4
SUKP25	10,506	34.4	**6.0**	34.3	38.5
SUKP26	11,209	31.8	**16.8**	32.4	36.3
SUKP27	11,325	38.9	**11.1**	37.9	41.3
SUKP28	10,220	25.7	**13.3**	28.3	28.3
SUKP29	10,238	28.8	**7.5**	24.4	24.4
SUKP30	10,381	33.6	**15.4**	30.6	31.6
Avg. (%)	–	29.1	**11.6**	28.5	30.4

3.4 *Comparison with I2PLS*

Our experiments show that VNS^1 and VNS^3 performed equally well, finding 27 out of 30 best known values (10 runs), with respective average gaps of 0.14% and 0.15%. For this reason, we select VNS^1 for the comparison with the state-of-the-art heuristic I2PLS. Table 6 shows the best values (and gaps to BKV) for I2PLS (100 runs, 500 s per run) and the best values (and gaps to BKV) for our GVNS (10 runs, $\max(m, n)$ seconds per run).

Table 4 GVNS results for 10 runs: best and average objective function values

	Best values (10 runs)				Average values (10 runs)				
Instance	VNS^1	VNS^2	VNS^3	VNS^4	VNS^1	VNS^2	VNS^3	VNS^4	BKV
SUKP01	12,045	12,045	12,045	12,045	12,045	12,045	12,045	12,045	**12,045**
SUKP02	13,283	13,283	13,283	13,283	13,245.7	13,283	13,283	13,283	**13,283**
SUKP03	14,044	14,044	14,044	14,044	14,044	14,011.9	14,011.9	14,033.3	**14,044**
SUKP04	13,696	13,696	13,696	13,696	13,696	13,696	13,696	13,695.9	**13,696**
SUKP05	13,521	13,521	13,521	13,521	13,485.9	13,509.3	13,509.3	13,509.3	**13,521**
SUKP06	12,522	12,384	12,522	12,522	12,401.4	12,380.6	12,394.6	12,456.6	**12,522**
SUKP07	11,568	11,568	11,568	11,568	11,550	11,546.8	11,549.8	11,550	**11,568**
SUKP08	11,563	11,563	11,563	11,563	11,533.9	11,563	11,513.6	11,547.7	**11,563**
SUKP09	12,817	12,817	12,817	12,784	12,775.5	12,742.9	12,754.7	12,755.6	**12,817**
SUKP10	10,600	10,600	10,600	10,600	10,476.1	10,521.9	10,466.8	10,506	**10,600**
SUKP11	11,484	11,484	11,484	11,484	11,474.2	11,474.2	11,469.3	11,469.3	**11,484**
SUKP12	11,665	11,643	11,665	11,665	11,570	11,512.6	11,577.8	11,622.6	**11,665**
SUKP13	11,321	11,321	11,321	11,321	11,149.8	11,183.8	11,210.3	11,152.5	**11,321**
SUKP14	11,771	11,771	11,771	11,771	11,754.6	11,656.2	11,715.9	11,754.6	**11,771**
SUKP15	11,249	11,249	11,249	11,249	11,125.6	11,062.4	11,174.2	11,169.6	**11,249**
SUKP16	12,369	12,369	12,369	12,369	12,354.5	12,369	12,369	12,369	**12,369**
SUKP17	12,479	12,479	12,479	12,479	12,397	12,294.5	12,376.5	12,397	**12,479**
SUKP18	13,508	13,508	13,508	13,508	13,393.8	13,346.4	13,467.6	13,467.6	**13,508**
SUKP19	11,298	11,298	11,298	11,298	11,249.7	11,298	11,183.5	11,255.6	**11,298**
SUKP20	14,215	14,215	14,215	14,215	13,929.9	13,967	14,180.8	14,009.9	**14,215**
SUKP21	12,317	12,317	12,317	12,317	12,140.8	12,177	12,281.1	12,246	**12,317**
SUKP22	11,802	11,802	11,802	11,763	11,667.6	11,661.5	11,532.3	11,533	**11,802**
SUKP23	12,411	12,411	12,500	12,411	12,343.8	12,366.7	12,357.5	12,346.5	12,607
SUKP24	11,585	11,585	11,585	11,585	11,521	11,553	11,473	11,507.3	11,858
SUKP25	10,506	10,506	10,506	10,506	10,407	10,506	10,379.7	10,421.3	**10,506**
SUKP26	11,209	11,209	11,209	11,209	11,015.4	11,013.8	11,169.2	10,956.8	**11,209**
SUKP27	11,325	10,679	11,325	11,325	10,942.4	10,679	11,002	10,752.7	**11,325**
SUKP28	10,220	10,220	10,220	10,220	10,123.9	10,206.4	10,184.6	10,110.2	**10,220**
SUKP29	10,217	10,217	10,194	10,217	10,141.3	10,158.2	10,078.8	10,060.3	10,238
SUKP30	10,381	10,381	10,381	10,381	10,230.7	10,168.2	10,264.1	10,246.9	**10,381**

We can observe that VNS^1 found 27 out of 30 best known values, while I2PLS found 26. The average gap is also lower for VNS^1, achieving 0.14% against 0.15% for I2PLS. We recognize that, although VNS^1 has performed better than I2PLS, the difference is tiny. Nevertheless, we emphasize that our GVNS is a much simpler approach than I2PLS, with virtually no parameters to tune, inspired by the LIMA principle.

Table 5 Time to best solutions

Instance	Time to best (seconds)			
	VNS^1	VNS^2	VNS^3	VNS^4
SUKP01	20.7	16.2	18.3	20.3
SUKP02	46.2	59.8	33	36.8
SUKP03	21.9	36.6	43.4	87.3
SUKP04	185.1	171.8	151.4	71.9
SUKP05	166.2	134.8	56.5	144.4
SUKP06	181.2	175.7	62.8	52.3
SUKP07	174.9	205.3	297.6	268.8
SUKP08	229.9	120.9	171.9	252.8
SUKP09	258.2	288.5	273.7	271.3
SUKP10	305.1	273.2	326.4	233.3
SUKP11	398.2	25.5	262.4	312.9
SUKP12	378.4	366.8	368.4	380.9
SUKP13	376.9	270.2	428.3	424.6
SUKP14	48.6	435.7	166.5	116.7
SUKP15	488.4	192.7	493.6	475.2
SUKP16	50.3	0.2	44.5	21.5
SUKP17	74.8	41.4	3.7	10.9
SUKP18	46.8	41.5	38.5	20.9
SUKP19	35	0.3	95.3	106.6
SUKP20	159.5	165.9	83.2	45.9
SUKP21	107.7	139.7	63.4	178.5
SUKP22	106.4	286.5	37.3	23.3
SUKP23	295	54.7	56.2	232.1
SUKP24	107.7	99.1	284.7	239.9
SUKP25	378.6	34	330.3	44.1
SUKP26	171.7	115.3	182.3	247.2
SUKP27	227.5	166	368.1	140.3
SUKP28	291.2	354.3	286.1	118.2
SUKP29	438.1	458.8	210.4	440.4
SUKP30	462.1	305.4	414.4	250.1

4 Conclusion

We have proposed an efficient and effective GVNS heuristic to solve the set-union knapsack problem (SUKP), a variant of the knapsack problem where each item is associated to multiple elements with different weights. We have proposed dynamic constructive heuristic strategies inspired by the combinatorial SUKP model, in order to provide initial solutions quickly. Our GVNS is a simple heuristic, inspired by the LIMA paradigm, that has been shown to be highly competitive with the state-of-the-art algorithms for the SUKP on a set of benchmark instances.

Table 6 Comparison of proposed GVNS with I2PLS

	I2PLS		VNS^1		
Instance	Best	Gap best	Best	Gap best	BKV
SUKP01	12,045	0.00%	12,045	0.00%	12,045
SUKP02	13,283	0.00%	13,283	0.00%	13,283
SUKP03	14,044	0.00%	14,044	0.00%	14,044
SUKP04	13,696	0.00%	13,696	0.00%	13,696
SUKP05	13,521	0.00%	13,521	0.00%	13,521
SUKP06	12,522	0.00%	12,522	0.00%	12,522
SUKP07	11,568	0.00%	11,568	0.00%	11,568
SUKP08	11,563	0.00%	11,563	0.00%	11,563
SUKP09	12,817	0.00%	12,817	0.00%	12,817
SUKP10	10,600	0.00%	10,600	0.00%	10,600
SUKP11	11,484	0.00%	11,484	0.00%	11,484
SUKP12	11,665	0.00%	11,665	0.00%	11,665
SUKP13	11,321	0.00%	11,321	0.00%	11,321
SUKP14	11,729	0.36%	11,771	0.00%	11,771
SUKP15	11,249	0.00%	11,249	0.00%	11,249
SUKP16	12,369	0.00%	12,369	0.00%	12,369
SUKP17	12,479	0.00%	12,479	0.00%	12,479
SUKP18	13,508	0.00%	13,508	0.00%	13,508
SUKP19	11,298	0.00%	11,298	0.00%	11,298
SUKP20	14,215	0.00%	14,215	0.00%	14,215
SUKP21	12,317	0.00%	12,317	0.00%	12,317
SUKP22	11,802	0.00%	11,802	0.00%	11,802
SUKP23	12,411	1.55%	12,411	1.55%	12,607
SUKP24	11,585	2.30%	11,585	2.30%	11,858
SUKP25	10,506	0.00%	10,506	0.00%	10,506
SUKP26	11,209	0.00%	11,209	0.00%	11,209
SUKP27	11,325	0.00%	11,325	0.00%	11,325
SUKP28	10,220	0.00%	10,220	0.00%	10,220
SUKP29	10,194	0.43%	10,217	0.21%	10,238
SUKP30	10,381	0.00%	10,381	0.00%	10,381

Potential avenues for future research include the exploration of MIP formulations and their integration into heuristic and exact methods that exploit the structure of the SUKP. In particular, inspired by the best known algorithms for the knapsack problem, approaches that combine dynamic programming and advanced relaxation techniques appear promising. Such algorithms would clearly benefit from being provided with very good feasible solutions. Hence, combining them with our GVNS heuristic is a promising avenue for future research.

Acknowledgments The work is partially supported by a "CHAIRE FRANCAISE à L'Université de l'Etat de Rio de Janeiro 2019" and the International Associate Laboratory "Operational Research and Computer Science in Transportation, Mobility and Logistics" between the LAMIH UMR CNRS 8201and the CIRRELT (Montréal, Canada). Igor M. Coelho also acknowledges the Brazilian funding agencies FAPERJ and CNPq for research grants employed during the development of this work.

References

Arulselvan, A. (2014). A note on the set union knapsack problem. *Discrete Applied Mathematics, 169*, 214–218.

Brimberg, J., Mladenović, N., Todosijević, R., & Urošević, D. (2017). Less is more: Solving the max-mean diversity problem with variable neighborhood search. *Information Sciences, 382*, 179–200.

Esteves, J. V., Coutinho, D., Schots, M., & Coelho, I. M. (2019). Is there an interplay between library usage and repository features? An analysis with regression models. In *Proceedings of the XXXIII Brazilian Symposium on Software Engineering* (pp. 407–416).

Fan, G., Wang, C., Wu, R., Xiao, X., Shi, Q., & Zhang, C. (2020). Escaping dependency hell: Finding build dependency errors with the unified dependency graph. In *Proceedings of the 29th ACM SIGSOFT International Symposium on Software Testing and Analysis*. Association for Computing Machinery, New York, ISSTA 2020 (pp. 463–474). https://doi.org/10.1145/3395363.3397388

Feng, Y., An, H., & Gao, X. (2019). The importance of transfer function in solving set-union knapsack problem based on discrete moth search algorithm. *Mathematics, 7*(1), 17.

Goldschmidt, O., Nehme, D., & Yu, G. (1993). On a generalization of the knapsack problem with applications to flexible manufacturing systems and database partitioning. Department of Management Science and Information Systems [University of Texas and Austin]

Goldschmidt, O., Nehme, D., & Yu, G. (1994). Note: On the set-union knapsack problem. *Naval Research Logistics, 41*(6), 833–842.

Hansen, P., Mladenović, N., Todosijević, R., & Hanafi, S. (2017). Variable neighborhood search: Basics and variants. *EURO Journal on Computational Optimization, 5*(3), 423–454.

He, Y., Xie, H., Wong, T. L., & Wang, X. (2018). A novel binary artificial bee colony algorithm for the set-union knapsack problem. *Future Generation Computer Systems, 78*, 77–86.

Kellerer, H., Pferschy, U., & Pisinger, D. (2004). Multidimensional knapsack problems. In *Knapsack problems* (pp. 235–283). Berlin: Springer.

Khuller, S., Moss, A., & Naor, J. S. (1999). The budgeted maximum coverage problem. *Information Processing Letters, 70*(1), 39–45.

Liao, B., Wang, X., Zhu, X., Yang, S., & Pardalos, P. M. (2020). Less is more approach for competing groups scheduling with different learning effects. *Journal of Combinatorial Optimization, 39*(1), 33–54.

Lu, S., Pei, J., Liu, X., Qian, X., Mladenovic, N., & Pardalos, P. M. (2020). Less is more: Variable neighborhood search for integrated production and assembly in smart manufacturing. *Journal of Scheduling, 23*, 649–664.

Mikić, M., Todosijević, R., & Urošević, D. (2019). Less is more: General variable neighborhood search for the capacitated modular hub location problem. *Computers & Operations Research, 110*, 101–115.

Mladenović, N., & Hansen, P. (1997). Variable neighborhood search. *Computers & Operations Research, 24*(11), 1097–1100.

Mladenović, N., Todosijević, R., & Urošević, D. (2016). Less is more: Basic variable neighborhood search for minimum differential dispersion problem. *Information Sciences, 326*, 160–171.

Mladenović, N., Alkandari, A., Pei, J., Todosijević, R., & Pardalos, P. M. (2020). Less is more approach: Basic variable neighborhood search for the obnoxious p-median problem. *International Transactions in Operational Research, 27*(1), 480–493.

Ozsoydan, F. B., & Baykasoglu, A. (2019). A swarm intelligence-based algorithm for the set-union knapsack problem. *Future Generation Computer Systems, 93*, 560–569.

Tang, C. S., & Denardo, E. V. (1988). Models arising from a flexible manufacturing machine, part ii: Minimization of the number of switching instants. *Operations Research, 36*(5), 778–784.

Taylor, R. (2016). Approximations of the densest k-subhypergraph and set union knapsack problems. arXiv preprint arXiv:161004935.

Wei, Z., & Hao, J. K. (2019). Iterated two-phase local search for the set-union knapsack problem. *Future Generation Computer Systems, 101*, 1005–1017.

Yang, X., Vernitski, A., & Carrea, L. (2016). An approximate dynamic programming approach for improving accuracy of lossy data compression by bloom filters. *European Journal of Operational Research, 252*(3), 985–994.

Common-Flow Formulations for the Diameter Constrained Spanning and Steiner Tree Problems

Luis Gouveia, Markus Leitner, and Ivana Ljubić

1 Introduction

We consider the diameter constrained variant of the Steiner tree problem on a graph (DCStTP). Formally, we are given an undirected graph $G = (V, E)$, with node set V, edge set E, edge costs $c_e \geq 0$, for all $e \in E$, and a diameter limit $D \in \mathbb{N}, 2 \leq D \leq |V| - 1$. The node set V is the disjoint union of terminal nodes T and potential Steiner nodes S. A feasible solution to the DCStTP is a subtree $G_S = (V_S, E_S)$ connecting all terminal nodes (also known as a Steiner tree, where $T \subset V_S$), whose diameter does not exceed D. Hence, the maximum number of edges between any pair of nodes in that tree is bounded from above by the value of D. For $D \in \{2, 3\}$, the problem can be solved in polynomial time. For $4 \leq D \leq |V| - 2$, the problem is NP-hard, even if we consider the spanning-tree variant of the problem (namely, $S = \emptyset$ and $T = V$), see Garey and Johnson (1979). The latter variant is known as the Diameter Constrained Spanning Tree problem on a graph, and we will refer to it as DCSpTP.

L. Gouveia
Departamento de Estatística e Investigação Operacional, Centro de Matemática, Aplicações Fundamentais e Investigação Operacional (CMAF-CIO), Universidade de Lisboa, Faculdade de Ciências, Lisbon, Portugal
e-mail: legouveia@fc.ul.pt

M. Leitner
Department of Operations Analytics, Vrije Universiteit Amsterdam, Amsterdam, Netherlands
e-mail: m.leitner@vu.nl

I. Ljubić (✉)
ESSEC Business School, Cergy-Pontoise, France
e-mail: ljubic@essec.edu

T. G. Crainic et al. (eds.), *Combinatorial Optimization and Applications*,
International Series in Operations Research & Management Science 358,
https://doi.org/10.1007/978-3-031-57603-4_3

There exist important applications of the DCStTP in telecommunications, data compression, or parallel computing (see, e.g., Bookstein and Klein (1991), Deo and Abdalla (2000), Santos et al. (2016)). In telecommunication networks, for example, the latency of a signal sent between a pair of nodes is directly proportional to the number of edges of the routing path. In networks that employ the so-called tree-multicasting, the maximum latency thus corresponds to the diameter of that tree (Vik et al., 2008).

Given a tree $G_S = (V_S, E_S)$, the *eccentricity* of a node v is the maximum number of edges on a path from v to any other node in G_S. The diameter of G_S is thus the maximum eccentricity over all nodes $v \in V_S$. Given a diameter D, the *center* of G_S is a node (if D is even) or a pair of adjacent nodes (if D is odd) of minimum eccentricity. Thus, a DCStTP can be seen as a Steiner tree rooted at an unknown center whose eccentricity (i.e., maximum number of hops to another node from G_S) H is bounded by the half of the diameter, i.e. $H = \lfloor D/2 \rfloor$. Many flow-based formulations have been proposed based on these concepts related to centers of trees and hop constraints. One of the first compact formulations for the DCStTP has been proposed by Achuthan et al. (1994). It is based on the Miller-Tucker-Zemlin (MTZ) modeling approach for trees, and it has been later improved by dos Santos et al. (2004). Formulations based on multi-commodity flows have been later studied by Gouveia and Magnanti (2003) and Gouveia et al. (2004, 2006). Compared to MTZ-based formulations, the multi-commodity-flow models typically provide much tighter linear programming (LP) bounds at the expense of involving a much larger number of flow variables. Most of these compact models are not (directly) suitable to be used for solving the DCStTP on very large graphs (either because of very weak lower bounds, or due to the prohibitive size of the underlying formulations). The most successful method for solving the DCSpTP is based on a layered graph reformulation in which the original problem is restated as a Steiner arborescence problem with additional constraints on a very large graph. Copies of the original nodes are created at each layer and the problem is solved using a branch-and-cut procedure (Gouveia et al., 2011). Most of the existing studies focus on the DCSpTP problem variant (notable exceptions are studies by Gouveia and Magnanti (2003), Gouveia et al. (2004, 2006)), hence slight model modifications are needed in order to address the more general problem variant, namely the DCStTP. The question on how to provide a valid formulation for the DCSpTP in the *natural space* of edge variables (using only one variable associated to each edge) remained open for some time. Gouveia et al. (2020) were the first to provide a valid model in the space of edge-variables using the concept of *circular jump* constraints. The authors provide a polyhedral study of the DCSpTP polytope and show that most of the proposed inequalities are not implied by the currently best-performing extended formulation based on layered graphs (Gouveia et al., 2011).

Heuristics for the DCSpTP have been investigated in Raidl and Julstrom (2003), Singh and Gupta (2007), Gruber et al. (2006), Binh et al. (2009), Lucena et al. (2010), Requejo and Santos (2009), Steitz (2015), Santos et al. (2016), and a constraint programming approach has been proposed by Noronha et al. (2010). Another problem variant, called the two-level diameter constrained spanning tree problem, has been introduced by Gouveia et al. (2015). Different flow-based formulations for the Steiner tree problem in graphs (STP) have, e.g., been considered in Maculan (1987), Bahiense et al. (2002, 2021), see also Ljubić (2021) for a recent survey on the STP.

Our contribution in this work is to introduce three mixed-integer linear programming (MILP) models for the DCStTP. We start with a MILP formulation that models a D-hop Steiner arborescence rooted at a randomly chosen terminal node. In order to derive valid formulations for the DCStTP, we then exploit the concept of triangles, i.e. we model diameter constrained Steiner trees induced by terminal subsets of size three. The first approach ensures that pairwise connections between any two terminals different than the root are not longer than D (cf. Sect. 2.1). For the second model, we start from the concept of common flows introduced by Polzin and Daneshmand (2001) to strengthen the multi-commodity flow formulation for Steiner trees. In the context of the DCStTP, we show that common flows are not only useful for strengthening the bounds, but provide an alternative way to impose diameter constraints (cf. Sect. 3.1). Finally, our third model is based on a new concept, to which we refer as *uncommon flow* (cf. Sect. 3.2). We analyse the strength of these models both theoretically and empirically, and investigate how the respective Benders reformulations influence the computational performance (cf. Sect. 4). As the numbers of variables and constraints of our formulations are already large, we refrain from combining the idea of common/uncommon flows with other disaggregation techniques such as position-indexed variables or layered graphs. Investigating the strength of such models could, however, be an interesting opportunity for future research.

This article is dedicated to Bernard Gendron whose scientific work has laid important foundations for many network optimization problems. Bernard made invaluable contributions in studying and improving standard *multi-commodity flow formulations* for various network design, location and routing problems (see, e.g., Gendron and Semet (2009), Contardo et al. (2013), Thiongane et al. (2015), Chouman et al. (2017)). Moreover, together with his co-authors, in his seminal works (Croxton et al., 2007; Gendron & Gouveia, 2017) Bernard has shown how *disaggregation* techniques can be applied to solve network flow problems with piecewise linear costs. The concepts of common and uncommon flows studied in our paper can be seen as an application of disaggregation techniques to multi-commodity flows, one of Bernard's favorite topics (Frangioni & Gendron, 2009).

2 Standard Flow-Based Model

We will address directed models, that is, we will view the Steiner tree as a subtree directed away from a node in T, e.g., node 0. To define a directed tree we need to "direct" the graph, that is, for each edge $\{i, j\}$ we also distinguish whether the edge is used in the direction from i to j (arc (i, j) is used) or is used the direction from j to i (arc (j, i) is used) and we denote by A the set of arcs defined in this way.

That way, we search for a Steiner arborescence directed away from 0, with a subset of remaining terminals as leaf nodes, and such that the diameter of the undirected counterpart of this arborescence does not exceed D.

In this section we start with a hop-constrained relaxation of the problem, based on the trivial property that if a tree has diameter D then the path between any given node (say 0) chosen as a root and any other terminal $t \in T \setminus 0$, has no more than D edges (or, in the case that we consider directed paths, it has no more than D arcs). Hence, any minimum-cost D-hop constrained Steiner arborescence rooted at 0, spanning all terminals from T, provides a valid lower bound to our problem. Indeed, let $\mathcal{T}$ be a subtree of G that spans all the terminals. If we fix a terminal node $0 \in T$ as a root node of the arborescence associated with $\mathcal{T}$, then the maximum number of hops between 0 and any other node in $\mathcal{T}$ is at most D. This property however does not guarantee that the diameter of $\mathcal{T}$ is at most D, if the condition is imposed *only* for a fixed node 0 (indeed, the diameter can be as large as $2D$). However, if we impose this property to hold for any choice of $0 \in T$, then it is not difficult to see that the resulting tree will have a diameter not larger than D.

In this article, we start from the multi-commodity flow model for the D-hop constrained Steiner arborescence, and we propose three different ways to extend this model so as to ensure that the diameter of the arborescence does not exceed D. The first model (see Sect. 2.1) is a formulation in which we model paths between any two pairs of terminals, the remaining two models (see Sects. 3.1 and 3.2) are based on the concepts of the common or uncommon flow, respectively.

To model the D-hop constrained problem, we consider the binary arc variables a_{ij}, for all $(i, j) \in A$, that indicate whether arc (i, j) is in a solution and the directed binary path variables f_{ij}^t, for all $t \in T \setminus \{0\}$ and all $(i, j) \in A$, indicating whether arc (i, j) is in the path from 0 to t. The model, denoted by (D-hop), reads as follows:

$$\min \sum_{(i,j)\in A} c_{ij} a_{ij} \tag{1a}$$

$$\text{s.t.} \sum_{(j,i)\in A} a_{ji} \begin{cases} = 1 & \text{if } i \in T \setminus \{0\} \\ = 0 & \text{if } i = 0 \\ \leq 1 & \text{otherwise} \end{cases} \qquad i \in V \tag{1b}$$

$$\sum_{(i,j)\in A} a_{ij} \geq \sum_{(j,i)\in A} a_{ji} \qquad i \in S \tag{1c}$$

$$\sum_{(i,j)\in A} f_{ij}^t - \sum_{(j,i)\in A} f_{ji}^t = \begin{cases} 1 & \text{if } i = 0 \\ 0 & \text{if } i \neq 0, t \\ -1 & \text{if } i = t \end{cases} \qquad t \in T \setminus \{0\}\,,\ i \in V \tag{1d}$$

$$\sum_{(i,j)\in A} f_{ij}^t \leq D \qquad t \in T \setminus \{0\} \qquad \text{(1e)}$$

$$0 \leq f_{ij}^t \leq a_{ij} \qquad t \in T \setminus \{0\}\ ,\ (i,j) \in A \qquad \text{(1f)}$$

$$a_{ij} \in \{0,1\} \qquad (i,j) \in A \qquad \text{(1g)}$$

Constraints (1b) are in-degree constraints, guaranteeing that the in-degree of each terminal node is one, except for the root node 0. The in-degree of each potential Steiner node is at most one. Inequalities (1c) state that for each potential Steiner node, the out-degree is greater than or equal to its in-degree. Constraints (1d) ensure that one unit of flow is sent from 0 to every other terminal node, and the capacity constraints (1f) ensure that the flow can be sent only along the arcs of the arborescence. Finally, inequalities (1e) guarantee that the number of arcs in any 0-t path is not larger than D. Also, a well known fact is that in the context of this model the flow variables can be defined as continuous, that is

$$f_{ij}^t \geq 0 \qquad t \in T \setminus \{0\}\ ,\ (i,j) \in A. \qquad (2)$$

Hence, the optimal solution of model (1) is a D-hop constrained minimum Steiner arborescence.

2.1 Flow-Based Model Based on Pairwise Terminal Connections

We now extend model (D-hop) using additional flow variables, denoted as g. The value of g_{ij}^{qt} ($q, t \in T \setminus \{0\}$, $q < t$, $(i,j) \in A$) is equal to one if and only if we can route one unit of flow from q to t along the edge $\{i, j\}$ (notice that we need to "ignore" the direction of the arc, as explained below). Hence, to ensure that the diameter of the arboresce whose incidence vector is given by a is not larger than D, the length of any q-t path should not exceed D. The resulting model, that we will denote by (SF), reads as follows:

$$\min \sum_{(i,j)\in A} c_{ij} a_{ij}$$

$$\text{s.t.} \sum_{(i,j)\in A} g_{ij}^{qt} - \sum_{(j,i)\in A} g_{ji}^{qt} = \begin{cases} 1 & \text{if } i = q \\ 0 & \text{if } i \neq q,t \\ -1 & \text{if } i = t \end{cases} \qquad \begin{array}{c} q,t \in T \setminus \{0\} \\ q < t\ ,\ i \in V \end{array}$$

(3a)

$$\sum_{(i,j)\in A} g_{ij}^{qt} \le D \qquad q,t \in T \setminus \{0\}\ ,\ q < t \tag{3b}$$

$$0 \le g_{ij}^{qt} + g_{ji}^{qt} \le a_{ji} + a_{ij} \qquad \begin{array}{r} q,t \in T \setminus \{0\} \\ q < t\ ,\ \{i,j\} \in E \end{array} \tag{3c}$$

(a, f) satisfy (1b)–(1g)

Constraints (3a) guarantee that one unit of flow is sent from each $q \in T$ to each $t \in T, q < t$. Inequalities (3b) ensure that each q-t path is of length at most D, and finally the capacity constraints (3c) impose that the flow g can only be sent along the arcs of the Steiner arborescence determined by a. We note that the direction of arcs and the direction of the flow g do not need to be identical. Consider an example given in Fig. 1a which demonstrates a Steiner arborescence rooted at 0 with terminals q and t. In this solution, the flow g^{qt} will use arcs a_{ij} in the opposite direction, i.e., $g_{ij}^{qt} = a_{ji}$ for all arcs between node x (the least common ancestor between q and t) and q, whereas $g_{ij}^{qt} = a_{ij}$ for all arcs connecting node x with t. This is why the capacities given on the right-hand-side have to be considered in an undirected fashion.

One may additionally consider constraints

$$0 \le g_{ij}^{qt} \le f_{ji}^{q} + f_{ij}^{t} \qquad q,t \in T \setminus \{0\}\ ,\ q < t\ ,\ (i,j) \in A \tag{4}$$

stating that the flow g can only be sent along arcs determined by the flow f. Since q is the destination node for f and a source node for g, if arc (i, j) is used along the q-t path, this arc corresponds to the opposite direction of the flow f^q. Similarly, since t is the destination node for both f and g, the orientation of the arc (i, j) will be identical for both of them.

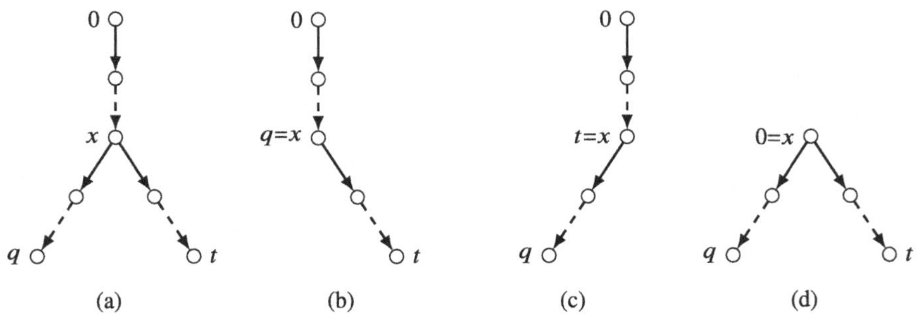

Fig. 1 Feasible triangle graphs for $(0, q, t)$. (**a**) Standard case, (**b**)–(**d**) special cases where one terminal node is the splitting node x. In either case $l_{0q} \le D$, $l_{0t} \le D$, and $l_{xq} + l_{xt} \le D$ must hold, where l_{ij} denotes the number of hops in a path from i to j

The LP relaxation of the previous model can be further enhanced by adding the following *bidirectional linking* constraints, motivated by Balakrishnan et al. (1989):

$$g_{ij}^{kt} + g_{ji}^{qt} \leq a_{ij} + a_{ji} \quad k, q, t \in T \setminus \{0\}, k < t, q < t, k \neq q, \{i, j\} \in E \tag{5a}$$

$$g_{ij}^{kq} + g_{ji}^{kt} \leq a_{ij} + a_{ji} \quad k, q, t \in T \setminus \{0\}, k < q, k < t, q \neq t, \{i, j\} \in E \tag{5b}$$

$$g_{ji}^{kq} + g_{ij}^{kt} \geq g_{ij}^{qt} \quad k, q, t \in T \setminus \{0\}, q < t, k < q, \{i, j\} \in E \tag{5c}$$

The first set of constraints states that we cannot have flow in opposite directions of a given edge going from a common origin to two different destinations. The second set has a similar symmetrical explanation. The third set "triangulates" the flow for a given pair of nodes in terms of a third node, that is, if arc (i, j) is in the path from q to t, then for any other node k, either it is in the path from k to t, or the reversed arc, (j, i), should be in the path from k to q.

The LP relaxation of this standard model can be further strengthened by adding the following set of constraints

$$f_{ij}^{q} + f_{ij}^{t} + f_{ji}^{q} + f_{ji}^{t} + g_{ij}^{qt} + g_{ji}^{qt} \leq 2(a_{ij} + a_{ji}) \quad \begin{array}{c} q, t \in T \setminus \{0\} \\ q < t \ , \ \{i, j\} \in E \end{array} \tag{6}$$

The validity of these constraints follows from the fact that the left hand-side of (6) cannot be greater than two. Observe that for a given pair $(i, j) \in A$, these constraints are a stronger version of an inequality that is obtained by adding (3c) for the same pair and four constraints (1f), respectively for the pairs (i, j) and (j, i) and destinations q and t. We will link constraints (6) with the uncommon flow formulation in Sect. 3.2.

Let us denote the models considered so far by (SF), (SF)+(4), (SF)+(4)–(5) and (SF)+(4)–(6). Let $v_{\text{LP}}(F)$ denote the value of the LP-relaxation of a given formulation F. We can derive the following result:

Proposition 1 *We have:*

$$v_{\text{LP}}(SF) \leq v_{\text{LP}}((SF) + (4)) \leq v_{\text{LP}}((SF) + (4) - (5)) \leq v_{\text{LP}}((SF) + (4) - (6)).$$

Moreover, there exist DCStTP instances for which the strict inequalities hold.

Instead of providing a proof of the second part of this result, we refer to our computational section where we provide some benchmark instances from the literature for which the strict inequalities between the proposed models hold.

3 Triangle-Based Models

The model discussed in the previous section is based on a characterization stating that a tree $\mathcal{T}$ has diameter at most D if any Steiner tree with two terminals contained in $\mathcal{T}$ has diameter at most D. In this section we extend the characterization by stating that a tree $\mathcal{T}$ has diameter at most D if any Steiner tree with three terminals contained in $\mathcal{T}$ has diameter at most D. This generalization follows the idea of strengthening a standard multi-commodity-flow-based model using common flows, as proposed by Polzin and Daneshmand (2001) for Steiner trees.

To see this connection, note that the standard multi-commodity flow models are characterized by including a set of variables and a system of constraints associated to each (or some of the) pair of nodes, essentially guaranteeing that there is path connecting the pair of nodes for which the systems are defined. The previous formulation exemplifies one such model. For each pair of nodes, the model includes sets of variables, either f and g, and the corresponding systems. Polzin and Daneshmand (2001) have proposed a formulation that generalizes this concept by also associating variables to triples of nodes $(0, q, t)$ that model the flow that is common to the paths from node 0 to node q and from node 0 to node t. The main idea of using these new variables is that by also including a new set of coupling constraints (relating new flow variables with standard flow variables and arc design variables) is to decrease the possibility of the standard flow from 0 to q and from 0 to t to split up and then to rejoin again (something that often happens in LP solutions of standard flow models even in problems where the constraints do not allow cycles in the integer feasible solutions, such as trees).

In this work we provide another modeling possibility of using the common flow variables, namely allowing the derivation of the diameter constraints for each pair of nodes $q, t \in T \setminus \{0\}, q < t$. In fact, as we shall shown next, for obtaining a valid model for the DCStTP we did not even need to add flow balancing constraints for the new flow variables. Figure 1 illustrates the four potential realizations of a triangle contained in $\mathcal{T}$ which is rooted at $0 \in T$ and also contains terminals $q, t \in T$. Next, to the standard case shown in Fig. 1a, any of the three terminals (i.e., 0, q or t) may be the splitting node x until which the paths from 0 to q and from 0 to t separate, see Fig. 1b, c, and d.

3.1 A Common Flow Realization

In the common flow variant, cf. Polzin and Daneshmand (2001), each triangle $(0, q, t)$, $\forall q, t \in T \setminus \{0\}, q < t$, is modeled using flow variables $f_{ij}^{q} \geq 0$, $f_{ij}^{t} \geq 0$, introduced in the model (D-hop), and additional *common flow* variables $m_{ij}^{qt} \geq 0$

for each $(i, j) \in A$. The latter describes the common part of flows f^q and f^t. The following model, denoted by (CF), provides a valid formulation for the DCStTP:

$$
\begin{aligned}
\min \ & \sum_{(i,j)\in A} c_{ij}a_{ij} && \\
\text{s.t. } & m_{ij}^{qt} \leq f_{ij}^{q} && q,t \in T\setminus\{0\}\ ,\ q<t\ ,\ (i,j)\in A && (7a)\\
& m_{ij}^{qt} \leq f_{ij}^{t} && q,t \in T\setminus\{0\}\ ,\ q<t\ ,\ (i,j)\in A && (7b)\\
& f_{ij}^{q} + f_{ij}^{t} \leq a_{ij} + m_{ij}^{qt} && q,t \in T\setminus\{0\}\ ,\ q<t\ ,\ (i,j)\in A && (7c)\\
& \sum_{(i,j)\in A} (f_{ij}^{q} + f_{ij}^{t} - 2m_{ij}^{qt}) \leq D && q,t \in T\setminus\{0\}\ ,\ q<t && (7d)\\
& (a, f) \text{ satisfy (1b)–(1g)} &&
\end{aligned}
$$

For each triple $(0, q, t)$, constraints (7a), (7b), and (7c) ensure that the common flow m_{ij}^{qt} on arc $(i, j) \in A$ is equal to one if both f_{ij}^{q} and f_{ij}^{t} are set to one. Constraints (7c) are the previously mentioned constraints that characterize the formulation of Polzin and Daneshmand (2001). These constraints allow us to write the diameter constraints without resorting to using the g flow. By restricting the number of arcs with flow f^q or f^t that are not part of their common flow constraints, inequalities (7d) ensure that the distance between q and t does not exceed D. Finally, observe that without these constraints, we would have a formulation for the D-hop constrained problem with a LP relaxation as tight as the LP relaxation of the formulation (D-hop).

Although not needed to obtain a valid formulation, we can still follow Polzin and Daneshmand (2001) and add, for each pair $q < t, q, t \in T \setminus \{0\}$, the additional flow conservation constraints for common flows m^{qt}:

$$
\sum_{(i,j)\in A} m_{ij}^{qt} - \sum_{(j,i)\in A} m_{ji}^{qt} \leq \begin{cases} 1 & \text{if } i = 0 \\ 0 & \text{if } i \neq 0 \end{cases} \qquad q,t \in T\setminus\{0\}\ ,\ q<t\ ,\ i\in V \qquad (8)
$$

We use (CF$^+$) to refer to model (CF)+(8). Polzin and Daneshmand (2001) show that their common flow model for the STP depends on the choice of the root node. Our computational experiments confirm that this is also the case for our (CF$^+$) model for the DCStTP.

3.2 An Uncommon Flow Realization

As observed previously, to ensure the diameter constraints between q and t, $q < t$, $q, t \in T \setminus \{0\}$, we need to restrict the total flow along those arcs on which either flow f^q or f^t, but not both of them, is used. Hence, we propose to use flow variables $u_{ij}^{qt} \geq 0$ modeling the *uncommon flow* (UF) of q and t.

$$\min \sum_{(i,j)\in A} c_{ij} a_{ij}$$

$$\text{s.t. } u_{ij}^{qt} \geq f_{ij}^q - f_{ij}^t \qquad q, t \in T \setminus \{0\}, \ q < t, \ (i,j) \in A \tag{9a}$$

$$u_{ij}^{qt} \geq f_{ij}^t - f_{ij}^q \qquad q, t \in T \setminus \{0\}, \ q < t, \ (i,j) \in A \tag{9b}$$

$$\sum_{(i,j)\in A} u_{ij}^{qt} \leq D \qquad q, t \in T \setminus \{0\}, \ q < t \tag{9c}$$

$$u_{ij}^{qt} \leq f_{ij}^q + f_{ij}^t \qquad q, t \in T \setminus \{0\}, \ q < t, \ (i,j) \in A \tag{9d}$$

$$u_{ij}^{qt} \leq 2a_{ij} - f_{ij}^q - f_{ij}^t \qquad q, t \in T \setminus \{0\}, \ q < t, \ (i,j) \in A \tag{9e}$$

$$(a, f) \text{ satisfy (1b)–(1g)}$$

For each triple $(0, q, t)$, constraints (9a), (9b), (9d) and (9e) ensure that the uncommon flow u_{ij}^{qt} on arc $(i, j) \in A$ is equal to one if exactly one of values f_{ij}^q and f_{ij}^t is equal to one. By restricting the number of arcs with uncommon flow u^{qt} using constraints (9c), we ensure that the distance between q and t does not exceed D.

Proposition 2 *The models (CF) and (UF) are equally strong in terms of their LP-relaxations, i.e., for all instances of the DCStTP, we have:*

$$v_{\text{LP}}(\text{CF}) = v_{\text{LP}}(\text{UF}).$$

Proof By using the following linear transformation

$$u_{ij}^{qt} = f_{ij}^q + f_{ij}^t - 2m_{ij}^{qt} \qquad (i,j) \in A, \ q, t \in T, q < t$$

we can transform any feasible LP-solution of the model (CF) into an LP-feasible solution of the model (UF), and vice versa. Under this equality, constraint (9d) corresponds to $m_{ij}^{qt} \geq 0$. Other equivalences are: (9a) corresponds to (7a) and (9b) corresponds to (7b); (9e) corresponds to (7c); (9c) corresponds to (7d). Finally, note that an inequality in the common-flow model that corresponds to $u_{ij}^{qt} \geq 0$, is the inequality that results from adding (7a) with (7b) for the same indices.

Additional flow conservation constraints for the uncommon flow can also be imposed:

$$\sum_{(i,j)\in A} u_{ij}^{qt} - \sum_{(j,i)\in A} u_{ji}^{qt} \geq \begin{cases} -1 & \text{if } i \in \{q,t\} \\ 0 & \text{otherwise} \end{cases} \quad q,t \in T \setminus \{0\}\ ,\ q < t\ ,\ i \in V. \tag{10}$$

We use (UF$^+$) to refer to the model (UF)+(10).

Proposition 3 *The models (CF$^+$) and (UF$^+$) are equally strong in terms of their LP-relaxations, i.e., for all instances of the DCSTP, we have:*

$$v_{\mathrm{LP}}(CF^+) = v_{\mathrm{LP}}(UF^+).$$

Proof By using the same linear transformation as in the proof of Proposition 2, we obtain

$$\begin{aligned} &\sum_{(i,j)\in A} u_{ij}^{qt} - \sum_{(j,i)\in A} u_{ji}^{qt} = \\ &\left(\sum_{(i,j)\in A} f_{ij}^{q} - \sum_{(j,i)\in A} f_{ji}^{q}\right) + \left(\sum_{(i,j)\in A} f_{ij}^{t} - \sum_{(j,i)\in A} f_{ji}^{t}\right) \\ &-2\left(\sum_{(i,j)\in A} m_{ij}^{qt} - \sum_{(j,i)\in A} m_{ji}^{qt}\right) \end{aligned} \tag{11}$$

for any $i \in V, q \in T \setminus \{0\},\ t \in T \setminus \{0,q\},\ q < t$. Let us now assume that (8) holds. We distinguish four cases:

1. $i \in V \setminus \{0, t, q\}$:

$$0 \geq 2\left(\sum_{(i,j)\in A} m_{ij}^{qt} - \sum_{(j,i)\in A} m_{ji}^{qt}\right) = 0 + 0 - \left(\sum_{(i,j)\in A} u_{ij}^{qt} - \sum_{(j,i)\in A} u_{ji}^{qt}\right).$$

2. $i = 0$:

$$2 \geq 2\left(\sum_{(i,j)\in A} m_{ij}^{qt} - \sum_{(j,i)\in A} m_{ji}^{qt}\right) = 1 + 1 - \left(\sum_{(i,j)\in A} u_{ij}^{qt} - \sum_{(j,i)\in A} u_{ji}^{qt}\right).$$

3. $i = t$:

$$0 \geq 2\left(\sum_{(i,j)\in A} m_{ij}^{qt} - \sum_{(j,i)\in A} m_{ji}^{qt}\right) = -1 + 0 - \left(\sum_{(i,j)\in A} u_{ij}^{qt} - \sum_{(j,i)\in A} u_{ji}^{qt}\right).$$

4. $i = q$:

$$0 \geq 2\left(\sum_{(i,j)\in A} m_{ij}^{qt} - \sum_{(j,i)\in A} m_{ji}^{qt}\right) = 0 - 1 - \left(\sum_{(i,j)\in A} u_{ij}^{qt} - \sum_{(j,i)\in A} u_{ji}^{qt}\right).$$

Hence, inequalities (10) are satisfied.

Similarly, using the same relationship given by (11), and by distinguishing the same four cases, one can show that inequalities (10) can be transformed into (8).

With Propositions 2 and 3 we have related the LP relaxations of the common and uncommon flow models. We present next an equality that relates the variables u of the uncommon flow model with the variables g of the (SF) model.

$$g_{ij}^{qt} + g_{ji}^{qt} = u_{ij}^{qt} + u_{ji}^{qt} \qquad q, t \in T \setminus \{0\}\ ,\ q < t\ ,\ \{i, j\} \in E \tag{12}$$

Equality (12) follows from the fact that an edge is used in one of the directions by the g flow model if and only if it is used by the uncommon flow u.

Consider the inequality (9e) for a given 4-tuple (i, j, q, t). If we add the same constraint for the tuple (j, i, q, t) we obtain:

$$f_{ij}^{q} + f_{ij}^{t} + f_{ji}^{q} + f_{ji}^{t} + u_{ij}^{qt} + u_{ji}^{qt} \leq 2(a_{ij} + a_{ji}) \quad q, t \in T \setminus \{0\}, q < t, \{i, j\} \in E \tag{13}$$

By using (12) we obtain (6). Hence, while constraints (6) strengthen the (SF) model, they are implied by the common/uncommon flow formulation.

Two observations are in order. First, we note that (6) is obtained after adding two constraints, (9e), from the uncommon flow model. Thus, before obtaining (6), we have weakened the LP relaxation of the uncommon flow model. Secondly, in the same way we have derived (6), we can derive other inequalities in the space of the f and g variables by combining other inequalities from the uncommon flow model. We leave the investigation of these inequalities for future research.

4 Computational Results

In this section, we discuss the results obtained in our computational study. All formulations have been implemented using julia 1.8.0 and JuMP 1.0 (Lubin et al., 2023) in combination with CPLEX 22.1 for solving MILPs. All experiments have been performed on a single core of a standard notebook with an i7-10510U CPU with 1.80 GHz and 16 GB RAM. We used benchmark instances with $|V| \in \{20, 30\}$ that are frequently used for testing approaches for hop- and distance-constrained tree problems (see, e.g., Gouveia et al. (2011, 2015)) in which nodes are randomly placed in a grid. We use the first instance from each set, define the first $|T|$ nodes ($|T| \in \{5, 10\}$) as terminals and use the first terminal as root node 0 in our formulations. We consider instances with random costs (TR) and Euclidean costs (TC, TE) and diameter values $D \in \{3, 4, 5\}$. The main difference between the latter two is the location of the first terminal node (in the center for TC and at the corner for TE).

Note that this study does not aim to improve the computational state-of-the-art for solving diameter-constrained Steiner trees which would likely by obtained by (straightforward) adaptations of the approach from Gouveia et al. (2011) originally proposed for spanning trees to the Steiner tree variant. Instead, we want to understand properties of and the differences between the formulations discussed in the previous sections and shed some light on the question whether using them within dedicated Benders decomposition approaches (that avoid explicitly considering the large number of flow variables) may be promising. Thus, we mainly focus on analyzing the lower bounds obtained from their LP-relaxations. In the second part of our study, we also discuss results obtained from solving the MILPs as well as on applying the automatic or annotated Benders decomposition method of CPLEX to them.

4.1 Comparison of Lower Bounds

Tables 1 and 2 provide LP-relaxation bounds and runtimes needed to solve the LP-relaxations, respectively. We consider the following formulations: (SF), (SF)+(4), (SF)+(4)–(5), (SF)+(4)–(6), and also (CF) and (CF^+). Given the results of Propositions 2 and 3, we can deduce the bounds for the models (UF) and (UF^+), respectively, which is why we do not explicitly report them. Each row in Tables 1 and 2 corresponds to a single instance. The first four columns in these tables indicate: the benchmark set (TR, TC, or TE), the number of nodes, the diameter D and the number of terminals $|T|$ in each instance, respectively. The value of the optimal solution is shown in column v_{IP} of Table 1, followed by the LP-relaxation bounds of each of the six models mentioned above. Bold values in Table 1 indicate the best obtained bounds (rounded to two decimal places). The times reported in Table 2 are runtimes (in seconds rounded to one decimal place) needed to solve the respective formulations after relaxing the binary variables.

The obtained results indicate that each of the constraints introduced in Sect. 2.1 helps in improving the quality of lower bounds of the standard (SF) model. We also observe that even the strongest model based on pairwise diameter constraints, namely (SF)+(4)–(6) can provide bounds which are weaker than the ones obtained by the (CF) model (see, e.g., instances TR-30-*-10). On the other hand, we do not

Table 1 LP-bounds. Bold values indicate the best obtained bounds per each instance

					$v_{LP}((SF)+\cdot)$				$v_{LP}(\cdot)$	
Set	$\|V\|$	D	$\|T\|$	v_{IP}	–	(4)	(4)–(5)	(4)–(6)	(CF)	(CF$^+$)
R	20	3	5	84	74.95	75.86	**76.00**	**76.00**	**76.00**	**76.00**
R	20	4	5	63	61.33	61.33	**61.67**	**61.67**	**61.67**	**61.67**
R	20	5	5	63	**55.00**	**55.00**	**55.00**	**55.00**	**55.00**	**55.00**
R	20	3	10	210	147.72	154.11	159.20	159.40	159.66	**159.78**
R	20	4	10	110	**110.00**	**110.00**	**110.00**	**110.00**	**110.00**	**110.00**
R	20	5	10	110	105.67	**106.00**	**106.00**	**106.00**	**106.00**	**106.00**
R	30	3	5	83	65.30	66.15	66.22	**66.75**	**66.75**	**66.75**
R	30	4	5	48	45.50	45.50	45.50	45.75	45.75	**46.00**
R	30	5	5	36	**36.00**	**36.00**	**36.00**	**36.00**	**36.00**	**36.00**
R	30	3	10	217	141.78	148.58	156.44	160.84	161.29	**161.40**
R	30	4	10	121	103.91	104.73	105.90	107.92	108.02	**108.21**
R	30	5	10	104	87.16	87.34	87.50	88.27	88.28	**88.30**
C	20	3	5	167	**167.00**	**167.00**	**167.00**	**167.00**	**167.00**	**167.00**
C	20	4	5	167	**167.00**	**167.00**	**167.00**	**167.00**	**167.00**	**167.00**
C	20	5	5	167	**167.00**	**167.00**	**167.00**	**167.00**	**167.00**	**167.00**
C	20	3	10	298	252.09	255.40	259.50	**261.00**	**261.00**	**261.00**
C	20	4	10	235	227.78	229.56	232.50	232.50	232.50	**233.13**
C	20	5	10	225	218.83	220.55	**221.59**	**221.59**	**221.59**	**221.59**
C	30	3	5	106	**104.00**	**104.00**	**104.00**	**104.00**	**104.00**	**104.00**
C	30	4	5	104	**104.00**	**104.00**	**104.00**	**104.00**	**104.00**	**104.00**
C	30	5	5	104	**104.00**	**104.00**	**104.00**	**104.00**	**104.00**	**104.00**
C	30	3	10	294	277.10	278.30	**279.25**	**279.25**	**279.25**	**279.25**
C	30	4	10	252	**252.00**	**252.00**	**252.00**	**252.00**	**252.00**	**252.00**
C	30	5	10	246	245.00	245.25	**245.33**	**245.33**	**245.33**	**245.33**
E	20	3	5	144	132.44	132.44	132.44	**132.50**	**132.50**	**132.50**
E	20	4	5	128	**128.00**	**128.00**	**128.00**	**128.00**	**128.00**	**128.00**
E	20	5	5	128	**128.00**	**128.00**	**128.00**	**128.00**	**128.00**	**128.00**
E	20	3	10	279	247.88	**249.00**	**249.00**	**249.00**	**249.00**	**249.00**
E	20	4	10	221	**221.00**	**221.00**	**221.00**	**221.00**	**221.00**	**221.00**
E	20	5	10	213	**213.00**	**213.00**	**213.00**	**213.00**	**213.00**	**213.00**
E	30	3	5	140	132.44	132.44	132.44	**132.50**	**132.50**	**132.50**
E	30	4	5	128	**128.00**	**128.00**	**128.00**	**128.00**	**128.00**	**128.00**
E	30	5	5	128	**128.00**	**128.00**	**128.00**	**128.00**	**128.00**	**128.00**
E	30	3	10	279	247.66	**249.00**	**249.00**	**249.00**	**249.00**	**249.00**
E	30	4	10	221	**221.00**	**221.00**	**221.00**	**221.00**	**221.00**	**221.00**
E	30	5	10	221	**213.00**	**213.00**	**213.00**	**213.00**	**213.00**	**213.00**

Table 2 Times (in seconds) for solving LP relaxations

				$t_{LP}((SF)+\cdot)$				$t_{LP}(\cdot)$	
Set	$\|V\|$	D	$\|T\|$	–	(4)	(4)–(5)	(4)–(6)	(CF)	(CF$^+$)
C	20	3	5	0.2	0.6	0.7	0.7	0.2	0.6
C	20	4	5	0.2	0.3	0.8	0.6	0.3	0.2
C	20	5	5	0.1	0.4	0.6	0.6	0.2	0.3
C	20	3	10	3.2	35.1	216.6	285.0	29.7	59.3
C	20	4	10	1.3	32.9	131.5	100.2	16.6	25.5
C	20	5	10	0.8	18.5	60.7	43.6	6.0	11.8
C	30	3	5	0.4	1.2	3.9	2.6	2.0	2.9
C	30	4	5	0.4	3.6	5.0	4.0	0.8	1.2
C	30	5	5	0.3	1.4	3.9	2.9	0.7	1.3
C	30	3	10	16.3	107.3	1818.6	1381.6	141.8	222.4
C	30	4	10	23.0	110.2	2649.6	1134.8	70.8	160.7
C	30	5	10	17.7	63.4	2502.0	1241.9	35.0	49.7
E	20	3	5	0.1	0.3	0.6	0.8	0.8	0.7
E	20	4	5	0.1	0.6	1.0	1.0	0.1	0.1
E	20	5	5	0.1	0.4	0.7	0.3	0.1	0.2
E	20	3	10	3.7	49.3	221.0	507.7	21.5	38.4
E	20	4	10	3.3	65.2	142.0	147.5	8.3	15.8
E	20	5	10	3.2	131.2	258.0	183.1	5.2	5.7
E	30	3	5	1.9	13.9	16.3	24.4	2.0	7.4
E	30	4	5	0.4	2.0	4.0	5.6	1.5	2.2
E	30	5	5	0.3	1.5	3.4	3.4	1.0	1.9
E	30	3	10	58.5	571.7	2159.3	1057.8	187.1	525.3
E	30	4	10	15.5	92.8	804.6	720.1	54.5	75.6
E	30	5	10	11.1	80.6	966.4	751.8	36.9	51.8
R	20	3	5	0.2	0.3	0.6	0.8	1.1	1.6
R	20	4	5	0.1	0.5	0.6	0.6	0.4	1.0
R	20	5	5	0.1	0.2	0.2	0.5	0.1	0.1
R	20	3	10	2.1	24.3	211.2	396.0	49.4	126.1
R	20	4	10	0.9	10.0	14.0	29.4	5.7	18.9
R	20	5	10	0.8	13.5	34.8	30.7	7.8	10.2
R	30	3	5	1.0	2.6	5.7	3.6	4.8	5.5
R	30	4	5	0.8	2.4	3.9	5.4	3.2	5.0
R	30	5	5	0.1	0.5	1.3	0.5	0.5	0.8
R	30	3	10	19.4	740.3	1683.0	1979.4	816.4	1696.6
R	30	4	10	13.8	97.4	1778.1	2243.6	302.6	491.8
R	30	5	10	12.2	66.8	2180.3	1881.9	101.3	297.7

have examples in which the bounds of the (CF) model are weaker than those of the (SF)+(4)–(6). Hence, it remains an open question whether the LP-relaxation bounds of the (CF) model theoretically dominate the LP-relaxation bounds of the (SF)-based models.

When focusing on the relative improvements of the lower bounds with respect to the basic reference value, which is $v_{\mathrm{LP}}(SF)$, we notice that these improvements are not very significant in most cases. While the two largest improvements are 13.8% (TR, $|V| = 30$, $|T| = 10$, $D = 3$) and 8.2% (TR, $|V| = 20$, $|T| = 10$, $D = 3$) they are below 3.5% in all other cases. Concerning the runtimes, we have several rather surprising results: the fastest model is the basic (SF) formulation, and adding constraints (4) may increase the runtimes by an order of magnitude. Addition of constraints (5) (or (5)–(6)) may further increase the runtimes by another order of magnitude. Given the relatively small improvement in the quality of LP-relaxation bounds achieved by adding these constraints, we conclude that there is a little value of adding these constraints within a branch-and-bound procedure. Indeed, it is more likely that MIP solvers will much faster improve the lower bounds by starting from the basic model (SF) while leveraging on branching strategies. When it comes to (CF) and (CF^+) formulations, we notice that quite often the runtimes are comparable to those of the (SF) model, but they can also be of an order of magnitude higher. Hence, the conclusion is the same: if we were to use the proposed models and solve them with an off-the-shelf solver, the basic model (SF) will be most likely the best performing one, as the runtime would be in favor of the (SF) model when resolving the trade-off between the quality of lower bounds and the size of the branching tree.

4.2 Branch-and-Benders Cut Implementations

We now turn our attention to alternative solution approaches. We namely consider branch-and-Benders-cut implementations for the proposed formulations and compare them with the traditional branch-and-bound approaches in which the LP-relaxations of the flow-based models are solved at every node of the branching tree. For both types of implementations, we use Cplex as off-the-shelf solver. Cplex parameter `Benders strategy` has to be set to 3, in order to activate the `automatic Benders` decomposition (i.e., the branch-and-Benders implementation), in which all flow variables are projected out from the respective model, and Benders feasibility and optimality cuts are generated in terms of a variables. We point out that in Cplex's implementation of branch-and-Benders-cut, all continuous variables are projected out and considered as part of the Benders subproblem. If the constraint matrix admits a block angular structure, the Benders LP subproblem is separated into multiple, smaller LPs, each of them is solved independently, and multiple Benders cuts are generated in each iteration (one per each subproblem). If, on the other hand, the subproblem is not separable, a single Benders cut is generated per each iteration. Thus, one can resort to `Benders strategy` equal to 2 (also known as `annotated Benders`) in order to explicitly keep certain continuous variables in the master problem (which may help Cplex to identify multiple separable subproblems).

In the Benders subproblem derived from the formulation (SF), the flow f is separable by each terminal. In addition, the flow g is separable by each pair (q, t)

of terminals, $q, t \in T \setminus \{0\}$, $q < t$, and hence the underlying LPs can be solved very fast. This separability property allows to add multiple violated Benders cuts in the same iteration and also reduces the number of iterations needed to converge to the LP-optimal solution. The remaining models unfortunately do not admit this property. The separability between f and g is destroyed after adding constraints (4) to (SF), since f and g flows can no longer be solved independently. Similar arguments apply to models (CF$^+$) and (UF$^+$) where all the model constraints link m, respectively u, variables with f, and hence, the separability property does not hold. This leads to a very poor performance of all models (except (SF)) when solved using the automatic Benders strategy. To circumvent this problem of non-separability, we resort to the annotated Benders strategy. In particular, we ensure that flow variables f stay in the master problem. As a consequence, the subproblems decompose per terminal pair $q, t \in T \setminus \{0\}$, $q < t$ for formulations (SF)+(4), (CF), (CF$^+$), (UF), and (UF$^+$). This modification comes at the cost of keeping additional $|T||A|$ variables in the master problem. On the contrary, the Benders subproblem remains non-separable for formulation (SF)+(4)–(6) due to the bidirectional forcing constraints (5). As we will see below, this separability property plays an important role when evaluating the efficiency of the proposed formulations, when it comes to branch-and-Benders-cut implementations.

Table 3 provides summary of obtained results, in which each row corresponds to a single instance. Both, branch-and-bound and branch-and-Benders-cut are given a timelimit of 1h. The table reports runtimes shown in seconds (if the instance is solved to optimality within the timelimit), or gaps (in %) at termination (shown in parenthesis). The gaps are computed relative to the best-known primal bound. A gap of 100% indicates that no feasible solution has been identified within the timelimit. We denote by $t_{\mathrm{IP}}(\cdot)$ the runtime of the branch-and-bound implementation, and by $t_{\mathrm{Ben}}(\cdot)$ the runtime of the branch-and-Benders-cut implementation. We compare the following formulations: the standard flow-based model (SF), and its two extensions, namely (SF)+(4) and (SF)+(4)–(6). In addition, we consider (CF$^+$) and (UF$^+$) formulations. We do not report results for using Benders decomposition on model (SF)+(4)–(6) as this approach failed to solve almost all instances (and resulted in very high optimality gaps) due to the non-decomposable subproblem leading to at most one Benders cut per iteration.

When comparing the solution times t_{IP} of five compact models, we observe a negative correlation between the LP solution times and the overall performance, i.e., the stronger LP-relaxation bounds do not pay off when the models are solved in their compact form. In particular, adding strengthening constraints (4)–(6) to the standard model (SF) does not help to reduce the overall runtimes. On the contrary, already by inserting only constraints (4) to (SF), for some of the instances we observe an order of magnitude increase in runtime. While only a single instance out of 36 instances remains unsolved within 1 h when (SF) model is used, this number increases to five, respectively 11, for the model (SF)+(4) and (SF)+(4)–(6), respectively. The models (CF$^+$) and (UF$^+$) perform similarly, and six, respectively, five instances remain unsolved within 1 h.

Table 3 IP solution times (values in parenthesis are gaps in percent after 1 h in case not solved)

				$t_{\mathrm{IP}}((SF)+\cdot)$			$t_{\mathrm{IP}}(\cdot)$		$t_{\mathrm{Ben}}(\cdot)$			
Set	$\|V\|$	D	$\|T\|$	–	(4)	(4)–(6)	(CF$^+$)	(UF$^+$)	(SF)	(SF)+(4)	(CF$^+$)	(UF$^+$)
C	20	3	5	0.2	0.5	0.4	0.4	0.6	0.3	0.2	0.2	0.3
C	20	4	5	0.1	0.7	0.4	0.5	0.3	0.2	0.1	0.2	0.2
C	20	5	5	0.2	0.3	0.4	0.2	0.3	0.2	0.1	0.2	0.2
C	20	3	10	2765.9	(8.7)	(100.0)	(11.1)	(15.4)	33.7	712.1	(9.1)	(8.2)
C	20	4	10	18.1	628.4	(100.0)	71.1	73	2.2	6.0	42	20.1
C	20	5	10	3.2	195.1	(100.0)	24.7	78.5	2.3	4.9	19.5	13.2
C	30	3	5	0.9	1.7	4.4	2.8	3.9	0.4	0.5	1.8	1.1
C	30	4	5	0.3	0.3	3.9	0.7	1.1	0.3	0.5	0.4	0.5
C	30	5	5	0.2	0.2	4.4	0.6	1	0.3	0.4	0.4	0.5
C	30	3	10	144.1	2794.6	(22.6)	(21.4)	1831.5	7.1	63.1	2087.3	1813.8
C	30	4	10	8.6	76.7	440.2	102.6	214.2	2.4	6.1	28.8	81.1
C	30	5	10	10.1	160.4	263	24.1	200.5	2.2	9.0	10.3	10.2
E	20	3	5	0.4	4.1	2.4	3	1.8	0.3	0.4	1.9	1.2
E	20	4	5	0.1	0.3	0.5	0.2	0.2	0.2	0.1	0.2	0.1
E	20	5	5	0.1	0.1	0.5	0.3	0.2	0.1	0.2	0.1	0.1
E	20	3	10	102.4	1686.8	(10.7)	948.4	1240.4	5.1	62.3	417.1	227.1
E	20	4	10	0.6	2.8	12.6	5.6	19.5	0.9	1.9	2.5	2.8
E	20	5	10	0.5	3.3	12	3.4	19.7	0.9	1.3	2.1	2.6
E	30	3	5	1.1	7.8	16.2	20.2	14.5	0.7	1.2	7.5	3.4
E	30	4	5	1.9	0.4	10.3	3.2	5.4	0.6	0.3	0.5	0.4
E	30	5	5	0.4	0.3	2.5	0.7	1.3	0.5	0.4	0.3	0.4
E	30	3	10	521.5	(5.1)	(100.0)	(8.1)	(7.1)	18.9	231.1	(5.1)	(7.9)
E	30	4	10	25.6	21.5	1697.9	122.9	395.4	1.6	6.2	16.5	37.9
E	30	5	10	5.0	34.9	382.4	63.8	466.6	1.9	6.8	13.3	22.4
R	20	3	5	0.5	3.2	2.9	3.7	3	0.3	0.3	2.3	2.5
R	20	4	5	0.1	0.6	1.6	1.6	0.6	0.1	0.1	0.6	0.5
R	20	5	5	0.4	3.5	3	1.7	3.6	0.2	0.3	0.3	0.3
R	20	3	10	2583.9	(27.4)	(67.0)	2979.4	2765.2	77.9	254.1	976.7	679.0
R	20	4	10	0.8	5.3	33.3	6.7	27.5	0.8	1.6	2.0	1.8
R	20	5	10	1.4	18.2	(2.7)	156.4	92	1	1.9	5.8	8.4
R	30	3	5	5.5	25.4	52.9	73.3	33.5	1.9	3.4	13.5	5.2
R	30	4	5	1.1	2.4	9.2	15.2	13.6	0.4	0.8	3.4	1.1
R	30	5	5	0.2	0.3	0.8	0.7	0.9	0.2	0.2	0.3	0.3
R	30	3	10	(37.9)	(44.0)	(100.0)	(100.0)	(38.5)	1571.6	(10.9)	(65.2)	(28.0)
R	30	4	10	2139.4	2586.5	(100.0)	(46.2)	(43.5)	12.7	133.5	2058.5	645.6
R	30	5	10	405.2	(9.7)	(100.0)	(21.7)	(18.4)	20.2	181.2	334.9	662.4

Comparing the branch-and-Benders-cut implementations versus their "IP counterparts", we observe significant speed-ups, especially for non-trivial instances. So, for example, using Benders cuts, all instances from our benchmark set can be solved to optimality by the model (SF) and the model (SF)+(4) solves all but one instance. For the remaining two models, (CF$^+$) and (UF$^+$), three instances remain unsolved. We observe that, when Benders cuts are used, the winner in terms of performance is the model (SF). This result is quite surprising, and we believe that several major factors explain such a performance: first, the model (SF) already provides strong lower bounds and the quality of these bounds is preserved when Benders cuts are dynamically separated. When comparing t_{IP} times, we have already seen that the minor relative improvement in the quality of lower bounds achieved by adding strengthening inequalities (or considering alternative formulations) does not pay off in the overall branch-and-bound context. Finally, the major reason why the Benders cuts do not make the models (SF)+(4), (SF)+(4)–(6), (CF^+) and (UF^+) competitive against (SF) lies in the fact that the size of the master problem is enlarged by a factor of $|T|$, both in terms of decision variables and constraints. Indeed, to maintain the separability property of the Benders subproblem, flow variables f are kept in the master problem, which significantly slows down every single LP-iteration of the master problem. On the other hand, leaving the variables f in the subproblem destroys the separability and makes the overall performance even worse (due to the excessive times needed to solve a single LP of the Benders subproblem, and slow convergence due to generation of a single cut per iteration).

To summarize, we conclude that Benders cuts significantly improve the performance of compact models, while preserving the quality of lower bounds. However, there is still a significant difference in the performance of branch-and-Benders-cut implementations, which does not only depend on the strength of the underlying model, but also on the separability of the respective Benders subproblems. If this property is not preserved, or is just partially preserved, the overall performance can be negatively affected.

5 Conclusions

In this work we have studied several flow-based formulations for modeling the spanning, respectively, Steiner tree problem whose diameter does not exceed a given parameter D. We have started from a problem relaxation in which we model a Steiner arborescence rooted at a randomly chosen terminal node and ensure that the length of each path between the root and any other terminal does not exceed D. We have considered three ways to extend this formulation in order to obtain a valid model. The first approach is a standard formulation, (SF), in which additional flow variables are introduced to model pairwise connections between the terminals. The model (SF) is then enhanced by several families of valid inequalities. The second approach is based on the concept of common flows which has been introduced to strengthen LP-relaxations for flow-based formulations of Steiner trees (Polzin & Daneshmand, 2001). We show that common flows can be used not only to strengthen the lower bounds, but also to enforce diameter constraints between pairs of terminals. Finally, we have introduced an alternative idea of *uncommon flows*, and shown that also the uncommon flows can be used to limit the pairwise distance between pairs of terminals.

We have compared the proposed models both theoretically and empirically. While the valid inequalities added to (SF) improve the quality of LP-relaxation bounds, and the models derived from common/uncommon flows in principle provide even stronger bounds, the runtimes needed to solve their LP-relaxations are prohibitive and do not make them applicable in practical context. In the second half of our empirical study we have examined the computational performance of the proposed formulations when they are embedded in a branch-and-Benders cut framework. We have seen that projecting out flow variables significantly improves the performance of compact models, while preserving the quality of lower bounds. However, we have observed a significant difference in the performance of branch-and-Benders cuts, as they are affected by the separability of the underlying Benders subproblems. When the separability does not or only partially holds, the overall performance is negatively affected. Hence, the best performing approach is derived from the model (SF) in combination with a branch-and-Benders cut implementation.

The results of our study can trigger further research on strong flow-based formulations for diameter constrained spanning/Steiner trees and related network design problems. Instead of solving LPs to derive Benders cuts, it would be interesting to investigate combinatorial procedures (in particular for the separation of integer infeasible points). In addition, departing from common/uncommon flows and along the lines of disaggregation techniques proposed in Croxton et al. (2007), Gendron and Gouveia (2017), we may obtain new models with even better LP-relaxation bounds. Disaggregations can be based on using position-indexed arc-variables as in Gouveia (1998), Gendron and Gouveia (2017), or on the choice of the root node as in Gouveia et al. (2006).

References

Achuthan, N. R., Caccetta, L., Caccetta, P., & Geelen, J. F. (1994). Computational methods for the diameter restricted minimum weight spanning tree problem. *Australasian Journal of Combinatorics, 10*, 51–72.

Bahiense, L., Maculan, N., & Sagastizábal, C. (2002). The volume algorithm revisited: Relation with bundle methods. *Mathematical Programming, 94*, 41–69.

Bahiense, L., Besso, A., Tostas, R., & Maculan, N. (2021). Using multiflow formulations to solve the Steiner tree problem in graphs. *RAIRO-Operations Research, 55*, S343–S350.

Balakrishnan, A., Magnanti, T. L., & Wong, R. T. (1989). A dual-ascent procedure for large-scale uncapacitated network design. *Operations Research, 37*(5), 716–740.

Binh, H. T. T, McKay, R. I., Hoai, N. X., & Nghia, N.D. (2009). New heuristic and hybrid genetic algorithm for solving the bounded diameter minimum spanning tree problem. In F. Rothlauf (Ed.), *Genetic and Evolutionary Computation Conference, GECCO 2009, Proceedings*, Montreal, July 8–12, 2009 (pp. 373–380). ACM.

Bookstein, A., & Klein, S. T. (1991). Compression of correlated bit-vectors. *Information Systems, 16*(4), 387–400.

Chouman, M., Crainic, T. G., & Gendron, B. (2017). Commodity representations and cut-set-based inequalities for multicommodity capacitated fixed-charge network design. *Transportation Science, 51*(2), 650–667.

Contardo, C., Cordeau, J., & Gendron, B. (2013). A computational comparison of flow formulations for the capacitated location-routing problem. *Discrete Optimization, 10*(4), 263–295.

Croxton, K. L., Gendron, B., & Magnanti, T. L. (2007). Variable disaggregation in network flow problems with piecewise linear costs. *Operations Research, 55*(1), 146–157.

Deo, N., & Abdalla, A. (2000). Computing a diameter-constrained minimum spanning tree in parallel. In G. Bongiovanni, R. Petreschi, & Gambosi G (Eds.), *Algorithms and complexity* (pp. 17–31). Berlin: Springer.

dos Santos, A. C., Lucena, A., & Ribeiro, C. C. (2004). Solving diameter constrained minimum spanning tree problems in dense graphs. In C. C. Ribeiro & S. L. Martins (Eds.), *Experimental and Efficient Algorithms, Third International Workshop, WEA 2004*, Angra dos Reis, May 25–28, 2004, Proceedings. Lecture notes in computer science (Vol. 3059, pp. 458–467). Berlin: Springer.

Frangioni, A., & Gendron, B. (2009). 0–1 reformulations of the multicommodity capacitated network design problem. *Discrete Applied Mathematics, 157*(6), 1229–1241.

Garey, M. R., & Johnson, D. S. (1979). *Computers and intractability: A guide to the theory of NP-completeness*. New York: W. H. Freeman.

Gendron, B., & Gouveia, L. E. N. (2017). Reformulations by discretization for piecewise linear integer multicommodity network flow problems. *Transportation Science, 51*(2), 629–649.

Gendron, B., & Semet. F. (2009). Formulations and relaxations for a multi-echelon capacitated location-distribution problem. *Computers & Operations Research, 36*(5), 1335–1355.

Gouveia, L. E. N. (1998). Using variable redefinition for computing lower bounds for minimum spanning and Steiner trees with hop constraints. *INFORMS Journal on Computing, 10*(2), 180–188.

Gouveia, L., & Magnanti, T. L. (2003). Network flow models for designing diameter-constrained minimum-spanning and Steiner trees. *Networks, 41*(3), 159–173.

Gouveia, L., Magnanti, T. L., & Requejo, C. (2004). A 2-path approach for odd-diameter-constrained minimum spanning and Steiner trees. *Networks, 44*(4), 254–265.

Gouveia, L. E. N., Magnanti, T. L., & Requejo, C. (2006). An intersecting tree model for odd-diameter-constrained minimum spanning and Steiner trees. *Annals of Operations Research, 146*(1), 19–39.

Gouveia, L., Simonetti, L., & Uchoa, E. (2011). Modeling hop-constrained and diameter-constrained minimum spanning tree problems as Steiner tree problems over layered graphs. *Mathematical Programming, 128*(1–2), 123–148.

Gouveia, L., Leitner, M., & Ljubić, I. (2015). The two-level diameter constrained spanning tree problem. *Mathematical Programming, 150*(1), 49–78.

Gouveia, L., Leitner, M., & Ljubić, I. (2020). A polyhedral study of the diameter constrained minimum spanning tree problem. *Discrete Applied Mathematics, 285*, 364–379.

Gruber, M., van Hemert, J. I., & Raidl, G. R. (2006). Neighbourhood searches for the bounded diameter minimum spanning tree problem embedded in a VNS, EA, and ACO. In M. Cattolico (Ed.), *Genetic and Evolutionary Computation Conference, GECCO 2006, Proceedings*, Seattle, July 8–12, 2006 (pp. 1187–1194). ACM.

Ljubić, I. (2021). Solving Steiner trees: Recent advances, challenges, and perspectives. *Networks, 77*(2), 177–204.

Lubin, M., Dowson, O., Garcia, J. D., Huchette, J., Legat, B., & Vielma, J. P. (2023). Jump 1.0: Recent improvements to a modeling language for mathematical optimization. 2206.03866

Lucena, A., Ribeiro, C. C., & Santos, A. C. (2010). A hybrid heuristic for the diameter constrained minimum spanning tree problem. *Journal of Global Optimization, 46*(3), 363–381.

Maculan, N. (1987). The Steiner problem in graphs. *North-Holland Mathematics Studies, 132*, 185–211.

Noronha, T. F., Ribeiro, C. C., & Santos, A. C. (2010). Solving diameter-constrained minimum spanning tree problems by constraint programming. *International Transactions in Operational Research, 17*(5), 653–665.

Polzin, T., & Daneshmand, S. V. (2001). A comparison of Steiner tree relaxations. *Discrete Applied Mathematics, 112*(1–3), 241–261.

Raidl, G. R., & Julstrom, B. A. (2003). Greedy heuristics and an evolutionary algorithm for the bounded-diameter minimum spanning tree problem. In *Proceedings of the 2003 ACM Symposium on Applied Computing* (pp. 747–752).

Requejo, C., & Santos, E. (2009). Greedy heuristics for the diameter-constrained minimum spanning tree problem. *Journal of Mathematical Sciences, 161*, 930–943.

Santos, A. C., Duhamel, C., & Andrade, R. (2016). Trees and forests. In R. Martí, P. Panos, & M. G. Resende (Eds.), *Handbook of heuristics* (pp. 1–27). Cham: Springer International Publishing.

Singh, A., & Gupta, A. K. (2007). Improved heuristics for the bounded-diameter minimum spanning tree problem. *Soft Computing, 11*, 911–921.

Steitz, W. (2015). New heuristic approaches for the bounded-diameter minimum spanning tree problem. *INFORMS Journal on Computing, 27*(1), 151–163.

Thiongane, B., Cordeau, J. F., & Gendron, B. (2015). Formulations for the nonbifurcated hop-constrained multicommodity capacitated fixed-charge network design problem. *Computers & Operations Research, 53*, 1–8

Vik, K. H., Halvorsen, P., & Griwodz, C. (2008). Multicast tree diameter for dynamic distributed interactive applications. In *IEEE INFOCOM 2008 - The 27th Conference on Computer Communications* (pp. 1597–1605). https://doi.org/10.1109/INFOCOM.2008.220

Models and Methods for Two-Level Facility Location Problems

Bernard Gendron, Paul-Virak Khuong, and Frédéric Semet

1 Introduction

This chapter aims to provide a synthesis of the research work carried out by B. Gendron and his co-authors on the two-level uncapacitated facility location problem (TUFLP). The motivation to work on this line of research came from a project with a multi-channel retailer in the early 2000s (Gendron et al., 2002; Gendron & Semet, 2009). The TUFLP is an extension of the uncapacitated facility location problem (UFLP) (Krarup & Pruzan, 1983). Solving an UFLP consists of selecting a set of depots from potential locations to minimize the sum of the fixed costs associated with each depot and of the transportation costs from any depot to each customer. In the TUFLP, the single set of locations is substituted with two tiers of locations (depots and satellites), and the path to each customer must start at a depot and pass through a satellite. In other words, the only arcs are between depots and satellites and between satellites and customers. A classical variant of the TUFLP is the two-level uncapacitated facility location problem with single assignment constraints (TUFLP-S). The TUFLP-S imposes the additional restriction that each satellite can be connected to at most one depot. These constraints appear

Bernard Gendron is deceased at time of publication.

B. Gendron · P.-V. Khuong
CIRRELT and Département d'informatique et de rechercher opérationnelle, Université de Montréal, Montréal, QC, Canada
e-mail: pvk@pvk.ca

F. Semet (✉)
Univ. Lille, CNRS, Inria, Centrale Lille, UMR 9189 CRIStAL, Lille, France
e-mail: frederic.semet@centralelille.fr

T. G. Crainic et al. (eds.), *Combinatorial Optimization and Applications*,
International Series in Operations Research & Management Science 358,
https://doi.org/10.1007/978-3-031-57603-4_4

in a number of applications, including transportation (Gendron & Semet, 2009) and telecommunications (Chardaire et al., 1999). For a large class of TUFLP instances for which the unique assignment constraints are not explicitly enforced, there exists an optimal solution that satisfies these constraints, due to the structure of the objective function (Chardaire et al., 1999).

In this chapter, we address two versions of the TUFLP-S: the *classical* TUFLP-S and the TUFLP-S with modular costs. In the classical TUFLP-S, the objective function includes fixed costs associated with the depots and the satellites, fixed costs for establishing connections between depots and satellites, and transportation costs from any depot to each customer, i.e., each path of the form depot-satellite-customer has a corresponding transportation cost. In the TUFLP-S with modular costs suggested to us by the industrial application, we have modular costs at depots and satellites associated with the number of product units handled and modular costs between depots and satellites associated with the number of vehicles used.

Multi-echelon location analysis is a central research theme in supply chain management reviewed by Melo et al. (2009) and Ortiz-Astorquiza et al. (2018). Most real-life applications lead to more challenging problems than the TUFLP, since they include multi-period and multi-commodity aspects or a complex cost function to optimize, see, e.g. Amrani et al. (2011), Gendron and Semet (2009), Huang et al. (2010), Thanh et al. (2010).

If we restrict ourselves to TUFLPs, there is a wide body of literature as illustrated by the surveys of Klose and Drexl (2005) and by Şahin and Süral (2007). Two main formulations are considered: arc-based (Kratica et al., 2014; Marín, 2007; Pirkul & Jayaraman, 1996, 1998) and path-based models (Barros & Labbé, 1994; Gao & Robinson, 1992; Kaufman et al., 1977; Ro & Tcha, 1984). Most authors propose exact solution methods based on strengthening the models with valid inequalities and facets (Aardal et al., 1996; Baïou & Barahona, 2014; Chardaire et al., 1999; Landete & Marín, 2009), on Benders decomposition (de Oliveira et al., 2020) and on Lagrangian relaxation (Barros, 1998; Marín & Pelegrín, 1999; Pirkul & Jayaraman, 1996, 1998). Some researchers compare relaxations for the arc-based and path-based formulations (Bloemhof-Ruwaard et al., 1994, 1996; Chardaire et al., 1999; Marín & Pelegrín, 1999). Considering the approximate solution of TUFLPs, heuristic methods based on Lagrangian or linear programming relaxations have been developed (Barros et al., 1998; Pirkul & Jayaraman, 1996, 1998), while other heuristic approaches rely on greedy strategies or on simple neighborhood search methods based on add, drop or exchange moves (Barros & Labbé, 1994; Mitropoulos et al., 2009; Narula & Ogbu, 1979; Zhang, 2006). To the best of our knowledge, there are very few attempts to solve TUFLPs with metaheuristics. Chardaire et al. (1999) describe a simulated annealing procedure that produces good solutions, while Ignacio et al. (2008) present a tabu search algorithm that is shown to be efficient compared to solving the mathematical program using CPLEX.

The remainder of this chapter is organized as follows. First, in Sect. 2, we recall the real-life problem at the origin of this research work introduced in Gendron et al. (2002), and give a path-based integer programming model which was proposed in Gendron and Semet (2009). Next, the two heuristic approaches to solve this problem described in Gendron et al. (2015) are presented in Sect. 3. Since this problem is a variant of the TUFLP-S, several formulations for this problem proposed in Gendron et al. (2017) are presented and compared in Sect. 4, followed by the description of solution methods developed (see Gendron et al., 2016) in Sect. 5.

2 A Real-Life TUFLP-S with Modular Costs

The motivation to study TUFLP originated in a project for a company that sells a wide range of products (clothing, electronics, and appliances) through its website, mail-order catalogs, and physical stores (Gendron et al., 2002; Gendron & Semet, 2009). One of the main logistical challenges it faces is to adapt its distribution system to demands that vary daily to meet a level of service (typically, a maximum delivery time of 24 hours) while minimizing the total cost. As a B2C company, consolidation is a major concern. Indeed, most orders contain few items, resulting in small to medium packages.

The company addresses this challenge through a multi-echelon distribution network with single-sourcing (see Fig. 1). Its operations are based in a few fixed primary facilities, which are central warehouses owned by the company. All transportation activities are subcontracted to carriers. A fleet of large vehicles delivers parcels from these primary facilities to secondary facilities, which are cross-docking terminals owned by carriers. Medium-size vehicles depart from these terminals and transfer parcels to smaller cross-docking locations (tertiary facilities), where the items are sorted into pre-assigned tours for final delivery to customers.

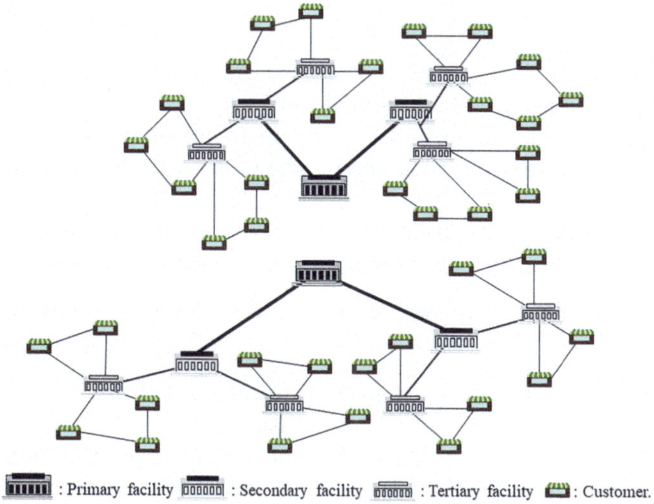

Fig. 1 Initial distribution network

The distribution network design problem is to determine the secondary and tertiary facilities to open and the number of trucks to connect the primary and secondary facilities and the secondary and tertiary facilities. Indeed, the company owns the primary facilities, and their locations are known. Moreover, it is assumed that the routes to serve the customers from a satellite are pre-computed.

Since there are no fixed costs associated with primary facilities and all products are available at each of them, each secondary facility can be assigned to its closed primary facility without losing optimality. Thus, the problem can be modeled as a TUFLP-S with modular costs. The transportation costs for individual connections are step functions that take into account the distance traveled and the capacity of vehicles. A fixed cost is associated with the use of each secondary facility. Secondary facilities incur costs for each batch of parcels sorted, which are also represented by step functions. To keep the vocabulary consistent with the following sections, secondary facilities will be referred to as depots and tertiary facilities as satellites. Gendron and Semet (2009) propose an arc-based and a path-based formulation for this real-life variant of the TUFLP-S. We present here only the path-based model, which outperforms the arc-based formulation. Indeed, even though the model size is larger, the linear relaxations are stronger and the better upper bounds are better when the path-based model is solved with CPLEX for a given CPU time. Let us first define the different sets of nodes:

I : set of potential sites to locate depots;
J : set of potential sites to locate satellites;
K : set of customers;
I_j : set of potential sites to locate depots connected to satellite $j \in J$;
J_i : set of potential sites to locate satellites connected to depot site $i \in I$;
K_j : set of customers connected to satellite $j \in J$;

The location and the transportation costs associated with the facilities and the customers are denoted as follows:

f_i : fixed cost for using and operating depot $i \in I$;
g_j : cost per Q product units for using and operating satellite $j \in J$;
d_i : transportation cost for using one vehicle to transport product units to depot $i \in I$ from its closest hub;
e_{ij} : transportation cost for using one vehicle from depot $i \in I$ to satellite $j \in J_i$;
c_{jk} : transportation cost between satellite $j \in J$ and customer $k \in K_j$.

The customer demands and the vehicle capacities are characterized by the following parameters:

n_k : number of product units to deliver to customer $k \in K$;
v_k : volume of product units to deliver to customer $k \in K$;
Q : capacity (in number of product units) of one batch of products handled at any satellite;
P : volumetric capacity of a vehicle transporting product units to any depot, from its closest hub;

R : volumetric capacity of a vehicle transporting product units from any depot to any satellite.

The path-based model involves six families of decision variables. Three families are associated with the design of the distribution system, while three families are required to compute the total cost. First, Y_i is a 0–1 variable which takes value 1 if depot $i \in I$ received product units from its closest hub; T_{ij} is a 0–1 variable which takes value 1 if product units are transported from depot $i \in I$ to satellite $j \in J$; X_{ijk} is a 0–1 variable which takes value 1 if product units are transported from depot $i \in I$ through satellite $j \in J_i$ to customer $k \in K_j$. Next, V_i is an integer variable that represents the number of trucks used between depot $i \in I$ and its closest hub; H_{ij} is an integer variable which represents the number of trucks used between depot $i \in I$ and satellite $j \in J_i$; U_j is an integer variable which represents the number of batches of products handled at satellite $j \in J$.

$$\min \sum_{i\in I} f_i\, Y_i + \sum_{i\in I} d_i\, V_i + \sum_{j\in J} g_j\, U_j$$

$$+ \sum_{i\in I}\sum_{j\in J_i} e_{ij}\, H_{ij} + \sum_{i\in I}\sum_{j\in J_i}\sum_{k\in K_j} c_{jk}\, X_{ijk} \qquad (1)$$

$$\text{s.t.} \sum_{i\in I}\sum_{j\in J_i} X_{ijk} = 1 \qquad k \in K \qquad (2)$$

$$\sum_{i\in I_j} T_{ij} \le 1 \qquad j \in J \qquad (3)$$

$$\sum_{j\in J_i}\sum_{k\in K_j} v_k X_{ijk} \le \Big(\sum_{k\in K} v_k\Big) Y_i \qquad i \in I \qquad (4)$$

$$X_{ijk} \le T_{ij} \qquad i \in I\,,\ j \in J_i\,,\ k \in K_j \qquad (5)$$

$$T_{ij} \le Y_i \qquad i \in I\,,\ j \in J_i \qquad (6)$$

$$\sum_{i\in I_j}\sum_{k\in K_j} n_k X_{ijk} \le Q U_j \qquad j \in J \qquad (7)$$

$$\sum_{j\in J_i}\sum_{k\in K_j} v_k X_{ijk} \le P V_i \qquad i \in I \qquad (8)$$

$$\sum_{k\in K_j} v_k X_{ijk} \le R H_{ij} \qquad i \in I\,,\ j \in J_i \qquad (9)$$

$$V_i \ge 0 \text{ and integer} \qquad i \in I \qquad (10)$$

$$U_j \ge 0 \text{ and integer} \qquad j \in J \qquad (11)$$

$$H_{ij} \geq 0 \text{ and integer} \qquad i \in I\ ,\ j \in J_i \tag{12}$$

$$Y_i \in \{0, 1\} \qquad i \in I \tag{13}$$

$$T_{ij} \in \{0, 1\} \qquad i \in I\ ,\ j \in J_i \tag{14}$$

$$X_{ijk} \in \{0, 1\} \qquad i \in I\ ,\ j \in J_i\ ,\ k \in K_j \tag{15}$$

In this model, the objective (1) represents the total cost for operating the distribution systems, i.e., the total handling cost and the total transportation costs between the facilities and customers. Constraints (2) enforce that each customer is served once. Constraints (3) ensure that each satellite is connected to at most one depot. Constraints (4), (5) and (6) are the linking constraints between the variables associated with the design of the distribution system. Constraints (7) count the number of batches processed in each satellite. Constraints (8) and (9) count the number of vehicles operated between the hubs and the depots and between the depots and the satellites. Note that constraints (4) are redundant in this formulation.

Gendron and Semet (2009) compare different relaxations of the arc-based and path-based models on data obtained from the company. The real-life network has 93 depots, 320 satellites, and 701 customers. Based on this data, 32 instances were generated, varying the size of the network, the costs at the depots and satellites, and the capacity of the vehicles used between them. As expected, the linear relaxation of the path-based formulation is stronger than that of the arc-based formulation. However, when we consider binary relaxations, i.e., only the integer requirements on the variables T_i, U_j and H_{ij} are relaxed, both models provide the same lower bound. Observing that there is no fixed cost associated with a connection between a depot and a satellite, the binary relaxation of the path-based model can be modeled as a UFLP. Computational experiments were performed to confirm the theoretical results. On the real-life instances, the smallest gaps between the lower bounds computed with a particular relaxation and the best-known upper bounds were obtained by solving the path-based formulation with CPLEX with an average value of 1.7% after 2 hours of CPU Time. Based on these results and motivated by the real application, a heuristic approach was designed, which is described in the next section.

3 Heuristic Approaches for the TUFLP-S with Modular Costs

Two heuristics were developed to solve the TUFLP-S with modular costs in Gendron et al. (2015). First, we propose a variant of the Variable Neighborhood Search (VNS) metaheuristic (Mladenović & Hansen, 1997; Hansen et al., 2019) called the Multilayer VNS (MLVNS). Second, we adapt the classical slope scaling heuristic.

Variable Neighborhood Search (VNS) is a metaheuristic based on the exploitation of neighborhoods. First, VNS orders the neighborhood structures and then iteratively considers them in that order. At each iteration, a new solution is obtained from the current best known solution by exploring the current neighborhood. Then, if a better solution is obtained, the local search continues considering the first neighborhood, otherwise the process continues from the best known solution by considering the next neighborhood.

MLVNS is a recursive variant of VNS based on partitioning the neighborhood structures into l_{max} layers. Each layer l has k^l_{max} neighborhood structures associated with it. As in the classical VNS, the neighborhoods grow in complexity with the layer index. For each layer l, a VNS is invoked. First, a solution x' is identified on a neighborhood $N_k(x)$ of the current solution x until a first improving solution is found. To define the subset of $N_k(x)$, we use m^k_{max} sub-neighborhood structures, resulting from a "natural" decomposition of N_k: $N^1_k, \ldots, N^{m^k_{max}}_k$. At each step m, a solution x' is generated by performing one move in $N^m_k(x)$. If $l = 1$, we assume that evaluating this candidate solution is easy. Otherwise, when $l > 1$, MLVNS is called recursively up to layer $l - 1$ to produce a final candidate solution x'. The generation of candidate solutions continues until x' improves upon x or $m > m^k_{max}$, i.e., all sub-neighborhood structures have been explored. Algorithm 1 summarizes the procedure.

Algorithm 1 MLVNS (x: current solution, l_{max}: number of layers)

for $l \leftarrow 1$ to l_{max} **do**
  Order the k^l_{max} neighborhood structures: $N_1, \ldots, N_{k^l_{max}}$
  repeat
    $k \leftarrow 1$
    repeat
      Order (randomly) the m^k_{max} sub-neighborhood structures: $N^1_k, \ldots, N^{m^k_{max}}_k$
      $m \leftarrow 1$
      repeat
        Generate $x' \in N^m_k(x)$
        if $l > 1$ **then**
          MLVNS($x', l - 1$)
        end if
        $m \leftarrow m + 1$
      until $(f(x') < f(x))$ OR $(m > m^k_{max})$
      if $f(x') < f(x)$ **then**
        $x \leftarrow x'$
        $k \leftarrow 1$
      else
        $k \leftarrow k + 1$
      end if
    until $k > k^l_{max}$
  until some termination condition is satisfied
end for

In our implementation of MLVNS, we first generate an initial solution by a greedy constructive procedure. Then, we design a MLVNS algorithm that includes three layers. In layer 1, four simple neighborhood structures are invoked. They are based on exchange or closing operations. More precisely, they aim at: (i) reassigning a customer from one satellite to another satellite; (ii) reassigning a satellite from one depot to another depot; (iii) closing a satellite; (iv) closing a depot. Note that the exploration of the last two neighborhoods relies on calls to the first two. In layer 2, two neighborhood structures are used, which consist in opening a depot or a satellite presently closed. In each case, the subneighborhood structures are obtained by scanning the set of closed facilities, i.e., depots or satellites. The satellites/customers are then assigned to the closest facilities while keeping the assignment of at least one satellite/customer at each depot/satellite. The move is completed after applying MLVNS at layer 1 and the final configuration is determined. Layer 3 includes neighborhood structures designed to diversify the search through some cost perturbations. Four types of perturbations are considered: (i) increasing the location cost for a selected subset of depots; (ii) increasing the location cost for a selected subset of satellites; (iii) increasing the transport costs from a depot belonging to a selected subset; (iv) increase the transport costs from a satellite belonging to a selected subset. Once the costs have been modified, MLVNS at layer 2 is applied, then the costs are restored and MLVNS at layer 2 is applied again.

To assess the efficiency of the proposed MLVNS approach, we implemented an adapted variant of the slope scaling heuristic. As mentioned above (see Sect. 2), if we relax the integer requirements on the variables T_i, U_j and H_{ij} in the formulation of the TUFLP-S with modular costs, the model reduces to a UFLP model. The slope scaling heuristic consists in first determining an optimal solution to the UFLP. Given the values of variables Y_i and X_{ijk}, a solution of the TUFLP-S with modular costs is derived, and then improved by calling layer 1 of the MLVNS heuristic. Then, the costs in the UFLP model are updated so that the objective function value is equal to the value obtained with the formulation of the TUFLP-S with modular costs if the solution of the UFLP model does not change. For example, let $(\tilde{Y}, \tilde{X})$ be the solution obtained after performing layer 1. Let $i \in I$ be a depot such that : $\sum_{j \in J_i} \sum_{k \in K_j} v_k \tilde{X}_{ijk} > 0$. When the integer requirement on V_i is relaxed, the corresponding value for V_i is

$$\tilde{V}_i = \frac{1}{P} \sum_{j \in J_i} \sum_{k \in K_j} v_k \tilde{X}_{ijk}$$

Then the cost of V_i in the binary relaxation of the model for the TUFLP-S with modular costs is updated as $d_i \lceil \tilde{V}_i \rceil \,/\, \tilde{V}_i$. The derived UFLP formulation is solved, and the process is repeated until the objective values of the UFLP model and the solution with layer 1 of the MLVNS are equal, or until layer 1 provides the same solution for two consecutive calls, or until a time limit is reached.

Gendron et al. (2015) compare the MLVNS with the slope scaling heuristic and with the solution of the path-based model for the TUFLP-S with modular costs with CPLEX on the testbed described in Gendron and Semet (2009) under a time limit of 24 minutes. Average gaps are computed relative to the lower bounds obtained by executing CPLEX for up to 2 hours. First, we analyzed the contribution of each layer of the MLVNS to its performance. When layer 1 only is invoked, the average gap varies from 8.9% to 14.2%. The gap ranges between 3.2% and 6.5% when layers 1 and 2 are applied. Last, when the MLVNS with the 3 layers is applied, it varies from 0.3% for the smallest instances to 4.9% for the full network with the original costs of the company. The MLVNS performs better than both alternative approaches. CPLEX identifies poor solutions with an average gap that reaches 114.7%. The slope scaling heuristic is also outperformed, although the difference is weaker on the full network with an average gap of 5.1%.

In the path-based model, constraints (7), (8) and (9) are introduced to capture the complex modular cost function. The constraints associated with the design of the network defined by constraints (2), (3), (4), (5) and (6) correspond to those of the TUFLP-S, which is considered in what follows.

4 Formulations for the TUFLP-S

The TUFLP is defined on a graph where the set of nodes is partitioned into three subsets: I, J and K. I is associated with the potential depot locations, J with the potential satellite locations, and K with customer locations. The costs are divided into fixed costs and transportation costs. Fixed costs are f_i for operating depot $i \in I$, g_j for operating satellite $j \in J$ and h_{ij} if satellite $j \in J$ is connected with depot $i \in I$. Finally, c_{ijk} is the transportation cost from $i \in I$ to customer $k \in K$ via satellite $j \in J$.

In Barros and Labbé (1994), the authors introduce a model for the TUFLP, which four types of decision variables. First, Y_i is a 0–1 variable which takes value 1 if depot $i \in I$ is open; Z_j is a 0–1 variable which takes value 1 if satellite $j \in J$ is open; T_{ij} is a 0–1 variable which takes value 1 if the connection between depot $i \in I$ to satellite $j \in J$ exists; X_{ijk} is a 0–1 variable which takes value 1 if customer $k \in K$ is served from depot $i \in I$ via satellite $j \in J$. Then, the TUFLP can be stated as formulation (G):

$$\min \sum_{i\in I} f_i\, Y_i \;+\; \sum_{j\in J} g_j\, Z_j \;+\; \sum_{i\in I}\sum_{j\in J} h_{ij}\, T_{ij} \;+\; \sum_{i\in I}\sum_{j\in J}\sum_{k\in K} c_{ijk}\, X_{ijk} \tag{16}$$

$$\text{s.t.} \quad \sum_{j\in J}\sum_{k\in K} X_{ijk} = 1 \qquad k \in K \tag{17}$$

$$X_{ijk} \leq T_{ij} \qquad (i, j, k) \in I \times J \times K \tag{18}$$

$$\sum_{j \in J} X_{ijk} \leq Y_i \qquad (i, k) \in I \times K \tag{19}$$

$$\sum_{i \in I} X_{ijk} \leq Z_j \qquad (j, k) \in J \times K \tag{20}$$

$$Y_i \in \{0, 1\} \qquad i \in I \tag{21}$$

$$Z_i \in \{0, 1\} \qquad i \in I \tag{22}$$

$$T_{ij} \in \{0, 1\} \qquad (i, j) \in I \times J \tag{23}$$

$$0 \leq X_{ijk} \leq 1 \qquad (i, j, k) \in I \times J \times K \tag{24}$$

In this formulation, constraints (17) ensure that every customer of K is served once. Constraints (18), (19), (20) are the linking constraints between the decision variables to compute the total fixed costs associated with paths. The binary requirements on the X_{ijk} variables are relaxed (24) since it is always possible to find an integer solution with the same cost when a solution with fractional values for X_{ijk} variables is obtained.

In the TUFLP-S, an additional constraint forces each satellite to be connected to at most one depot. A *weak* model (Gendron et al., 2017) is obtained by adding the following constraints to (G):

$$\sum_{i \in I} T_{ij} \leq 1 \qquad j \in J \tag{25}$$

Note that constraints (25) are the same constraints as (3) in TCLDP. They can be viewed as generalized upper bound (GUB) constraints. This model can be improved by observing that Z_j is equal to 1 if and only if it exists $i' \in I$ such that $T_{i'j}$ is equal to 1. This leads to the constraints:

$$\sum_{i \in I} T_{ij} = Z_j \qquad j \in J \tag{26}$$

which are stronger than (25) and lead to the so-called *strong* formulation (S) of the TUFLP-S:

$$\min \sum_{i \in I} f_i Y_i + \sum_{j \in J} g_j Z_j + \sum_{(i,j) \in I \times J} h_{ij} T_{ij} + \sum_{(i,j,k) \in I \times J \times K} c_{ijk} X_{ijk}$$

$$\text{s.t.} \sum_{(i,j) \in I \times J} X_{ijk} = 1 \qquad k \in K$$

$$\sum_{i \in I} T_{ij} = Z_j \qquad j \in J$$

$$X_{ijk} \leq T_{ij} \qquad (i, j, k) \in I \times J \times K$$

$$\sum_{j \in J} X_{ijk} \leq Y_i \qquad (i, k) \in I \times K$$

$$Y_i \in \{0, 1\} \qquad i \in I$$

$$Z_i \in \{0, 1\} \qquad i \in I$$

$$T_{ij} \in \{0, 1\} \qquad (i, j) \in I \times J$$

$$0 \leq X_{ijk} \leq 1 \qquad (i, j, k) \in I \times J \times K$$

Note that in (S), constraints (20) are removed since they are redundant. Then, we have the following result:

Proposition 1 (Gendron et al. (2017)) *The linear relaxation of (S) provides a lower bound which is greater than or equal to that of the weak model (G) with (25). This inequality can be strict.*

Other formulations of the TUFLP-S can be easily obtained from (S). A model (S_p) with fewer decision variables can be obtained by projecting out variables Z_j using constraints (26) and adding the constraints (25). Then, the integrality requirements on variable Y_i can be relaxed in models (S) and (S_p) once we add the following constraints:

$$T_{ij} \leq Y_i \qquad (i, j) \in I \times J \tag{27}$$

Thus, two alternative formulations (S^C) and (S_p^C) are derived. Note that the linear relaxation of the four models (S), (S_p), (S^C) and (S_p^C) provided the same bound (Gendron et al., 2017).

The four models plus the weak model were compared using CPLEX on a large testbed, including industrial instances derived from the data provided by the company for the TCLDP (see Sect. 2) and artificial instances obtained using the generator proposed by Kochetov and Ivanenko (2005) for the UFLP. These two sets have different characteristics with respect to the objective function. For industrial instances, the fixed costs associated with the satellites and with the connections between depots and satellites are set to zero, while the fixed costs associated with satellites have high values for artificial instances. Thus, the best results were obtained for industrial instances with (S_p) and for artificial instances with (S^C). The formulation (S) is a good compromise overall when a priority rule for branching is defined on variables Y_i for industrial instances and on variables Z_j for artificial instances.

5 Lagrangian-Based Methods for the TUFLP-S

We now present the exact and approximate solution methods proposed to the TUFLP-S (Gendron et al., 2016). The main idea in developing the Lagrangian-based method was to recover a property already exploited in the relaxations of (TCLDP). As mentioned in Sect. 2, the binary relaxation of (TCLDP) can be modeled as a UFLP. Here, we aim to define a Lagrangian relaxation for a TUFLP-S formulation, which leads to define UFLPs as subproblems. To do so, we considered the model (S_p^C). (S_p^C) is the formulation (S) (see Sect. 4), where variables Z_j are projected out, constraints (25) are added and the binary requirements on Y_i are relaxed. When constraints (18) are relaxed, the Lagrangian subproblem $S_p^C(\lambda)$ is as follows, where λ are the Lagrange multipliers:

$$\min \sum_{i \in I} f_i Y_i + \sum_{(i,j) \in I \times J} \Big(g_j + h_{ij} - \sum_{k \in K} \lambda_{ijk}\Big) T_{ij} + \sum_{(i,j,k) \in I \times J \times K} (c_{ijk} + \lambda_{ijk}) X_{ijk}$$

$$\text{s.t.} \quad \sum_{(i,j) \in I \times J} X_{ijk} = 1 \qquad k \in K$$

$$\sum_{i \in I} T_{ij} \leq 1 \qquad j \in J$$

$$\sum_{j \in J} X_{ijk} \leq Y_i \qquad (i,k) \in I \times K$$

$$T_{ij} \leq Y_i \qquad (i,j) \in I \times J$$

$$0 \leq Y_i \leq 1 \qquad i \in I$$

$$T_{ij} \in \{0,1\} \qquad (i,j) \in I \times J$$

$$0 \leq X_{ijk} \leq 1 \qquad (i,j,k) \in I \times J \times K$$

In the TUFLP-S, all variables are binary. Therefore, we can impose that variables Y_i take 0–1 values. Then, we can relax the binary requirements on variable T_{ij}, since an integer solution can be easily derived from any optimal fractional solution. Finally, we can observe that all but one variable X_{ijk} can be eliminated for a given pair (i,k). Indeed, any optimal solution can be transformed into an equivalent solution where the only nonzero variable X_{ijk} for (i, k) is X_{ij*k} with $c_{ij*k} + \lambda_{ij*k} = \min_{j \in J}(c_{ijk} + \lambda_{ijk})$. By defining new binary decision variables: $W_{ik} = X_{ij*k}$ and parameters : $d_{ik} = c_{ij*k} + \lambda_{ij*k}$ for all $(i, k) \in I \times K$, we get the compact formulation of the Lagrangian subproblem $S_p^C(\lambda)$:

$$
\begin{aligned}
\min \sum_{i \in I} f_i Y_i + \sum_{(i,j) \in I \times J} \Big(g_j + h_{ij} - \sum_{k \in K} \lambda_{ijk}\Big) T_{ij} \quad + \sum_{(i,k) \in I \times K} d_{ik} W_{ik} & \\
\text{s.t.} \sum_{i \in I} W_{ik} = 1 \qquad & k \in K \\
\sum_{i \in I} T_{ij} \le 1 \qquad & j \in J \qquad (25) \\
W_{ik} \le Y_i \qquad & i \in I \\
T_{ij} \le Y_i \qquad & (i, j) \in I \times J \\
Y_i \in \{0, 1\} \qquad & i \in I \\
0 \le T_{ij} \le 1 \qquad & (i, j) \in I \times J \\
0 \le W_{ik} \le 1 \qquad & (i, k) \in I \times K
\end{aligned}
$$

This compact model is a formulation of an UFLP since GUB constraints (25) can be transformed into an equality by adding an artificial depot with zero assignment cost for each satellite.

Different approaches, such as the subgradient and bundle methods, were considered to solve the Lagrangian dual. However, they obtained tight lower bounds only when several iterations were performed, which was costly from a computational time point of view. Therefore, we used a simpler heuristic to obtain lower bounds. This approach consists in first solving the linear relaxation of (S_p^C), then the Lagrange multipliers are set to the values of the dual variables of constraints (18), and finally the Lagrangian subproblem $S_p^C(\lambda)$ is solved. This heuristic is fast and provides lower bounds, at least as good as the linear relaxation, since the Lagrangian subproblem does not have the integrality property.

A primal heuristic was devised from the solution of the Lagrangian subproblem obtained to obtain a feasible solution for the TUFLP-S. This heuristic alternatively solves two UFLPs, providing feasible solutions for the TUFLP-S. First, based on the solution of the subproblem, the set of open depots is fixed, and a UFLP, in which a connection between a depot and a satellite is viewed as a facility, is solved. Next, the connections between the satellites and the customers are fixed, and a UFLP, where the cost of assigning a satellite to a depot includes the total cost of serving the customers via the satellite from the depot, is solved. These two problems are solved to optimality repeatedly until no improvement of the best-known solution is observed.

These lower and upper bounds were embedded in an LP-based branch-and-bound algorithm. Two specific branching schemes were implemented. The first one is based on the GUB constraints (25). For satellite j, two branches are defined either $\sum_{i \in I} T_{ij} = 1$ or $\sum_{i \in I} T_{ij} = 0$. The second branching scheme implements a polytomic branching strategy based on the same constraint. As many child nodes as variables T_{ij} in the sum are generated with $T_{ij} = 1$ and an additional node where the sum is set to zero. The most effective branching strategy tested is to first branch with the GUB scheme and use the polytomic scheme if necessary.

Reliability branching (Achterberg et al., 2005) was implemented to perform the branching choice among the candidate branches. The main adaptation was related to the evaluation of unreliable branches, i.e., branches for which the pseudocosts cannot be evaluated. The unreliable branches are sorted according to the sum of absolute complementary slackness violations in the Lagrangian subproblem. Satellites corresponding to the larger values are considered first.

The efficiencies of the LP-based branch-and-bound algorithm and the primal heuristic were assessed on the testbed described in Sect. 4. On the industrial instances, all models presented in Sect. 4 provide the same gaps to optimal values at the root node. The Lagrangian relaxation approach reduced these gaps significantly (by around 50%), and the Lagrangian-based heuristic provides solutions with an optimality gap of 0.12% on average. Considering the branch-and-bound algorithms, the best results were achieved using only the polytomic branching. However, the approach was not competitive with the solution of S_p^C with CPLEX, even though the number of explored nodes was significantly reduced (more than 14 times on average). On artificial instances, large gaps are obtained at the root node with all models described in Sect. 4. Unfortunately, the Lagrangian bounding procedure did not improve the values obtained with the linear relaxation of S_p^C. However, the branch-and-bound algorithm with the branching strategy combining the GUB scheme and the polytomic branching was able to solve all instances in less than 1'600 seconds on average while CPLEX on formulation S_p^C required up to 4'800 seconds on average. This difference in computation time is related to the number of nodes explored in the branch-and-bound tree, which is more than 50 times larger for the CPLEX solver. The implementation of reliability branching based on information derived from solving the Lagrangian subproblem explains this reduction in the number of nodes.

6 Conclusions

This chapter summarizes the research work of B. Gendron with his co-authors over a period of 15 years on the TUFLP-S. An industrial application in freight transportation drove the interest in this problem. This led to address the TUFLP-S with modular costs. The contribution was twofold. First, formulations of the classical TUFLP-S, on the one hand, and of the TUFLP-S with modular costs, on the other hand, were compared, both theoretically and experimentally, on a testbed including artificial and industrial instances. Second, heuristic and exact methods have been designed, and the computational results show their efficiencies. These results could be improved by developing an adaptive branching strategy choice for solving the models as well as for the specialized branch-and-bound algorithm. The latter would benefit to a great extent from a specialized algorithm for solving the UFLP. Regarding the industrial application, the solutions obtained with the MLVNS are within 5% of the best lower bounds obtained by solving the path-based model. A promising research project is the development of exact algorithms based on decomposition methods applied to the path-based model. Note that one of the main challenges lies in the modular cost structure of the objective function. We have shown that industrial instances can be solved easily when considered with the classical linear cost function.

The planned research avenues include the study of extensions to the multi-level facility location problems and to the multiperiod version. An interesting extension of the TUFLP-S with modular costs relies on the observation that the distribution network results from contracts between the company and carriers for longhaul and last-mile delivery services. Designing the network to meet time-varying demand while considering contracting with carriers is a promising research topic addressed recently in Clavijo López et al. (2024).

Acknowledgments We are thankful to the anonymous referee whose comments helped us write a better paper. We are grateful to our master students and engineers who participated in these research projects.

References

Aardal, K., Labbé, M., Leung, J., & Queyranne, M. (1996). On the two-level uncapacitated facility location problem. *INFORMS Journal on Computing, 8*(3), 289–301.

Achterberg, T., Koch, T., & Martin, A. (2005). Branching rules revisited. *Operations Research Letters, 33*(1), 42–54.

Amrani, H., Martel, A., Zufferey, N., & Makeeva, P. (2011). A variable neighborhood search heuristic for the design of multicommodity production–distribution networks with alternative facility configurations. *OR Spectrum, 33*, 989–1007.

Baïou, M., & Barahona, F. (2014). A polyhedral study of a two level facility location model. *RAIRO-Operations Research, 48*(2), 153–165.

Barros, A. I., Dekker, R., & Scholten, V. (1998). A two-level network for recycling sand: A case study. *European Journal of Operational Research, 110*(2), 199–214.

Barros, I. (1998). *Discrete and fractional programming techniques for location models*. Dordrecht: Kluwer Academic Publishers.

Barros, I., & Labbé, M. (1994). A general model for the uncapacitated facility and depot location problem. *Location Science, 2*, 173–191.

Bloemhof-Ruwaard, J. M., Salomon, M., & Van Wassenhove, L. N. (1994). On the coordination of product and by-product flows in two-level distribution networks: Model formulations and solution procedures. *European Journal of Operational Research, 79*(2), 325–339.

Bloemhof-Ruwaard, J. M., Salomon, M., & Van Wassenhove, L. N. (1996). The capacitated distribution and waste disposal problem. *European Journal of Operational Research, 88*(3), 490–503.

Chardaire, P., Lutton, J. L., & Sutter, A. (1999). Upper and lower bounds for the two-level simple plant location problem. *Annals of Operations Research, 86*, 117–140.

Clavijo López, C., Crama, Y., Pironet, T., & Semet, F. (2024). Multi-period distribution networks with purchase commitment contracts. *European Journal of Operational Research, 312*(2), 556–572.

de Oliveira, P. B., Contreras, I., de Camargo, R. S., & de Miranda Júnior, G. (2020). A comparison of separation routines for benders optimality cuts for two-level facility location problems. *Expert Systems with Applications, 141*, 112928.

Gao, L. L., & Robinson, Jr. E. P. (1992). A dual-based optimization procedure for the two-echelon uncapacitated facility location problem. *Naval Research Logistics, 39*(2), 191–212.

Gendron, B., & Semet, F. (2009). Formulations and relaxations for a multi-echelon capacitated location–distribution problem. *Computers & Operations Research, 36*(5), 1335–1355.

Gendron, B., Semet, F., & Strozyk, C. (2002). Adaptive distribution systems. In *TRANSTECH: Transport Technology Product & Process Innovation Management, Presses Universitaires de Valenciennes*, Valenciennes (pp. 36–42)

Gendron, B., Khuong, P. V., & Semet, F. (2015). Multilayer variable neighborhood search for two-level uncapacitated facility location problems with single assignment. *Networks, 66*(3), 214–234.

Gendron, B., Khuong, P. V., & Semet, F. (2016). A lagrangian-based branch-and-bound algorithm for the two-level uncapacitated facility location problem with single-assignment constraints. *Transportation Science, 50*(4), 1286–1299.

Gendron, B., Khuong, P. V., & Semet, F. (2017). Comparison of formulations for the two-level uncapacitated facility location problem with single assignment constraints. *Computers & Operations Research, 86*, 86–93.

Hansen, P., Mladenović, N., Brimberg, J., & Pérez, J. A. M. (2019). *Variable neighborhood search*. Berlin: Springer.

Huang, Y., Chen, C. W., & Fan, Y. (2010). Multistage optimization of the supply chains of biofuels. *Transportation Research Part E: Logistics and Transportation Review, 46*(6), 820–830.

Ignacio, A. A. V., Ferreira Filho, V. J. M., & Galvão, R. D. (2008). Lower and upper bounds for a two-level hierarchical location problem in computer networks. *Computers & Operations Research, 35*(6), 1982–1998.

Kaufman, L., Eede, M. V., & Hansen, P. (1977). A plant and warehouse location problem. *Journal of the Operational Research Society, 28*(3), 547–554.

Klose, A., & Drexl, A. (2005). Facility location models for distribution system design. *European Journal of Operational Research, 162*(1), 4–29.

Kochetov, Y., & Ivanenko, D. (2005). Computationally difficult instances for the uncapacitated facility location problem. *Metaheuristics: Progress as Real Problem Solvers, 32*, 351–367.

Krarup, J., & Pruzan, P. M. (1983). The simple plant location problem: Survey and synthesis. *European Journal of Operational Research, 12*(36–81), 41.

Kratica, J., Dugošija, D., & Savić, A. (2014). A new mixed integer linear programming model for the multi level uncapacitated facility location problem. *Applied Mathematical Modelling, 38*(7), 2118–2129.

Landete, M., & Marín, A. (2009). New facets for the two-stage uncapacitated facility location polytope. *Computational Optimization and Applications, 44*(3), 487.

Marín, A. (2007). Lower bounds for the two-stage uncapacitated facility location problem. *European Journal of Operational Research, 179*(3), 1126–1142.

Marín, A., & Pelegrín, B. (1999). Applying lagrangian relaxation to the resolution of two-stage location problems. *Annals of Operations Research, 86*, 179–198.

Melo, M. T., Nickel, S., & Saldanha-Da-Gama, F. (2009). Facility location and supply chain management–A review. *European Journal of Operational Research, 196*(2), 401–412.

Mitropoulos, P., Giannikos, I., & Mitropoulos, I. (2009). Exact and heuristic approaches for the locational planning of an integrated solid waste management system. *Operational Research, 9*, 329–347.

Mladenović, N., & Hansen, P. (1997). Variable neighborhood search. *Computers & Operations Research, 24*(11), 1097–1100.

Narula, S. C., & Ogbu, U. I. (1979). An hierarchal location–allocation problem. *Omega, 7*(2), 137–143.

Ortiz-Astorquiza, C., Contreras, I., & Laporte, G. (2018). Multi-level facility location problems. *European Journal of Operational Research, 267*(3), 791–805.

Pirkul, H., & Jayaraman, V. (1996). Production, transportation, and distribution planning in a multi-commodity tri-echelon system. *Transportation Science, 30*(4), 291–302.

Pirkul, H., & Jayaraman, V. (1998). A multi-commodity, multi-plant, capacitated facility location problem: Formulation and efficient heuristic solution. *Computers & Operations Research, 25*(10), 869–878.

Ro, H. B., & Tcha, D. W. (1984). A branch and bound algorithm for the two-level uncapacitated facility location problem with some side constraints. *European Journal of Operational Research, 18*(3), 349–358.

Şahin, G., & Süral, H. (2007). A review of hierarchical facility location models. *Computers & Operations Research, 34*(8), 2310–2331.

Thanh, P. N., Péton, O., & Bostel, N. (2010). A linear relaxation-based heuristic approach for logistics network design. *Computers & Industrial Engineering, 59*(4), 964–975.

Zhang, J. (2006). Approximating the two-level facility location problem via a quasi-greedy approach. *Mathematical Programming, 108*, 159–176.

Facility Location: A Guide to Modeling and Solving Complex Problem Variants via Lagrangian Relaxation Heuristics

Sanjay Dominik Jena

1 Introduction

Facility location problems aim at finding the optimal locations to place facilities that are required to satisfy customer demand. Among the most simple and classical problem variants is the *Capacitated Facility Location Problem*, which considers as input a set of customer demands, along with a set of candidate facility locations, production capacities and transportation costs. The problem consists in selecting a subset of the candidate locations to construct production facilities, from which the customer demand is then served. As such, the problem balances the trade-off between investment and operational costs. While constructing too many facilities is likely to be costly, too few facilities may result in large (and therefore expensive) transportation distances to serve customer demands.

Facility location problems are among the most studied NP-hard combinatorial optimization problems, given that they realistically represent the planning contexts in many application domains, including production planning, telecommunication (see, e.g. Chardaire & Sutter, 1996), health care (see, e.g. Vahidnia et al., 2009), education (see, e.g. Antunes et al., 2009) and forestry (see, e.g. Jena et al., 2015b). While initial facility location problems were rather simpler, the degree of realism to which facility location problems have been modeled has drastically increased over the years. Planning problems nowadays may involve the selection of production equipment and capacity, multiple time-periods, the adjustment of production capacity over time, the relocation of facilities, multiple commodities, refined cost functions, the representation of parameter uncertainty and, in the context of more complex supply chains, distributions networks spanning over several echelons.

S. D. Jena (✉)
CIRRELT and School of Management, Université du Québec à Montréal, Montréal, QC, Canada
e-mail: jena.sanjay-dominik@uqam.ca

T. G. Crainic et al. (eds.), *Combinatorial Optimization and Applications*,
International Series in Operations Research & Management Science 358,
https://doi.org/10.1007/978-3-031-57603-4_5

Even though multi-purpose optimization solvers have greatly evolved, the complexity and scale of the planning problems have outgrown their capacity to solve such problems in sufficiently short computing times. As a remedy, mathematical decomposition has been successfully applied to a variety of combinatorial optimization problems. In the context of facility location problems, Lagrangian relaxation has shown to be an efficient approach, if applied correctly. However, given the variety of facility location problems, developing an appropriate optimization model and the right ingredients required for a Lagrangian heuristic can be a challenging task. This chapter aims at guiding the reader through the process of developing both an appropriate optimization model and the corresponding Lagrangian heuristic. To this end, we focus on the development of strong optimization models, enabling the Lagrangian solutions to provide strong bounds. We will here focus on Mixed-integer Programming models, which have been by far the predominant modelling paradigm in this domain, particularly due to its flexibility. We then explain how Lagrangian relaxation can be applied and how its ingredients can be tailored to the specific problem variant at hand.

Contributions and Scope Given the diversity of problem variants and Lagrangian heuristics proposed in the last decades, this work aims at providing a systematic guide on how to model complex facility location problems and solve them by means of Lagrangian heuristics. As such, it synthesizes content from three principal references (Jena et al., 2016, 2017; Šťádlerová et al., 2023), but also provides a few new suggestions that have not yet been considered in the literature (e.g., a formulation to model piecewise linear cost functions that does not make assumptions on the shape of the function).

While this work focuses on modeling many complex planning characteristics (as will be reviewed in Sect. 2), it restricts to the case of multi-sourcing and single-echelon problems. As such, we do not discuss the modeling or resolution of problems with multiple echelons or general transportation networks in supply chains (see, e.g., Allen et al., 2022). We therefore also exclude hub-location problems, which have been successfully tackled by Lagrangian relaxations (see, e.g., Contreras et al., 2009). Finally, it is worthwhile mentioning that Lagrangian Relaxation has also been applied to location problems based on discrete-choice models. A general framework is presented by Pacheco Paneque et al. (2022).

Outline The structure of this chapter is as follows. Section 2 first reviews the related literature on Lagrangian relaxation for Facility Location problems, highlighting the evolution of problem complexity over time. Section 3 then defines the general single-echelon planning problem, along with the extensions that are considered throughout the paper, namely, complex production cost functions, modular facility structures, facility relocation and parameter uncertainty. Along with such definitions, this section introduces a general problem formulation and the required model extensions. Section 4 then reviews Lagrangian Relaxation and its components (i.e., the Lagrangian subproblem, the Lagrangian Dual, and the generation of feasible upper bounds) in the context of the base formulation. The following subsections then explain how these components can be adapted to the various problem variants. Finally, Sect. 5 concludes the chapter.

2 A Review of Lagrangian Relaxation for Facility Location

Facility location problems have been solved by Lagrangian Relaxation since over 30 years. To the best of our knowledge, up to today, Beasley (1993) has been the only work proposing a framework to apply Lagrangian Relaxation to location problems. The authors considered the classical Capacitated Facility Location Problem and explored the two principal relaxations, which, until today, remain the most promising: relaxing the capacity constraints, or relaxing the demand constraints. Relaxing the capacity constraints mostly makes sense when tackling a simple location problem, in which no additional capacity decisions are involved. The resulting subproblem is an uncapacitated location problem, which itself is NP-hard. As a result, relaxing the demand constraints is a more popular approach, which allows, in most cases, to decompose the problem by candidate facility locations.

Table 1 reviews a selection of papers that applied Lagrangian relaxation to some sort of facility location problem. In particular, it illustrates the characteristics of the tackled facility location problems. Throughout the abbreviations used to define the characteristics of the planning problem, lower-case letters indicate that the characteristic results in an easier-to-solve problem, while upper-case letters indicate that the problem becomes more difficult to solve (typically, since it becomes more complex).

The first column refers to the planning horizon: a problem has either a **(s)**ingle or **(M)**ultiple time-periods. The second column refers to the network structure over which commodities are routed. We here focus on **(s)**ingle-echelon problems, i.e., those with one layer of location decisions (here, facilities), additional to the layer of customers that have to be served. For the sake of completeness, we also include a few works with **(M)**ultiple layers (including, for example, a layer of warehouses) and those with a **(G)**eneral network structure. The following column indicates whether a **(s)**ingle or **(M)**ultiple commodities are considered. The next set of columns indicates what kind of capacity restrictions facilities are subject to. If none of the subcolumns is checked, then the problem is uncapacitated. Capacitated problems either have a **(f)**ixed pre-defined capacity or allow to select the capacity from a pre-defined set of capacity **(L)**evels. Some works also allow to install **(M)**ultiple facilities at the same location. Finally, in multi-period settings, the capacity of facilities may be **(E)**xpanded or **(R)**educed once, or even be **(A)**djusted several times throughout the planning horizon. Problems may also involve some particular **(I)**ndustrial capacity constraints in order to represent the application context more realistically. The next column indicates whether customers can be served by several facilities or only by a single one, also known as **(m)**ulti- and **(S)**ingle-sourcing, respectively. The latter involves that allocation variables are binary, which significantly complicates the solution of the problem. Single-

Table 1 Characteristics of location problems tackled by Lagrangian relaxation

Paper	Time per.	Eche-lons	Comm-odities	Capacity constr.							Sour-cing	Cost func.	Relo-cation	Uncer-tainty	Relaxed constr.
				f	L	M	E	R	I	A					
Guignard-Spielberg and Kim (1983)	s	s	s	✓							S	s			S
Barcelo et al. (1990)	s	s	s	✓							S	s			D
Sridharan (1991)	s	s	s	✓							m	s			D
Beasley (1993)	s	s	s	✓							m	s			D v C
Shulman (1991)	M	s	s	✓			✓				m	s			D
Chardaire and Sutter (1996)	M	s	s								m	s			D
Holmberg and Ling (1997)	s	s	M		✓						m	C			D
Agar and Salhi (1998)	s	s	s	✓							S	s			
Hinojosa et al. (2000)	M	M	M	✓							m	s			D
Correia and Captivo (2003)	s	s	s		✓						m	s			D
Wu et al. (2006)	s	s	M	✓		✓					m	s			D
Correia and Captivo (2006)	s	s	s		✓						S	s			D
Hinojosa et al. (2008)	s	M	M	✓							m	s			D & F
Li et al. (2009)	s	s	M	✓							m	s			C
Diabat et al. (2011)	M	s	s								S	s			D
Görtz and Klose (2012)	s	s	s	✓							m	s			D
Ghodsi (2012)	s	s	s	✓							m	s		✓	C & N
Gendron et al. (2016)	s	M	s								m	s			C
Jena et al. (2016)	M	s	M		✓		✓	✓	✓	✓	m	s	✓		D & R
Jena et al. (2017)	M	s	M		✓		✓			✓	m	s			D
Marín et al. (2018)	M	s	s	✓						✓	m	s		✓	D
Cabezas et al. (2021)	s	s	s								m	s		✓	D
Kadri et al. (2022)	s	G	M	✓							m	s			F
Štádlerová et al. (2023)	M	s	M		✓		✓				m	C		✓	D

sourcing also complicates the solution via Lagrangian Relaxation, since the demand allocation for each facility in the Lagrangian subproblem cannot be simply solved as a transportation problem. The function used to describe the operational cost function (including, e.g., maintenance, production or transportation costs) is indicated in the next column: either it is a **(s)**imple linear cost-function, or it is more **(C)**omplex, often involving piecewise linear functions. The next two columns indicate whether the problem allows for facility relocation and the representation of parameter uncertainty. Finally, the last set of columns indicates the constraints relaxed within the Lagrangian decomposition: **(D)**emand constraints, **(C)**apacity constraints, **(F)**low conservation constraints (only in multi-echelon problems), **(R)**elocation linking constraints (only when relocation is allowed) or **(N)**on-anticipativity constraints (only in the case of stochastic programming).

While not exhaustive, Table 1 aims at illustrating how both the complexity of the planning problems and their corresponding Lagrangian relaxations have evolved over time. Indeed, the considered planning problems demonstrate that Lagrangian Relaxation has been consistently applied to new and more complex problem variants. In particular, complicating features, such as multiple time-periods, multiple commodities and more complex cost functions have been tackled more commonly in recent years. Most of the works have assumed predefined fixed production capacities for the facilities. In the last 20 years, more complex variants have been considered and solved, including the choice of the capacity level and the possibility of several facilities at the same location. Capacity expansion, reduction, and the adjustment of capacities along time are more recent integrations. Most problems also focus on the multi-sourcing problem variant, which is easier to solve than the single-sourcing variant (and also more common in the real planning practices). The proposed Lagrangian algorithms mostly relax the demand constraints. When problem variants are particularly complex, such as those including general network structures, relocation or parameter uncertainty, the corresponding complicating constraints often have also been relaxed. Literature on problem variants integrating such characteristics is rather recent.

3 Problem Definition and Mathematical Formulations

In the here considered planning problem, a number of facilities has to be placed over a planning horizon, discretized into a number of time periods, to satisfy customer demands for several commodities. Facilities can be constructed at different capacity levels (selected from a finite set of available capacity levels), incurring different construction costs, different production capacities and different per-unit production costs. Demand quantities for each of the commodities are defined for each of the customer nodes and have to be served from the available facilities, incurring transportation costs. Facilities may be constructed at any time period. Further, existing facilities may expand or reduce capacity level, therefore changing their capacity level at the various time periods. Given that demand quantities may vary over time, this allows for more dynamic planning solutions that can adapt to geographical demand changes. Changing capacity levels incurs costs. The objective is to find a facility capacity schedule that serves all customer demands at minimum total costs.

This general planning problem is modelled in Sect. 3.1, generalizing a variety of facility location problems. We also discuss the modelling elements that render this formulation stronger than alternative models. Section 3.2 then illustrates how the use of the model parameters results in different special cases.

The following subsections then extend the general problem by additional planning features. For the sake of concision, the definition of such planning features and the corresponding formulations are presented in the same section. Specifically, Sect. 3.3 discusses the use of more complex (specifically, piecewise linear) objective functions. Section 3.4 models complex facility structures, which may be partially and temporarily activated or closed. Section 3.5 focuses on a model extension that enables facility relocation. Finally, Sect. 3.6 elaborates on the representation of parameter uncertainty and the integration of different scenarios via 2-stage stochastic programming.

3.1 A General Model

This section first reviews a general formulation (Jena et al., 2017) that encompasses a variety of different facility location problems, including those with multiple commodities, multiple time-periods, multiple capacity levels and allowing to adjust the capacity level throughout the planning horizon.

Facilities may be constructed at candidate facility locations indicated by a set J. A facility is constructed at one of the eligible capacities, given within a discrete set $L = \{0, 1, \ldots, \overline{\ell}\}$ of possible capacity levels, where level 0 indicates that no facility exists. Customer demand points are given by set I. The planning horizon contains a total of $\overline{t}$ time periods, included in set $T = \{1, 2, \ldots, \overline{t}\}$. Without loss of generality, we assume that the beginning of period $t+1$ corresponds to the end of period $t \in T$. In addition, we make the assumption that all facility openings, closings and capacity changes are implemented at the beginning of a time period. Customer demands may be specified for different commodities, given within set P. The demand of customer $i \in I$ for commodity $p \in P$ in period $t \in T$ is denoted by d_{ip}^{t}, while the cost to serve one unit of commodity $p \in P$ from facility $j \in J$ operating at capacity level $\ell \in L$ to customer $i \in I$ during period $t \in T$ is denoted by $g_{i\ell p}^{jt}$. The capacity of a facility of size $\ell \in L$ at location $j \in J$ is given by u_{ℓ}^{j} (with $u_{0}^{j} = 0$). The cost matrix $f_{\ell_1 \ell_2}^{jt}$ describes the combined cost to change the capacity level of a facility at location $j \in J$ from $\ell_1 \in L$ to $\ell_2 \in L$ at the beginning of period $t \in T$ and to operate the facility at capacity level $\ell_2 \in L$ throughout time period $t \in T$. Note that if $\ell_1 = \ell_2$, in most applications, there are no capacity change costs, but only operating costs.

Furthermore, we let $\ell^j \in L$ be the initial capacity level of an existing facility at location $j \in J$.

The mathematical formulation uses binary variables $y^{jt}_{\ell_1\ell_2}$ equal to 1 if and only if the facility at location $j \in J$ changes its capacity level from $\ell_1 \in L$ to $\ell_2 \in L$ at the beginning of period $t \in T$. For a facility $j \in J$ open at capacity level $\ell \in L$, set $L^-(j,t,\ell) \subseteq L$ defines the capacity levels to which the capacity may be changed at the beginning of period $t \in T$. In a similar fashion, set $L^+(j,t,\ell) \subseteq L$ defines the capacity levels from which the capacity may be changed to level $\ell \in L$ at facility $j \in J$ at the beginning of period $t \in T$. Depending on the application context, only a limited set of capacity changes may be eligible in practice. As will be explained in the next section, a careful definition of sets L^- and L^+ therefore enables the model to represent different problem variants. The continuous allocation variables $x^{jt}_{i\ell p}$ denote the fraction of the demand of customer $i \in I$ for commodity $p \in P$ in period $t \in T$ that is served from a facility of size $\ell \in L$ located at $j \in J$. The resulting model is known as the *Generalized Modular Capacities (GMC)* formulation:

$$\min \sum_{j\in J}\sum_{t\in T}\sum_{\ell_1\in L}\sum_{\ell_2\in L^-(j,t,\ell_1)} f^{jt}_{\ell_1\ell_2} y^{jt}_{\ell_1\ell_2} + \sum_{i\in I}\sum_{j\in J}\sum_{\ell\in L}\sum_{p\in P}\sum_{t\in T} g^{jt}_{i\ell p} d^t_{ip} x^{jt}_{i\ell p} \tag{1}$$

$$\text{s.t.} \sum_{j\in J}\sum_{\ell\in L} x^{jt}_{i\ell p} = 1 \qquad i\in I,\ p\in P,\ t\in T \tag{2}$$

$$\sum_{i\in I}\sum_{p\in P} d^t_{ip} x^{jt}_{i\ell p} \le \sum_{\ell_1\in L^+(j,t,\ell)} u^j_\ell y^{jt}_{\ell_1\ell} \qquad j\in J,\ \ell\in L,\ t\in T \tag{3}$$

$$\sum_{\ell_1\in L^+(j,t-1,\ell)} y^{j(t-1)}_{\ell_1\ell} = \sum_{\ell_2\in L^-(j,t,\ell)} y^{jt}_{\ell\ell_2} \qquad j\in J,\ \ell\in L,\ t\in T\setminus\{1\} \tag{4}$$

$$\sum_{\ell_2\in L^-(j,t,\ell^j)} y^{j1}_{\ell^j\ell_2} = 1 \qquad j\in J \tag{5}$$

$$x^{jt}_{i\ell p} \le \sum_{\ell_1\in L} y^{jt}_{\ell_1\ell} \qquad i\in I,\ j\in J,\ \ell\in L,\ p\in P,\ t\in T \tag{6}$$

$$x^{jt}_{i\ell p} \ge 0 \qquad i\in I,\ j\in J,\ \ell\in L,\ p\in P,\ t\in T \tag{7}$$

$$y^{jt}_{\ell_1\ell_2} \in \{0,1\} \qquad j\in J,\ \ell_1\in L,\ \ell_2\in L^-(j,t,\ell),\ t\in T \tag{8}$$

The objective function (1) minimizes the total cost for changing the capacity levels, maintaining open facilities and serving the demand. Constraints (2) are the demand constraints for the customers. Constraints (3) are the capacity constraints at the facilities. For each location $j \in J$, the model tracks the current capacity level throughout the planning horizon. Constraints (4) link the capacity change variables in consecutive time periods, while Constraints (5) initialize the sizes of the facilities at the beginning of the planning horizon, ensuring that exactly one capacity level is selected. We may refer to such constraints as capacity flow conservation constraints, given that, for each facility location, a flow of 1 unit is conserved throughout the network to specify the capacity level defined over the various time periods.

Constraints (6), called the *Strong Inequalities (SI)*, ensure that no demand can be served from a facility of size $\ell \in L$ at period $t \in T$ if no such facility is available during period $t \in T$. Since the capacity constraints are also enforcing the same requirements, the SIs are redundant to the MIP model. Similar inequalities, which can be seen as a special case of flow cover inequalities (Padberg et al., 1983), are used in many facility location and network design problems (e.g., Gendron, 2011; Chouman et al., 2017). In some simple location problems, such SIs are also facet-defining (see, e.g., Cornuejols & Thizy, 1982). In general facility location and network design formulations, they tend to drastically improve the LP relaxation bound.

Figure 1 illustrates the network that, for a given facility, manages the open capacity level at each of the time-periods. A flow of 1 unit is sent through the network by activating binary variables $y^{jt}_{\ell_1\ell_2}$. Again, activating variable $y^{jt}_{\ell_1\ell_2}$, i.e., selecting arc $(\ell_1\ell_2)$, with $\ell_1 = \ell_2$ indicates that the capacity level remains at that level throughout time period t, while an activated arc with $\ell_1 \neq \ell_2$ indicates a capacity level change. As such, a path through this network represents a capacity opening schedule for that facility. The use of such detailed variables may seem overly complicated on first sight, as there is a large number of binary variables if L is large. A more classical formulation would consider the simpler binary variables

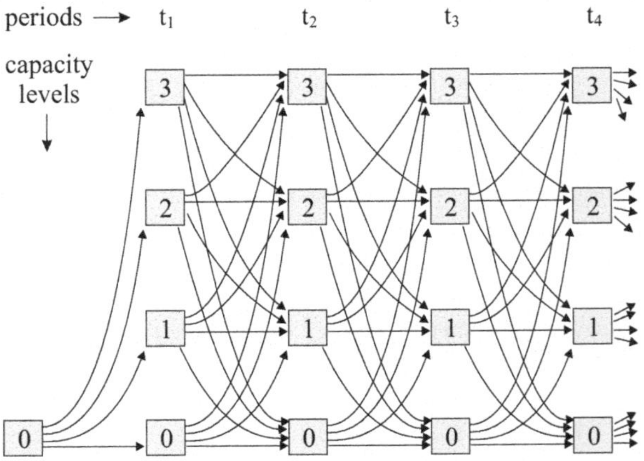

Fig. 1 Network flow structure of capacity changes for a given facility

y_ℓ^{jt} that take value 1 if the facility $j \in J$ has capacity level $\ell \in L$ at throughout time period $t \in T$. Such formulation requires additional binary variables indicating the expansion or the reduction of ℓ capacity levels. Both formulations are compared in Jena et al. (2015a). While the formulation using variables with a single ℓ-index consists of less variables, it tends to provide a larger integrality gap (i.e., it provides a weaker LP relaxation) than the GMC formulation. The main reason being that, in the GMC formulation, the OF coefficients $f_{\ell_1\ell_2}^{jt}$ incorporate the costs for both changing the capacity level and for maintaining the facility open at that level throughout the time-period. Constraints (4) are defined for each $\ell \in L$. This is not possible in the simpler formulation, where the LP relaxation solution may mix construction and maintenance variables with lowest costs. When developing a Lagrangian heuristic, one is therefore particularly interested in strong formulations which may allow the Lagrangian subproblem to provide stronger lower bounds.

Another advantage of using the more complex GMC formulation lies in its ability to represent more refined capacity change costs. While the simpler formulation represents only the costs to increase or decrease the capacity level by ℓ levels, the GMC formulations can encode more details within $f_{\ell_1\ell_2}^{jt}$, allowing, for example, to have different costs for a capacity expansion from level 1 to 3 then for an expansion from level 2 to 4, even though, in both cases, the capacity is increased by two levels.

3.2 Special Cases

The formulation presented above may be unnecessarily complex for several applications. By careful design of the sets used within this formulation, several special cases can be modelled. In particular, the single-commodity variant is modelled when set P holds only one commodity. The single-period problem variant is modelled when set T has only a single time-period. In this case, the set of eligible capacity levels for facility $j \in J$ is solely defined by $L^-(j, t = 0, \ell)$. If the problem is uncapacitated, i.e., facilities are assumed to have infinite capacity, Constraints (3) are not required. The strong inequalities (6) are then sufficient to ensure that only opened facilities can be used. Finally, if L holds only one capacity level in addition to capacity level 0, the model represents a problem variant with fixed capacity levels.

In the general case, if $L^- = L$ and $L^+ = L$, the formulation allows to expand and reduce from any capacity level to any other capacity level. Jena et al. (2015a) explore two problem variants as special cases. In one problem variant, facilities can be constructed at any capacity level. Once constructed, they can be temporarily closed when idle in order to avoid high maintenance costs. However, the available capacity at a given location cannot be changed. This problem variant can be modelled by adding artificial capacity levels that represent facilities that are temporarily closed, along with the corresponding arcs that represent the closing and reopening. The other problem variant allows for expanding and reducing capacity throughout the planning horizon. Here, a temporary closing is not possible, and one needs to completely shut down the facility (to level 0) in order to completely avoid maintenance costs. Both problem variants, as well as many more, can be modelled by appropriately defining L^- and L^+.

Finally, if minimum production quantities are required at the facilities, the total production quantity may exceed the amount transferred to the customers. Such excess capacity may incur penalties, which can be easily integrated into the model (see, e.g., Štádlerová et al., 2023).

3.3 Accounting for Complex Cost Functions

The facility location problem considered above makes the standard assumption that, for each capacity level $\ell \in L$, production costs are linear in function of the total production quantity, i.e., the per-unit production costs is constant for each level $\ell \in L$. Specifically, for a facility $j \in J$ and time period $t \in T$, parameter $g_{i\ell p}^{jt}$ defines both the per-unit production cost at capacity level $\ell \in L$ and the per-unit distribution costs for product $p \in P$ to customer $i \in I$. In practice, however, such costs may not be linear. In this case, a more accurate modeling of the cost function is particularly important for production costs (see, e.g., Christensen & Klose, 2021), which tend to be of higher order than transportation costs. In our case, such production costs may be defined differently for each capacity level. For example in the case of economies scale, the per-unit production costs would be a concave function.

Štádlerová et al. (2022, 2023) consider a location problem in which equipment is placed to capture hydrogen. The type and quantity of capturing equipment installed defines the operating costs. As such, the proposed model uses different capacity levels, and for each level, a concave function is defined, representing the capturing costs for the total amount of hydrogen captured at that location. This cost function is approximated by a piecewise linear function. The total production costs are exemplified in Fig. 2. Subfigure (a) illustrates the total production costs F that

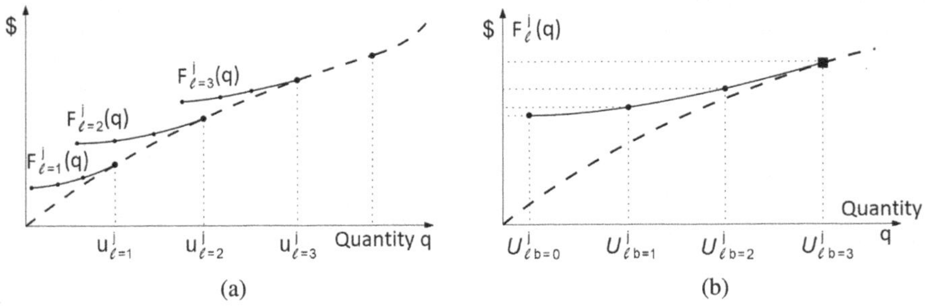

Fig. 2 Exemple of a piecewise-linear production cost function (adapted from Štádlerová et al., 2023). (**a**) Separate functions for different capacity levels. (**b**) Piecewise-linear cost function for a specific capacity level $\ell \in L$

depend on both the selected capacity level $\ell \in L$ and the total production quantity q at that location. Subfigure (b) exemplifies a piecewise-linear function for a given capacity level $\ell \in L$: specifically, 3 line pieces defined by 4 breakpoints. The piecewise linear function, specified for each capacity level, does not only indicate the maximum production quantity possible at that capacity level, but also a minimum production requirement. Note that the dashed line is a lower bounding trend line that has only been drawn for illustrative purposes. While the authors propose a formulation tailored to the concave shape of the function, we will now show how any continuous piecewise linear function can be used (i.e., the end of each linear piece is identical to the beginning of the next linear piece).

We separate the operational costs into two parts. Parameter $g_{i\ell p}^{jt}$ now exclusively refers to the per-unit transportation costs to distribute product $p \in P$ from facility $j \in J$ operating at level $\ell \in L$ to customer $i \in I$ during time period $t \in T$. In addition, we separately define the production costs by a cost function indicating the total costs to produce a quantity of q units. Specifically, for each $\ell \in L, \ell \geq 1$ and facility $j \in J$, we define a piecewise linear function $F_\ell^j(q)$ that determines the costs to produce a total quantity of q units at facility $j \in J$ operating at capacity level $\ell \in L$. This piecewise linear function consists of $\overline{b}_\ell^j$ line pieces, where $b = 1, \ldots, \overline{b}_\ell^j$ are the breakpoints of the production quantity defining the piecewise linear function. For each line piece $b \in [1, \overline{b}_\ell^j]$, the production quantity at the beginning of the piece is given by $U_{\ell(b-1)}^j$, while the end of the piece is given by $U_{\ell b}^j$. As such, for a facility $j \in J$ operating at level $\ell \in L$, $U_{\ell(b=0)}^j$ is the minimum production level allowed, while $u_\ell^j = U^j_{\ell(b=\overline{b}_\ell^j)}$ is the maximum production level allowed.

Let $\overline{\mu}_{j\ell bt}$ be a binary variable that takes value 1 if facility $j \in J$ operating at level $\ell \in L$ produces a total quantity that falls into line piece b during time period $t \in T$. Further, let continuous variables $\mu_{j\ell bt}^1$ and $\mu_{j\ell bt}^2$ represent the linear combination between the start and the end of the selected line piece such that this linear combination corresponds to the total production quantity. The GMC model can be modified as follows in order to account for such complex production cost functions by adding the following constraints, where $[\overline{b}_\ell^j]$ denotes the sequential set $\left\{1, \ldots, \overline{b}_\ell^j\right\}$:

$$\sum_{b \in [\overline{b}_\ell^j]} \overline{\mu}_{j\ell bt} = \sum_{\ell_1 \in L} y_{\ell_1 \ell}^{jt} \qquad j \in J \;,\; \ell \in L \;,\; t \in T \tag{9}$$

$$\mu_{j\ell bt}^1 + \mu_{j\ell bt}^2 = \overline{\mu}_{j\ell bt} \qquad j \in J \;,\; \ell \in L \;,\; b \in [\overline{b}_\ell^j] \;,\; t \in T \tag{10}$$

$$\sum_{i \in I} \sum_{p \in P} x_{i\ell p}^{jt} = \sum_{b \in [\overline{b}_\ell^j]} U_{\ell(b-1)}^j \mu_{j\ell bt}^1 + U_{\ell b}^j \mu_{j\ell bt}^2 \qquad j \in J\ ,\ \ell \in L\ ,\ t \in T \tag{11}$$

$$\overline{\mu}_{j\ell bt} \in \{0, 1\} \qquad j \in J\ ,\ \ell \in L\ ,\ b \in [\overline{b}_\ell^j]\ ,\ t \in T \tag{12}$$

$$\mu_{j\ell bt}^1, \mu_{j\ell bt}^2 \geq 0 \qquad j \in J\ ,\ \ell \in L\ ,\ b \in [\overline{b}_\ell^j]\ ,\ t \in T \tag{13}$$

Constraints (9) ensure that a line piece is only selected if a corresponding facility operating at the correct capacity level exists. Constraints (10) set the linear combination between the two breakpoints defining the selected line piece. Constraints (11) require that the total production at a facility equals the production represented by the linear combination of the two breakpoints. Constraints (12) and (13) are the domain constraints.

The total production costs can now be computed based on the linear combination for the selected breakpoints. Hence, the following term has to be added to the Objective Function (1), computing the total production costs as the linear combination between the two breakpoints on the selected line-piece:

$$+ \sum_{j \in J} \sum_{\ell \in L} \sum_{b \in [\overline{b}_\ell^j]} \sum_{t \in T} F_\ell^j (U_{\ell(b-1)}^j) \mu_{j\ell bt}^1 + F_\ell^j (U_{\ell b}^j) \mu_{j\ell bt}^2$$

A similar technique can be applied to model more complex transportation costs, if required.

3.4 Accounting for Modular Facility Structures

In specific applications, exclusively defining the capacity level may not be sufficient to accurately represent a more complex structure of the production facility. This is the case, for example, if a facility is composed by several modules (see, e.g., Wu et al., 2006; Alarcon-Gerbier & Buscher, 2022), each of which may have different capacities and production costs. Expanding (or reducing) the total production capacity at a location then requires a more refined modelling.

Jena et al. (2016) consider location decisions for forestry camps, each of which is composed of different hosting units. The total capacity given by different combinations of such units can be represented as capacity levels within the GMC model. However, in this planning context, idle units can further be temporarily closed independently from each other to avoid unnecessary maintenance costs. The variables used in the corresponding model therefore tracks two types of information at each location: the existing capacity level and the currently open capacity level.

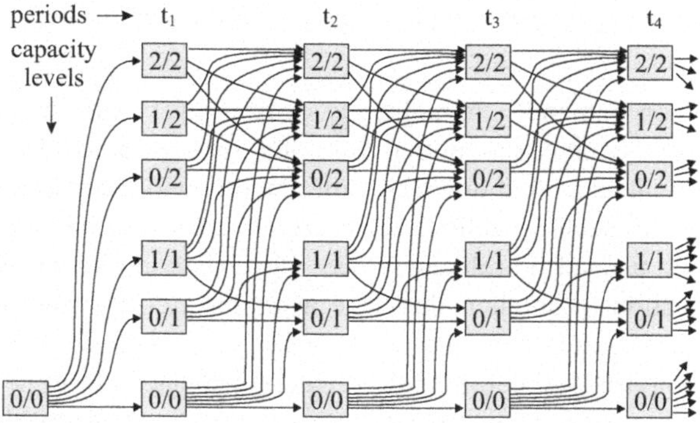

Fig. 3 Network flow structure to manage partial facility closing and reopening. Each node indicates the level of open and existing capacity (Jena et al., 2016)

In particular, the authors use variables $y^{jt}_{\ell_1\ell_2 n_1 n_2}$ that take value 1 if the facility at location $j \in J$ changes its existing capacity from level $n_1 \in L$ to level $n_2 \in L$ at the beginning of time period $t \in T$, while its available (open) capacity changes from level $\ell_1 \in L$ to level $\ell_2 \in L$. As a consequence, a total of $n_2 - \ell_2$ capacity levels are assumed to be temporarily unavailable (closed) during that time period. The total costs for the set of changes are encoded in the more complex parameter $f^{jt}_{\ell_1\ell_2 n_1 n_2}$, which includes the cost to change the existing capacity level, the change of open capacity and the cost to maintain the facility open at level $\ell_2 \in L$ throughout the time period. In order to account for such complex decisions in the above defined GMC model, variables $y^{jt}_{\ell_1\ell_2}$ have to be replaced by $\sum_{\ell_1 \in L} \sum_{\ell_2 \in L} y^{jt}_{\ell_1\ell_2 n_1 n_2}$.

The resulting network structure to manage the open and existing capacity levels at a location is exemplified in Fig. 3. For the sake of simplicity, this figure only illustrates the expansion of existing capacity levels (and not the reduction).

Within the model, capacity conservation Eqs. (4) and (5) have to be replaced by the new capacity conservation constraints, which track the individual changes for each $n \in L$ and $\ell \in L$:

$$\sum_{\ell_1 \in L^+(j,t-1,\ell)} \sum_{n_1=\ell_1,\ldots,\overline{\ell}} y^{j(t-1)}_{\ell_1 \ell n_1 n} = \sum_{\ell_2 \in L^-(j,t,\ell)} \sum_{n_2=\ell_2,\ldots,\overline{\ell}} y^{jt}_{\ell\ell_2 n n_2}$$

$$j \in J \ , \ n \in L \backslash \{0\} \ , \ \ell = 1, \ldots, n \ , \ t \in T \backslash \{1\} \qquad (14)$$

$$\sum_{\ell_2 \in L^-(j,t=0,\ell^j)} \sum_{n_2=\ell_2,\ldots,\overline{\ell}} y^{j1}_{(\ell_1=0)\ell_2(n_1=\ell^j)n_2} = 1 \qquad j \in J \qquad (15)$$

Specifically, Eq. (15) specifies that, at the beginning of the planning horizon, a facility at location $j \in J$ initially has no open capacity levels, even if a facility already exists at capacity level ℓ^j. Further note that the capacity flow conservation constraints (14) could be formulated using only one constraint for each location $j \in J$ and time period $t \in T$. However, using separated constraints for each existing level $n \in L$ and open level $\ell \in L$ tends to significantly strengthen the LP relaxation.

3.5 Accounting for Facility Relocation

When demand geographically evolves over time, the initial locations of facilities may become suboptimal for later time periods. It may then be economically beneficial to shut down an entire facility and reconstruct it at another location, even though such operations are costly. In specific application contexts, entire facilities may instead be relocated (see, e.g., Alarcon-Gerbier & Buscher, 2022) from one location to another, therefore avoiding a costly shut-down and reconstruction. A specific example is given by Jena et al. (2015b, 2016), where forestry camps, composed by mobile trailers, can be relocated closer to the new logging regions.

From a modeling perspective, two different approaches can be taken to integrate facility relocation, visualized in Fig. 4. A relocation can be modelled as a direct arc between an origin and destination location, as exemplified in Sub-figure (a). Such modelling allows for representing costs that are tailored to the specific origin-destination pair, e.g., based on the relocation distance. However, it also results in a large number of relocation decision variables. As an alternative, a facility may first be moved to an artificial hub-node, as illustrated in Sub-figure (b), and then be directed to its destination location. The latter approach assumes that the cost of facility relocation primary depends on the size of the facility, while the cost difference due to varying distances between the origin and destination locations are negligible.

We now explicitly model the latter option. To this end, binary variables $w^-_{jt\ell}$ and $w^+_{jt\ell}$ are added to the model, which take value 1 if a facility of capacity size $\ell \in L$ is removed from and relocated to, respectively, location $j \in J$ at the beginning of time period $t \in T$.

These relocation variables have to be added to the capacity conservation constraints (4), which then write as follows $j \in J$ $\ell \in L$ $t \in T \setminus \{1\}$:

$$\sum_{\ell_1 \in L^+(j,t-1,\ell)} y^{j(t-1)}_{\ell_1 \ell} + w^+_{jt\ell} = \sum_{\ell_2 \in L^-(j,t,\ell)} y^{jt}_{\ell \ell_2} + w^-_{jt\ell} \tag{16}$$

If the relocation of an existing facility at the first time-period is allowed, the corresponding $w^-_{jt\ell j}$ variable needs to be added to the left-hand side of Eq. (5). The

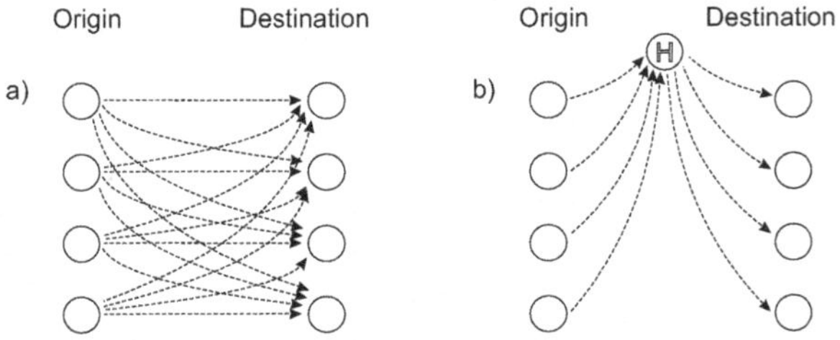

Fig. 4 Two possible network flow architectures to model facility relocation (Jena, 2014)

following equalities ensure that for each facility of size $\ell \in L$ relocated to a location, there is exactly one facility of the same size removed from a location:

$$\sum_{j \in J} w^{+}_{jt\ell} = \sum_{j \in J} w^{-}_{jt\ell} \qquad \ell \in L \setminus \{0\}, \ , \ t \in T \setminus \{1\}. \tag{17}$$

Finally, the relocation costs have to be added to the objective function. Assume that c^R_ℓ denotes the costs to relocate a facility of capacity size $\ell \in L$ from one location to another. The following term has to be added to the objective function (1):

$$+\sum_{j \in J} \sum_{\ell \in L} \sum_{t \in T} \frac{c^R_\ell}{2} w^{+}_{jt\ell} + \sum_{j \in J} \sum_{\ell \in L} \sum_{t \in T} \frac{c^R_\ell}{2} w^{-}_{jt\ell}.$$

Here, the relocation costs have been equally split onto the incoming and outgoing relocation variables. Given that Equality (17) ensures that any outgoing facility matches an incoming facility of the same size, the relocation cost can also be associated to only one of the two variables. However, in the case of a subsequent Lagrangian relaxation, splitting the costs over the two variables is preferred to ensure that both variables imply a usage cost.

If the relocation distance between specific location pairs has to be modelled, a single binary variable $w_{j_1 j_2 t\ell}$ with more specific relocation costs $c^R_{\ell j_1 j_2}$ can be used instead, indicating that the respective facility is relocated from location $j_1 \in J$ to location $j_2 \in J$. The capacity change network, illustrated in Fig. 1, then uses variables that are directly used in Eq. (4). As a consequence, Equalities (17) are no longer required. The use of such binary variables with more detailed information does, however, not improve the strength of the LP relaxation (Jena, 2014).

Note that the formulation above allows for reducing or expanding capacity (using the y variables) once the facility has arrived at the new location. The presented formulations also make the assumption that relocation is instantaneous. If relocation is assumed to take one (or more) time-period(s), the $w^{+}_{jt\ell}$ variables have to be added to the capacity conservation constraints (16) of a later occurring time-period. The formulations discussed above can also be easily adapted to the case where only a part (e.g., certain units) of the facility is relocated.

3.6 *Accounting for Uncertainty via 2-Stage Stochastic Programming*

Parameter uncertainty in facility location problems typically concerns the customer demands or the costs to provide available capacity and to serve the customers. While deterministic problem variants aim at minimizing the total costs required to satisfy the entire demand, the latter becomes ill defined when demand is uncertain. In this case, the main paradigm classes to deal with such uncertainty include robust optimization (see, e.g., Ben-Tal et al., 2009) and stochastic optimization (see, e.g., Birge & Louveaux, 2011). Roughly speaking, the former aims at minimizing the total costs required to satisfy the worst possible demand outcome, while the latter aims at minimizing the expected costs required to satisfy the possible demand outcomes. We here focus on the case of the latter, where we minimize the expected costs and demand uncertainty is assumed to be well represented by a set of demand scenarios. Such a case, where capacity decisions have to be made in the first stage, while demand allocation is done in the second stage, has more recently been handled by Lagrangian relaxation in the literature (see, e.g., Marín et al., 2018; Štádlerová et al., 2023). When optimizing expectation, it is rarely beneficial to aim at satisfying the entire demand of extreme cases that have a low probability of occurrence. It is therefore common to quantify the penalties induced by shortfall demand units (also known as the recourse).

In a dynamic planning context, the set of eligible first-stage decisions directly depends on the application context, as well as the flow of information and decisions. Specifically, some of the capacity decisions (i.e., the opening of facilities at some time periods and, potentially, capacity adjustments) may have to be made in the first stage, while others can be postponed to the second stage, once demand quantities are more reliably known. The set of first-stage decisions defines the problem variant. At one extreme of the problem variants, all capacity decisions have to be made in the first stage. At another extreme, only a few facility opening decisions (e.g., those for the first time period) have to be made in the first stage, while the remainder can be tailored to the demand scenarios in the second stage. Such a flexible problem variant particularly makes sense if the planning is carried out dynamically in a rolling horizon fashion. For example, Štádlerová et al. (2023) requires that all facility openings, no matter at which time period, are decided in the first stage, while capacity expansions can be made in the second stage.

For the scope of this chapter, it is impossible to cover all problem variants that may result from the various definitions of the first-stage decisions. We therefore here focus on the simplest case of 2-stage stochastic programming, where capacity decisions y are made in the first stage. At that moment, only a probability distribution over the demand is available. We then assume that demand quantities are revealed in the second stage. Here, capacity decisions remain the same, but demand allocation can be tailored to the revealed demand. We also assume that the demand probabilities are sufficiently well approximated by a set of scenarios S, with ρ_s being the probability that scenario $s \in S$ will occur. For each scenario $s \in S$, d_{ip}^{st} represents the demand quantity for commodity $p \in P$ requested by customer $i \in I$ during time period $t \in T$. While other parameters may also be uncertain and depend on $s \in S$, for the sake of illustration, we here restrict to demand uncertainty. We assume a penalty cost in the case that the available capacity falls short of the requested demand. To this end, let f_p^S be the cost of being one unit short of demand for commodity $p \in P$. The objective is then to minimize the expected total costs required to satisfy the various demand scenarios.

Demand allocation decisions are tailored to the specific demand scenario $s \in S$ in the second stage. Specifically, let $x_{i\ell p}^{sjt}$ be the proportion of demand d_{ip}^{st}, satisfied from the facility at location $j \in J$ open at capacity level $\ell \in L$. If minimum production quantities at the facilities are required in the problem, potential excess capacity will have to be defined separately for each demand scenario (see, e.g., Štádlerová et al., 2023).

We adapt the GMC formulation as follows. We define continuous variables $\tilde{x}_{ip}^{st}$ as the under-served unit amount of the demand of customer $i \in I$ for commodity $p \in P$ at time period $t \in T$ in scenario s. The corresponding penalties for shortfall quantities are added to Objective Function (1):

$$+\sum_{i \in I} \sum_{p \in P} \sum_{t \in T} \sum_{s \in S} \rho_s f_p^S \tilde{x}_{ip}^{st}$$

Furthermore, term $d_{ip}^t x_{i\ell p}^{jt}$ in Objective Function (1) is replaced by $\sum_{s \in S} d_{ip}^{st} x_{i\ell p}^{sjt}$ to account for the scenario-dependent demand.

Demand constraints (2), capacity constraints (3) and SIs (6) are then replaced by their scenario-dependent counterparts (18), (19) and (20), respectively. Domain constraints (21) are further added for the shortfall variables.

$$\sum_{j \in J} \sum_{\ell \in L} x_{i\ell p}^{sjt} + \tilde{x}_{ip}^{st} = 1 \qquad i \in I\,,\ p \in P\,,\ t \in T\,,\ s \in S \tag{18}$$

$$\sum_{i \in I} \sum_{p \in P} d_{ip}^{st} x_{i\ell p}^{sjt} \leq \sum_{\ell_1 \in L^+(j,t,\ell)} u_\ell^j y_{\ell_1 \ell}^{jt} \qquad j \in J\,,\ \ell \in L\,,\ t \in T\,,\ s \in S \tag{19}$$

$$x_{i\ell p}^{sjt} \leq \sum_{\ell_1 \in L} y_{\ell_1 \ell}^{jt} \qquad i \in I\,,\ j \in J\,,\ \ell \in L\,,\ p \in P\,,\ t \in T\,,\ s \in S \tag{20}$$

$$0 \leq \tilde{x}_{ip}^{st} \leq 1 \qquad i \in I\,,\ p \in P\,,\ t \in T\,,\ s \in S. \tag{21}$$

As previously mentioned, such formulation assumes that facilities do not have minimum production requirements. For such a case, we refer to Sect. 3.3, which shows how to model piece-wise linear functions defined independently for each capacity level. Those piece-wise linear functions also implicitly bound the allowable minimum and maximum production quantities. Under uncertain demand, additional variables, representing the produced excess capacity, will have to be added. Such production excess will then have to be penalized in the objective function. We refer the reader to the work of Štádlerová et al. (2023) as an example for such a case.

4 Solution via Lagrangian Relaxation

Lagrangian Relaxation, also referred to as Lagrangian decomposition, belongs to the class of mathematical decomposition methods. As such, it has been developed on the premise of exploiting the structure of the constraint coefficient matrix of the optimization model. While each method has its advantages and disadvantages, Lagrangian decomposition tends to be particularly promising when a large part of the constraint matrix is organized in a block structure, i.e., most of the constraints have non-zero coefficients organized in blocks, while only a few constraints link these blocks. The latter type of constraints, those that have non-zero coefficients simultaneously in several blocks, may be called the "complicating constraints". Lagrangian relaxation aims at relaxing the complicating constraints, and transfers them into the objective function by penalizing their violation multiplied by weights, called the *Lagrangian dual values* (or, multipliers) of the relaxed constraints. The resulting optimization problem is called the *Lagrangian subproblem*. Given that the complicating constraints have been eliminated from the constraint matrix, the problem can be decomposed into independent subproblems. Ideally, solving each of these subproblems is now easy and can be performed by a specialized algorithm. Section 4.1 relaxes the demand constraints, defining the corresponding Lagrangian subproblem. Then, the efficient solution of the subproblem is discussed.

The solution of the Lagrangian subproblem is, however, unlikely to be feasible to the original problem. Most likely, the relaxed constraints are not satisfied. Theoretically, a set of Lagrangian dual values exists such that the solution to the Lagrangian subproblem is also feasible to the original problem. This set of optimal Lagrangian dual values is the result of solving the so-called *Lagrangian dual problem*, which optimizes over the space of Lagrangian dual values. This optimization problem is non-smooth in general, which makes it hard to find the optimal Lagrangian dual values (see, e.g., Frangioni, 2005). Instead, one aims at converging to a set of Lagrangian dual value such that the violation of the constraints is as small as possible. Iterative algorithms to solve the Lagrangian dual problem are discussed in Sect. 4.2.

Throughout the iterative process of improving the Lagrangian duals, the Lagrangian solution may be used to derive *solutions that are feasible* to the original problem. Naturally, the challenge here lies within the reconstitution of the relaxed constraints. As such, the employed process is an heuristic, which will be discussed in Sect. 4.3.

Within each of the Sects. 4.1–4.3, we first address the case of the general GMC formulation. Then, the adjustments necessary to account for more complex problem variants (see Sects. 3.3–3.6) are discussed.

We finally note that, given that the Lagrangian subproblem is a relaxation of the original (minimization) problem, the optimal solution to any Lagrangian subproblem constitutes a *lower bound* to the original problem. A Lagrangian heuristics therefore typically produces both upper bounds (i.e., feasible solutions) and lower bounds. Unless this procedure is embedded in an exact algorithm (such as Branch-and-Bound), it remains an heuristic. In order to provide a bound on the quality of the found upper bounds, it is important to find the best possible lower bound. This lower bound is typically impacted by the strength of the underlying formulation. This reiterates the importance of using a formulation that is as strong as possible, as previously emphasized in Sect. 3. For an in-depth discussion of Lagrangian Relaxation itself, as well as valuable related insights, we refer the reader to the excellent introduction of Guignard (2003).

4.1 Solution of the Lagrangian Subproblem

For Facility Location problems, two different relaxations have typically been considered: relaxing the demand constraints (2) or relaxing the capacity constraints (3). The latter has been successfully attempted by several authors (Beasley, 1993; Li et al., 2009; Ghodsi, 2012; Gendron et al., 2016). However, the resulting subproblem may still be hard to solve. While relaxing the capacity constraints may make sense in some particular situations, it is unlikely to provide competitive results when some of the key features of the problem involve the facility capacities, such as several capacity levels and capacity adjustment over time (as it is the case with the GMC).

We therefore here decide to relax the demand constraints (2) within the GMC formulation (see Sect. 3.1), which are the only constraints that provide a link among the different candidate facility locations. Such a relaxation also allows for a more unified approach that can be adjusted to the various extensions discussed throughout Sects. 3.3–3.6.

Let α be the vector of Lagrange multipliers of demand constraints (2). After relaxing these constraints, transferring their violation into the objective function and rearranging its terms, we obtain the following Lagrangian subproblem:

$$
\begin{aligned}
L(\alpha) = \min & \sum_{j\in J}\sum_{\ell_1\in L}\sum_{\ell_2\in L^-(j,t,\ell)}\sum_{t\in T} f_{\ell_1\ell_2}^{jt} y_{\ell_1\ell_2}^{jt} \\
& + \sum_{i\in I}\sum_{j\in J}\sum_{\ell\in L}\sum_{p\in P}\sum_{t\in T}(g_{i\ell p}^{jt} d_{ip}^t - \alpha_{ip}^t) x_{i\ell p}^{jt} + \sum_{i\in I}\sum_{p\in P}\sum_{t\in T}\alpha_{ip}^t \\
\text{s.t. } & (3)\text{–}(8).
\end{aligned}
$$

In order to simplify the representation, let $\tilde{c}_{i\ell p}^{jt} = g_{i\ell p}^{jt} d_{ip}^t - \alpha_{ip}^t$ denote the modified costs for the $x_{i\ell p}^{jt}$ variables. Given that the demand constraints have been relaxed, the Lagrangian subproblem can be separated into one subproblem for each candidate facility location $j \in J$. The objective function of the Lagrangian subproblem can then be written as $L(\alpha) = \sum_{j\in J} L_j(\alpha) + \sum_{i\in I}\sum_{p\in P}\sum_{t\in T}\alpha_{ip}^t$. Here, $L_j(\alpha)$

corresponds to the part of the objective function specific to candidate location $j \in J$, written as follows:

$$
\begin{aligned}
L_j(\alpha) = \min & \sum_{\ell_1 \in L} \sum_{\ell_2 \in L^-(j,t,\ell_1)} \sum_{t \in T} f^{jt}_{\ell_1 \ell_2} y^{jt}_{\ell_1 \ell_2} + \sum_{i \in I} \sum_{\ell \in L} \sum_{p \in P} \sum_{t \in T} \tilde{c}^{jt}_{i\ell p} x^{jt}_{i\ell p} \\
\text{s.t.} & \sum_{i \in I} \sum_{p \in P} d^t_{ip} x^{jt}_{i\ell p} \le \sum_{\ell_1 \in L^+(j,t,\ell)} u^j_\ell y^{jt}_{\ell_1 \ell} && \ell \in L\,,\; t \in T \\
& \sum_{\ell_1 \in L^+(j,t-1,\ell)} y^{j(t-1)}_{\ell_1 \ell} = \sum_{\ell_2 \in L^-(j,t,\ell)} y^{jt}_{\ell \ell_2} && \ell \in L\,,\; t \in T \setminus \{1\} \\
& \sum_{\ell \in L^-(j,t,\ell^j)} y^{j1}_{\ell^j \ell} = 1 \\
& x^{jt}_{i\ell p} \le \sum_{\ell_1 \in L^+(j,t,\ell)} y^{jt}_{\ell_1 \ell} && i \in I\,,\; \ell \in L\,,\; p \in P\,,\; t \in T \\
& x^{jt}_{i\ell p} \ge 0 && i \in I\,,\; \ell \in L\,,\; p \in P\,,\; t \in T \\
& y^{jt}_{\ell_1 \ell_2} \in \{0, 1\} && \ell_1 \in L\,,\; \ell_2 \in L^-(j,t,\ell_1)\,,\; t \in T.
\end{aligned}
$$

In the following sections, we will elaborate on how to solve the Lagrangian subproblem above, as well as those derived from the other proposed problem extensions.

4.1.1 Solving the Subproblem for Each Candidate Location

Subproblem $L_j(\alpha)$ can be solved independently for each facility location $j \in J$ (see, e.g. Jena et al., 2017). Each subproblem corresponds to the capacity planning over the planning horizon at that given location $j \in J$, which can be computed by solving a shortest-path problem defined on the acyclic time-capacity graph, as exemplified in Fig. 1. The arc costs to change from one capacity level to another at the beginning of each time-period includes the costs to change capacity, maintain the facility open at the new capacity level, as well as the demand allocation to the customers (considering the modified costs $\tilde{c}^{jt}_{i\ell p}$). Specifically, the shortest-path network for facility location $j \in J$ is defined by nodes (ℓ, t), $\ell \in L, t \in T$. The arc costs representing a capacity change from level $\ell_1 \in L$ to level $\ell_2 \in L$ at the beginning of period $t \in T$, maintenance costs at level $\ell_2 \in L$ and the optimal demand allocation is then given by $f^{jt}_{\ell_1 \ell_2} + \hat{L}^\alpha_j(\ell, t)$. Here $\hat{L}^\alpha_j(\ell, t)$ represents the optimal demand allocation at a facility located at $j \in J$ and open at level $\ell \in L$ during time period $t \in T$ under multipliers α, and will be defined next.

Computing the Optimal Demand Allocation Computing the optimal demand allocation with costs $\hat{L}^\alpha_j(\ell, t)$ for each time period $t \in T$ and capacity level

$\ell \in L$ minimizing the modified costs $\tilde{c}_{i\ell p}^{jt} = g_{i\ell p}^{jt} d_{ip}^{t} - \alpha_{ip}^{t}$ is equivalent to solving a continuous knapsack problem. This can be done in polynomial time. First, all demand nodes d_{ip}^{t} with non-positive allocation costs are sorted in increasing order of their adjusted per-unit transportation costs, given by $\tilde{c}_{i\ell p}^{jt}/d_{ip}^{t}$. Then, demand is allocated (respecting the sorted sequence) until either the demand of a customer $i \in I$ and commodity $p \in P$ is fully met or the capacity u_{ℓ}^{j} is filled.

The resulting shortest-path problem can be solved by standard algorithms, such as the Dijkstra algorithm, a minimum cost-flow network solver or Dynamic Programming. As mentioned earlier, the solutions obtained from solving the Lagrangian subproblem are likely infeasible to the original problem. In order to obtain feasible solutions that constitute an upper bound to the problem, one may aim at "repairing" the Lagrangian solutions. This will be discussed in Sect. 4.3.

Solution Under Special Cases The solution of the subproblem $L_j(\alpha)$ may simplify under some of the special cases discussed in Sect. 3.2. Naturally, having a single commodity instead of several commodities still requires to solve a shortest-path problem as described above. In contrast. in the case of a single-period problem, the optimal capacity level can easily be found by inspecting the costs for each level. If the problem is uncapacitated, the demand allocation described above requires to allocate to the facility all demand nodes with non-positive allocation costs. Finally, if set L only has a single capacity level in addition to level 0 and a facility cannot be closed once opened, the optimal time-period to open the facility can be identified by inspection: for each time-period, one considers the costs to open the facility and maintain it open until the end of the planning horizon. If, however, facilities may be closed as well, or in the case of any other more complex capacity change rules induced by sets L^- and L^+, the subproblem can be easily solved as a shortest-path problem (as previously described).

4.1.2 Solution in the Case of Complex Cost Functions

When piecewise linear cost functions are used for the production costs that depend on the facility's capacity level $\ell \in L$, the Lagrangian subproblem is still be decomposed as previously shown. However, in order to find the optimal production schedule for each facility location, the demand allocation has to be adjusted. Specifically, the demand allocation, solved as a continuous knapsack problem, now has piecewise linear costs (see, e.g., Christensen & Klose, 2021). The objective to be minimized now also accounts for the piecewise-linear production costs given through the linear combination between μ^1 and μ^2, while g now exclusively refers to the distribution costs. As such, the continuous knapsack with piecewise-linear production costs minimizes the following term:

$$\sum_{j\in J}\sum_{\ell\in L}\sum_{t\in T}\sum_{b\in[\overline{b}_{\ell}^{j}]} F_{\ell}^{j}(U_{\ell(b-1)}^{j})\mu_{j\ell bt}^{1} + F_{\ell}^{j}(U_{\ell b}^{j})\mu_{j\ell bt}^{2} + \tilde{c}_{i\ell p}^{jt},$$

where $\tilde{c}^{jt}_{i\ell p} = g^{jt}_{i\ell p} d^t_{ip} - \alpha^t_{ip}$. Let

$$h^j_{\ell bt} = \frac{F^j_\ell(U^j_{\ell b}) - F^j_\ell(U^j_{\ell(b-1)})}{U^j_{\ell b} - U^j_{\ell(b-1)}}$$

be the per-unit production costs at facility $i \in I$ that produces at capacity level $\ell \in L$ and breakpoint b of its piecewise linear cost function during time period $t \in T$. The marginal costs to produce and serve one additional unit while having selected line piece b on capacity level $\ell \in L$ is computed as $m^j_{\ell bt} = \tilde{c}^{jt}_{i\ell p} + h^j_{\ell bt}$.

We define one piecewise linear continuous knapsack for each capacity level $\ell \in L$ and time-period $t \in T$. Customer demands have to be added line piece by line piece. To this end, for a given $\ell \in L$ and $t \in T$, customer demands d^t_{ip} are sorted in non-decreasing order of their per-unit distribution costs $\tilde{c}^{jt}_{i\ell p}/d^t_{ip}$. Given that the distribution costs do not depend on b, the sorted sequence of customer-commodity pairs is the same no matter the line piece b. Starting with $b = 1$, customer demands are then allocated in that order either until the marginal cost is positive (i.e., $m^j_{\ell bt} > 0$) or until the capacity limit $U^j_{\ell b}$ of the current line piece b is reached. In the case of the latter, the next highest line piece $b = 2, \ldots, \overline{b}$ is considered.

If this iterative procedure reaches the maximum capacity $U^j_{\ell\overline{b}}$ of the last line piece $\overline{b}$, the facility cannot serve all profitable customer demands. On the other hand, if the minimum production capacity $U^j_{\ell(b=0)}$ has not been met by adding customers with marginal costs $m^j_{\ell bt} \leq 0$, customers with positive marginal costs have to be considered until the minimum production capacity is reached.

4.1.3 Solution in the Case of Modular Facility Structures

In the context of modular facility structures that can be modelled by tracking both the existing capacity level and the capacity level that is currently open (i.e., available for production), the solution approach for the Lagrangian subproblem outlined in Sect. 4.1.1 can be extended. In this case, the shortest path problem is defined on a more complex capacity-time network, as illustrated in Fig. 3.

Each node (ℓ, n) now represents the actual capacity that is available for use and the total capacity that exists at a given location. The arc costs to transition from one node to another additionally need to consider the costs to temporarily close or reopen a part of the existing capacity. Specifically, the shortest-path network for facility location $j \in J$ is defined by nodes (ℓ, n, t) with $\ell \in L, n = \ell, \ldots, q, t \in T$. As before, the arc costs represent a change from existing level $n_1 \in L$, open at level $\ell_1 \in L$, to existing level $n_2 \in L$ open at level $\ell_2 \in L$ at the beginning of period $t \in T$, as well as the maintenance costs at level $\ell_2 \in L$ and the optimal demand allocation. The arc transitioning costs are therefore given by $f^{jt}_{\ell_1\ell_2 n_1 n_2} + \hat{L}^\alpha_j(\ell, t)$, where $\hat{L}_j^{\alpha}(\ell, t)$ represents the costs of the optimal demand allocation as defined and computed in Sect. 4.1.1. As before, the resulting shortest path problem can be solved, for example, by dynamic programming. For further details concerning the solution procedure in the case of modular facility structures, we refer to Jena et al. (2016).

4.1.4 Solution in the Case of Facility Relocation

If the context of the planning problem allows for facility relocation, both the demand constraints (2) and the relocation hub constraints (17) are considered complicating, since they sum over different facility locations. Two possibilities can be considered: either only relax the demand constraints or relax both sets of constraints.

In the former case, the problem cannot be decomposed by candidate facility location. One may add to the subproblem aggregated demand constraints, enforcing that the total production capacity available at each time-period is sufficient to satisfy the entire demand of that time-period. However, the problem cannot be solved combinatorially and has to be solved by a general-purpose MIP solver. The solution of the subproblem may therefore be rather slow. While the obtained solutions may give insights about beneficial facility relocations and yield interesting lower bounds, to the best of our knowledge, literature has not yet reported on competitive approaches for this type of relaxation.

In contrast, the latter case, relaxing both sets of contraints, has been successfully used in the literature. Here, the subproblem can be decomposed by facility locations (see Jena et al., 2016), each of which can be solved as a shortest-path problem in a similar fashion as previously discussed. In this case, additional Lagrangian multipliers β are added when relaxing the relocation hub constraints (17), and the following term is added in the objective function, indicating the violation of the relaxed constraints:

$$+\sum_{\ell \in L\setminus\{0\}} \sum_{t \in T\setminus\{1\}} \left(\sum_{j \in J} w_{jt\ell}^{-} - \sum_{j \in J} w_{jt\ell}^{+} \right) \beta_{\ell t}$$

Given that all constraints linking the various facility locations have been relaxed, the resulting subproblem can be solved in a similar way as outlined at the beginning of Sect. 4.1, additionally considering the above terms for variables w^{-} and w^{+} in the objective function. Specifically, the capacity-change network illustrated in Fig. 1 should now include arcs representing relocation variables $w_{jt\ell}^{-}$ and $w_{jt\ell}^{+}$, weighted by the costs $\beta_{\ell t}$ and $-\beta_{\ell t}$, respectively.

Note that the Lagrangian solutions obtained may not satisfy the relocation constraints, i.e., the number of removed facilities may not match the number of arriving facilities. The solutions must be "repaired" to be feasible to the original problem, which will be outlined in Sect. 4.3.3. Finally, we also note that a location problem that combines both facility relocation and modular facility structures requires a more subtle analysis in order to solve the Lagrangian subproblem. We refer the reader to the work of Jena et al. (2016) who have solved such a problem by means of Lagrangian relaxation.

4.1.5 Solution in the Case of Uncertainty via Scenarios

When parameter-uncertainty is explicitly acknowledged in the model via demand scenarios, for the case where the entire capacity schedule has to be decided in the first stage, the solution of the Lagrangian subproblem changes only marginally. Specifically, scenario-dependent demand constraints (18) are relaxed, and Lagrangian multipliers α_{ipts} are used, now also depending on scenario $s \in S$.

After the relaxation of demand constraints (18), the objective function of the Lagrangian subproblem $L_j(\alpha)$ additionally contains the following term, given that the shortfall variables $\tilde{x}_{ip}^{st}$ are part of the relaxed constraints:

$$+\sum_{i \in I} \sum_{p \in P} \sum_{s \in S} \sum_{t \in T} (\rho_s f_p^S - \alpha_{ipts}) \tilde{x}_{ip}^{st}.$$

To solve the Lagrangian subproblem, the shortest path problem (see Sect. 4.1.1) throughout the time-capacity network can still be applied, given that the entire capacity schedule is assumed to remain the same for all demand scenarios. Two changes have to be made to compute the costs at each node. First, the objective function of the Lagrangian solution now has to take into consideration the term above for shortfall variables $\tilde{x}_{ip}^{st}$. Second, the costs for the demand allocation within each of the time-capacity nodes within this network have to consider the different demand scenarios. To this end, for each time period $t \in T$ and capacity level $\ell \in L$, the continuous knapsack has to be solved for each demand scenario s.

Given the demand shortfall variables $\tilde{x}_{ip}^{st}$ only appeared in the relaxed demand constraints (18), but not in the remaining constraints, they are now constrained only by their domain constraints (21). As such their solution values do not impact the solution values of the demand allocation variables x. The continuous knapsack can therefore be solved as previous explained in Sect. 4.1.1. In order to obtain valid lower bounds from the Lagrangian solution, one still requires to solve for the values of $\tilde{x}_{ip}^{st}$, which can be computed by inspection (i.e., they are set either to 0 or to the corresponding customer demand).

Different Subsets of Second-Stage Decisions As noted in Sect. 3.6, we here assume that all capacity decisions have to be made within the first stage of the two-stage stochastic optimization problem. In many applications, this is not the case. For example, in a dynamic context in which the optimization model is reoptimized at each time-period, only capacity decisions for the first (i.e., the next upcoming) time-period may have to be made in the first stage, while the remainder can be approximated in the second stage and will only be taken in the upcoming executions of the optimization model. In such cases, the Lagrangian subproblem for each facility location cannot be solved simply by solving a shortest path-problem. Instead, one needs to distinguish decisions that are made within the first stage and those that are made in the second stage, independently for each scenario. A shortest path problem then has to be solved separately for each scenario and for a given set of first-stage decisions. For a facility location that features both uncertainty-based scenarios and complex production cost functions, and that defines certain capacity decisions within the second stage of the stochastic optimization problem, we refer to the work of Šťádlerová et al. (2023), who explain in detail the use of Lagrangian relaxation to solve such a problem.

4.2 Solution of the Lagrangian Dual Problem

For any given Lagrange multiplier α, the solution provided by the Lagrangian subproblem $L(\alpha)$ provides a lower bound to the original problem. The best possible lower bound z^* is obtained by solving the so-called Lagrangian dual problem:

$$z^* = \max L(\alpha).$$

In theory, there exists a set of multipliers α such that the solution of the Lagrangian subproblem is also feasible to the original problem, i.e., the relaxed constraints are not violated. In practice, however, these multipliers are rather difficult to find, as the Lagrangian function $L(\alpha)$ is generally non-differentiable, resulting in a non-smooth optimization problem (see, e.g., Frangioni, 2005). Iterative approaches, aiming at converging to multipliers that maximize the Lagrangian dual, and therefore minimize the implicit violation of the relaxed constraints, has been shown to be an effective approach. These approaches can be divided into two classes: subgradient methods, which consider one subgradient at each iteration and bundle methods, which make use of a subset of subgradients. Both classes will be discussed below.

4.2.1 Subgradient Method

Subgradient methods iteratively adjust the Lagrangian multipliers into the direction that allows for a smaller violation of the relaxed constraints. They consist of two main ingredients: the subgradient direction and the stepsize. In the case where the demand constraints (2) have been relaxed, for multipliers α_{ip}^t at iteration k, a subgradient direction with components γ_{ip}^{kt} can be easily computed as the violation of the relaxed constraints when considering the solution x of the current Lagrangian subproblem:

$$\gamma_{ip}^{kt} = 1 - \sum_{j \in J} \sum_{\ell \in L} x_{\ell p}^{ijt} \qquad i \in I\ ,\ p \in P\ ,\ t \in T.$$

For all subgradient directions, a common stepsize is used, representing the amount each Lagrangian multipliers will be adjusted into the subgradient direction. A common approach, suggested by Held et al. (1974) and successfully used in succeeding works (see, e.g., Shulman, 1991; Sridharan, 1991; Correia & Captivo, 2003; Jena et al., 2017), is to choose the stepsize λ^k at iteration k depending on how close the current solution value (which constitutes a lower bound) is to the best known upper bound $\hat{Z}$, scaled by scalar δ^k and the magnitude of all current subgradient directions:

$$\lambda^k = \delta^k \frac{\widehat{Z} - L^k(\alpha)}{\sum_{i \in I} \sum_{p \in P} \sum_{t \in T} (\gamma_{ip}^{kt})^2}.$$

The Lagrangian multipliers used within the next iteration are then updated as follows:

$$\alpha_{ip}^{(k+1)t} = \alpha_{ip}^{kt} + \lambda^k \gamma_{ip}^{kt} \qquad i \in I \ , \ p \in P \ , \ t \in T.$$

The performance and convergence may highly vary among problem instances and is known to be sensitive to the values of its parameters. For the scalar, it has been found beneficial (see, e.g., Jena et al., 2017) to use an initial value of δ^0 and a subsequent halving of the value every 25 consecutive iterations without improvement of the lower bound. Lagrangian multipliers α are typically initialized with 0.

Subgradient for Relaxed Relocation Constraints If the problem accounts for relocation, and the corresponding relocation linking constraints (17) have been relaxed, the subgradient method has to take into consideration the subgradients for the multipliers β. At the k-th iteration, a subgradient can be computed as follows, with variables w^- and w^+ fixed to the values found in the solution to the Lagrangian subproblem:

$$\mu_{\ell t}^k = \sum_{j \in J} w_{jt\ell}^- - \sum_{j \in J} w_{jt\ell}^+ \qquad \ell \in L \backslash \{0\} \ , \ t \in T.$$

The computation of the step size has to be extended as follows:

$$\lambda^k = \delta^k \frac{\widehat{Z} - L^k(\alpha)}{\sum_{i \in I} \sum_{p \in P} \sum_{t \in T} (\gamma_{ip}^{kt})^2 + \sum_{\ell \in L \backslash \{0\}} \sum_{t \in T} (\mu_{\ell t}^k)^2}.$$

Finally, the Lagrangian multipliers for the relaxed relocation linking constraints, used within the next iteration are then updated as follows:

$$\beta_{\ell t}^{(k+1)} = \beta_{\ell t}^k + \lambda^k \mu^{\ell t} \qquad \ell \in L \backslash \{0\} \ , \ t \in T.$$

Subgradient in the Case of Demand Uncertainty As the subgradient is computed as the violation of the relaxed constraint, it now has to account for shortfall variables $\tilde{x}$ in each of the scenarios $s \in S$, computed as follows:

$$\gamma_{ip}^{kts} = 1 - \sum_{j \in J} \sum_{\ell \in L} x_{\ell p}^{ijt} - \tilde{x}_{ip}^{st} \qquad i \in I\ ,\ p \in P\ ,\ t \in T\ ,\ s \in S.$$

The stepsize λ^k for the next iteration k now has to sum over all squared subgradients as before, but additionally including those for the different scenarios:

$$\lambda^k = \delta^k \frac{\widehat{Z} - L^k(\alpha)}{\sum_{i \in I} \sum_{p \in P} \sum_{t \in T} \sum_{s \in S} (\gamma_{ip}^{kts})^2}.$$

Finally, the computation of the next Lagrangian multipliers α is now also carried out separately for each scenario $s \in S$:

$$\alpha_{ip}^{(k+1)ts} = \alpha_{ip}^{kts} + \lambda^k \gamma_{ip}^{kts} \qquad i \in I\ ,\ p \in P\ ,\ t \in T\ ,\ s \in S.$$

4.2.2 Bundle Methods

An alternative to classical subgradient methods are so-called Bundle methods, which use a subset $B \subseteq \{1, \ldots, k\}$ of the previous multiplier-subgradient tuples $< L(\alpha^m), \gamma^m >$ from any previous iteration $m \in [1, k]$, where k is the current iteration. In contrast to the subgradient method explained above, those methods do not rely on an upper bound. We will here focus on two bundle methods that have been successfully applied to solve complex facility location problems.

Aggregated Bundle Method We illustrate the aggregated bundle method proposed by Frangioni and Gallo (1999) and Frangioni (2005), which has been successfully applied in several location (see, e.g., Jena et al., 2016, 2017) and network design problems (see, e.g., Frangioni & Gorgone, 2014). The method aims at finding a linear combination θ among the available subgradients by solving the following quadratic optimization problem:

$$\min \left\{ \frac{1}{2} \left\| \sum_{m \in B} \gamma^s \theta^m \right\|^2 + \frac{1}{R} E_B \theta \ ;\ \sum_{m \in B} \theta^m = 1\ ,\ \theta \geq 0 \right\},$$

where $E_B = L(\alpha) + \gamma(\hat{\alpha} - \alpha) - L(\hat{\alpha})$ is the linearization error from the current point $\hat{\alpha}$. Further, R is the corresponding trust region parameter, scaling the penalty related to the distance from from the current point. The tentative ascent direction is then computed by the convex combination of the subgradients, using the convex multipliers θ. Alternatively, the dual problem can be solved to compute the ascent direction, or directly the new point. Frangioni and Gallo (1999) elaborate on these relationships in detail.

The solution values for θ^m do not only derive the next ascent direction, but may also be useful to construct feasible integer solutions. Given that high values of θ^m indicate that certain subgradients are judged more promising then others, one may want to pay close attention to the corresponding multipliers and the Lagrangian solution resulting from them. This relationship is briefly discussed in Sect. 4.3.5.

Boxstep Method Another method that has been successfully used to solve the Lagrangian dual for complex location problems (see, e.g., Schütz et al., 2009; Štádlerová et al., 2023) is the boxstep method (Marsten et al., 1975). The method directly computes the set of the next Lagrangian multipliers $\alpha^{(k+1)}$ by solving the following linear program, where $L^k = L(\alpha^k) - \sum_{i\in I}\sum_{p\in P}\sum_{t\in T}\alpha_{ip}^{kt}\gamma_{ip}^{kt}$:

$$
\begin{aligned}
&\max \phi \\
&\text{s.t. } \phi \le L^m + \sum_{i\in I}\sum_{p\in P}\sum_{t\in T}\alpha_{ip}^{(k+1)t}\gamma_{ip}^{mt} && m = 1,\dots,k \\
&\alpha_{ip}^{(k+1)t} \le \alpha_{ip}^{kt} + \Delta_{ip}^{kt} && i \in I\ ,\ p \in P\ ,\ t \in T \\
&\alpha_{ip}^{(k+1)t} \ge \alpha_{ip}^{kt} - \Delta_{ip}^{kt} && i \in I\ ,\ p \in P\ ,\ t \in T \\
&\phi \in \mathbb{R}\ ,\ \alpha_{ip}^{(k+1)t} \in \mathbb{R}
\end{aligned}
$$

The first set of constraints are the cutting planes that outer approximate the Lagrangian solution objective. The second and third sets of constraints are the box-step constraints, ensuring that the next Lagrangian multiplier is not too distant from the current one. Here, Δ_{ip}^{kt} defines the box, specific to each multiplier α_{ip}^{kt}. Typically, the box decreases in size as the algorithm converges. For example, Štádlerová et al. (2023) propose to decrease the boxsize by a predefined percentage whenever α_{ip}^{kt} changes its sign from one iteration to the next. Further, the authors also reset the box size if the multipliers do not change for three consecutive iterations in order to allow for escaping local optimal. Note that the ideal initial box size depends on the magnitude of the multiplier values and should be selected such that the convergence is the fastest.

If several demand scenarios are used (as suggested in Sect. 3.6), the right-hand side of the first constraint of the LP of the boxstep method has to be adjusted such that the Lagrange multipliers are also summed over all demand scenarios $s \in S$. Similarly, the two remaining constraints have to be defined for each demand scenario $s \in S$.

4.3 Generation of Feasible Upper Bounds

At each iteration, the Lagrangian heuristic may make the attempt to construct an upper bound to the problem, which corresponds to a feasible solution to the original problem. When the subgradient method is used to solve the Lagrangian dual problem, this upper bound also directly impacts the computation of the step size and therefore the convergence of the algorithm.

The generation of good upper bounds may be time consuming. If the Lagrangian relaxation is used within another optimization framework, e.g., within a branch-and-bound algorithm to provide stronger lower bounds, the quality of the upper bounds may be less important. In contrast, if the main objective is to produce high quality feasible solutions that can be used in practical planning, then the generation of such solutions should be given much care.

In both cases, Lagrangian heuristics typically aim at exploiting the solutions obtained from the Lagrangian subproblem in each of the iterations. These solutions are unlikely to be feasible, given that certain constraints have been relaxed (in our case, the demand constraints, and possibly the relocation linking constraints). However, when a set of "good" multipliers are used, the violation of the constraints may be small, and the provided solution may be close to a feasible one, or at least contain certain information that is helpful to approximate the optimal solution to the original problem. It therefore makes sense to attempt to "repair" such solutions such that they become feasible to the original problem, as will be outlined throughout Sects. 4.3.1–4.3.4. While such heuristics may be sufficient and provide high quality solutions for more simple problem variants, they are unlikely to find close-to optimal solutions in more complex problem variants. In this case, techniques aiming at improving the feasible solutions found may be employed, which will be discussed in Sect. 4.3.5.

4.3.1 Repairing Infeasibility of the Lagrangian Solutions

As previously explained, given that the demand constraints have been relaxed, the demand allocation from facilities to customers as proposed by the solution of the Lagrangian subproblem is most likely infeasible to the original problem. We now aim at repairing such infeasibility in order to obtain a valid upper bound.

A feasible solution to the here considered problem consists of two components:

- *A facility opening schedule* given by $\ell(j,t) \in L$, indicating a capacity level for each facility location $j \in J$ and each time period $t \in T$, as suggested by the Lagrangian solution. This opening schedule may be appropriate to efficiently serve demands in the original problem. However, given that demand constraints have been relaxed, capacity may also either be short or in excess at certain time periods. To ensure feasibility, additional capacity may have to be opened at certain time periods (either by opening new facilities or by increasing capacity at existing facilities). In addition, excess capacity may be reduced to further reduce maintenance costs and therefore improve the planning solution.

- *A demand allocation* from facilities to customers. The Lagrangian solution provides a tentative demand allocation, which may either exactly meet customer demands d_{ip}^t, under-serve (i.e., the flow sent from facilities does not cover the entire demand) or over-serve them (i.e., too much flow is being sent). Two solutions are common to correct unsatisfied demand: either the tentative demand allocation is repaired such that all demand is met exactly, or the tentative demand is ignored and demand is allocated optimally from scratch.

An attempt to "repair" one component of the Lagrangian solution will inevitably affect the other component, as the available capacity impacts the demand allocation and the demand allocation limits the possible capacity schedules. There are multiple ways how to approach this interplay. Given that computational efficiency is a key concern within the generation of upper bounds, below we consider a greedy approach to ensure sufficient capacity and repair the demand allocation. The demand allocation may be reoptimized afterwards, if required.

Greedy Capacity Increase and Demand Correction We first focus on the greedy approach proposed by Jena et al. (2017) (and original proposed by Shulman (1991) for a simpler problem variant) that aims at correcting missing production capacity, along with a greedy correction of under- or over-served demand. Let Σ_1, Σ_2 and Σ_3 denote three subsets, containing the demand triplets $< i, p, t >$ that are exactly met, over-served and under-served, respectively. They are defined as follows:

$$\Sigma_1 = \left\{ < i, p, t >: \sum_{j \in J} x_{i\overline{\ell(j,t)}p}^{jt} = 1 \right\}, \quad \Sigma_2 = \left\{ < i, p, t >: \sum_{j \in J} x_{i\overline{\ell(j,t)}p}^{jt} > 1 \right\}$$

$$\text{and } \Sigma_3 = \left\{ < i, p, t >: \sum_{j \in J} x_{i\overline{\ell(j,t)}p}^{jt} < 1 \right\}.$$

Based on this tentative demand allocation, the greedy approach proceeds as follows:

1. *Reduce redundant demand allocation*: more flow than required may currently be sent to certain customers. In an attempt to reduce redundant flows that is the most expensive, one may first sort, for each demand triplet $< i, p, t > \in \Sigma_2$, the allocation from facilities $j \in J$ in decreasing order of their allocation costs $g_{i\ell p}^{jt}$. Excess demand allocation is then iteratively reduced from one facility at a time until the requested demand quantity is met exactly.
2. *Increase capacity* in the case of capacity shortfall: if the total available capacity is insufficient to cover the total demand at certain time periods, the available capacity has to be increased. A naive approach may identify facilities already opened at some other time-periods and increase the capacity for the critical time-period to the same capacity level. If this is not possible, any random facility may be opened. More elaborate approaches may want to consider the transportation costs to under-served customers.

3. *Complete missing demand allocation*: for each demand triplet $< i, p, t > \in \Sigma_3$, feasible allocations from facilities $j \in J$ are first sorted in increasing order of their allocation costs $g_{i\ell p}^{jt}$ (considering, of course, only facilities $j \in J$ with left-over capacity at time period $t \in T$). Missing demand is then iteratively increased from one facility at a time until the requested demand is met exactly.
4. *Reduce capacity* in the case of capacity excess: once the demand allocation is feasible, a corresponding optimal capacity schedule may be computed, just as done when solving the Lagrangian subproblem, but considering only capacity levels that are sufficiently high to serve the allocated demand.

More Elaborate Approaches While the approach above may not provide optimal demand allocation, it is quite fast and can therefore be carried out at each iteration. Alternatively, once sufficient production capacity is ensured, an optimal demand allocation may be derived relatively quickly. This corresponds to a transportation problem in a bipartite graph, with production capacity as supply and customers as demand nodes. While this may be solved with several algorithms, from an implementation standpoint, an efficient way is to use specialized network solvers (such as the network algorithm from CPLEX), which can be reoptimized quickly with updated supply, demand and transportation costs.

In simple problem variants, the above greedy approach may be sufficient to generate upper bounds of sufficiently high quality (see, e.g. Jena et al., 2017). In more complex problem variants, the trade-off between different time-periods, costs for capacity adjustment, maintenance and demand allocation becomes too complex. Even more elaborate efforts to increase capacity may therefore not be effective when considered isolated from the other decisions, and therefore unnecessarily increase computing time. In such circumstances, a wiser choice may be to aim at improving existing feasible solutions within an improvement phase, where the interplay between these decisions may be jointly considered (see, e.g. Jena et al., 2016). One possibility is the construction of a restricted MIP model, which will be explained in Sect. 4.3.5.

Upper Bound Generation for Special Cases The outlined upper bound procedure is naturally also valid for special cases of the GMC. In particular, with a single time-period or predefined capacities (as opposed to capacity levels), the procedures described above simplify trivially and significantly. Finally, note that the approaches outlined above also exactly apply to one of the proposed extensions. Specifically, in the case of modular facility structures (see Sect. 3.3), the capacity opening schedule is then given by a pair $(\ell(j,t), n(j,t))$ with $\ell \in L, n \in L$, indicating the available and existing capacity levels for each facility location $j \in J$ and each time period $t \in T$.

4.3.2 Repairing in the Case of Complex Cost Functions

As a result of relaxing the demand constraints, the optimal demand allocation has been computed separately for each of the candidate facility locations by solving a piecewise-linear continuous knapsack. When generating feasible upper bounds, the demand allocation has to be solved by taking into consideration all available facilities at once. Given that the total production costs $F_\ell^j(U_{\ell q}^j)$ now depend on the total production quantity q at that facility (i.e., it has a piecewise-linear cost function), the optimal demand allocation cannot be computed by solving a transportation problem (or a minimum cost flow problem). As such, Lagrangian heuristics for planning problems that contain such cost functions resort to general purpose MIP solvers to solve for the optimal demand allocation (see, e.g., Štádlerová et al., 2023).

4.3.3 Repairing in the Case of Facility Relocation

When relocation linking constraints have been relaxed, the greedy approach presented above can be adapted by adding a step carried out at the beginning of the procedure. This step aims at matching the proposed incoming and outgoing facility relocations of the different capacity sizes at each time period. Specifically, for each facility $j' \in J$ for which the Lagrangian solution suggests an outgoing relocation of size $\ell \in L$ at time period $t \in T$ (i.e., $w^-_{j't\ell} = 1$), one needs to find another location $j'' \in J, j' \neq j''$ with an incoming relocation decisions of the same size and at the same time period (i.e., $w^+_{j''t\ell} = 1$). While some of the proposed incoming and outgoing relocation decisions can be adopted into a feasible solution, some of those decisions cannot be matched. Any procedure to carry out such matching should therefore ensure that the final capacity schedule remains coherent and feasible, and consider the resulting costs of the matching. Specifically, failing to match an incoming or outgoing relocation suggested by the Lagrangian solution results in an incoherent capacity opening schedule at that facility. Assume that the capacity opening schedule suggested by the Lagrangian solution has been adopted. Further assume that some of the incoming and outgoing relocation suggestions for facilities of the same size and at the same time period have been matched. Specific attention has to be given to two cases:

- An incoming relocation suggestion to location j at period t is not matched: Not matching a suggested incoming relocation at time period t implies that no facility is available, even though the opening schedule (i.e., the y variables) at location j assumes the subsequent availability of such a facility. This incoherence can be "repaired" by constructing a facility of the corresponding size at period t to obtain a coherent opening schedule.
- An outgoing relocation suggestion from location j at period t is not matched: Similarly, not matching a suggested outgoing relocation at time period t implies that the facility is still available at that location, even though the capacity

schedule assumes that there is none. This incoherence can be "repaired" by shutting down the facility at period t to obtain a coherent opening schedule.

A greedy approach (similar to the one used by Jena et al., 2016) may iterate through the time periods, count the number of matches of each size $\ell \in L, \ell \geq 1$ at each time period, and then arbitrarily select some of those facilities to select these matches. If the costs for facility shutdown and construction are not the same at all locations, outgoing relocations may be prioritized for locations where shutdown is expensive, and incoming relocation may be prioritized for locations where construction is expensive.

Optimal Matching A more complex approach, which, to the best of our knowledge, has not yet been considered in the literature, would be to optimize these matching decisions. The corresponding matching may be formulated as a restricted variant of the original MIP formulation, where only a fraction of decisions is used. Specifically, demand allocation is ignored. The resulting problem is a min-cost network flow problem, which aims at deciding whether the suggested outgoing $w^-_{jt\ell}$ and incoming $w^+_{jt\ell}$ relocations should be adopted, and which capacity expansions and reductions should be used in case one of the suggested relocations is not adopted. Capacity expansion decisions are only possible when an incoming capacity relocation has been suggested, and capacity reduction decisions are only possible when an outgoing capacity relocation has been suggested.

While such an optimal matching may be carried out computationally efficiently, it has to be noted that the relocation decisions, along with certain capacity changes may also be reoptimized in an improvement phase, as exemplified in Sect. 4.3.5.

4.3.4 Repairing in the Case of Uncertainty via Scenarios

The upper bound procedure outlined in Sect. 4.3.1 can easily be adapted to the case where a set of demand scenarios is used, but all capacity decisions are taken in the first stage of the stochastic planning problem. Specifically, the costs for the demand allocation has to be computed over the given demand scenarios according to their probabilities. While, ideally, all demand scenarios are considered in order to obtain a realistic estimation of the expected demand allocation costs when modifying the capacity schedule, a subset can be used in order to speed up the procedure. For instance, Štádlerová et al. (2023) use four reference scenarios to approximate the expected demand allocation costs, corresponding to those that have maximum, minimum, mean and medium total demand among all scenarios. In particular, the scenario with maximum total demand ensures that the designed capacity schedule has sufficient capacity to satisfy all demand.

4.3.5 Improvement Strategies

The feasible solutions derived from the Lagrangian solutions may be of sufficiently high quality when the problem is not too complex (see, e.g., Shulman, 1991; Correia & Captivo, 2003; Wu et al., 2006). In more complex problem variants, the synergies and interaction between the various decisions may be too high, and the derived solutions may not be of sufficient quality. In this case, improvement heuristics can be employed to improve upon existing solutions. A variety of improvement heuristics has been proposed in the literature, including:

- Local and tabu searches: starting from a specific feasible solution, search procedures, exploring neighborhoods tailored to the problem structure, can be employed to iteratively improve the solution (see, e.g., Correia & Captivo, 2006; Hinojosa et al., 2008; Li et al., 2009).
- Iterative Matheuristics such as Local Branching (Fischetti & Lodi, 2003): starting from a feasible solution, a maximum number of variables may change their solution variables in order to allow for iterative improvements.
- Restricted MIP models: considering only a subset of the original decisions may sufficiently reduce the problem size such that it can be solved quickly (see, e.g., Jena et al., 2017).

In contrast to the restricted MIP, the first two types of heuristics are iterative procedures. Matheuristics and the restricted MIP have the advantage that search neighbourhoods do not have to be tailored to the problem structure. All of such methods are typically time-intensive. The feasible solution on which such methods are applied should therefore be chosen carefully. They may, for example, be employed at each iteration at which a new best upper bound has been found, in order to further polish this bound. In the simplest case, the improvement method is added as a second phase once the Lagrangian relaxation has sufficiently converged.

Restricted MIP Model While tailored improvement heuristics may be very efficient, restricted MIP models have the advantage that they are flexible, easy to implement and consider the complex synergies among the different types of decisions. In other words, they may be able to consider neighbourhoods that are manually hard to design. Selecting the decisions available in the restricted model is a delicate trade-off: making available too few decisions may not allow for a significant improvement, while making available too many decisions may make the resulting model hard to solve. In order to define a restricted MIP to the problem at hand, one may first identify the part of the capacity schedule decisions that should be fixed (i.e., the facilities, along with their capacity levels for certain time periods). For locations and time periods where the schedule is not fixed, one then defines the available capacity levels, as well as the eligible capacity transitions (i.e., the eligible changes between different capacity levels). Clearly, it is crucial to identify decisions that may be part of the optimal solution to the original problem. Luckily, in this regard, the iterative process of the Lagrangian Relaxation may be helpful. Two options can be considered (Jena et al., 2017):

1. Consider decisions selected by the **Lagrangian solutions**: Assume that a total of n^{iter} number of iterations has been performed. Let $n^C_{j\ell t}$ be the number of Lagrangian solutions where capacity level $\ell \in L$ has been selected at location $j \in J$ and time period $t \in T$. Further, let L^R_{jt} be the set of capacity levels that will be made available in the restricted MIP at location $j \in J$ and time period $t \in T$. **Capacity opening decisions may be fixed**, if they appear in at least $100 \times pFix$ percent of all Lagrangian solutions, where $pFix$ is a predefined parameter from]0, 1]. Specifically, a capacity at location $j \in J$ and time period $t \in T$ is fixed to $\ell \in L$, i.e., $L^R_{jt} = \{\ell\}$, if $n^C_{j\ell t}/n^{iter} \geq pFix$. As **available capacity levels** at each location $j \in J$ and period $t \in T$, one may use those that have been the most popular. That is, if L^R_{jt} has not been fixed in the previous step, it can be composed by the n^S capacity levels that have appeared most often in the Lagrangian solutions, where n^S is a predefined parameter. Finally, the set of **eligible capacity transitions** (corresponding to variables $y^{jt}_{\ell_1\ell_2}$) may, in the most generous case, be defined as all transitions among the defined sets of available capacity levels, i.e., $\ell_1 \in L^R_{jt}$, $\ell_2 \in L^R_{j(t+1)}$ $j \in J, t \in T$.
2. Decisions endorsed by the aggregated **Bundle method**: the aggregated bundle method (see Sect. 4.2.2) provides a weight θ^m for each of the previous Lagrangian multipliers and solutions $m \in B$. This weight can be seen as a probability that the corresponding multipliers, and hence the associated solution, lead to a good opening schedule. This information can be exploited in various ways. A straightforward approach computes the probability $\tilde{y}^{jt}_\ell = \sum_{m \in B} \theta^m \overline{y}^{mjt}_\ell$, where $\overline{y}^{mjt}_\ell$ takes value 1 if the Lagrangian solution of Bundle member m selects capacity level $\ell \in L$ for location $j \in J$ at time period $t \in T$. A restricted MIP model can then be constructed: for each location $j \in J$ and time period $t \in T$, a capacity decision is fixed if $\tilde{y}^{jt}_\ell \geq pFix$. If a decision is not fixed, the set L^R_{jt} of available capacity levels is composed of the n^S capacity levels with highest probabilities $\tilde{y}^{jt}_\ell$.

5 Conclusions

Even though facility location problems have been investigated since over 50 years, they still remain an active component of contemporary research, given their strategic importance in most supply chains. In this chapter, we have reviewed the chronological development of solution methods based on Lagrangian relaxation to tackle facility location problems, which have become increasingly complex over time.

We have reviewed a strong formulation, that allows for multiple time periods, multiple commodities, multiple capacity levels and the adjustment of capacity along time. We have also explicitly addressed extensions that account for complex production cost functions, modular structures of facilities, and the incorporation of parameter uncertainty. As such, we have reviewed most of the relevant features in modern facility location planning problems. We have then shifted focus to the efficient solution of such problem variants via Lagrangian relaxation: specifically, the solution of the Lagrangian subproblem, the solution of the Lagrangian dual problem and the generation of feasible upper bounds. As such, we have provided a systematic guide to modeling and solving complex facility location problems, that will, hopefully, be useful to practitioners and researchers that aim at solving problem variants composed of any combination of the here discussed problem characteristics.

While Lagrangian relaxation has certainly received less than attention than other decomposition methods in recent years, more recent literature suggests that this approach remains powerful even when faced with complex problem variants. While such problems are likely to become even more complex in the future in order to account for more details of the practical planning problems, certain contemporary applications demonstrate the need to integrate location problems in more complex planning contexts. For example, the design of carbon-dioxide transportation and storage networks translates into an integrated location and network design problem, where both the capturing and storage facilities, as well as the pipelines, have to be selected from a set of available (capacity) sizes. Such problems remain challenging to solve. Specifically, in this problem class, demand quantities are not specified separately for each node as in classical facility location, or between origin-destination node-pairs as in classical network design. Instead, one typically aims at attaining a single global target level for the entire system, no matter where the carbon is extracted and stored (see, e.g., Middleton & Bielicki, 2009). In such case, the popular strong inequalities, which have been key to developing strong formulations, become ineffective, which makes it difficult to obtain strong bounds. Research on such problems is eagerly needed, and is therefore attractive from an academic and a practical point of view.

Acknowledgments I am thankful to the two anonymous referees whose comments helped to improve this paper. Further, I am grateful to my collaborators and co-authors who have contributed to shaping many of the here presented developments throughout our joint research.

References

Agar, M. C., & Salhi, S. (1998). Lagrangean heuristics applied to a variety of large capacitated plant location problems. *Journal of the Operational Research Society, 49*(10), 1072–1084.

Alarcon-Gerbier, E., & Buscher, U. (2022). Modular and mobile facility location problems: A systematic review. *Computers and Industrial Engineering*, 108734.

Allen, R. C., Avraamidou, S., Butenko, S., & Pistikopoulos, E. N. (2022). Solution strategies for integrated distribution, production, and routing problems arising in modular manufacturing. Technical Report.

Antunes, A. P., Berman, O., Bigotte, J., & Krass, D. (2009). A location model for urban hierarchy planning with population dynamics. *Environment and Planning A, 41*(4), 996–1016.

Barcelo, J., Hallefjord, A., Fernandez, E., & Jörnsten, K. (1990). Lagrangean relaxation and constraint generation procedures for capacitated plant location problems with single sourcing. *OR Spektrum, 12*(2), 79–88.

Beasley, J. E. (1993). Lagrangean heuristics for location problems. *European Journal of Operational Research, 65*(3), 383–399.
Ben-Tal, A., El Ghaoui, L., & Nemirovski, A. (2009). *Robust optimization* (Vol. 28). Princeton University Press.
Birge, J. R., & Louveaux, F. (2011). *Introduction to stochastic programming*. Springer Science & Business Media.
Cabezas, X., García, S., Martin-Barreiro, C., Delgado, E., & Leiva, V. (2021). A two-stage location problem with order solved using a lagrangian algorithm and stochastic programming for a potential use in covid-19 vaccination based on sensor-related data. *Sensors, 21*(16), 5352.
Chardaire, P., & Sutter, M.C.A.C. (1996). Solving the dynamic facility location problem. *Networks, 28*(2), 117–124.
Chouman, M., Crainic, T. G., & Gendron, B. (2017). Commodity representations and cutset-based inequalities for multicommodity capacitated fixed-charge network design. *Transportation Science 51*(2), 650–667.
Christensen, T. R. L., & Klose, A. (2021). A fast exact method for the capacitated facility location problem with differentiable convex production costs. *European Journal of Operational Research, 292*(3), 855–868.
Contreras, I., Díaz, J. A., & Fernández, E. (2009). Lagrangean relaxation for the capacitated hub location problem with single assignment. *OR Spectrum, 31*, 483–505.
Cornuejols, G., & Thizy, J. M. (1982). Some facets of the simple plant location polytope. *Mathematical Programming, 23*(1), 50–74. https://doi.org/10.1007/BF01583779. http://www.springerlink.com/index/10.1007/BF01583779.
Correia, I., & Captivo, E. M. (2006). Bounds for the single source modular capacitated plant location problem. *Computers & Operations Research, 33*(10), 2991–3003.
Correia, I., & Captivo, M. E. (2003). A Lagrangean heuristic for a modular capacitated location problem. *Annals of Operations Research, 122*(1), 141–161.
Diabat, A., Richard, J. P., & Codrington, C. W. (2011). A Lagrangian relaxation approach to simultaneous strategic and tactical planning in supply chain design. *Annals of Operations Research, 203*(1), 55–80.
Fischetti, M., & Lodi, A. (2003). Local branching. *Mathematical Programming, 98*, 23–47.
Frangioni, A. (2005). About Lagrangian methods in integer optimization. *Annals of Operations Research, 139*(1), 163–193.
Frangioni, A., & Gallo, G. (1999). A bundle type dual-ascent approach to linear multicommodity min-cost flow problems. *INFORMS Journal on Computing, 11*(4), 370–393.
Frangioni, A., & Gorgone, E. (2014). Bundle methods for sum-functions with "easy" components: Applications to multicommodity network design. *Mathematical Programming, 145*(1), 133–161.
Gendron, B. (2011). Decomposition methods for network design. *Procedia - Social and Behavioral Sciences, 20*, 31–37.
Gendron, B., Khuong, P. V., & Semet, F. (2016). A lagrangian-based branch-and-bound algorithm for the two-level uncapacitated facility location problem withguin single-assignment constraints. *Transportation Science, 50*(4), 1286–1299.
Ghodsi, G. (2012). A lagrangian relaxation approach to a two-stage stochastic facility location problem with second-stage activation cost. Master's Thesis, University of Waterloo.
Görtz, S., & Klose, A. (2012). A simple but usually fast branch-and-bound algorithm for the capacitated facility location problem. *INFORMS Journal on Computing, 24*(4), 597–610.
Guignard, M. (2003). Lagrangean Relaxation. *Top, 11*(2), 151–228.
Guignard-Spielberg, M., & Kim, S. (1983). A strong Lagrangian relaxation for capacitated plant location problems. Technical Report, Department of Statistics.
Held, M., Wolfe, P., & Crowder, H. P. (1974). Validation of subgradient optimization. *Mathematical Programming, 6*(1), 62–88.
Hinojosa, Y., Puerto, J., & Fernández, F. (2000). A multiperiod two-echelon multicommodity capacitated plant location problem. *European Journal of Operational Research*, 123(2), 271–291.

Hinojosa, Y., Kalcsics, J., Nickel, S., Puerto, J., & Velten, S. (2008). Dynamic supply chain design with inventory. *Computers & Operations Research, 35*(2), 373–391.
Holmberg, K., & Ling, J. (1997). A Lagrangean heuristic for the facility location problem with staircase costs. *European Journal of Operational Research, 97*(1), 63–74.
Jena, S. D. (2014). Dynamic facility location with modular capacities: Models, algorithms and applications in forestry. Ph.D Thesis, Université de Montréal.
Jena, S. D., Cordeau, J. F., & Gendron, B. (2015a). Dynamic facility location with generalized modular capacities. *Transportation Science, 49*(3), 484–499.
Jena, S. D., Cordeau, J. F., & Gendron, B. (2015b). Modeling and solving a logging camp location problem. *Annals of Operations Research, 232*, 151–177.
Jena, S. D., Cordeau, J. F., & Gendron, B. (2016). Solving a dynamic facility location problem with partial closing and reopening. *Computers & Operations Research, 67*, 143–154.
Jena, S. D., Cordeau, J. F., & Gendron, B. (2017). Lagrangian heuristics for large-scale dynamic facility location with generalized modular capacities. *INFORMS Journal on Computing, 29*(3), 388–404.
Kadri, A., Koné, O., & Gendron, B. (2022). A lagrangian heuristic for the multicommodity capacitated location problem with balancing requirements. *Computers & Operations Research, 142*, 105720.
Li, J., Chu, F., & Prins, C. (2009). Lower and upper bounds for a capacitated plant location problem with multicommodity flow. *Computers & Operations Research, 36*(11), 3019–3030.
Marín, A., Martínez-Merino, L. I., Rodríguez-Chía, A. M., & Saldanha-da Gama, F. (2018). Multi-period stochastic covering location problems: Modeling framework and solution approach. *European Journal of Operational Research, 268*(2), 432–449.
Marsten, R. E., Hogan, W. W., & Blankenship, J. W. (1975). The boxstep method for large-scale optimization. *Operations Research, 23*(3), 389–405.
Middleton, R. S., & Bielicki, J. M. (2009). A scalable infrastructure model for carbon capture and storage: Simccs. *Energy Policy, 37*(3), 1052–1060.
Pacheco Paneque, M., Gendron, B., Sharif Azadeh, S., & Bierlaire, M. (2022). A lagrangian decomposition scheme for choice-based optimization. *Computers & Operations Research, 148*, 105985.
Padberg, M. W., Van Roy, T. J., & Wolsey, L. A. (1983). Valid linear inequalities for fixed charge problems. *Operations Research, 33*(4), 842–861.
Schütz, P., Tomasgard, A., & Ahmed, S. (2009). Supply chain design under uncertainty using sample average approximation and dual decomposition. *European Journal of Operational Research, 199*(2), 409–419.
Shulman, A. (1991). An algorithm for solving dynamic capacitated plant location problems with discrete expansion sizes. *Operations Research, 39*(3), 423–436.
Sridharan, R. (1991). A lagrangian heuristic for the capacitated plant location problem with side constraints. *Journal of the Operational Research Society, 42*(7), 579–585.
Štádlerová, Š., Aglen, T. M., Hofstad, A., & Schütz, P. (2022). Locating hydrogen production in norway under uncertainty. In: *Computational logistics: 13th international conference, ICCL 2022, Barcelona, Spain, September 21–23, 2022, Proceedings* (pp. 306–321). Springer.
Štádlerová, Š., Jena, S.D., & Schütz, P. (2023). Using lagrangian relaxation to locate hydrogen production facilities under uncertain demand: a case study from norway. *Computational Management Science, 20*(1), 10.
Vahidnia, M. H., Alesheikh, A. A., & Alimohammadi, A. (2009). Hospital site selection using fuzzy AHP and its derivatives. *Journal of Environmental Management, 90*(10), 3048–3056.
Wu, L., Zhang, X., & Zhang, J. (2006). Capacitated facility location problem with general setup cost. *Computers & Operations Research, 33*(5), 1226–1241.

Bin Packing Methodologies for Capacity Planning in Freight Transportation and Logistics

Maria Elena Bruni, Teodor Gabriel Crainic, and Guido Perboli

1 Introduction

Capacity planning is an umbrella term which may be synthetically defined as the process of determining the appropriate mix and quantity of resources to meet (predicted) varying levels of demand. It is a general concept encountered in a very broad spectrum of human activities, particularly in production (manufacturing, services, energy, etc.), project management, human resource management, telecommunication, transportation, and logistics. We focus on *Consolidation-based Freight Transportation and Logistics* (CFTL) in this chapter, particularly on the planning of the *supply capacity* required to satisfy known or predicted, customer or internally-generated demand.

Consolidation in CFTL is the process of combining several shipments, potentially with different origins and destinations, and loading them into the same vehicle or container for their complete or partial journey. Consolidation aims to reduce the unit shipment cost and the journey time, benefiting all parties involved. Railways, less-than-truckload motor carriers, shipping companies moving containers on oceans, seas, rivers, and canals, postal services and express couriers are prime

M. E. Bruni (✉)
CIRRELT and University of Calabria, Rende CS, Italy
e-mail: mariaelena.bruni@unical.it

T. G. Crainic
CIRRELT and Département d'analytique, opérations et technologies de l'information, École des sciences de la gestion, Université du Québec à Montréal, Montréal, QC, Canada
e-mail: TeodorGabriel.Crainic@cirrelt.net

G. Perboli
CIRRELT and Politecnico di Torino, Turin, Italy
e-mail: guido.perboli@polito.it

T. G. Crainic et al. (eds.), *Combinatorial Optimization and Applications*,
International Series in Operations Research & Management Science 358,
https://doi.org/10.1007/978-3-031-57603-4_6

examples of consolidation-based carriers moving a large and valuable part of the world trade over short, medium, long, and intercontinental distances. Innovative “new" CFTL concepts, e.g., City Logistics, Physical Internet, and Synchromodality, are also based on consolidation. Such activities require terminals to perform consolidation/deconsolidation of goods and, eventually, of vehicles (into/out of convoys, e.g., rail, road, and barge trains) as well as to store the goods for varying periods.

Freight transportation may be briefly described as the interplay between *shippers* and *carriers*. The former makes up the multicommodity, origin-destination *demand* side of the system, while the latter *supply* the mono/multi/intermodal transportation and terminal resources, services, and capacity required to satisfy the demand.

Each stakeholder, be it a shipper or carrier, faces planning challenges to procure and secure or to plan and provide capacity, respectively. The particular challenges and the most appropriate Operations Research methodology to address them vary according to the type of firm and the planning horizon considered. All of them share, however, concerns regarding the adequate representation of how the contemplated freight flows will use the planned capacity, in terms of operational and economic feasibility and efficiency. Yet, standard formulations with classical capacity constraints do not adequately represent many physical and operational attributes of freight storage and movement capacity units, resulting in rough approximations of how demand flows use available capacity.

Bin Packing increasingly appears as a more appropriate methodology to address the challenge of the adequate representation of the interplay between the demand flows and the storage and movement capacity. Packing offers a gateway to infuse CFTL capacity planning models with novel nuances and attributes. It brings a micro-level perspective by delving into a more efficient allocation of resources, e.g., vehicle and container fleets, minimizing empty spaces within them, and reducing transportation costs and environmental impacts. This granular approach ensures that resources are better utilized, minimizing waste and enhancing cost-effectiveness.

The amount of contributions found in the literature on this topic is still limited. Our objective is to start filling this gap in knowledge. We identify issues and challenges, define new problem settings, and propose research directions.

The chapter is organized as follows. Section 2 recalls the main packing problems and models. Section 3 identifies the main categories of stakeholders and their major capacity requirements. Section 4 reviews the limited literature on CFTL capacity planning with packing considerations. Section 5 discusses research perspectives. We conclude in Sect. 6.

2 Packing Problems and Models

Given a set of objects and a set of loading units, packing problems are related to the choice of the objects to put into each loading unit and the way to accommodate the selected objects into their respective loading units, in order to optimize a specified objective function. A given subset of objects assigned to a specific loading unit is said to be *loadable* if the objects and their accommodation into the loading units respect one or more constraints, called *loading* or *packing constraints*.

In the basic packing problem setting, objects and loading units are mono-dimensional, i.e., the geometry of both objects and loading units is disregarded, and one attribute only is associated with them. The attribute generally associated with the objects is *weight* or *volume*, while that of loading units is often named *capacity*. This attribute is related to the packing aspect only, i.e., it is used to define the amount of space available in the loading units and the amount of space used inside the loading units by the objects. Let $\mathcal{I}$ be a set of objects and $\mathcal{J}$ a set of loading units. Let w_i be the volume (weight) of the object $i \in \mathcal{I}$ and W_j be the capacity of loading unit $j \in \mathcal{J}$, respectively. A set of objects $\mathcal{I}_j \subseteq \mathcal{I}$ is then defined as loadable into a loading unit $j \in \mathcal{J}$ in this basic packing setting, if the sum of the volume of the objects is less or equal to the capacity of the loading unit, yielding the following packing constraint, often referred to as the *knapsack constraint*, for each loading unit in $\mathcal{J}$:

$$\sum_{i \in \mathcal{I}_j} w_i \leq W_j. \tag{1}$$

Two main families of packing problems may be defined, depending on the objective function considered: the *Knapsack* (KP) and the *Bin Packing* (BPP) problems.

In the KP, one is given a set of objects, characterized by volume and profit, and one loading unit of finite capacity, called knapsack. The aim is to find the subset of objects that can be accommodated into the knapsack, such that the total profit of the loaded objects is maximized, subject to the knapsack constraint (1) (Martello & Toth, 1990).

In the basic BPPs, a set of objects, called *items*, must be loaded into a set of equal-sized loading units, called *bins*, such that the number of bins used is minimized. All the objects must be loaded and each item must be loaded into exactly one bin (Crainic et al., 2007b,a). As for the KP, the packing constraint requires that, for every bin, the sum of the volume of the objects assigned to it is less or equal to its capacity.

The Multiple Knapsack Problem (MKP) is the extension of the KP where one is given a set of knapsacks of finite capacity, instead of a single one. No cost is associated with the knapsacks, and the objective is the maximization of the total profit of the subset of items chosen to be loaded (Pisinger, 1999).

The Variable Sized Bin Packing Problem (VSBPP) is a generalization of the BPP, proposed Friesen and Langston (1986), which introduces bin types. In the VSBPP, the cost of the bins is strongly correlated to the volume of the bin. A more interesting variant, without correlation between the volumes and the costs of the bins, is discussed in Crainic et al. (2011b) and is named Variable Cost and Size Bin Packing Problem (VCSBPP). All the items must be compulsorily loaded in both the VSBPP and VCSBPP.

Several problems can be modeled using this broad setting, from classical operational issues related to the loading of vehicles, trains, and ships, to the less classic loading of space vehicles, transportation, logistics, telecommunication, and planning of systems (Zhao et al., 2016; Alba Martínez et al., 2013; Crainic et al., 2007c; Colaneri et al., 2003), by appropriately defining the items, the bins, the objective function, and the set of packing constraints.

2.1 Multiple Attribute Packing Problem Extensions

The basic, mono-dimensional, packing problems have been extended by adding specific attributes to objects and loading units constraints, or both. We now present the main extensions of the packing problem defined in the literature to introduce new attributes to objects and loading units and the main references. According to the majority of the literature, we refer to objects as items and to loading units as bins.

We analyze these extensions along four axes: additional packing constraints or attributes, temporal requirement, uncertainty of one or more attributes, and presence of multiple stakeholders.

2.1.1 Additional Packing Constraints or Attributes

Multi-constrained packing problems address settings where both items and bins are characterized by two or more packing attributes. Most contributions on this topic in the literature have addressed the KP case, that is, the Multi-Constrained Knapsack Problem (MCKP). Items and knapsack are then characterized by two or more packing attributes, and, as in the KP, the items have a profit and the objective is to find the subset of items that maximizes the sum of the loaded item profits. The main difference lies in the packing constraints, as the subset must satisfy a knapsack constraint for each packing attribute.

Multi-constrained problems have received little attention in transportation and logistics. The MCKP has been used for mobile-communications planning, where it models the management of different quality-of-service levels (Osei-Mensah et al., 2022; García Llinás and Nagi, 2015).

A second family of multi-attribute packing problems considers the geometry of the items and bins. They are called K-Dimensional (KD) Packing Problems, where K is the number of geometrical dimensions associated with the items and the bins. The main difference with the previous multi-constrained packing problems lies in the fact that the knapsack constraints are not sufficient to ensure the feasibility of the packing of the items. Actually, the items in a KD packing problem must be accommodated into the bins in such a way that their geometry does not overlap with that of the other items in the same bin. Thus, a feasible packing is defined by associating a specific position to each item inside the loading space of the bin.

According to the regularity of the shapes of items and bins (boxed or irregular), the number of dimensions (2, 3, or more) and the presence of additional constraints (e.g., layered loadings, rotation of the items, items in fixed position, presence of constraints on the stability of the packing in terms of the position of the centre of gravity of the packing, and the presence of fragile items), several *K*D packing-problem variants may be defined (Baldi et al., 2012b; Colaneri et al., 2003). For the sake of brevity, we recall the main results for the most used case only, i.e., the loading of boxed-shaped items into boxed-shaped bins. Concerning the number of dimensions, most of the existing literature concentrates on the 2D and 3D cases when examining dimensionality. However, studying cases with more than 3 dimensions offers both theoretical understanding and real-world applicability. First, analyzing how properties and algorithms behave as the number of dimensions grows provides useful academic insights. Second, practical situations that require consideration of more than three dimensions occur, for example, when incorporating time or an additional variable for which one needs to avoid overlapping, such as in crane loading and unloading processes (Ghirardi et al., 2011).

A first attempt to model *K*D Packing Problems is due to Gilmore and Gomory (1965). The authors propose a representation given by the enumeration of all the *patterns*, i.e., all subsets of items that could be accommodated into a bin, given the problem constraints. The huge number of patterns that can be defined from a given set of items makes the approach appropriate for column-generation approaches (Gilmore & Gomory, 1965; Baldacci & Boschetti, 2007).

Beasley (1985) consider a formulation for 2D packing based on the discretization of the bin surface into $p \times q$ rectangles. The bottom-left corner of each item was then placed on the bottom-left corner of a rectangle. A similar representation was introduced by Hadjiconstantinou and Christofides (1995), except that instead of explicitly partitioning the bin into rectangles, they limited the set of coordinates each item could assume to p and q values. In both cases, the number of variables grows with the accuracy of the discretization. Consequently, both representations are principally used to compute upper bounds through Lagrangian relaxation and subgradient optimization.

An approach often used for 2D-packing building consists of combining procedures designed for mono-dimensional problems, namely *shelf* (or *layer*) methods (Chung et al., 1982; Berkey & Wang, 1987). The items are first sorted and packed into "shelves" with sizes equal to the width of the bin. The problem then reduces to solving a mono-dimensional packing instance. Indeed, a 2D packing can be obtained by placing the shelves into the container according to the solution of a mono-dimensional packing problem, where the size of the items equals the depth of the shelves and the size of the mono-dimensional bins equals the depth D of the two-dimensional bin. The same approach can also be used to build 3D packings. Build first two-dimensional shelves by using any 2D algorithm and, then, arrange them into three-dimensional bins by solving a mono-dimensional packing problem, where the size of the items equals the height of the shelves and the size of the bins equals the height of the 3D bin. When the 2D shelves are also built according to the shelf approach, the method is known as *wall-building* (Pisinger, 2002). The drawback of the shelf approach is that it introduces guillotine cuts on the depth and height of the two and three-dimensional bins, respectively, leading to their under-utilization.

Martello et al. (2000) define *Corner Points* as the non-dominated locations where an item can be placed into an existing packing. In two dimensions, Corner Points are located where the envelope of the items in the bin changes from vertical to horizontal. Corner Points on the three-dimensional envelope can be found applying the two-dimensional algorithm for each distinct value of the height of the bin defined by the lower and upper terminal lines of each item.

Fekete and Schepers (2004) propose a graph-theoretical approach for the characterization K-Dimensional Packing Problems. The authors consider the relative positions of the items in a feasible packing and defined a graph describing the item "overlapping" according to the projection of the items on each orthogonal axis. The authors prove necessary conditions on the interval graphs to define a feasible packing. Combined with good heuristics for dismissing infeasible subsets of items, this characterization was used to develop a two-level tree search. According to computational results, mainly limited to 2D problems, this strategy outperforms previous methods. The representation suffers from several limitations, e.g., the difficulty of handling additional packing constraints, such as fixing the position of one or more items, and the presence of multiple equivalent graph representations of the same packing of items (Crainic et al., 2009).

Extending the Corner Point concept, Crainic et al. (2007c) define the Extreme Points, providing a better exploitation of the volume of the bins. The basic idea is to efficiently find the points where an item can be added to an existing packing. Computational results show that the proposed method outperforms all the other constructive heuristics for both 2-D and 3-D Bin Packing Problems, Container Packing (the variant of the 3D KP where the profits of the items are their volume), and the Balanced 3D Knapsack Problem (the 3D Knapsack Problem with constraints on the stability of the bin and packing in terms of the position of the centre of gravity of the packing) (Crainic et al., 2007c; Perboli et al., 2011; Baldi et al., 2012b). Extreme Points are broadly used in the literature due to their computational efficiency and the possibility to be adapted to incorporate additional constraints.

2.1.2 Temporal Requirements

The introduction of time-related considerations and characteristics for items and bins makes up a very relevant and challenging family of problem-setting extensions, related to resource sharing in the on-demand economy, subcontracting, and co-operative frameworks, such as synchromodality (Perboli et al., 2017), City Logistics (Crainic et al., 2021b), and Many-to-One-to-Many systems (Taherkhani et al., 2022).

Initiated recently, work in this area started from the VCSBPP, the problem is defined on a representative-for-planning schedule length divided into periods by equally-distant time moments.

Liu et al. (2023) consider an extension of the VCSBPP where the items have a time-window during which they must be loaded. A subset of items assigned to a bin is then considered loadable if the total volume of the items does not exceed the bin capacity and the intersection of the items time windows is non-empty.

Crainic et al. (2021a) introduce a mono-dimensional problem variant where items and bins display several time-related attributes. A bin, if selected, should be used within an availability interval. Similarly, an item is characterized by an availability time moment, a delivery due date, as well as earliest and latest delivery time moments. The authors define time-dependent item-to-bin assignment costs and propose two formulations, for the myopic (single period) and multi-period variants, aimed at minimizing the total system cost, which comprises the sum of the fixed costs of the selected bins and the total item-to-bin assignment cost, using both regular and spot-market bins. Very efficient heuristic methods are proposed to tackle the problem.

Perboli et al. (2021) extend the results in Perboli and Rosano (2019) by introducing time-dependency in the VCSBPP. A new variant is proposed, namely the Variable Cost and Size Bin Packing Problem with Time-Dependent Costs (VCSBPP-TD), where the characteristics associated with the bins are time-dependent. A period-specific item-to-bin assignment cost combines the various costs associated with the time characteristics of items and bins. The authors propose a mixed-integer programming formulation and heuristics to solve realistic instances.

2.1.3 Uncertainty

Item and bin attributes are deterministic in most packing problem settings found in the literature. Yet, the assumption that all the problem parameters are deterministically known is quite unrealistic for many cases of actual planning applications, where the exact values associated with some parameters are not known in advance, and partial information only is available at the planning stage. The situation is evolving, however, and research is increasingly considering various sources of uncertainty.

Two paradigms are primarily found in the literature considering uncertainty in packing problems, namely, chance-constraints (Charnes & Cooper, 1959) and two-stage stochastic programming (Birge & Louveaux, 2011). While the first approach has been predominantly used in the context of surgery scheduling (Lamiri et al., 2008; Bruni et al., 2015) and virtual machine consolidation in cloud services (Martinovic et al., 2022), the second appears in logistics capacity planning. Stochastic versions of packing problems typically assume that the volume of the objects is uncertain, while the capacity of the loading unit is a known parameter. A notable exception is the work of Crainic et al. (2024), which is discussed at the end of the subsection.

Song et al. (2014) present a chance-constrained version of the KP, where both the item size and the knapsack capacity are random variables. Chance constraints are imposed on the packing, requiring that the knapsack capacity limits are satisfied with a given probability. Focusing on the special case in which the uncertainty is represented by a finite number of scenarios, a deterministic equivalent model is derived and solved by a coefficient-strengthening procedure. Wang et al. (2021) consider a generalization of Song et al. (2014), where the bin capacity is deterministically known and the item random size follows a discrete distribution. In the proposed setting and formulation, chance constraints are imposed separately on each bin to guarantee that the capacity is not exceeded with a given probability. Two equivalent reformulations are derived and a tailored solution approach is presented.

Cohen et al. (2019) consider the bin packing with chance constraints under the assumption that the item sizes are independent random variables with a continuous Gaussian probability distribution. The deterministic equivalent reformulation of the problem is also tailored to address the case of unknown probability distributions. The distributionally robust version of the model is derived assuming that the item size probability distributions belong to a family of distributions with a known mean and covariance matrix.

Zhang et al. (2020) consider two versions of the chance-constrained bin packing problem. Specifically, the first version is a two-stage stochastic mixed-integer program, which assumes that the discrete distribution of item sizes is known. The second version is a distributionally robust problem, which assumes again that the distributions of item sizes belong to a family of distributions with known mean and covariance matrix.

Perboli et al. (2012) and Perboli et al. (2014) consider the uncertainty due to the oscillations of the cost associated with bin loading. In both cases, the random profit is defined as the sum of a deterministic term and a stochastic component with an unknown probability distribution. In both papers, the stochastic problems are formulated as two-stage stochastic programming models. The authors also show, under mild assumptions regarding the random parameters and by applying the extreme-value theory, how to derive a deterministic model able to replicate the first-stage decisions with an error of less than 1% from the optimal solution of the original two-stage model.

Crainic et al. (2016, 2014a) propose the Stochastic Variable Cost and Size Bin Packing Problem (SVCSBPP), an extension of the VCSBPP considering uncertainty in both supply and demand. Relative to supply, the authors consider that the future availability and cost of bins, of various types, may be different from those known when the selection decisions are taken, including bins to be used when unfeasibility issues arise. On the demand side, the problem setting considers the number and dimension of items, which will need to be packed into available bins. The authors propose a two-stage integer stochastic programming formulation, with integer decision variables in the second stage modelling the selection of additional bins.

Given that state-of-the-art mixed-integer programming solvers struggle to solve even small instances of SVCSBPP formulations (Crainic et al., 2011a, 2014b), the authors propose an efficient meta-heuristic based on progressive-hedging (Rockafellar & Wets, 1991) ideas. The authors also propose a new consensus method aimed at addressing symmetry issues, i.e., the existence of several values of the model decision variables representing the same solution, which afflict many problem settings including packing ones. This enhancement, combined with the efficacy of the heuristics presented in Crainic et al. (2011b), enabled to tackle large-sized instances with a negligible computational effort, providing optimal solutions to all the instances for which proven optima were known.

Crainic et al. (2024) consider an extension of the SVCSBPP and VCSBPP problem settings, of significant practical relevance, addressing broader uncertainty issues and the reliability of the future supply. The authors explicitly address the possibility that “certain" bins, i.e., those contracted at decision time (first stage of the two-stage stochastic formulation) may yet suffer from a loss of capacity when needed in the future (due to, e.g., delays, mechanical failures, accidents, etc.). The authors propose a two-stage stochastic programming model for this SVCSBPP with Capacity Loss. They then successfully apply and further enhance the progressive-hedging meta-heuristic of Crainic et al. (2016), and show the effectiveness and value of the methodology on applications to long-haul and urban freight transportation planning.

2.1.4 Multiple Stakeholders

Packing problems traditionally considered the point of view of one stakeholder only. An interesting direction of research addresses how to incorporate the conjoint points of view of multiple stakeholders. The current literature focuses on cases where a leader makes certain decisions, and one or more followers optimize their own objectives based on the leader's choices. The followers' decisions, in turn, influence the leader's objective and the feasibility of their decisions. The primary goal is to identify the optimal decision from the leader's perspective and bilevel optimization is the methodology of choice.

In the context of packing, the literature primarily focuses on KPs. Buchheim et al. (2022) addresses a bilevel continuous knapsack problem where the leader controls only the capacity (bins), while the followers select the items to optimize their objective function. Solving this problem without uncertainty is relatively straightforward, making it an ideal starting point for exploring the increased difficulty that arises when uncertainty is introduced in bilevel optimization. The authors consider the continuous version of the problem only, where the follower solves a continuous knapsack problem.

Another variant of this problem, where the follower solves a binary knapsack problem instead, is introduced in Dempe and Richter (2000) and further explored in Brotcorne et al. (2009). Both articles propose pseudo-polynomial algorithms for this problem. Caprara et al. (2014) further investigates these variants of bilevel KPs, including cases where the leader and follower control subsets of items and pack them into a shared knapsack of fixed capacity (Mansi et al., 2012), or where each has its own knapsack but chooses items from a common set (Tahernejad et al., 2020). Algorithms for the continuous version of the problem in Tahernejad et al. (2020) can be found in Carvalho et al. (2018) and Fischer and Woeginger (2020).

Qiu and Kern (2015) examine a variation of the problem where an item can be selected by both the leader and the follower, resulting in a change in its profit (either an increase or decrease). Finally, Ozaltin et al. (2010) considers a stochastic version of the bilevel KP described in Dempe and Richter (2000), assuming that the followers' capacity depends on both the leader's decision and an uncertain value drawn from a discrete probability distribution.

As discussed in Caprara et al. (2014), these particular KPs are more manageable compared to the standard BPP. Carvalho et al. (2018) presents a polynomial algorithm for the bilevel KP variant where the follower solves a continuous knapsack problem, while Pferschy et al. (2019) address a bilevel KP where the leader has control over the weights of a subset of the follower's items. Finally, Della Croce and Scatamacchia (2020) consider the bilevel KP with interdiction constraints, an extension of the classic 0–1 knapsack problem formulated as a Stackelberg game with two agents, a leader and a follower, that choose items from a common set and hold their own private knapsacks. Bounds and an exact method are derived.

2.2 A General Packing Model

Packing problems stem from the KP or the BPP. We present a general model that joins the characteristics of the KP, the BPP, and generalizations derived from them. In what follows, we describe this generalized packing problem, present two formulations, and show how the packing problems presented earlier in the chapter can be seen as special cases. The problem and models are defined in the basic setting, i.e., deterministic, one packing attribute, and a single knapsack packing constraint.

The Generalized Bin Packing Problem (GBPP) considers a set of items characterized by volume and profit and sets of bins of various types characterized by volume and cost. A subset of the items, which we call *compulsory*, must be loaded, while a selection has to be made among the *non-compulsory* ones. The objective is to select the non-compulsory items to load and the bins into which to load the compulsory and the selected non-compulsory items in order to minimize the total net cost. The latter is given by the difference between the total cost of the used bins and the total profit of the non-compulsory loaded items.

Let $\mathcal{I}$ be the set of items, and w_i and p_i the volume and profit, respectively, of item $i \in \mathcal{I}$. Let $\mathcal{I}^{\mathrm{C}} \subseteq \mathcal{I}$ be the subset of compulsory items and $\mathcal{I}^{\mathrm{NC}} = \mathcal{I} \setminus \mathcal{I}^{\mathrm{C}}$ the subset of non-compulsory items. Let $\mathcal{J}$ be the set of available bins, $\mathcal{T}$ the set of bin types, and $\sigma(j) = t \in \mathcal{T}$ the type of bin $j \in \mathcal{J}$. Define for each bin type $t \in \mathcal{T}$, its cost C_t and volume W_t, as well as the minimum L_t and the maximum U_t number of bins of that type that may be selected. Finally, denote $U \leq \sum_{t \in \mathcal{T}} U_t$ the maximum number of bins that we are allowed to select.

The item-to-bin packing rules of the GBPP are stated as follows:

- All items in I^{C} must be loaded;

- For all used bins, the sum of the volumes of the items loaded into a bin must be less than or equal to the bin volume;
- The number of bins used for each type $t \in \mathcal{T}$ must be within the lower and upper availability limits L_t and U_t;
- The total number of used bins cannot exceed the total number of available bins U.

Infeasibility may arise when the available bins are not sufficient to load all compulsory items. To address this issue, we add a special bin v of volume $W_v = \sum_{i \in \mathcal{I}^{\mathrm{C}}} w_i$, thus able to load all compulsory items, and set its cost C_v to a value much higher than the costs of the initial bin types in order to discourage its use, e.g., $C_v \gg \sum_{t \in \mathcal{T}} C_t$.

Consider the following decision variables:

- Bin selection binary variables y_j equal to 1 if bin $j \in \mathcal{J}$ is used, 0 otherwise;
- Item-to-bin assignment binary variables x_{ij} equal to 1 if item $i \in \mathcal{I}$ is loaded into bin $j \in \mathcal{J}$, 0 otherwise.

The first model of the GBPP considers disaggregated item-to-bin assignment decisions and, thus, it is also named *assignment model*. It can be formulated as follows:

$$\min \sum_{j \in \mathcal{J}} C_{\sigma(j)} y_j - \sum_{j \in \mathcal{J}} \sum_{i \in \mathcal{I}^{\mathrm{NC}}} p_i x_{ij} \tag{2}$$

$$\text{s.t.} \sum_{i \in \mathcal{I}} w_i x_{ij} \leq W_{\sigma(j)} y_j, \qquad j \in \mathcal{J}, \tag{3}$$

$$\sum_{j \in \mathcal{J}} x_{ij} = 1, \qquad i \in \mathcal{I}^{\mathrm{C}}, \tag{4}$$

$$\sum_{j \in \mathcal{J}} x_{ij} \leq 1, \qquad i \in \mathcal{I}^{\mathrm{NC}}, \tag{5}$$

$$\sum_{j \in \mathcal{J}: \sigma(j)=t} y_j \leq U_t, \qquad t \in \mathcal{T}, \tag{6}$$

$$\sum_{j \in J: \sigma(j)=t} y_j \geq L_t, \qquad t \in \mathcal{T}, \tag{7}$$

$$\sum_{j \in \mathcal{J}} y_j \leq U, \tag{8}$$

$$y_j \in \{0, 1\}, \qquad j \in \mathcal{J}, \tag{9}$$

$$x_{ij} \in \{0, 1\}, \qquad i \in \mathcal{I}, j \in \mathcal{J}. \tag{10}$$

The objective function (2) minimizes the total net cost of the packing, given by the difference between the total cost of the used bins and the total profit of the selected non-compulsory items. The profit of the compulsory items is not considered in the objective function because it is a constant. We choose to present the minimization version to follow the tradition of packing problems presented in the literature. The equivalent formulation obtained by maximizing the total net profit (total profit minus total cost) recalls the knapsack problem setting.

Constraints (3) have the double effect of linking the item-assignment and bin-selection decisions and enforcing the capacity of used bins. Constraints (4) and (5) ensure that each compulsory and not-compulsory item is loaded into exactly one and at most one bin, respectively. Constraints (6) and (7) enforce the maximum and minimum number of available bins of each type, while constraints (8) limit the total number of selected bins. Constraints (9) and (10) enforce the integrality nature of the decision variables.

Model (2)–(10) involves a polynomial number of variables and constraints and it is based on an explicit definition of the assignment of items to bins (Baldi et al., 2012a). It is unsuitable for developing efficient algorithms, however, due to the significant symmetry of the solutions following from the item-to-bin assignment variables, which is typical of packing models.

A *set covering formulation* of the GBPP is based on enumerating all the possible combinations of items that can fit into a bin while satisfying the packing constraints. Those combinations of items are called in the literature *patterns*, *feasible patterns* or *loading patterns* (Delorme et al., 2016). Thus, a loading pattern k_t, for a bin of type $t \in \mathcal{T}$, is described by a vector a_{k_t} of indicator functions $a^i_{k_t}$, $i \in \mathcal{I}$, $k_t \in \mathcal{K}_t$, $t \in \mathcal{T}$, such that $a^i_{k_t} = 1$ if item i may packed into pattern k_t, 0 otherwise. $\mathcal{K}_t = \{k_t : k_t \text{ pattern for bin type } t \in \mathcal{T}\}$ is the set of all feasible patterns for bin type $t \in \mathcal{T}$. The cost of pattern k_t is then the difference between the cost of the bin type t and the total profit of non-compulsory items in that pattern:

$$c_{k_t} = C_t - \sum_{i \in \mathcal{I}^{\text{NC}}} p_i a^i_{k_t}. \tag{11}$$

We define the bin loading pattern selection variables λ_{k_t} equal to 1 if pattern $k_t \in \mathcal{K}_t$ is used, 0 otherwise. The set covering formulation of the GBPP may then be written as follows:

$$\min \sum_{t \in \mathcal{T}} \sum_{k \in \mathcal{K}_t} c_{k_t} \lambda_{k_t} \tag{12}$$

$$\text{s.t.} \sum_{t \in \mathcal{T}} \sum_{k \in \mathcal{K}_t} a^i_{k_t} \lambda_{k_t} = 1, \qquad i \in \mathcal{I}^{\text{C}}, \tag{13}$$

$$\sum_{t \in \mathcal{T}} \sum_{k \in \mathcal{K}_t} a^i_{k_t} \lambda_{k_t} \leq 1, \qquad i \in \mathcal{I}^{\text{NC}}, \tag{14}$$

$$\sum_{k \in \mathcal{K}_t} \lambda_{k_t} \leq U_t, \qquad t \in \mathcal{T}, \tag{15}$$

$$\sum_{k \in \mathcal{K}_t} \lambda_{k_t} \geq L_t, \qquad t \in \mathcal{T}, \tag{16}$$

$$\sum_{t \in \mathcal{T}} \sum_{k \in \mathcal{K}_t} \lambda_{k_t} \leq U, \tag{17}$$

$$\lambda_{k_t} \in \{0, 1\}, \qquad k_t \in \mathcal{K}_t, t \in \mathcal{T}. \tag{18}$$

The objective function (12) minimizes the total cost of the selected bin loading patterns. Since the feasibility of the item-to-bin assignment is guaranteed by the packing feasibility of the loading patterns, respecting the packing constraints (3), the latter is not required anymore. Constraints (13)–(17) have the same meaning as (4)–(8), and (18) are the integrality constraints.

The set covering model (12)–(18) has the advantage of separating the feasibility phase from the optimality one. The pattern generation addresses the feasibility phase, and the model is only devoted to finding an optimal combination of patterns. Due to the exponential number of variables, this formulation is suitable for Column Generation and Branch-and-Price approaches (Baldi et al., 2014, 2019).

Whilst the two models provide the same optimal solution, the strength of their continuous relaxations is generally different. In particular, the lower bound of the GBPP obtained by optimizing the continuous relaxation of the set covering formulation is not weaker than the one obtained by optimizing the formulation (2)–(10) (Baldi et al., 2012a).

It is easy to show the GBPP generalizes classic packing and knapsack problems:

- Bin Packing Problem: Set $|\mathcal{T}| = 1$, with $C_{\sigma(j)} = 1$, $j \in \mathcal{J}$, and $\mathcal{I}^{\text{NC}} = \emptyset$, i.e., all items must be loaded. Constraints (5) - (7) are then redundant and the objective function becomes the minimization of the number of bins;
- Knapsack Problem: Set $|\mathcal{T}| = 1$, $|\mathcal{J}| = 1$, with $C_{\sigma(j)} = 0$, $j \in \mathcal{J}$, and $\mathcal{I}^{\text{C}} = \emptyset$;
- Multiple Knapsack Problem: Set $|\mathcal{T}| > 1$, $|\mathcal{J}| = m$, with $C_{\sigma(j)} = 0$, $j \in \mathcal{J}$, where m is the number of knapsacks, and $\mathcal{I}^{\text{C}} = \emptyset$;
- Variable Cost and Size Bin Packing Problem and Variable Sized Bin Packing Problem: Considering $\mathcal{I}^{\text{NC}} = \emptyset$ and the presence of several bin types, one obtains the VCSBPP and the VSBPP when the cost of a bin type is or is not strongly correlated to its volume, respectively. Notice that, an equivalent formulation for the VCSBPP can be obtained by imposing $\mathcal{I}^{\text{C}} = \emptyset$ and setting the item profit higher than the cost of any bin type, $p_i > \max_{j \in \mathcal{J}} C_{\sigma(j)}$, which makes any bin profitable even when only one item is loaded into it.

3 Stakeholders and Capacity Planning in CFTL

This section is dedicated to the presentation of the *capacity-planning* concept for the main stakeholder types in consolidation-based freight transportation and logistics.

Recall that, the movements of freight, and of the vehicles and convoys performing them, together with the freight transiting or being stored for some time in terminals, result from the local, regional, national, continental, and worldwide trade generated by the needs and requirements of people, and, hence, of the full array of public and private organizations characterizing the human society (e.g., Crainic et al., 2018). It is thus natural and widespread to summarily describe CFTL as illustrated in Fig. 1, that is, as the interplay between *shippers*, generating the demand for transport (and storage), and *carriers*, supplying the facilities and services answering this demand.

The figure also recalls that the CFTL industry and stakeholders evolve in an environment shaped by many other agents and events, e.g., governmental institutions, setting up policies, laws, and rules, while offering subsidies and levying taxes and fees; ancillary service providers of material and human resources, information and decision-support technologies, as well as a large gamut of other services; people, who create demand as customers, work for the institutions and industries involved, and put pressure on all other stakeholders by voting as citizens, vociferating through social networks, and being part of groups and associations within the civil society; as well as the micro and major events, both human and nature generated, characterizing our human society (Perboli et al., 2018a; Crainic et al., 2018; Neirotti et al., 2014). We do not address these aspects in this chapter.

Stakeholders have their own goals and, generally, make independent plans and decisions in an uncertain environment. They are interconnected, however, through numerous interactions and interdependencies, in particular through the buyer-seller relationships, understandings, and contracts (Perboli et al., 2018b, 2014). The capacity-planning issues follow from these relationships.

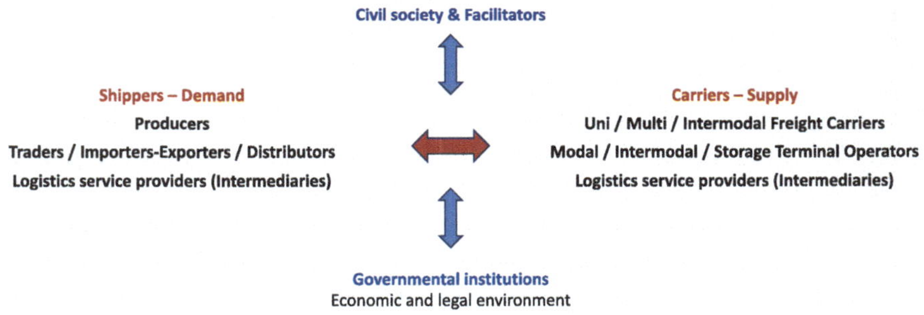

Fig. 1 Transportation and logistics main stakeholders

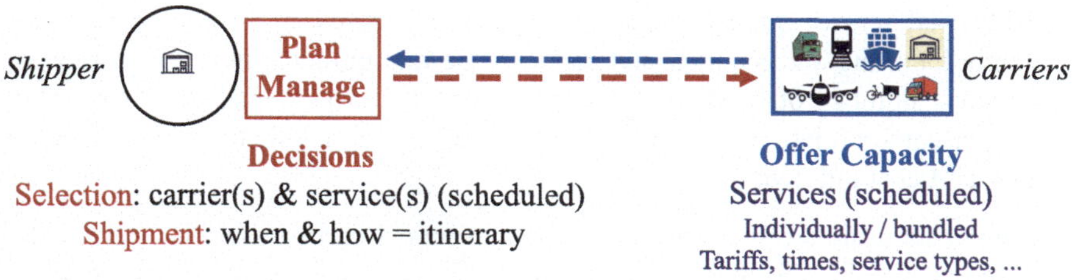

Fig. 2 Shipper case, selecting from many carrier offers

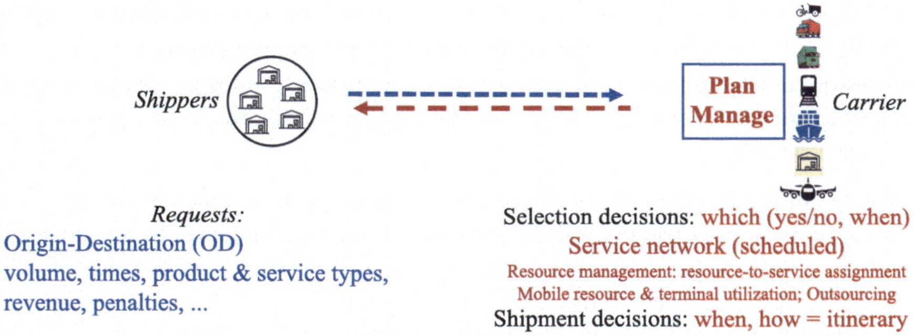

Fig. 3 Carrier case, planning its operations to answer many shipper requests

We now focus on the two main categories of CFTL stakeholders, shippers (demand) and carriers (supply). Figures 2 and 3 illustrate the interactions from the points of views of shipper and carrier, respectively.

Producers, traders, brokers, importers/exporters, whole- and retail buyers, sellers, and distributors of goods may be described as *shippers*. They send goods to other private or public organizations or to customers, either directly or by outsourcing part or all of these decisions to providers of logistics services, also known as Party Logistics (PL) firms. Such *intermediaries* may then be classified as shippers as well.

Shippers are typically involved in organizing shipments and selecting how to move (and store, when appropriate) freight in the most suitable way. They are thus determining the *demand* for transportation and logistics activities.

Shippers request, from one or several carriers, time and cost-efficient transport and storage services with the associated capacity. The requests, which may be repetitive for the duration of a shipper-carrier contract, are defined by:

- Specific origin-destination location pairs over given geographic areas and networks;
- Given or predicted volumes of freight measured in numbers of tons, packages, pallets, containers, etc.;
- Temporal attributes for moving or storing activities, e.g., availability time, due date, acceptable delays (with penalties), etc.;

- Additional service-quality requirements and objectives, e.g., specifications on freight handling and loading, and service-type requirements in terms of modes, vehicles, number of modal/service transfers, etc., which may be involved in shipper-carrier relations. To ease the presentation, we include them in the classic time-and-cost efficiency measures.

Shipper requests are met by modal and intermodal carriers, facility (terminals, warehouses, depots, fulfilment and distribution centers, etc.) operators, and intermediaries, which we group under the generic *carrier* term. Given demand predictions and, possibly, firm contracts with customers, carriers need to plan their operations and resources, that is, the means and capacity to, on the one hand, perform profitably and efficiently, while, on the other hand, satisfying the shipper demand.

Consolidation-based modal and intermodal carriers propose capacity over a network of *services* covering a given geographical area and time horizon. A service is characterized by:

- Origin, destination, and physical route (potentially with intermediary stops);
- Specific frequency (number of departures within a typical operation period, e.g., a day, a week, a month) or schedule;
- Service type, e.g., travel time, speed, type of vehicle, etc;
- Tariffs, varying with the type of product, distance, type of service, etc.

Similar service quality, duration, and tariff characteristics also qualify the storage-capacity offers, as well as the offers made by intermediaries.

Regarding intermediaries, present in both groups, it is important to note that xPL firms (specifically, x^{th}-party logistics service providers, where x equals 3, 4, or 5, depending on the level of service offered) typically do not possess transportation resources and do not manage the service network and capacity. (They might have ownership or oversight of storage and consolidation facilities, however.) Hence, xPLs behave as shippers when dealing with carriers to secure capacity on their services while becoming carriers when negotiating with their shipper customers to secure freight to move or store using the secured capacity.

This is also the case for the integrated planning and management activities of City Logistics, Physical Internet, and Synchromodal systems. Taherkhani et al. (2022) identify all such intermediaries as multi-stakeholders or Many-to-One-to-Many systems (*M1M*). Fig. 4 illustrates the shipper-carrier interactions from the point of view of an intermediary, identified as an Intelligent Decision-Support Platform (Taherkhani et al., 2022).

All the activities of all stakeholders mentioned above require that the capacity to store and transport be available when and where needed. Hence, capacity planning takes place at all levels of decision-making, strategic, tactical, and operational. We discuss these issues and the role packing methodologies may play in Operations Research-based decision support methods in the next sections.

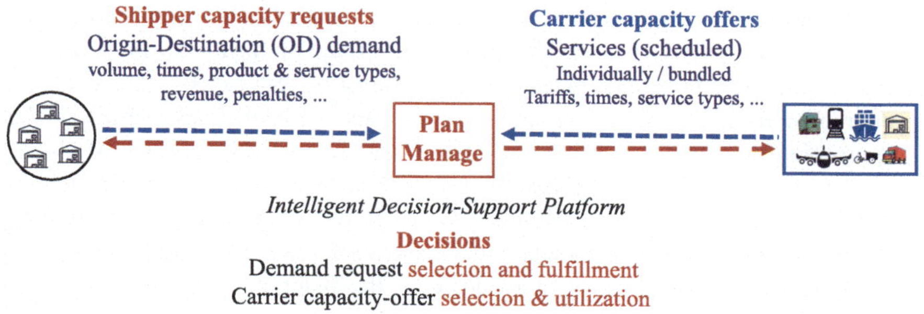

Fig. 4 Intermediary case, selecting and matching many shipper requests and carrier offers

4 CFTL Capacity Planning with Packing Considerations

In this section, we discuss the capacity-related issues faced by the two main categories of CFTL stakeholders, shippers and carriers at the strategic and tactical-operational planning levels. *Strategic*, long-term, capacity planning refers to the foundation for an efficient, effective, and profitable functioning of the system over an extended period, and very often concerns significant investments in securing resources and capacity. *Tactical*, medium-term, capacity planning is essential for consolidation-based freight transportation and logistics. The tactical plan defines, and hopefully optimizes, the policies and rules governing the utilization of capacity provided by a restricted set of resources, owned or to be contracted, in order to efficiently and profitably perform for the next season of operations. In all cases, planning may concern a specific geographical location or terminal, a group of locations over a given region, or the network made up of terminals and the transportation infrastructure and services connecting them.

Shippers address various capacity-planning issues to secure the transportation and logistics infrastructure and services to satisfy known or forecast market demand.

- *Strategic Planning*. In terms of transportation and logistics long-term planning, shippers assess their logistics network, from suppliers to end customers. The, hopefully optimal, location and size of their own facilities, such as warehouses and distribution centres, is part of this process, the issue concerning also the downsizing or closing of some of the existing facilities. When the shippers own transportation means, e.g., motor vehicles, decisions regarding the composition and sizing of the fleet belong to strategic planning as well. Furthermore, most shippers must also secure transportation and logistics services and capacity from external suppliers, the carriers as defined previously. Long-term contracts, normally accounting for the estimated performance characteristics of the carrier (e.g., service capability and reliability) in addition to offered resources and tariffs, are generally negotiated and decided within strategic-planning activities.
- *Tactical Planning* may concern securing, for a short-to-medium planning horizon, transportation or storage capacity (or both) to support procurement and

sales processes, bring in resources, or distribute products. Decisions may relate, as mentioned above, to a single location, a group of locations, or particular customer-related networks. When resources and sufficient capacity are available internally, these decisions are mainly concerned with allocating them to the appropriate locations, processes, product lines, or customer groups. Medium-term contracts must be negotiated otherwise with carriers for transportation services for given lines or customer groups, or for additional storage capacity in some terminals, in order to secure the capacity one expects to need for regular operations addressing regular demand, over the planning horizon considered. Adjusting the tactical plan, e.g., adjusting the facility and transportation-service capacities and fine-tuning the supplier selections, to respond to changing market conditions as well as demand and supply fluctuations, is also part of tactical planning but for shorter, operational, planning horizons.

Carriers aim for an efficient allocation of transportation capacity to meet specific shipment requests, control costs, and acquire resources.

- *Strategic Planning* involves locating, improving, or expanding (closing or downgrading) terminals to ensure efficient access to key markets. It also concerns decisions related to the composition (addressing major technological and legal upheavals is part of this aspect), dimension, and acquisition (including replacing or upgrading existing resources) of the fleet of power units (tractors, locomotives and vehicles (containers, trailers, vessels, planes), and their distribution among the regions serviced to meet anticipated demand. All those material resources require people to handle and maintain them. Strategic human resources management is therefore part of the process and aims to determine the amplitude and qualifications of each component of the workforce for the same planning horizon used for material resources. This strategic foresight ensures carriers can compete effectively in the market.
- *Tactical Planning* concerns, as for all other stakeholders, adjusting/adapting the resources available following strategic decisions, and operations to the conditions and regular demand estimated for the duration of the forthcoming medium-term planning horizon. An illustration of such processes is the yearly decision North American railroads take concerning the number of intermodal railcars of each type they will rent for the operations of the following year. Such decisions are often part, or are evaluated through, the main tactical-planning process of setting up the operation (also known in various fields as transportation or load) plan for the following season of activity. The plan includes the services and the resources to support them, with respective routes, schedules, and characteristics (Sect. 3), as well as the itineraries to efficiently and profitably move the forecast regular demand using those services and resources.

Table 1 depicts the current status of the literature on CFTL Capacity Planning with packing considerations. It classifies the main contributions by scope of the study (Single node, i.e., a terminal, group of terminals, or a corridor, or the entire Network), stakeholder type (Shipper or Carrier), and level of planning (Strategic or

Table 1 Literature on capacity planning with packing concerns

	Single node		Network	
	Strategic	Tactic	Strategic	Tactic
Shipper	Crainic et al. (2024) Crainic et al. (2016) Crainic et al. (2014a)	Crainic et al. (2021a) Crainic et al. (2016) Crainic et al. (2014a)	None	None
Carrier	Bruni et al. (2024)	Bruni et al. (2023) Perboli et al. (2021) Baldi et al. (2019) Perboli et al. (2012) Baldi et al. (2012a)	None	Hewitt and Lehuédé (2023) Côté et al. (2017)

Tactic). It is noteworthy that, in many cases, the relevant methodology is the same for both tactical and strategic planning, differences being mainly observed in the detail of the data representations. We consequently include each reference in all the locations corresponding to the applications it describes.

4.1 Shipper Capacity Planning

All shipper-related contributions we are aware of address single-node issues, presenting models and methods relevant for both strategic and tactical planning.

In seminal papers by Crainic et al. (2016, 2014a), the authors make initial attempts to address capacity planning problems in strategic and tactical applications. They introduce the VCSBPP and the presence of multiple bin types with costs uncorrelated with their volumes, and its stochastic counterpart SVCSBPP. The focus of these papers is the perspective of shippers needing to book their capacity in advance, while the uncertainty is related to the demand. In particular, Crainic et al., 2016 conduct analyses on the impact of demand uncertainty and container availability in different industrial settings, demonstrating how employing a stochastic model can provide a forecast for the appropriate capacity to pre-book in various operational scenarios. This includes situations such as container scarcity and high prices, which are currently prevalent in the post-COVID-19 era.

Crainic et al. (2024) open a new research avenue by explicitly considering broader uncertainty issues and the reliability of the future supply. Enlarging the SVCSBPP to address the possibility that the capacity contracted at decision time (first stage of the two-stage stochastic formulation) may not be fully available in the future (due to, e.g., delays, mechanical failures, accidents, etc.), the authors show the effectiveness and value of the methodology on applications to long-haul and urban freight transportation planning.

Applications of packing problems to tactical decisions for shippers can be found in Crainic et al. (2014a, 2016, 2021a).

Crainic et al. (2014a, 2016) share the same packing model for both the strategic and tactical levels. The authors show how, while the packing problem structure remains the same at the two decision levels, the information knowledge and flow differ, yielding different ways of using the same packing methodology.

Inspired by the tactical-planning challenges of M1M corridor-based integrated logistics systems that support freight exchanges between two regions, e.g., South-West Asia and Canada, Crainic et al. (2021a) conduct a thorough analysis of corridor-based logistic systems. In the context of freight logistics, a corridor refers to a designated transportation route or pathway that connects two or more regions or cities. It serves as a strategic transportation link for the movement of goods and cargo between different areas. Corridors are typically established based on geographic, economic, and infrastructure considerations. This research extends the packing methodology incorporating time-dependent packing decisions. It also provides valuable insights into the effects of planning, specifically capacity planning, on promoting sustainable solutions without imposing undue burdens on small and medium enterprises or undermining the competitiveness of local economies. By examining the impacts of capacity planning within a cooperative framework, they contribute to the advancement of sustainable practices in the field of transportation and emphasize the potential benefits of fostering collaboration among industry players.

4.2 Carrier Capacity Planning

Most carrier-related contributions we identified in the literature address single-node issues at tactical-planning level. Network-wide optimization is still in its infancy, but presents a rich set of modelling, algorithmic, and application challenges we discuss in Sect. 5.

With respect to the contributions addressing tactical planning issues in single-node problem settings, Baldi et al. (2012a) focus on considering bin costs and item profits simultaneously, proposing the GBPP. Perboli et al. (2012) then address the stochastic version of the GBPP, where the source of uncertainty is given by the item profits. Baldi et al. (2019) address the planning problem of a last-mile parcel city-delivery service, by considering different delivery options differing in cost and service quality. The carriers must decide which delivery options to select for the service to minimize the total cost and maximize the total service quality. The problem is modelled as a GBPP with item profits varying according to the bin.

Capacity planning issues in urban delivery and city logistics are further highlighted in Perboli and Rosano (2019), where the VCSBPP is proposed as a method to assist the management of a mixed-type fleet of vehicles, including cargo-bikes, electric vans, and standard vans. The VCSBPP is also proposed as the main optimization model in a simulation-optimization decision support system proposed by Perboli et al. (2018b) to select delivery modes in an urban area. The authors analyze urban delivery systems, considering not only operational and economic sustainability, but also environmental sustainability aspects, aiming for more comprehensive and sustainable urban delivery strategies.

Perboli et al. (2021) further explore the context of multi-tier urban logistics, proposing a last-mile parcel delivery model based on the usage of shared urban consolidation facilities (satellites) and a heterogeneous fleet of vehicles. The model is formulated as a VCSBPP-TD, where the planning horizon is divided into discrete time periods to capture time-dependent variations in delivery costs and satellite capacities, as well as customer time preferences. More in detail, given a set of customer orders, each with a particular volume to be delivered within a particular time slot (interval), the goal is to define delivery plans that minimize the total cost, which accounts for satellite and vehicle usage and transportation. The analysis of real data from Turin shows that the satellite-based system significantly reduces costs (up to 57%) compared to a single-echelon delivery system. The approach is more beneficial when the customer delivery time slots are unknown. The analysis also shows that the lack of understanding of customer time-slot preferences can increase delivery costs substantially (up to 300%).

Lastly, Bruni et al. (2023) explore the same problem setting as Crainic et al. (2016) and Perboli and Rosano (2019), but from the perspective of online retailers and e-commerce. To meet the computational requirements of online systems, which demand solutions in nearly real-time, the authors introduce a machine-learning approach for capacity-planning forecasting. The results obtained using the machine-learning approach are comparable in quality with those provided by a stochastic programming model but with a much lower computational effort. This makes machine learning more suitable for online systems, such as those of Amazon or Alibaba.

All the papers mentioned above tackle problems arising in tactical planning. To the best of our knowledge, the only paper dealing with strategic-level issues is proposed by Bruni et al., 2024. Moreover, this paper addresses the planning problem of a group of nodes representing satellite depots. The authors propose a bi-level version of the strategic pricing definition problem in a two-tiered city logistics system, where a group of transshipment points is used in the second tier to perform last-mile deliveries. The problem is presented as a multi-depot variant of the VCSBPP, where the tariffs for the use of satellites are defined in the upper level, considering the reaction of the carriers in the lower level. The numerical analysis highlights the potential of satellite-based solutions to achieve both economic sustainability and the promotion of environmentally friendly delivery modes.

There are very few papers in the CFTL capacity planning literature incorporating packing considerations within a network setting (Iori et al., 2007; Côté et al., 2014, 2017, 2020; Alonso et al., 2022; Hewitt & Lehuédé, 2023).

The largest stream of research comes from the integration of packing constraints into vehicle routing problems and models, within operational-planning contexts (Iori & Martello, 2010, 2013). These papers deal with different variants of routing problems, including location-routing, inventory-routing, and multi-echelon routing. One of the first problems to be studied in this area has been the Capacitated VRP with Two-dimensional Loading constraints (2L-CVRP), which is a variant of the

Capacitated VRP (CVRP), where rectangular-shaped items have to be transported and loading constraints have to be satisfied (see Côté et al., 2017, for a literature survey on the topic).

Iori et al. (2007) propose the first exact algorithm for a 2L-CVRP where single-customer routes are not allowed and all the available vehicles have to be routed. They model the problem by using the classical two-index vehicle flow model for the CVRP (Toth & Vigo, 2002), adding to it so-called *infeasible-path constraints*. The latter remove one or more routes, which are considered infeasible, that is, routes for which there is no feasible arrangement of the customer's items to be served by the same route inside the associated vehicle. The proposed solution method is based on a Branch-and-Cut (B&C) algorithm that iteratively calls an inner branch-and-bound for the solution of the loading sub-problem. This modelling approach is extended by Côté et al. (2020) to a stochastic variant of the 2L-CVRP, where the size and weight of some items are not known with certainty when the routes are planned. The authors show that their method can also be applied to the deterministic counterpart of the stochastic 2L-CVRP addressed by Iori et al. (2007).

Côté et al. (2017) present a mixed integer formulation of an integrated routing-packing problem and propose a Branch-and-Check method integrating a packing feasibility phase. If a routing solution violates the loading constraints, then the infeasible-path constraints are added as cuts to the original model. The authors consider several types of loading constraints and compare how the different solutions change with respect to the packing constraints and how much this impacts the computational effort.

Alonso et al. (2022) introduce a model for vehicle routing where a set of items is associated with each customer. The items have a boxed shape and a weight. Split delivery is allowed, i.e., the items of the same customer can be assigned to different vehicles. The authors present a mixed-integer model, where the packing constraints are explicitly included, and the position of the centre of gravity and the weight on the axles of the vehicles are taken into account.

It is noticeable that the above-mentioned modelling approaches may be extended to other variants of vehicle routing. Also noticeable is the fact that all these problem settings and models assume that the bins, i.e., vehicles, are given and no bin selection is considered. This may very well be sufficient for homogeneous-fleet routing cases but is a shortcoming in other contexts, e.g., when the choices for each route is to be made among vehicles of different capacities, with or without compartments.

Hewitt and Lehuédé (2023) propose a new formulation for a basic variant of the Scheduled Service Network Design (SSND) problem, where each origin-destination demand is characterized by volume and time windows at origin (availability) and destination (due date), services stand for the possibility to move vehicles directly between two consolidation terminals, any number of vehicles may be dispatched on each service at any time, at a unit-vehicle cost, demand itineraries are thus sequences of consolidation terminals, and each demand must follow a single itinerary. Consolidation is possible, and desirable to reduce the number of vehicles used, when multiple shipments dispatch on the same terminal-to-terminal service at the same time. The objective is to minimize the total vehicle cost while satisfying the demand.

The authors propose a formulation based on the concept of *consolidation* on physical links at given dispatching moments. A consolidation associated to an arc/service is a set of demands which may move on the physical link at the same time. Given the somewhat simple service structure, the set of consolidations which can be associated to each physical link may be generated a priori. Consequently, more complex packing considerations and requirements may be accounted for. Furthermore, in that setting, consolidations may be defined on the physical network, without the need for a time-space network formulation. This concept yields a compact formulation for problem settings of somewhat limited dimensions that outperforms a more classic time-space network formulation with knapsack capacity constraints. In a very interesting development, the authors also present a hybrid formulation for large problem dimensions, where the dispatch times on arcs are modelled differently, as consolidations or as time-space knapsacks, for each time interval (period) of the given planning duration, depending on the number of consolidations which may be defined for the interval.

Reviewing the literature one observes that most contributions address vehicle routing problems, which present somewhat simpler network and operation structures than, e.g., long-haul service network design. The integration of packing considerations going further than the classic knapsack constraint is then also simpler than in the SSND case, as illustrated in Hewitt and Lehuédé (2023), where packing is performed *a priori* and not within the SSND formulation. Even in the routing cases, however, the packing constraints are generally treated as subproblems, e.g., verifying the packing feasibility of the loads assigned to a vehicle in the current solution and generating valid inequalities to eliminate those solutions when packings are not feasible.

We also notice that the capacity units are assumed to exist and be associated to each movement. There is no consideration of the selection of the appropriate bins in terms of attributes and total cost. One thus fails to capture an important aspect of the capacity planning process. Since the goal of any packing problem is to either pack a single loading unit as densely as possible or pack all objects using as few loading units as possible, this aspect is of paramount importance, in particular in SSND problem settings when one needs to select the mode or the carrier or to manage heterogeneous vehicle fleets.

5 Research Perspectives

This section outlines key research frontiers that can enhance the integration of packing methodologies into CFTL capacity planning models and methods. We explore these avenues along three axes. The first two axes are methodological and consider the integration of packing constraints in network capacity planning and the generalization of packing problems. The other one is related to the use of packing methodologies in the application domains and focuses on the research on richer environments with multi-stakeholders and the inclusion of new decision-makers and stakeholders.

5.1 Integration of Packing Constraints in Network Capacity Planning

As illustrated in Table 1, the current literature on capacity planning with packing considerations is predominantly focused on the optimization of a single node, largely disregarding the network-related characteristics of transportation and logistics systems. Very few papers consider tactical planning problems on a group of locations or on a network. Clearly, fully considering the complexity of a CFTL network within the realm of packing methodology brings methodological, modelling and computational, challenges. These challenges are related to replacing the standard, aggregated, knapsack link-capacity constraints of network-optimization problems, with more complex packing ones.

The challenge comes from the nature of most network flow problems, in which the demand flow may be split among several paths from origin to destination, yielding a finite but large (when flows are required to be integer) or infinite (continuous flow case) number of potential items to pack on each link. Moreover, the packing takes place on all links (nodes, eventually), globally constrained by the total capacity and by the network flow-conservation constraints. Generalizing the consolidation concept of Hewitt and Lehuédé (2023) to such general network settings is already a challenging endeavour.

Integrating packing constraints into general SSND problems, with multi-stop services and the management of heterogeneous and substituable vehicle fleets, presents a truly major challenge. Consider, for example, that managing resources, e.g., vehicles of various capacities involves selecting for each service leg (arc) the appropriate number of capacity units. While this is a true packing problem characteristic, it complexifies significantly the SSND formulation, both in the circulation of the capacity units, which may support several services, and in the transfer of flows from one service leg to the following one on the itinerary.

To the best of the authors' knowledge, only one paper makes a step toward the incorporation of packing constraints into a networked transport system. In particular, Flamand et al. (2023) combine two classical problems, the transportation problem and the VSBPP, with the aim of optimally shipping and packing, from supply nodes to demand nodes, multiple types of commodities on different types of vehicles. The objective function accounts for the transportation cost as well as for the cost of acquiring the capacity resources, which in this case are vehicles of different types. Packing constraints ensure that the total weight of the commodities that are loaded on each vehicle does not exceed its capacity. The paper assumes integer flows, only marginally addressing the intrinsic complexity of packing constraints. Moreover, the topology of the network considered is very simplified, being the transportation problem defined on a bipartite network.

These observations led to the identification of a research venue of particular interest in integrating packing aspects on the network. There is a need for new models and methods to incorporate network-wise the families of packing problems already in use in capacity planning. In particular, Service Network Design and Packing can be combined in a unified formulation to better represent several realistic settings in multi-echelon urban delivery systems and complex supply chains. An approach integrating Service Network Design and Packing opens avenues both from an academic and industrial point of view. On the academic side, the integrated models are much more difficult to solve from a computational viewpoint, because the two problems, Service Network Design and Packing, both have a highly degenerate combinatorial structure. Finding solution methods, both exact and meta-heuristic, for integrated formulations will pose significant algorithmic challenges, especially as the system dimensions increases (Crainic et al., 2018, 2021b). Considering the practical usage of this integrated approach, as observed in Côté et al. (2017), the main benefits will be solutions presenting a different structure and better quality, both in terms of total cost, number of vehicles routed and load factors achieved.

5.2 *Intensify the Research on Generalizations of Packing Problems*

As highlighted in Sect. 2, the existing literature on cutting and packing predominantly focuses on studying variations of KP and BPP. These problems hold significant importance in various applications and offer a favourable combinatorial structure, enabling researchers to derive theoretical insights and develop efficient solution methods. However, capacity planning in CFTL needs more complex generalizations of packing problems, where many fundamental properties become weak or even invalid (Crainic et al., 2016, 2018; Baldi et al., 2019). It is crucial to explore new models capable of effectively addressing variants involving multiple attributes, like costs, profits, quality and time (Crainic et al., 2021b). Another challenging venue is given by the integration of revenue management in packing problems. Indeed, capacity planning integrated with a revenue management vision can be used to validate the integration of new operational practices and technologies and evaluate their impacts on transportation systems. This approach can help in optimizing revenue generation while ensuring efficient use of the capacity (Perboli et al., 2018a; Fadda et al., 2018).

Current research has not paid much attention to the role of uncertainty in packing problems, mainly focusing on random demand fluctuations within the two-stage stochastic programming paradigm. This modelling paradigm fits very well the majority of tactical and strategical planning problems, ranging from the long-haul shipping to last-mile delivery (Perboli et al., 2012; Crainic et al., 2016, 2024), since at the planning stage the future values of the problem parameters are seldom known with certainty. However, the assumption that the distribution of the random parameters is readily available, which is required to apply this approach, is quite stringent. Very often, only a limited number of samples or scenarios are available from historical data. The appropriateness of the approach is, hence, strongly conditioned by the type and amount of information available about the uncertainty, which is also a characteristic of the problem at hand. Different representations for the uncertainty should be explored, considering, for instance, emerging research on distributional ambiguity, and appropriate modelling paradigms devised.

Besides demand, other important problem characteristics may be considered uncertain. The literature that considers the uncertainty about the time at which demands arise or about time interval availability of the supply is still extremely scarce, highlighting relevant research opportunities in the context of multi-period and time-dependent models under uncertainty.

There is a huge lack of research on uncertainty on the supply side. As a matter of fact, the capacity of loading units is indeed characterized by exogenous uncertainty driven by both global and internal uncontrollable factors. The former may depend on changing supply chain strategies in response to market dynamics (for instance, for shared facilities) or disruptive events (for instance, for container ports). Internal factors may be related to the absence of information or poor management; the capacity of a satellite can be only roughly estimated in advance because carriers do not know precisely when the freight will arrive and how much time will be needed to start delivering it to customers. Considering and facing these uncertainties is fundamental to appropriately adopt and evaluate different strategic/tactical flexibility approaches for securing capacity. In this respect, a diversified gamut of options is available, like outsourcing transport services with occasional carriers or forming alliances with other stakeholders to share terminal capacity, or establishing long-term contracts. When, to what extent, and at which cost flexibility options at strategic and tactical levels should be considered in response to uncertainty is a key research question yet to be addressed. Advances in this research stream will enable researchers and practitioners to develop strategies to build flexibility and gain competitive advantage.

5.3 Richer Environments: Multi-Stakeholders, Cooperation, and Competition

Traditionally, public transit, personal transportation, and freight have been modeled and optimized independently, even as new active transport modes, business models, and connected/automated vehicles emerge. However, an integrated perspective across all transportation systems is crucial at regional and national levels. Research on such an integrated vision has been limited since initial examples in the 1990s (Crainic et al., 2018; Crainic & Florian, 2008; Jourquin & Beuthe, 2006). In concepts like hyperconnected logistics systems (e.g. city logistics, physical internet, synchromodality) and Logistics 5.0 capacity planning play a central role (ALICE Consortium, 2015; Bektaş et al., 2017; Crainic and Montreuil, 2016; Perboli et al., 2020). However, current research on these areas remains sparse, with overly simplified system representations. In this context, packing methodologies have classical applications in CFTL capacity planning from shipper and carrier viewpoints, but they also hold promise for new economic paradigms like sharing economy logistics, e-commerce, demand-driven supply chains, Lean Production, and Lean Business (Fadda et al., 2018; Perboli et al., 2022; Fadda et al., 2022; Perboli et al., 2018a). This requires adapting packing problems to capture issues and characteristics unique to these emerging settings and business models.

Research on capacity planning problems involving multiple stakeholders sharing common resources represents a promising and underexplored direction. The literature on multi-stakeholder packing problems is still limited and needs further attention, particularly regarding modeling the complex interactions and resource sharing between different stakeholders such as urban satellite depots or spare capacity in on-demand services (Giusti et al., 2018; Bruni et al., 2024; Crainic et al., 2024, 2021a). The vast majority of existing research examines capacity planning from a single stakeholder perspective, neglecting critical aspects of cooperation, competition, integration, and synchronization between stakeholders. Recent works have made preliminary attempts to account for the presence of multiple operators in shared facilities (Perboli et al., 2021; Bruni et al., 2024), but more research is needed to thoroughly explore frameworks for understanding stakeholders' competitive and cooperative relationships, how to enable greater integration between stakeholders, and how to synchronize disparate stakeholders' objectives and constraints (Perboli et al., 2017; Crainic et al., 2021b).

Another important research direction is incorporating the perspectives of new stakeholder groups into capacity planning, such as policymakers and institutional stakeholders. Packing problems could help these groups plan capacity for shared resources and validate policy decisions, as demonstrated in urban logistics and mobility planning (see the project for the development of the new Logistics and Mobility Plan to be implemented in 2025 of the Regional Government of Piedmont, Perboli et al., 2021; Perboli & Rosano, 2019) and in the planning of urban electric vehicles charging infrastructures in European cities (INCIT-EV Consortium, 2017). However, packing problems must be enriched to fully represent the specific priorities of policymakers and institutions, including business interactions between stakeholders (cooperation and/or competition), environmental sustainability, and social equity (De Marco et al., 2017; Crainic et al., 2021b). Research integrating packing problems into policy and institutional planning contexts remains very limited.

6 Conclusions

Capacity planning is a critical management activity in a broad spectrum of human activities, and in consolidation-based freight transportation and logistics, in particular. Insufficient capacity can have detrimental effects on performance, customer service, revenue, and market share. Yet, much research and development work is still needed to properly represent the physical and operational attributes of storage and movement capacity units and the adequate interplay between them and demand flows, into Operations Research-based models and methods targeting the planning of the supply capacity required to satisfy known or predicted, customer or internally-generated demand.

Following recent contributions to the literature, we identify *packing* as an appropriate methodology to address this representation issue. Developing packing-enriched Operations Research-based planning models and methods for the various stakeholders involved in consolidation-based freight transportation and logistics activities is challenging, however, and very little has been undertaken so far.

This chapter aims to start filling these gaps. We recalled the main packing problems and models and proposed a general formulation encompassing them. We then identified the main categories of stakeholders with their major capacity planning issues and reviewed the limited available literature on the topic.

The chapter concluded with a discussion of three important research avenues, namely, the integration of packing considerations and constraints into network-wide planning models and methods, the intensification of research on multi-attribute packing problems and applications incorporating revenue management and uncertainty, and the modelling of the requirements of multi-stakeholder, hyper-connected systems operating under various cooperation and sharing mechanisms.

Pursuing an agenda focused on these research directions can significantly expand the scope and impact of packing methods. The ability to formulate packing problems that integrate rich operational details into strategic decision-making will be invaluable as logistics systems grow more complex, global, and technology-driven. However, realizing this potential requires dedicated efforts to formulate more comprehensive packing models while considering the increased computational challenges.

Acknowledgments As the entire book, this chapter is dedicated to the loving memory of Bernard. We miss him. While working on this paper, the third author was Head of the Urban Mobility and Logistics Systems (UMLS) initiative of the interdepartmental Center for Automotive Research and Sustainable mobility (CARS) at the Politecnico di Torino, Turin, Italy, while working on this paper, the second author was Adjunct Professor, Department of Computer Science and Operations Research, Université de Montréal, Canada. We gratefully acknowledge the financial support provided by the Canadian Natural Sciences and Engineering Research Council (NSERC) through its Discovery and Collaborative Research and Development grant programs, as well as by the Fonds de recherche du Québec through their Teams and CIRRELT infrastructure grants.

References

Alba Martínez, M. A., Clautiaux, F., Dell'Amico, M., & Iori, M. (2013). Exact algorithms for the bin packing problem with fragile objects. *Discrete Optimization, 10*(3), 210–223.

ALICE Consortium (2015). WG2 Corridors, Hubs and Synchromodality. https://www.etp-logistics.eu/?page_id=292, last check: 13.05.2013.

Alonso, M., Martinez-Sykora, A., Alvarez-Valdes, R., & Parreño, F. (2022). The pallet-loading vehicle routing problem with stability constraints. *European Journal of Operational Research, 302*(3), 860–873.

Baldacci, R., & Boschetti, M. A. (2007). A cutting plane approach for the two-dimensional orthogonal non guillotine cutting stock problem. *European Journal of Operational Research, 183*, 1136–1149.

Baldi, M. M., Crainic, T. G., Perboli, G., & Tadei, R. (2012a). The generalized bin packing problem. *Transportation Research Part E: Logistics and Transportation Review, 48*(6), 1205–1220.

Baldi, M. M., Perboli, G., & Tadei, R. (2012b). The three-dimensional knapsack problem with balancing constraints. *Applied Mathematics and Computation, 218*, 9802–9818.

Baldi, M. M., Crainic, T. G., Perboli, G., & Tadei, R. (2014). Branch-and-price and beam search algorithms for the variable cost and size bin packing problem with optional items. *Annals of Operations Research, 222*, 125–141.

Baldi, M. M., Manerba, D., Perboli, G., & Tadei, R. (2019). A generalized bin packing problem for parcel delivery in last-mile logistics. *European Journal of Operational Research, 274*(3), 990–999.

Beasley, J. E. (1985). An exact two-dimensional non-guillotine cutting stock tree search procedure. *Operations Research, 33*, 49–64.

Bektaş, T., Crainic, T. G., & Van Woensel, T. (2017). From managing urban freight to smart city logistics networks. In: K. Gakis, & P. Pardalos (Eds.), *Networks design and optimization for smart cities, series on computers and operations research* (Vol. 8, pp. 143–188) World Scientific Publishing.

Berkey, J. O., & Wang, P. Y. (1987). Two dimensional finite bin packing algorithms. *Journal of the Operational Research Society, 38*, 423–429.

Birge, J. R., & Louveaux, F. (2011). *Introduction to stochastic programming* (2nd ed.). Springer.

Brotcorne, L., Hanafi, S., & Mansi, R. (2009). A dynamic programming algorithm for the bilevel knapsack problem. *Operations Research Letters, 37*(3), 215–218.

Bruni, M. E., Beraldi, P., & Conforti, D. (2015). A stochastic programming approach for operating theatre scheduling under uncertainty. *IMA Journal of Management Mathematics, 26*(1), 99–119.

Bruni, M. E., Fadda, E., Fedorov, S., & Perboli, G. (2023). A machine learning optimization approach for last-mile delivery and third-party logistics. *Computers and Operations Research, 157*.

Bruni, M. E., Khodaparasti, S., & Perboli, G. (2024). A bi-level approach for last-mile delivery with multiple satellites. *Transportation Research Part C: Emerging Technologies, 160*, 104495.

Buchheim, C., Henke, D., & Irmai, J. (2022). The stochastic bilevel continuous knapsack problem with uncertain follower's objective. *Journal of Optimization Theory and Applications, 194*(2), 521–542.

Caprara, A., Carvalho, M., Lodi, A., & Woeginger, G. J. (2014). A study on the computational complexity of the bilevel knapsack problem. *SIAM Journal on Optimization, 24*(2), 823–838.

Carvalho, M., Lodi, A., & Marcotte, P. (2018). A polynomial algorithm for a continuous bilevel knapsack problem. *Operations Research Letters, 46*(2), 185–188.

Charnes, A., & Cooper, W. W. (1959). Chance-constrained programming. *Management Science, 6*(1), 73–79.

Chung, F. K. R., Garey, M. R., & Johnson, D. S. (1982). On packing two-dimensional bins. SIAM - *Journal of Algebraic and Discrete Methods, 3*(1), 66–76.

Cohen, M. C., Keller, P. W., Mirrokni, V., & Zadimoghaddam, M. (2019). Overcommitment in cloud services: Bin packing with chance constraints. *Management Science, 65*(7), 3255–3271.

Colaneri, L., Croce, F. D., Perboli, G., & Tadei, R. (2003). A heuristic procedure for rack configuration in the space vehicle accommodation problem. In: T. A. Ciriani, G. Fasano, S. Gliozzi, & R. Tadei (Eds.), *Operations research in space and air* (pp. 27–42). Springer US.

Côté, J. F., Gendreau, M., & Potvin, J. Y. (2014). An exact algorithm for the two-dimensional orthogonal packing problem with unloading constraints. *Operations Research, 62*(5), 1126–1141.

Côté, J. F., Guastaroba, G., & Speranza, M. G. (2017). The value of integrating loading and routing. *European Journal of Operational Research, 257*(1), 89–105.

Côté, J. F., Gendreau, M., & Potvin, J. Y. (2020). The vehicle routing problem with stochastic two-dimensional items. *Transportation Science, 54*(2), 453–469.

Crainic, T. G., & Florian, M. (2008). National planning models and instruments. *INFOR: Information Systems and Operational Research, 46*(4), 299–308.

Crainic, T. G., & Montreuil, B. (2016). Physical internet enabled hyperconnected city logistics. In: E. Taniguchi, & R. G. Thompson (Eds.), *9th international conference on city logistics, June 2015, Tenerife, Spain, Elsevier, Transportation research procedia* (Vol. 12, pp. 383–398)

Crainic, T. G., Perboli, G., Pezzuto, M., & Tadei, R. (2007a). Computing the asymptotic worst-case of bin packing lower bounds. *European Journal of Operational Research, 183*, 1295–1303.

Crainic, T. G., Perboli, G., Pezzuto, M., & Tadei, R. (2007b). New bin packing fast lower bounds. *Computers and Operations Research, 34*, 3439–3457.

Crainic, T. G., Perboli, G., & Tadei, R. (2007c). Extreme point-based heuristics for three-dimensional bin packing. *INFORMS Journal on Computing, 20*, 368–384.

Crainic, T. G., Perboli, G., & Tadei, R. (2009). Ts2pack: a two-level tabu search for the three-dimensional bin packing problem. *European Journal of Operational Research, 195*, 744–760.

Crainic, T. G., Fu, X., Gendreau, M., Rei, W., & Wallace, S. W. (2011a). Progressive hedging-based metaheuristics for stochastic network design. *Networks, 58*(2), 114–124.

Crainic, T. G., Perboli, G., Rei, W., & Tadei, R. (2011b). Efficient lower bounds and heuristics for the variable cost and size bin packing problem. *Computers and Operations Research, 38*(11), 1474–1482.

Crainic, T. G., Gobbato, L., Perboli, G., Rei, W., Watson, J. P., & Woodruff, D. L. (2014a). Bin packing problems with uncertainty on item characteristics: An application to capacity planning in logistics. *Procedia - Social and Behavioral Sciences, 111*, 654–662.

Crainic, T. G., Hewitt, M., & Rei, W. (2014b). Scenario grouping in a progressive hedging-based meta-heuristic for stochastic network design. *Computers and Operations Research, 43*, 90–99.

Crainic, T. G., Gobbato, L., Perboli, G., & Rei, W. (2016). Logistics capacity planning: A stochastic bin packing formulation and a progressive hedging meta-heuristic. *European Journal of Operational Research, 253*(2), 404–417.

Crainic, T. G., Perboli, G., & Rosano, M. (2018). Simulation of intermodal freight transportation systems: A taxonomy. *European Journal of Operational Research, 270*, 401–418.

Crainic, T. G., Djeumou Fomeni, F., & Rei, W. (2021a). Multi-period bin packing model and effective constructive heuristics for corridor-based logistics capacity planning. *Computers and Operations Research, 132*, 105308.

Crainic, T. G., Perboli, G., & Ricciardi, N. (2021b). City logistics. In: T. G. Crainic, M. Gendreau, & B. Gendron (Eds.), *Network design with applications in transportation and logistics* (Chap. 16, pp. 507–537). Springer.

Crainic, T. G., Perboli, G., Rei W., Rosano M., & Lerma V. (2024). Capacity planning with uncertainty on contract fulfillment. *European Journal of Operational Research, 314*(1), 152–175.

De Marco, A., Mangano, G., Zenezini, G., Cagliano, A. C., Perboli, G., Rosano, M., & Musso, S. (2017). Business modeling of a city logistics ict platform. In: *Computer software and applications conference (COMPSAC), 2017 IEEE 41st annual* (pp. 783–789)

Della Croce, F., & Scatamacchia, R. (2020). An exact approach for the bilevel knapsack problem with interdiction constraints and extensions. *Mathematical Programming, 183*(1–2), 249–281.

Delorme, M., Iori, M., & Martello, S. (2016). Bin packing and cutting stock problems: Mathematical models and exact algorithms. *European Journal of Operational Research, 255*(1), 1–20.

Dempe, S., & Richter, K. (2000). Bilevel programming with knapsack constraints. *Central European Journal of Operations Research, 8*(2), 93–107.

Fadda, E., Perboli, G., & Tadei, R. (2018). Customized multi-period stochastic assignment problem for social engagement and opportunistic iot. *Computers and Operations Research, 93*, 41–50.

Fadda, E., Perboli, G., Rosano, M., Mascolo, J. E., & Masera, D. (2022). A decision support system for supporting strategic production allocation in the automotive industry. *Sustainability, 14*.

Fekete, S. P., & Schepers, J. (2004). A combinatorial characterization of higher-dimensional orthogonal packing. *Mathematics of Operations Research, 29*(2), 353–368.

Fischer, D., & Woeginger, G. J. (2020). A faster algorithm for the continuous bilevel knapsack problem. *Operations Research Letters, 48*(6), 784–786.

Flamand, T., Iori, M., & Haouari, M. (2023). The transportation problem with packing constraints. *Computers & Operations Research, 157*, 106278.

Friesen, D. K., & Langston, M. A. (1986). Variable sized bin packing. *SIAM Journal on Computing, 15*, 222–230.

García Llinás, G. A., & Nagi, R. (2015). Network and QoS-based selection of complementary services. *IEEE Transactions on Services Computing, 8*(1), 79–91.

Ghirardi, M., Perboli, G., & Sasia, D. (2011). Maximizing the throughput of multimodal logistic platforms by simulation-optimization: The duferco case study. In: *2011 IEEE international conference on automation science and engineering* (pp. 52–57).

Gilmore, P. C., & Gomory, R. E. (1965). Multistage cutting problems of two and more dimensions. *Operations Research, 13*, 94–119.

Giusti, R., Manerba, D., Perboli, G., Tadei, R., & Yuan, S. (2018). A new open-source system for strategic freight logistics planning: The SYNCHRO-NET optimization tools. In: *Transportation research procedia* (Vol. 30, pp. 245–254).

Hadjiconstantinou, E., & Christofides, N. (1995). An exact algorithm for general, orthogonal, two-dimensional knapsack problems. *European Journal of Operational Research, 83*(1), 39–56.

Hewitt, M., & Lehuédé, F. (2023). New formulations for the scheduled service network design problem. *Transportation Research Part B: Methodological, 172*, 117–133.

INCIT-EV Consortium (2017). INCIT-EV Home Page. https://www.incit-ev.eu/, last check: 13.05.2013.

Iori, M., & Martello, S. (2010). Routing problems with with loading constraints. *TOP, 18*(1), 4–27.

Iori, M., & Martello, S. (2013). An annotated bibliography of combined routing and loading problems. *Yugoslav Journal of Operations Research, 23*(3), 311–326.

Iori, M., Salazar-González, J. J., & Vigo, D. (2007). An exact approach for the vehicle routing problem with two-dimensional loading constraints. *Transportation Science, 41*(2), 253–264.

Jourquin, B., & Beuthe, M. (2006). A decade of freight transport modeling with virtual networks: Aquired experiences and new challenges. In: A. Reggiani, & P. Nijkamp (Eds.), *Spatial dynamics, networks and modelling* (pp. 181–200). Edward Elgar Publishing.

Lamiri, M., Xie, X., Dolgui, A., & Grimaud, F. (2008). A stochastic model for operating room planning with elective and emergency demand for surgery. *European Journal of Operational Research, 185*(3), 1026–1037.

Liu, Q., Cheng, H., Tian, T., Wang, Y., Leng, J., Zhao, R., Zhang, H., & Wei, L. (2023). Algorithms for the variable-sized bin packing problem with time windows. *Computers & Industrial Engineering, 155*, 107175.

Mansi, R., Alves, C., de Carvalho, J. M.V., & Hanafi, S. (2012). An exact algorithm for bilevel 0–1 knapsack problems. *Mathematical Problems in Engineering 2012*, 1–23.

Martello, S., & Toth, P. (1990). *Knapsack problems - algorithms and computer implementations*. John Wiley & Sons.

Martello, S., Pisinger, D., & Vigo, D. (2000). The three-dimensional bin packing problem. *Operations Research, 48*(2), 256–267.

Martinovic, J., Hähnel, M., Scheithauer, G., & Dargie, W. (2022). An introduction to stochastic bin packing-based server consolidation with conflicts. *TOP 30*(2), 296–331.

Neirotti, P., De Marco, A., Cagliano, A. C., Mangano, G., & Scorrano, F. (2014). Current trends in smart city initiatives: Some stylised facts. *Cities, 38*, 25–36.

Osei-Mensah, E., Thabet, S. K. S., Bamisile, O., Asiedu-Ayeh, E., Agbesi, V. K., & Li, J. (2022). Energy saving in 5g substations using edge caching for adaptive immersive media streaming. In: *2022 4th Asia energy and electrical engineering symposium (AEEES)* (pp. 47–51).

Ozaltin, O. Y., Prokopyev, O. A., & Schaefer, A. J. (2010). The bilevel knapsack problem with stochastic right-hand sides. *Operations Research Letters, 38*(4), 328–333.

Perboli, G., & Rosano, M. (2019). Parcel delivery in urban areas: Opportunities and threats for the mix of traditional and green business models. *Transportation Research Part C: Emerging Technologies, 99*, 19–36.

Perboli, G., Crainic, T. G., & Tadei, R. (2011). An efficient metaheuristic for multi-dimensional multi-container packing. In: *Proceedings of the IEEE international conference on automation science and engineering* (pp. 563–568)

Perboli, G., Tadei, R., & Baldi, M. M. (2012). The stochastic generalized bin packing problem. *Discrete Applied Mathematics, 160*(7–8), 1291–1297.

Perboli, G., Tadei, R., & Gobbato, L. (2014). The multi-handler knapsack problem under uncertainty. *European Journal of Operational Research, 236*(3), 1000–1007.

Perboli, G., Musso, S., Rosano, M., Tadei, R., & Godel, M. (2017). Synchro-modality and slow steaming: New business perspectives in freight transportation. *Sustainability, 9*(10), 1843.

Perboli, G., Musso, S., & Rosano, M. (2018a). Blockchain in logistics and supply chain: A lean approach for designing real-world use cases. *IEEE Access, 6*, 62018–62028.

Perboli, G., Rosano, M., Saint-Guillain, M., Rizzo, P., & Perboli, G. (2018b). Simulation-optimisation framework for city logistics: An application on multimodal last-mile delivery. *IET Intelligent Transport Systems, 12*, 262–269.

Perboli, G., Fedorov, S., & Rosano, M. (2020). The European concept of smart city: A taxonomic analysis. In: *Proceedings of the IEEE 44th annual computers, software, and applications conference (COMPSAC)* (pp. 1725–1730). IEEE.

Perboli, G., Brotcorne, L., Bruni, M. E., & Rosano, M. (2021). A new model for last-mile delivery and satellite depots management: The impact of the on-demand economy. *Transportation Research Part E: Logistics and Transportation Review, 145*, 102184.

Perboli, G., Rosano, M., & Wei, Q. (2022). A simulation-optimization approach for the management of the on-demand parcel delivery in sharing economy. *IEEE Transactions on Intelligent Transportation Systems, 23*(8), 10570–10582.

Pferschy, U., Nicosia, G., & Pacifici, A. (2019). A stackelberg knapsack game with weight control. *Theoretical Computer Science, 799*, 149–159.

Pisinger, D. (1999). An exact algorithm for large multiple knapsack problems. *European Journal of Operational Research, 114*(3), 528–541.

Pisinger, D. (2002). Heuristics for the container loading problem. *European Journal of Operational Research, 141*, 382–392.

Qiu, X., & Kern, W. (2015). Improved approximation algorithms for a bilevel knapsack problem. *Theoretical Computer Science, 595*, 120–129.

Rockafellar, R. T., & Wets, R. J.-B. (1991). Scenarios and policy aggregation in optimization under uncertainty. *Mathematics of Operations Research, 16*(1), 119–147.

Song, Y., Luedtke, J. R., & Küçükyavuz, S. (2014). Chance-constrained binary packing problems. *INFORMS Journal on Computing, 26*(4), 735–747.

Taherkhani, G., Bilegan, I., Crainic, T. G., Gendreau, R., & Rei, W. (2022). Tactical capacity planning in an integrated multi-stakeholder freight transportation system. *OMEGA, 110*, article 102628.

Tahernejad, S., Ralphs, T. K., & DeNegre, S. T. (2020). A branch-and-cut algorithm for mixed integer bilevel linear optimization problems and its implementation. *Mathematical Programming Computation, 12*(4), 529–568.

Toth, P., & Vigo, D. (2002). An overview of vehicle routing problems. In: P. Toth, & D. Vigo (Eds.), *The vehicle routing problem* (pp. 1–26). SIAM.

Wang, S., Li, J., & Mehrotra, S. (2021). Chance-constrained multiple bin packing problem with an application to operating room planning. *INFORMS Journal on Computing, 33*(4), 1661–1677.
Zhang, Z., Denton, B. T., & Xie, X. (2020). Branch and price for chance-constrained bin packing. *INFORMS Journal on Computing, 32*(3), 547–564.
Zhao, X., Bennell, J. A., Bektaş, T., & Dowsland, K. (2016). A comparative review of 3d container loading algorithms. *International Transactions in Operational Research, 23*, 287–320.

Models for Network Flow and Network Design Problems with Piecewise Linear Costs

Bernard Fortz, Bernard Gendron, and Luis Gouveia

1 Introduction

Network flow problems are fundamental optimization problems that involve finding the optimal flow of resources through a network, subject to constraints such as capacity limits and cost considerations. A natural extension is multicommodity network flow problems, where there may be multiple types of resources flowing through the network, each with its own cost function. These models are also useful when considering origin-destination based flows, where it is important that the flow served to a node comes from a particular other node in the network. For example, in a telecommunications network, there may be multiple types of data traffic flowing through the network, each with a different cost function based on the bandwidth and other parameters. Often the complex structure of the cost of using a particular network edge or link can be modeled (or approximated) using a piecewise linear function. This means that the cost of using the link changes at certain thresholds or

B. Fortz (✉)
HEC - Management School of the University of Liège, Liège, Belgium

Computer Science Department, Université libre de Bruxelles (ULB), Brussels, Belgium

INRIA, Université de Lille, Lille, France
e-mail: bernard.fortz@uliege.be

B. Gendron
CIRRELT and DIRO, Université de Montréal, Montréal, QC, Canada
e-mail: bernard.gendron@cirrelt.net

L. Gouveia
Departamento de Estatística e Investigação Operacional, Faculdade de Ciências da Universidade de Lisboa and CMAFcIO - Centro de Matemática, Aplicações Fundamentais e Investigação Operacional, Lisbon, Portugal
e-mail: legouveia@fc.ul.pt

T. G. Crainic et al. (eds.), *Combinatorial Optimization and Applications*,
International Series in Operations Research & Management Science 358,
https://doi.org/10.1007/978-3-031-57603-4_7

breakpoints, and may not increase at a constant rate. For example, in transportation planning, the cost of using a road may be modelled as a piecewise linear function of the traffic volume on the road. As the traffic volume increases, the cost of using the road may increase at a faster rate, due to congestion or other factors.

Efficiently solving network design and multicommodity network flow problems with piecewise linear costs can be challenging, especially when these cost functions are non-convex. Bernard Gendron has made significant contributions on solving this class of problems, namely on developing models and algorithms that can effectively handle non-convex cost functions, allowing for more accurate and efficient resource allocation in these types of networks. Several references to Bernard's contributions will appear on this review.

In this chapter, we focus on mixed-integer linear models for variants of multicommodity network flow problems with piecewise linear costs. The formulations presented in this chapter are, in general, extended formulations of large size. The formulations will presented in the context of three related problems, the continuous multicommodity network flow problem, the integer multicommodity network flows problem and the unsplittable multicommodity network flow problem. The first problem is a relaxation of the second one, which in turn is a relaxation of the third one.

The presentation of this chapter intends to review Bernard Gendron's contributions, also emphasizing which modelling techniques are needed to better solve the more difficult versions. We observe and emphasize that our work is a tour of known modelling approaches that might be helpful for a researcher wanting to study these or related problem. To the best of our knowledge, this chapter is the first review paper to systematically present models for different cases that have been studied in the literature, unifying notation and presenting the main similarities and differences between the models. However, it is not our intention to make a computational study. We refer the reader to Croxton et al. (2007), Frangioni and Gendron (2009), and Gendron and Gouveia (2017) and the recent survey by Frangioni and Gendron (2021) for algorithmic approaches based on columns and rows generation or Lagrangian relaxation to solve the models presented in this chapter efficiently.

In addition to the large scale models presented here, a different perspective is followed by some authors who focus on direct approaches to solve network flow problems with piecewise linear costs with specialized algorithms, see e.g. Murthy and Helgason (1994) and Pinto and Shamir (1994).

After an introduction of the notation and the definition of the main concepts used throughout the chapter in Sect. 2, we review models for multicommodity network flows with piecewise linear costs, starting in Sect. 3 with the case of continuous flows. In Sect. 4, we extend these models to the case where flow have to be integral, while Sect. 5 covers the case where flows need to follow a single path for each commodity. We conclude in Sect. 6.

2 Notation and Definitions

Let $G = (V, A)$ be a directed graph, with a set of nodes V, and a set of arcs A. Consider as well a set of commodities K, where each commodity $k \in K$ has a given origin o_k, a destination d_k, and a demand of d_k units of flow to be routed from o_k to d_k. In order to simplify the writing of classical flow conservations that will appear in all our formulation, we define, for each node $i \in V$ and commodity $k \in K$, λ_i^k such that $\lambda_{o_k}^k = 1$, $\lambda_{d_k}^k = -1$, and $\lambda_i^k = 0$ for every $i \neq \{o_k, d_k\}$. For any proper subset $S \subsetneq V$, $S \neq \emptyset$, we define the forward cut $\delta^+(S)$ associated to S as the set of arcs leaving S, i.e. $\delta^+(S) = \{(i, j) \in A : i \in S, j \notin S\}$, and the backward cut $\delta^-(S)$ as the set of arcs entering S, i.e. $\delta^-(S) = \{(i, j) \in A : i \notin S, j \in S\}$. When $S = \{i\}$, we write $\delta^+(i)$ and $\delta^-(i)$ instead of $\delta^+(\{i\})$ and $\delta^-(\{i\})$.

Each arc $a \in A$ has an associated cost function $g_a(l_a)$ of the load flowing through the arc, l_a. This cost function is piecewise linear, with the segments being indexed by the finite set $S_a = \{1, 2, ..., |S_a|\}$. Each segment $s \in S_a$ has a lower and upper bound on the flow, represented by the breakpoints b_a^{s-1} and b_a^s. If finite, the breakpoint of the last segment of each arc $a \in A$, $b_a^{|S_a|}$, can be interpreted as the capacity of the arc. However, the case where $b_a^{|S_a|} = \infty$ also holds. A segment is also characterized by a slope c_a^s and an intercept f_a^s. In conclusion, the cost function g_a is such that $g_a(l_a) = f_a^s + c_a^s l_a$ if $b_a^{s-1} < l_a \leq b_a^s$ for $s \in S_a$. We assume that the function is lower semi-continuous, i.e. $g(l) \leq \liminf_{l' \to l} g(l')$ and that $g(0) = 0$. Figure 1 illustrates this notation. When the objective function is convex, these values are such that $c_a^1 \geq 0$, $c_a^s > c_a^{s-1}$ and $f_a^s \leq 0$, $f_a^s < f_a^{s-1}$. Moreover, when the cost function is continuous, $b_a^s c_a^s + f_a^s = b_a^s c_a^{s+1} + f_a^{s+1}$ must hold.

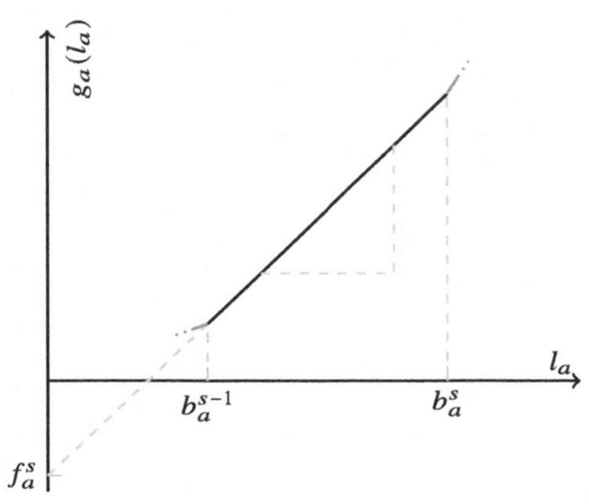

Fig. 1 Notation for each segment of $g_a(l_a)$

3 Continuous Multicommodity Network Flows

In this section, we focus on problems where flows are continuous. The classical multicommodity flow problem in this case can be formulated as a linear program, as described in Sect. 2, and is polynomially solvable. We start with a basic multicommodity formulation for the case when flows are continuous and the cost function is convex. Then, we consider the case when costs are not convex, point out why the previous model cannot be used directly, and as a consequence introduce the classical multiple choice formulation. Finally, based on the fact the previous formulation has a weak LP relaxation, the concept of disaggregation is introduced to lead to a formulation with a much tighter LP bound.

3.1 *Convex Costs and a Basic Multicommodity Flow Model*

The easiest case of multicommodity network flow problems with piecewise linear costs arises when flow are continuous and the cost function is convex. In this case, the problem can be formulated by explicitly introducing a variable g_a for each arc $a \in A$ to represent the cost $g_a(l_a)$ associated with the load of arc a. As the piecewise linear functions are convex, it is sufficient to impose that g_a is the maximum of the linear functions defining the cost for each segment. This approach was among others used by Fortz and Thorup (2000).

$$\min \sum_{a \in A} g_a \tag{1a}$$

$$\text{s.t.} \quad \sum_{a \in \delta^+(i)} x_a^k - \sum_{a \in \delta^-(i)} x_a^k = \lambda_i^k, \qquad i \in V, k \in K, \tag{1b}$$

$$l_a = \sum_{k \in K} d^k x_a^k, \qquad a \in A, \tag{1c}$$

$$g_a \geq f_a^s + c_a^s l_a, \qquad a \in A, s \in S_a, \tag{1d}$$

$$0 \leq x_a^k \leq 1, \qquad a \in A, k \in K, \tag{1e}$$

$$l_a \geq 0, \qquad a \in A, s \in S_a, \tag{1f}$$

$$g_a \geq 0, \qquad a \in A. \tag{1g}$$

In this formulation, (1b) are the flow conservation constraints, (1c) define the load on each arc and (1d) bounds the cost on each arc from below. Constraints (1e) to (1g) define the bounds on the variables. As for the basic multicommodity flow problem, this is a linear programming model that can be solved in polynomial time.

3.2 The Non-Convex Case and the Multiple Choice Model

Unfortunately, when costs are not convex and there are multiple commodities, the problem becomes NP-hard (Fortz et al., 2017). Indeed, the difficulty lies in the identification of the segment to which the load of an arc belongs. A classical approach to handle this, is to extend the previous model with a multiple choice structure, leading to the so-called multiple choice model that was first introduced in the context of multicommodity flow problems with piecewise linear costs by Balakrishnan and Graves (1989). We introduce binary variables y_a^s, with $y_a^s = 1$, if arc $a \in A$ contains a non-zero flow on segment $s \in S_a$, and $y_a^s = 0$ otherwise. For the sake of simplicity, if arc a contains a non-zero flow on segment $s \in S_a$, we say that arc a is on segment $s \in S_a$. We also define continuous variables l_a^s, that indicate the load going through arc $a \in A$ on segment $s \in S_a$. The problem can then be formulated as follows.

$$\min \quad \sum_{a \in A} \sum_{s \in S_a} (f_a^s y_a^s + c_a^s l_a^s) \tag{2a}$$

$$\text{s.t.} \quad \sum_{a \in \delta^+(i)} x_a^k - \sum_{a \in \delta^-(i)} x_a^k = \lambda_i^k, \qquad i \in V, k \in K, \tag{2b}$$

$$\sum_{s \in S_a} y_a^s \le 1, \qquad a \in A, \tag{2c}$$

$$\sum_{s \in S_a} l_a^s = \sum_{k \in K} d^k x_a^k, \qquad a \in A, \tag{2d}$$

$$b_a^{s-1} y_a^s \le l_a^s \le b_a^s y_a^s, \qquad a \in A, s \in S_a, \tag{2e}$$

$$0 \le x_a^k \le 1, \qquad a \in A, k \in K, \tag{2f}$$

$$y_a^s \in \{0, 1\}, \qquad a \in A, s \in S_a, \tag{2g}$$

$$l_a^s \ge 0, \qquad a \in A, s \in S_a. \tag{2h}$$

Constraints (2b) are flow conservation constraints. Constraint sets (2c)–(2e) identify the segment each arc is on. Naturally, only a single segment per arc may be selected (2c). The choice of segment is implied by the load flowing through the respective arc. This load is given by constraints (2d) and its value is assigned to one of the variables l. To ensure that only the appropriate load variable is non-zero, in (2e) we either bound l_a^s by the breakpoints of the corresponding segment, if $y_a^s = 1$, or we force it to zero, if $y_a^s = 0$.

Note that in the case of convex cost functions, Fortz et al. (2017) showed the Linear Programming (LP) relaxation of the basic formulation (1) has integer y_a^s variables, and the basic formulation (1) becomes equivalent to the multiple choice formulation (2) in the sense that they have the same optimal objective value.

Essentially, this means that in the convex costs case, there is no gain in augmenting the original basic model with the multiple choice structure.

In the general (non-convex) case, however, the value of the optimal solution of the LP relaxation of the multiple choice formulation (2) can be arbitrarily far from the optimal solution, even for a single commodity. To illustrate this, consider a toy network with only two nodes v and w, a single arc a from v to w and a unit demand between v and w. Obviously, the only feasible solution is to send one unit of flow on a. Consider an arbitrarily large value $c > 0$. If the cost function is given by two segments, with breakpoints $b_a^0 = 0$, $b_a^1 = 1$ and $b_a^2 = 2$, intercepts $f_a^1 = 0$ and $f_a^2 = 2c$, and slopes $c_a^1 = c$ and $c_a^2 = -c$, the optimal solution of (2a)–(2h) is obtained with $y_a^1 = y_a^2 = 0.5$ and $l_a^1 = 0$, $l_a^2 = 1$ with a total cost of 0, while the optimal solution of the problem has objective value c which can be made arbitrarily large.

3.3 The Disaggregated Formulation

In order to strengthen the LP relaxation of Model (2), Croxton et al. (2003, 2007) use disaggregation, a common technique to strengthen the LP relaxation of MIPs. We present here a slightly adapted version of it, to keep notation consistent. Let us introduce variables x_a^{ks} to represent the proportion of commodity k routed on arc $a \in A$ on segment $s \in S_a$. These new variables are linked to those in the multiple choice formulation (2) by the relations

$$l_a^s = \sum_{k \in K} d^k x_a^{ks} \qquad a \in A, s \in S_a, \tag{3}$$

$$x_a^k = \sum_{s \in S_a} x_a^{ks} \qquad a \in A, k \in K. \tag{4}$$

and a straightforward substitution leads to

$$\min \quad \sum_{a \in A} \sum_{s \in S_a} \left(f_a^s y_a^s + c_a^s \sum_{k \in K} d^k x_a^{ks} \right) \tag{5a}$$

$$\text{s.t.} \quad \sum_{a \in \delta^+(i)} \sum_{s \in S_a} x_a^{ks} - \sum_{a \in \delta^-(i)} \sum_{s \in S_a} x_a^{ks} = \lambda_i^k, \qquad i \in V, k \in K, \tag{5b}$$

$$\sum_{s \in S_a} y_a^s \leq 1, \qquad a \in A, \tag{5c}$$

$$b_a^{s-1} y_a^s \leq \sum_{k \in K} d^k x_a^{ks} \leq b_a^s y_a^s, \qquad a \in A, s \in S_a, \tag{5d}$$

$$\sum_{s \in S_a} x_a^{ks} \leq 1, \qquad k \in K, a \in A, \tag{5e}$$

$$y_a^s \in \{0, 1\}, \qquad a \in A, s \in S_a. \tag{5f}$$

This formulation has the same LP relaxation value as formulation (2), but the introduction of disaggregated variables x_a^{ks} in the model allows an enhancement based on the observation that the flow on arc a can be on segment s if and only if $y_a^s = 1$. This observation leads to the following valid inequalities:

$$x_a^{ks} \leq y_a^s, \qquad k \in K, a \in A, s \in S_a. \tag{5g}$$

To see the effect of this enhancement on the LP relaxation of the original model, let us consider again the toy example described before. Since there is a single commodity ($K = \{1\}$), the LP relaxation of this model corresponds to the vector of disaggregated variables $x_a^{11} = 0$, $x_a^{12} = 1$. But this vector violates the new valid inequality (5g) as $x_a^{12} = 1 > 0.5 = y_a^2$. On the other hand, it is quite easy to see that the optimal solutions of the LP relaxation of (5) are all the convex combinations of $x_a^{11} = y_a^1 = 1; x_a^{12} = y_a^2 = 0$ and $x_a^{11} = y_a^1 = 0; x_a^{12} = y_a^2 = 1$, with optimal objective function value c, and the LP relaxation is integral. Intuitively, the gain obtained by (5g) is due to the fact that these constraints impose the value of variable y_a^s to be at least equal to the maximum (over k) of the values of variables x_a^{ks} whereas the previous constraint (5d) states that the value of the same variable should be at least equal to a weighted average variables x_a^{ks}.

This clearly shows that the LP relaxation of the strong disaggregated multiple choice formulation (5) can be much better than the LP relaxation of the original multiple choice formulation (2). Croxton et al. (2007) show that the LP relaxation of the disaggregated multiple choice formulation (5) approximates the cost function on each arc with its lower convex envelope, which is the best one can hope for with linear constraints.

Note that the disaggregated multiple choice formulation also strenghtens the LP bound in the convex case, but this reformulation is only possible in the presence of binary y_a^s variables. The same technique cannot be applied to the basic model (1), hence justifying furthermore the interest of the multiple choice model.

4 Integer Multicommodity Network Flows

Gendron and Gouveia (2017) study the same problem with the additional restriction that the flow on each arc is integer. The basic formulation (1) (in the convex case) and the multiple choice formulation (2) (for the general case) immediately extend to this case by adding the restriction that variables l_a or l_a^s are integer. The disaggregated formulation (5) can also be used if variables l_a are re-introduced in the model as integer variables, together with linking constraints (3). These formulations have a very weak LP relaxation, as their continuous counterparts. Moreover, as we replace continuous variables by integer variables, the continuous flow models are relaxations of the integer ones, indicating that these latest problems are more difficult to solve.

4.1 The Point-Based Model

In order obtain stronger models, Gendron and Gouveia (2017) propose several formulations, based on discretization. The first one leads to the so-called Point-Based Model. This model is based on the assumption that the flow in each arc is integer, allowing the definition of a set of possible values the flow an arc a can carry, $Q_a = \{1, 2, \ldots, u_a\}$ where u_a is an upper bound on the maximal flow that can be sent on a in a feasible solution. Note that in the uncapacitated case considered here, we might take $u_a = \sum_{kinK} d^k$, but this approach also allows to model cases with explicit arc capacities. Sets Q_a can also be partitioned by segments, by defining $Q_a^s = \{q \in \mathcal{Z} : b_a^{s-1} < q \leq b_a^s\}$. The discretization of the the value of the flow on any arc allows the introduction of new binary variables y_a^q such that $y_a^q = 1$ if and only if the total flow on arc a is equal to q.

Observe that these new variables are linked to original variables y_a^s and l_a^s in (2) by the following relations:

$$y_a^s = \sum_{q \in Q_a^s} y_a^q, \qquad s \in S_a, a \in A,$$

$$\sum_{s \in S_a} l_a^s = \sum_{q \in Q_a} q y_a^q, \qquad a \in A.$$

Substituting in (2) leads immediately to the new formulation

$$\min \sum_{a \in A} \sum_{s \in S_a} \sum_{q \in Q_a^s} (f_a^s + c_a^s q)\, y_a^q \tag{6a}$$

$$\text{s.t.} \sum_{a \in \delta^+(i)} x_a^k - \sum_{a \in \delta^-(i)} x_a^k = \lambda_i^k, \qquad i \in V, k \in K, \tag{6b}$$

$$\sum_{q \in Q_a} y_a^q \leq 1, \qquad a \in A, \tag{6c}$$

$$\sum_{q \in Q_a} q y_a^q = \sum_{k \in K} d^k x_a^k, \qquad a \in A, \tag{6d}$$

$$0 \leq x_a^k \leq 1, \qquad a \in A, k \in K, \tag{6e}$$

$$y_a^q \in \{0, 1\}, \qquad a \in A, q \in Q_a. \tag{6f}$$

As far as we know, this discretization technique was first introduced in Gouveia (1995). One of the motivations for this technique, as noted in the earlier work, is also seen here, namely that transformed constraints (5d) become redundant with the new variables and therefore are not needed in the new model. Also it can be easily shown that the LP bounds of the disaggregated multiple choice formulation (5) and of the discretized point-based formulation (6) are equal. The LP relaxation of this model can be enhanced in two different ways described next, that, when combined together, lead to the formulation with the strongest LP bound.

4.2 Disaggregating the Point-Based Formulation

As noticed in Gendron and Gouveia (2017), the disaggregation technique used in the previous section can also be used together with the discretization technique used to derive the previous model. However, the load variables l_a^s must be included and defined as integer, and constraints (3) need to be added to the disaggregated multiple choice formulation (5). Based on this observation, they propose to apply the same ideas of disaggregation to the point-based model.

To do this, let us introduce variables x_a^{kq} to represent the proportion of commodity k routed on arc $a \in A$ if the total flow on arc a is exactly q (i.e., $\sum_{kinK} d^k x_a^k = q$). These new variables are linked to those in formulation (6) by the relations

$$x_a^k = \sum_{q \in Q_a} x_a^{kq}, \qquad a \in A, k \in K, \tag{7}$$

$$q y_a^q = \sum_{k \in K} d^k x_a^{kq}, \qquad a \in A, q \in Q_a. \tag{8}$$

Substituting and adding strong linking constraints then leads to

$$\min \sum_{a \in A} \sum_{s \in S_a} \sum_{q \in Q_a^s} \left(f_a^s y_a^q + c_a^s \sum_{k \in K} d^k x_a^{kq} \right) \tag{9a}$$

$$\text{s.t.} \sum_{a \in \delta^+(i)} \sum_{q \in Q_a} x_a^{kq} - \sum_{a \in \delta^-(i)} \sum_{q \in Q_a} x_a^{kq} = \lambda_i^k, \qquad i \in V, k \in K, \tag{9b}$$

$$\sum_{q \in Q_a} y_a^q \leq 1, \qquad a \in A, \tag{9c}$$

$$\sum_{q \in Q_a} q y_a^q = \sum_{k \in K} d^k \sum_{q \in Q_a} x_a^{kq}, \qquad a \in A, \tag{9d}$$

$$x_a^{kq} \leq y_a^q, \qquad a \in A, q \in Q_a, k \in K, \tag{9e}$$

$$x_a^{kq} \in \{0, 1\}, \qquad a \in A, k \in K, q \in Q_a. \tag{9f}$$

Note that in this last formulation, we have relaxed the fact that y_a^q need to be defined as binary variables as this is implied by the fact that x_a^{kq} are defined as binary variables combined with constraints (9c) and (9d).

4.3 Adding Chvátal-Gomory Rank 1 Inequalities

Another improvement to the point-based formulation (6) can be obtained by considering rounded cut-set inequalities, described next. Multiplying each flow conservation constraint (9b) by the corresponding demand d^k, summing over K and substituting using (9d) leads to

$$\sum_{a \in \delta^+(i)} \sum_{q \in Q_a} q y_a^q - \sum_{a \in \delta^-(i)} \sum_{q \in Q_a} q y_a^q = \sum_{k \in K} \lambda_i^k d^k, \; i \in V. \tag{10}$$

Summing over all nodes in a proper subset $S \subsetneq V$, $S \neq \emptyset$, leads to the following set of (redundant) constraints, called point-based cut-set equations:

$$\sum_{a \in \delta^+(S)} \sum_{q \in Q_a} q y_a^q - \sum_{a \in \delta^-(S)} \sum_{q \in Q_a} q y_a^q = D(S), \; \emptyset \neq S \subsetneq V, \tag{11}$$

where $D(S) = \sum_{i \in S} \sum_{k \in K} \lambda_i^k d^k$ is the net supply accross cut S.

Although these point-based cut-set equations do not improve the LP relaxation of the model, Chvátal-Gomory rank 1 inequalities derived from Eq. (11) turn out to be very effective in improving the corresponding LP relaxation.The idea is to divide Eq. (11) by each possible integer value in the disagreggated set, then round up or down, leading to:

$$\sum_{a \in \delta^+(S)} \sum_{q \in Q_a} \left\lceil \frac{q}{p} \right\rceil y_a^q - \sum_{a \in \delta^-(S)} \sum_{q \in Q_a} \left\lfloor \frac{q}{p} \right\rfloor y_a^q \geq \left\lceil \frac{D(S)}{p} \right\rceil, \emptyset \neq S \subsetneq V, p \in Q_a, \tag{12a}$$

$$\sum_{a\in\delta^+(S)}\sum_{q\in Q_a}\left\lfloor\frac{q}{p}\right\rfloor y_a^q - \sum_{a\in\delta^-(S)}\sum_{q\in Q_a}\left\lceil\frac{q}{p}\right\rceil y_a^q \le \left\lfloor\frac{D(S)}{p}\right\rfloor, \emptyset \ne S \subsetneq V, p \in Q_a, \tag{12b}$$

This idea has been used previously in other works using discretization, see e.g., Gouveia and Saldanha-da Gama (2006), Correia et al. (2010), and Gouveia and Moura (2012). Similar to the results in these works, the results reported in Gendron and Gouveia (2017) indicate that inequalities (12) considerably improve the LP bound of the point-based model (6). The reported results also show that strengthened with inequalities (12), the disaggregated point-based formulation (9) provides the strongest lower bound of all formulations presented in this section.

5 Unsplittable Multicommodity Network Flows

An important extension of the models above is to impose that all the flow for a commodity follows the same path. This arises for example in the context of telecommunications networks, see, e.g., the models proposed by Papadimitriou and Fortz (2014a,b), in the context of a complex multi-period design and routing problem. Lower bounds resulting from the LP relaxation of the problem are very weak, and part of this weakness is due to the piecewise linear objective function combined with single-path routing. This can be explained furthermore by observing that the relaxation obtained by relaxing single-path routing is exactly the continous version of the problem studied in Sect. 3, and both have the same weak LP relaxation.

Again, models for the continuous case (see Sect. 3) can easily be adapted to this extension by imposing that variables x_a^k are defined as binary. Fortz et al. (2017) extend the models from Croxton et al. (2003, 2007), using disaggregation to build a strong model similar to Model (5) where, in addition, x_a^{ks} are also defined as binary.

Exploiting the fact that variables x_a^{ks} are binary, Fortz et al. (2017) present two model enhancements. First, since flows are unsplittable, an arc being traversed by a given commodity cannot be on a segment whose upper breakpoint is smaller than the demand flow. Therefore, some x_a^{ks} variables can be fixed to zero (or removed from the model):

$$x_a^{ks} = 0, \qquad a \in A, k \in K, s \in S_a : b_a^s < d^k, \tag{13}$$

Secondly, since x_a^{ks} and y_a^s are binary, coefficients in the right-side inequality in (5d), namely $b_a^{s-1} y_a^s \le \sum_{k\in K} d^k x_a^{ks}$, can be tightened, leading to valid inequalities

$$b_a^{s-1} y_a^s \le \sum_{k\in K} \min\left(d^k, b_a^{s-1}\right) x_a^{ks}, \qquad a \in A, s \in S_a. \tag{14}$$

Fixing variables to zero (13) is a major improvement since, in addition to reducing the size of the model to be solved, it also strenghtens the LP bound. Similarly, coefficients tightening like the one proposed with (14) is well-known to have a major impact on LP bounds and helps reduce considerably the computation times needed to solve the model. The important message here is that these enhancements would not be possible without applying disaggregation.

Fortz et al. (2017) have also shown that the problem is polynomially solvable if there is a single commodity, i.e., when $|K| = 1$, as it reduces to a single path problem, and that the problem is NP-hard if $|K| > 1$. They also showed that the LP relaxation of their strengthened formulation always has an optimal solution with x_a^{ks} taking binary values when $|K| = 1$.

6 Conclusion

Bernard Gendron has made tremendous contributions in modeling and solving large scale transportation and network design problems, with a significant focus on variants of the fixed-charge multi-commodity flow problem. Besides contributions on modelling the problem and variants described in this chapter, other chapters of this book illustrate the careful and innovative ways in which Bernard has developed and applied decomposition techniques such as Lagrangian relaxation or Benders decomposition to these models.

Our objective in this chapter was to emphasize the importance of methodological advances in developing models for these problems. Finding an appropriate (strong) formulation is a necessary and important step to make the development of decomposition algorithms relevant and effective in practice.

We proposed a systematic review on mixed-integer linear models for multicommodity network flow problems with piecewise linear costs, unifying notation and presenting the main similarities while pointing out differences between the models. We have reviewed formulations and enhancements in the context of several variants of the problem, with convex and non-convex costs, and with and without integer flows. The approach taken was to start from the most basic version of continuous multicommodity network flow problems with piecewise linear costs and restrict progressively the set of feasible solutions, by imposing first integer flows, then unsplittable flows.

The review points out and emphasizes relevant model contributions by Bernard on this topic. In particular, the ideas of (variable and constraint) disaggregation, allowing to develop structural valid inequalities or to eliminate variables in a more efficient pre-processing step, are illustrated.

Due to the large size of these disaggregated models that involve a huge number of variables and constraints and thus, are difficult to solve directly by available ILP packages, over the last years, the contributions of Bernard were going in the direction of applying efficient decomposition techniques to solve large-scale instances in an efficient way. There is plenty of room for future research on the subject: on the one hand, advances are still needed to solve large scale instances of these models exactly. On the other hand, these techniques of disaggregation used as a tool to better catch the structure of the underlying problem could be applied to other classes of problems.

References

Balakrishnan, A., & Graves, S. C. (1989). A composite algorithm for a concave-cost network flow problem. *Networks, 19*(2), 175–202.

Correia, I., Gouveia, L., & Saldanha-da Gama, F. (2010). Discretized formulations for capacitated location problems with modular distribution costs. *European Journal of Operational Research, 204*(2), 237–244.

Croxton, K. L., Gendron, B., & Magnanti, T. L. (2003). A comparison of mixed-integer programming models for nonconvex piecewise linear cost minimization problems. *Management Science, 49*(9), 1268–1273.

Croxton, K. L., Gendron, B., & Magnanti, T. L. (2007). Variable disaggregation in network flow problems with piecewise linear costs. *Operations Research, 55*(1), 146–157.

Fortz, B., & Thorup, M. (2000). Internet traffic engineering by optimizing ospf weights. In: *INFOCOM 2000. Nineteenth annual joint conference of the IEEE computer and communications societies. Proceedings* (Vol. 2, pp. 519–528). IEEE.

Fortz, B., Gouveia, L., & Joyce-Moniz, M. (2017). Models for the piecewise linear unsplittable multicommodity flow problems. *European Journal of Operational Research, 261*(1), 30–42.

Frangioni, A., & Gendron, B. (2009). 0–1 reformulations of the multicommodity capacitated network design problem. *Discrete Applied Mathematics 157*(6), 1229–1241. Reformulation Techniques and Mathematical Programming.

Frangioni, A., & Gendron, B. (2021). Piecewise linear cost network design. In: T. G. Crainic, M. Gendreau, & B. Gendron (Eds.), *Network design with applications to transportation and logistics* (pp. 167–185). Springer International Publishing.

Gendron, B., & Gouveia, L. (2017). Reformulations by discretization for piecewise linear integer multicommodity network flow problems. *Transportation Science, 51*(2), 629–649.

Gouveia, L. (1995). A 2n constraint formulation for the capacitated minimal spanning tree problem. *Operations Research, 43*(1), 130–141.

Gouveia, L., & Saldanha-da Gama, F. (2006). On the capacitated concentrator location problem: a reformulation by discretization. *Computers & Operations Research, 33*(5), 1242–1258.

Gouveia, L., & Moura, P. (2012). Enhancing discretized formulations: the knapsack reformulation and the star reformulation. *Top, 20*, 52–74.

Murthy, R. V., & Helgason, R. V. (1994). A direct simplex algorithm for network flow problems with convex piecewise linear costs. *Optimization Methods and Software, 4*(3), 191–207.

Papadimitriou, D., & Fortz, B. (2014a). Methods for time-dependent combined network design and routing optimization. In: *IEEE global communications conference, GLOBECOM 2014, Austin, TX, USA, December 8–12, 2014* (pp. 1303–1309).

Papadimitriou, D., & Fortz, B. (2014b). Time-dependent combined network design and routing optimization. In: *IEEE international conference on communications, ICC 2014, Sydney, Australia, June 10–14, 2014* (pp. 1124–1130)

Pinto, Y., & Shamir, R. (1994). Efficient algorithms for minimum-cost flow problems with piecewise-linear convex costs. *Algorithmica, 11*, 256–277.

New Formulations for the Scheduled Service Network Design Problem with Piecewise Linear Costs

Mike Hewitt and Fabien Lehuédé

1 Introduction

The planning of consolidation-based transportation networks involves defining the moves of vehicles in a hub network in order to satisfy a given transportation demand. In this context, a set of demands, or *commodities*, have to be transported from their respective origins to their destinations in the network. Each commodity has a quantity. A release date can be provided to specify at what date and time the commodity is available at its origin and a due date can specify the latest date at which it can be delivered to its destination. The network is organized to efficiently serve multiple commodities which are consolidated in vehicles. Network design problems then typically consists in determining one or several paths for each commodity in the network as well as determining the number of vehicles required on each link of the network as well as the time at which each vehicle starts (its *dispatch time*) if release dates and due dates are involved, such that the sum of transportation cost is minimized.

Such planning operations involve solving challenging optimization problems to minimize the transportation costs. Different problem variants exist depending, for example, on the type of company making the decision (typically carriers, industries or third-party logistics companies), on the decision level (tactical or more operational) or even of the type of service (FTL: full-truckload or LTL:less-than-

M. Hewitt (✉)
Department of Information Systems and Supply Chain Management, Quinlan School of Business, Loyola University Chicago, Chicago, IL, USA
e-mail: mhewitt3@luc.edu

F. Lehuédé
IMT Atlantique, LS2N, UMR CNRS, Nantes, France
e-mail: fabien.lehuede@imt-atlantique.fr

T. G. Crainic et al. (eds.), *Combinatorial Optimization and Applications*,
International Series in Operations Research & Management Science 358,
https://doi.org/10.1007/978-3-031-57603-4_8

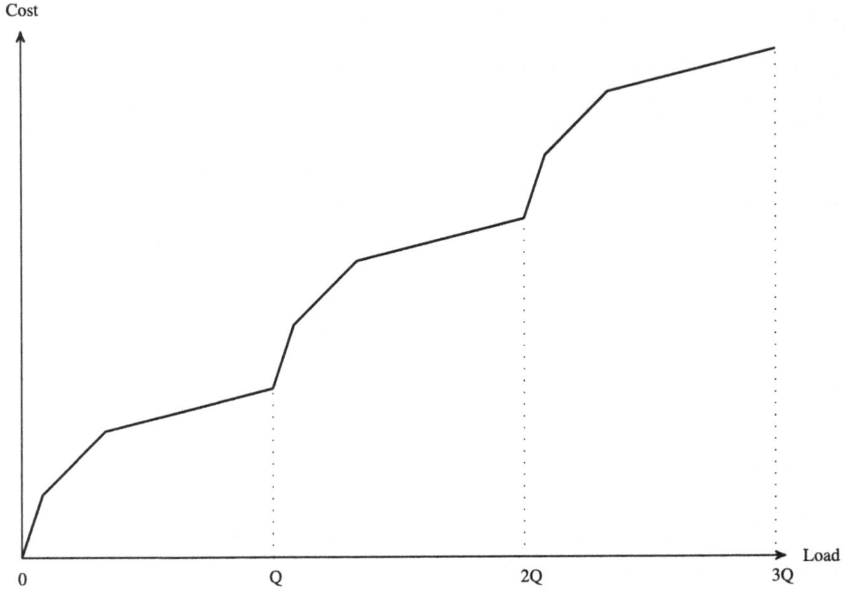

Fig. 1 Typical LTL rate on a transportation network link for a homogeneous fleet of vehicles of capacity Q

truckload) in the case of motor carriers. The most studied in operations research literature are modeled as Service Network Design (SND) or Scheduled Service Network Design (SSND) problems, both reviewed recently in Crainic and Hewitt (2021), (within the book Crainic et al., 2021, co-edited by our dear friend Bernard). In the same book, the chapter of Bakir et al. (2021) on motor carrier services, underlines that a large number of problems, from flow planning to vehicle routing and scheduling could be classified as network design problems in this domain.

In this chapter, we are particularly interested in variants of the SSNDP that integrates piecewise-linear costs. This type of cost, illustrated in Fig. 1, is met in numerous contexts in the planning of transportation networks. One of the first application is the modeling of economies of scale related to the transported quantity when a shipper seeks to optimize its distribution (Balakrishnan & Graves, 1989; Croxton et al., 2003). It is the case in particular in a collaborative context where multiple shippers organize the pooling of their transport operations (Chabot et al., 2018; Tang et al., 2019; Arslan et al., 2020; Lai et al., 2022). Accordingly, numerous models have been proposed to integrate piecewise-linear cost functions in network design application (see e.g. Balakrishnan and Graves (1989), Croxton et al. (2007), Frangioni and Gendron (2009), Gendron and Gouveia (2017), and Fortz et al. (2017)). Frangioni and Gendron (2021) present a recent synthesis of models and methods for piecewise-linear costs optimization in network design applications.

Considering the SSNDP, it is classically modeled by discretizing time and designing a so-called Time-Expanded Network (TEN) (Ford & Fulkerson, 1958,

1962). In such a network, nodes and arcs encode both physical and temporal attributes. In an SSNDP context, a node represents a physical location and period of time. An arc represents a physical transportation move dispatched from its origin during one period of time for arrival at its destination at another period of time that reflects (at least) the travel time associated with the physical transportation move. Further, as a node models both a location and period of time, where and when a commodity is available for pickup can be modeled with the appropriate choice of origin node for that commodity in a TEN. A similar statement can be made regarding the choice of destination node for a commodity.

With the construction of such a network, the SSNDP can be modeled as a Fixed-Charge Multicommodity Network Flow Problem. Subsequently, techniques for modeling piecewise-linear costs for network flow problems can then be applied to the SSNDP. However, a TEN is typically constructed in an enumerative fashion in which each physical location and transportation move is duplicated for each period of time to be modeled. This renders solving large-scale instances of the SSNDP, either with piecewise linear costs or per-vehicle transportation costs, formulated on a TEN, challenging to solve for commercial solvers.

One approach to mitigating this challenge is the dynamic discretization discovery algorithm, which generates such networks dynamically (Boland et al., 2017; Hewitt, 2019). Another approach is to avoid formulating on a TEN completely. However, doing so necessitates a new mechanism for modeling a *consolidation*. Namely, commodities that dispatch on the same physical transportation move at the same time (or period of time) and thus can potentially travel in the same vehicle. With a TEN this is captured by commodity flows that travel on the same arc.

Recently, (Hewitt & Lehuédé, 2022, 2023) proposed an alternative formulation of the SSNDP that foregoes a TEN and instead models with an *a priori* enumeration of possible consolidations. In particular, Hewitt and Lehuédé (2023) show that realistic size instances of this formulation of the SSNDP can be solved by a MIP solver without having to implement an elaborate algorithm. The paper also states that piecewise-linear cost functions can be easily integrated in the model but no comparison of this model to the literature or experiments of this model has been performed yet.

As a result, this chapter presents the TEN and consolidation based formulations of the SSNDP with piecewise linear costs. It proposes a comparison of the performance of the two models. This comparison is performed on instances derived from a real carrier network and demand. We show that, on these instances, the new consolidation based formulation outperforms the classical TEN model in terms of number of instances solved, solving time and optimality gap for instances that are not solved within a 2 hours time limit.

2 The Piecewise-Linear Scheduled Service Network Design Problem: Time-Expanded Network Formulation

Let us consider a set of terminals, modeled by a set of nodes $\mathcal{N}$, and the transportation links between these terminals, modeled by a set of arcs $\mathcal{A}$. Each arc $(i, j) \in \mathcal{A}$ in this network is associated with a travel duration τ_{ij} for the vehicles that can be dispatched on this link. The corresponding network is called the *flat network* and it is denoted by $\mathcal{D} = (\mathcal{N}, \mathcal{A})$.

The transportation demand is modeled by a set of shipments $k \in \mathcal{K}$, each associated with a size q_k (in the same unit as vehicle capacity), an origin $o_k \in \mathcal{N}$ and a destination $d_k \in \mathcal{N}$. The date at which the shipment is available at its origin is the *release date*, denoted by e_k, while the latest delivery date is the *due date*, denoted by l_k. In addition, we assume that a set P_k is given to model the possible paths from o_k to d_k of a shipment k in the network.

Related to these definitions: we denote by $\mathcal{K}_{ij}$, the set of shipments k for which there exist a path $p \in P_k$ that contains the arc (i, j); we let $P(i, j)_k \subseteq P_k$ denote the set of paths that use arc (i, j); and we denote by $[\alpha_{ij}^{kp}, \beta_{ij}^{kp}]$ the time window in which a shipment $k \in \mathcal{K}$ can dispatch on arc (i, j) if it takes path p according to the shipment release date and due dates as well as the travel times on the path. Here, *dispatch time* denotes the time at which the shipment leaves the origin of the related arc. It can also refer to the time at which the vehicle that transports this shipment starts travelling on this arc. Unloading, sorting or loading times can be integrated on the arc travel times.

For each arc $(i, j) \in \mathcal{A}$, a piecewise-linear cost function $g_{ij}(u_{ij})$ can be defined according to the load u_{ij} on the arc. This function is determined by a finite set of segments $S_{ij} = \{1, \ldots, |S_{ij}|\}$. Each segment $s \in S_{ij}$ is defined by a lower and upper bound $[b_{ij}^{s-1}, b_{ij}^{s}]$ related to the load u_{ij}, an intercept f_{ij}^{s} and a slope c_{ij}^{s}, such that $g_{ij}(u_{ij}) = f_{ij}^{s} + c_{ij}^{s} u_{ij}$ if $u_{ij} \in [b_{ij}^{s-1}, b_{ij}^{s}]$. This cost structure and notation is illustrated on Fig. 2.

We define the PLSSNDP as the problem of selecting one path for each shipment, determining a feasible dispatch time for each arc on this path and determining the number of necessary vehicles on the network to perform the shipments transportation such that the overall transportation cost is minimized.

To model this problem, the TEN formulation is based on a discretization of the time horizon in T periods, denoted by $\mathcal{T} = \{1, \ldots, T\}$. Then the TEN $\mathcal{D}_{\mathcal{T}} = (\mathcal{N}_{\mathcal{T}}, \mathcal{A}_{\mathcal{T}})$ is defined with: $\mathcal{N}_{\mathcal{T}}$ denoting the node set $\{(i, t), i \in \mathcal{N}, t \in \mathcal{T}\}$ that models the activities at terminal i during period t, and the set $\mathcal{A}_{\mathcal{T}}$ which models traveling and waiting operations in the network. More explicitly, $\mathcal{A}_{\mathcal{T}}$ is composed of traveling arcs of the form $((i, t), (j, t'))$, $(i, t), (j, t') \in \mathcal{N}_{\mathcal{T}}, i \neq j, t' = \min(t'' \in \tau | t'' - t \geq \tau_{ij})$, which model the travel from a terminal i at time t to a terminal j at time t', and waiting arcs of the form $((i, t), (i, t + 1))$, $(i, t), (i, t + 1) \in \mathcal{N}_{\mathcal{T}}$, which model a shipment that is waiting at a terminal i from period t to period $t + 1$. Note that we do not consider any cost on waiting arcs.

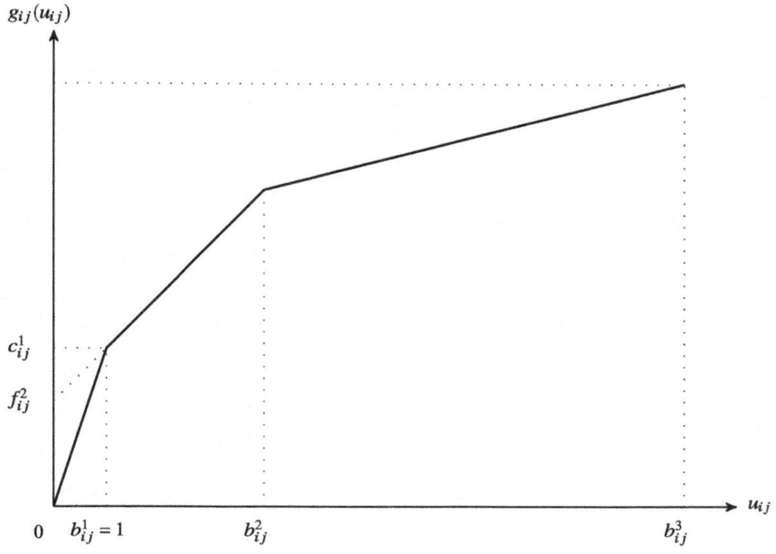

Fig. 2 Piecewise-linear cost function notation

Based on this TEN, we define four sets of decision variables. First, for all shipments and paths, we define a binary variable v^k_p, equal to one if the path $p \in \mathcal{P}_k$ is selected for shipment $k \in \mathcal{K}$. Second, the binary variables $x^{ktt'}_{ij}, k \in \mathcal{K}, ((i,t),(j,t')) \in \mathcal{A}_{\mathcal{T}}$ denote whether shipment k takes arc $((i,t),(j,t'))$. Third, for each arc $((i,t),(j,t')) \in \mathcal{A}_{\mathcal{T}}$, we define the binary variable $y^{stt'}_{ij}$ which is equal to one if the load on $((i,t),(j,t'))$ falls within $[b^{s-1}_{ij}, b^s_{ij}]$, $s \in S_{ij}$ and last, $z^{stt'}_{ij} \in \Re^+$, which models the load on this arc if it falls within segment $s \in S_{ij}$.

Given these decision variables, the SSNDP can be formulated as follows. We label this formulation *TEN*($\mathcal{D}$):

$$z_{TEN(\mathcal{D})} = \min \sum_{((i,t),(j,t')) \in \mathcal{A}_{\mathcal{T}}} \sum_{s \in S_{ij}} (f^s_{ij} y^{stt'}_{ij} + c^s_{ij} z^{stt'}_{ij}) \tag{1}$$

$$s.t. \sum_{p \in P_k} v^k_p = 1, \qquad k \in \mathcal{K}, \tag{2}$$

$$\sum_{p \in P(i,j)_k} v^k_p = \sum_{(t,t'):((i,t),(j,t')) \in \mathcal{A}_{\mathcal{T}}} x^{ktt'}_{ij}, \qquad (i,j) \in \mathcal{A}, k \in \mathcal{K}_{ij}, \tag{3}$$

$$\sum_{((i,t),(j,t'))\in\mathcal{A}_{\mathcal{T}}} x_{ij}^{ktt'} - \sum_{((j,t'),(i,t))\in\mathcal{A}_{\mathcal{T}}} x_{ji}^{kt't} = \begin{cases} 1 & (i,t)=(o_k,e_k) \\ -1 & (i,t)=(d_k,l_k) \\ 0 & \text{o.w.,} \end{cases}$$

$$(i,t)\in\mathcal{N}_{\mathcal{T}}, k\in\mathcal{K}, \quad (4)$$

$$\sum_{k\in\mathcal{K}_{ij}} q_k x_{ij}^{ktt'} = \sum_{s\in S_{ij}} z_{ij}^{stt'}, \quad ((i,t),(j,t'))\in\mathcal{A}_{\mathcal{T}}, \quad (5)$$

$$b_{ij}^{s-1} y_{ij}^{stt'} \le z_{ij}^{stt'} \le b_{ij}^{s} y_{ij}^{stt'}, \quad ((i,t),(j,t'))\in\mathcal{A}_{\mathcal{T}}, s\in S_{ij}, \quad (6)$$

$$\sum_{s\in S_{ij}} y_{ij}^{stt'} \le 1, \quad ((i,t),(j,t'))\in\mathcal{A}_{\mathcal{T}}, \quad (7)$$

$$v_p^k \in \{0,1\}, \quad k\in\mathcal{K}, p\in P(i,j)_k, \quad (8)$$

$$x_{ij}^{ktt'} \in \{0,1\}, \quad ((i,t),(j,t'))\in\mathcal{A}_{\mathcal{T}}, k\in\mathcal{K},. \quad (9)$$

$$y_{ij}^{stt'} \in \{0,1\}, \quad ((i,t),(j,t'))\in\mathcal{A}_{\mathcal{T}}, s\in S_{ij}, \quad (10)$$

$$z_{ij}^{stt'} \ge 0, \quad ((i,t),(j,t'))\in\mathcal{A}_{\mathcal{T}}, s\in S_{ij}. \quad (11)$$

The objective function (1) minimizes the sum of piecewise-linear costs. Constraints (2) state that one path should be assigned to each shipment. Constraints (3) tie the chosen path to the arcs of the network. Constraints (4) model the balancing of shipment flows at their respective origin, destination and intermediate terminals. Constraints (5) connect the flow variables of the piecewise-linear cost model to the SSNDP load variables. Constraints (6) state that a segment s is used for an arc $((i,t)(jt'))$ if the load on this arc falls within the segment bounds. Constraints (7) state that only one segment can be selected for the piecewise-linear cost function on one arc. Constraints (8)–(11) define the decision variable definition domains.

This model is referred to as the *Multiple Choice Model* in the literature (Balakrishnan & Graves, 1989; Vielma et al., 2010). Like network design problems in general, this formulation can be strengthened with the following valid inequalities:

$$x_{ij}^{ktt'} \le \sum_{s\in S_{ij}:b_{ij}^s\ge q_k} y_{ij}^{stt'}, \quad ((i,t),(j,t'))\in\mathcal{A}_{\mathcal{T}}, k\in\mathcal{K}. \quad (12)$$

These constraints reflect the observation that if shipment k flows on arc $((i,t),(j,t'))$ then because of the binary nature of the x variables, a segment must be chosen for that arc that has an upper bound at least as great as the demand of shipment k.

3 The Consolidation Based Formulation

The consolidation based formulation has been recently proposed in (Hewitt & Lehuédé, 2022, 2023) for the SSNDP and the service network scheduling problem, a variant where the path of each commodity is know. This model presumes an *a priori* enumeration of consolidations on each arc $(i, j) \in \mathcal{A}$, wherein a *consolidation* is defined as a set of shipments that dispatch on the same physical move at the same time. Regarding notation for this formulation, we let Ω_{ij} define the set of feasible shipment consolidations on arc (i, j). Thus, an element $\omega \in \Omega_{ij}$ is a subset of commodities (i.e., $\omega \subseteq \mathcal{K}$) that have a non-zero intersection of their time windows on arc $(i, j) \in \mathcal{A}$. For $(i, j) \in \mathcal{A}, \omega \in \Omega_{ij}$, we let ϕ_ω^k indicate whether shipment $k \in \mathcal{K}$ is contained in consolidation ω. For a more extensive description and illustration of the concept of consolidation, we refer the interested reader to Hewitt and Lehuédé (2022).

Since all shipment consolidations are enumerated, the consolidation based formulation allows us to calculate the cost associated with transporting consolidation $\omega \in \Omega_{ij}$ on $(i, j) \in \mathcal{A}$ independently of the dispatch time of the consolidation. This cost is denoted by c_ω for all $\omega \in \Omega_{ij}$ and defined as $c_\omega = g_{ij}(\sum_{k \in \omega} q_k)$. Hence, the formulation integrates the piecewise linear cost function before optimization.

We define the binary decision variable δ_ω to indicate whether the consolidation $\omega \in \Omega_{ij}$ is chosen. Like the formulation *TEN*($\mathcal{D}$), the choice of physical routing of shipments is done with the binary variables $v_p^k \in \{0, 1\}, p \in P_k, k \in \mathcal{K}$. The decision variable γ_{ij}^k represents the time at which shipment k dispatches on arc $(i, j) \in \mathcal{A}$. The consolidation-based formulation we present, which we label *CONS*($\mathcal{D}$), is as follows:

$$z_{CONS(\mathcal{D})} = \min \sum_{(i,j) \in \mathcal{A}} \sum_{\omega \in \Omega_{ij}} c_\omega \delta_\omega \tag{13}$$

$$s.t. \sum_{p \in P_k} v_p^k = 1, \qquad k \in \mathcal{K}, \tag{14}$$

$$\sum_{\omega \in \Omega_{ij}} \phi_\omega^k \delta_\omega = \sum_{p \in P(i,j)_k} v_p^k, \qquad (i, j) \in \mathcal{A}, k \in \mathcal{K}_{ij}, \tag{15}$$

$$\gamma_{ij}^k - \gamma_{ij}^{k'} \leq T(1 - \sum_{\omega \in \Omega_{ij}} \phi_\omega^k \phi_\omega^{k'} \delta_\omega), \qquad (i, j) \in \mathcal{A}, k, k' \in \mathcal{K}_{ij}, \tag{16}$$

$$\sum_{p \in P(i,j)_k} \alpha_{ij}^{kp} v_p^k \leq \gamma_{ij}^k, \qquad (i,j) \in \mathcal{A}, k \in \mathcal{K}_{ij}, \tag{17}$$

$$\gamma_{ij}^k \leq \sum_{p \in P(i,j)_k} \beta_{ij}^{kp} v_p^k, \qquad (i,j) \in \mathcal{A}, k \in \mathcal{K}_{ij}, \tag{18}$$

$$\sum_{(o_k,j) \in \mathcal{A}} \gamma_{o_k j}^k \geq e_k, \qquad k \in \mathcal{K}, \tag{19}$$

$$\sum_{(i,d_k) \in \mathcal{A}} (\gamma_{id_k}^k + \tau_{id_k}(\sum_{p \in P(i,d_k)_k} v_p^k)) \leq l_k, \qquad k \in \mathcal{K}, \tag{20}$$

$$\sum_{(i,j) \in \mathcal{A}} (\gamma_{ij}^k + \tau_{ij}(\sum_{p \in P(i,j)_k} v_p^k)) \leq \sum_{(j,i) \in \mathcal{A}} \gamma_{ji}^k, \qquad j \in N, k \in \mathcal{K}, \tag{21}$$

$$v_p^k \in \{0,1\}, \qquad k \in \mathcal{K}, p \in P(i,j)_k, \tag{22}$$

$$\delta_\omega \in \{0,1\}, \qquad \omega \in \Omega_{ij}, (i,j) \in \mathcal{A}, \tag{23}$$

$$\gamma_{ij}^k \geq 0, \qquad k \in \mathcal{K}, (i,j) \in \mathcal{A}. \tag{24}$$

The objective function (13) minimizes the sum of transportation costs. Constraints (14) state that a path should be assigned to each commodity. Constraints (15) indicates that if the path assigned to a commodity k uses an arc $(i, j) \in \mathcal{A}$, then this commodity must be part of a selected consolidation on (i, j). Constraints (16) state that two commodities that belong to the same selected consolidation on an arc $(i, j) \in \mathcal{A}$ are dispatched at the same time on (i, j). Constraints (17) and Constraints (18) state that a commodity cannot be dispatched before the opening and after the end of its time window on each arc of its selected path. Constraints (19) enforce the departure of a commodity from its origin after its release date and constraints (20) make sure each commodity is delivered at its destination by its due date. Precedences on a commodity path, integrating travel times, is modeled by Constraints (21). Constraints (22) and (23) define the domains of the variables.

Regarding enumerating consolidations, one method is to generate them on an arc by arc basis. Doing so relies on the concept of a *maximal consolidation*, a consolidation that is not contained in any other consolidation for that arc. With such a method, for a given arc $(i, j) \in \mathcal{A}$, time window(s) can be derived for when each commodity can dispatch on (i, j). From these time windows, maximal consolidations can be determined. The set Ω_{ij} can then be constructed by adding to it each maximal consolidation and each of its subsets. While such a method is not polynomial, its run time depends greatly on the sizes of the maximal consolidations.

Consolidation-Pruning Hewitt and Lehuédé (2023) present a condition under which consolidations can be removed from *CONS*($\mathcal{D}$) without rendering all optimal solutions infeasible. The logic behind that observation can be adapted to a similar observation for the case of non-linear costs. Namely, that when a consolidation ω can be partitioned into subsets, $\omega_i, i = 1, \ldots, m$ such that the $c_\omega \geq \sum_{i=1}^m c_{\omega_i}$

then consolidation ω can be removed from $CONS(\mathcal{D})$ without rending all optimal solutions infeasible.

4 Computational Analysis

In this section, we computationally evaluate the solvability of the two formulations. Before presenting the results of our computational study, we describe the instances on which it is based as well as the computational environment in which experiments were run.

4.1 Instances

We consider a subset of the set of instances used in Hewitt and Lehuédé (2023). That set consists of randomly generated instances that model a portion of the network of a United States-based LTL carrier that consists of 25 terminals ($|\mathcal{N}| = 25$) and 530 physical moves, or services, between terminals ($|\mathcal{A}| = 530$).

We consider instances that vary in the number of shipments, or the size of the set $\mathcal{K}$. In these instances, all commodities have at most ten potential paths (i.e. $|\mathcal{P}_k| \leq 10, k \in \mathcal{K}$). To identify the set $\mathcal{P}_k$ for shipment k, we enumerated all paths from o_k to d_k to identify a set of candidate paths. We then removed from this candidate set all paths for which the total time required to travel that path exceeded the time available to deliver the shipment. Formally, we removed path p for commodity k if $\sum_{(i,j)\in p} \tau_{ij} > l_k - e_k$. Next, we removed from the candidate set, paths that exhibited one of two criteria, both of which were inspired by discussions with practitioners. The first criterion is paths that visit more than three terminals other than the origin or destination. The second factor is paths with distance greater than 150% the distance between origin and destination terminals, when such a direct move is possible. Then, from the remaining set of candidate paths, we included in $\mathcal{P}_k$ the ten shortest with respect to travel time. If the candidate set did not contain ten paths then we included in $\mathcal{P}_k$ all candidate paths.

Regarding transportation costs on arc (i, j), we define piecewise linear costs based on a given vehicle capacity of Q and carrier-provided transportation cost C_{ij}^{FTL} for a full truckload on arc (i, j). To represent cost functions with an increasing number of segments in our experiments, we model that the capacity of an individual vehicle, Q, is divided into M segments of identical lengths. Hence, $b_{ij}^1 = Q/M$, $b_{ij}^2 = 2Q/M$, $b_{ij}^M = Q$. To model economies of scales, we define a cost F_{ij} as the maximal cost of the first segment: $g_{ij}(b_{ij}^1) = F_{ij}$. Hence, the first segment has a slope $c_{ij}^1 = F_{ij}/b_{ij}^1 = MF_{ij}/Q$. Then the second segment has a slope $c_{ij}^2 = F_{ij}/b_{ij}^2 = MF_{ij}/2Q$, the third segment has a slope $c_{ij}^3 = MF_{ij}/3Q, \ldots$, the Mth segment has a

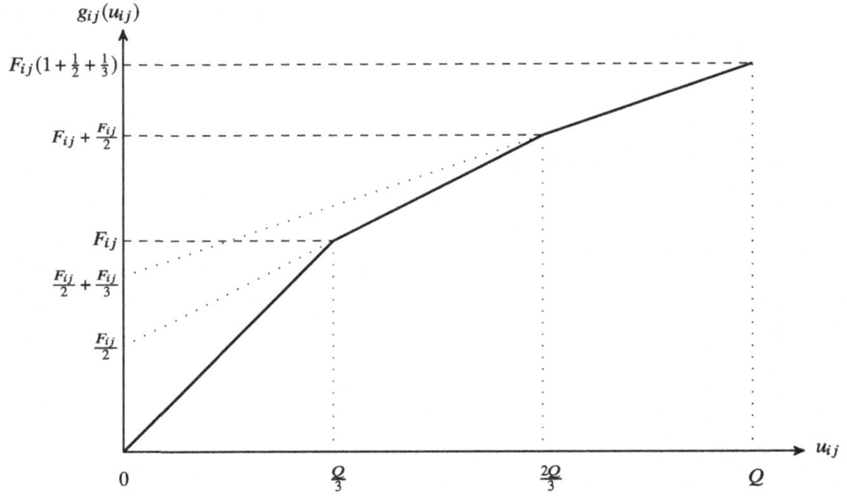

Fig. 3 First M segments of a piecewise-linear cost function defined for our experiments for vehicles of capacity Q

slope $c_{ij}^M = F_{ij}/Q$. The intersect of the first segment is $f_{ij}^1 = 0$. Then to ensure that the function is continuous, the intersect of each segment $s \in \{2, \dots, M\}$ is defined as $f_{ij}^s = F_{ij} \times \sum_{k=2}^{s} 1/k$. To calibrate the cost function such that $g_{ij}(Q) = C_{ij}^{FTL}$, we define $F_{ij} = C_{ij}^{FTL}/\sum_{k=1}^{M} 1/k$, depending on the number of required segments M. This function is illustrated on Fig. 3. It is repeated for loads that are greater than one truck with the same M slopes.

Regarding the time dimension of the instances, we consider a planning horizon that corresponds to a five-day week. We presume each travel time, τ_{ij}, is a multiple of a half-hour and each available and due time is a multiple of an hour. We note these assumptions are based on the data from the carrier network that motivated this research. As such, instances of the TEN formulation are created based on a discretization of time into half-hour intervals, as this is sufficiently granular to accurately model the temporal aspects of these instances. Consequently, instances of the TEN formulation are based on a time horizon consisting of $T = 240$ periods.

We presume a vehicle capacity of $Q = 22{,}000$ pounds and consider divisions of that capacity into three, four, and five segments of identical lengths. We classify an instance by the number of shipments, $|\mathcal{K}|$, and capacity division. We consider classes with 100, 150, or 200 numbers of shipments and five instances per class, leaving a total of 45 instances in our experimental study. Instances are available upon request.

4.2 Computational Setting

We next describe the computational environment and settings used in our experimental study. We first note that each experiment was run on the same computer, which is equipped with 200 GB of RAM and 64 Intel Xeon Gold 6130 CPU processors operating at 2.10GHz. This computer runs the Ubuntu distribution of the Linux operating system. All formulations and code to instantiate them were implemented in Python 3.7 (VanRossum & Drake, 2010). All code was run on a single thread for 2 hours, or, 7200 seconds. All optimization models were also solved with CPLEX 22.10 (Studio-CPLEX, 2022) to a 1% tolerance. In all experiments CPLEX 22.10 was configured to use a single thread. All CPLEX parameters not mentioned were left at their default values.

4.3 Analysis

In our computational analysis we seek to understand the following issues. The first is which of the two proposed formulations is more amenable to solution by a MIP solver. The second is to what extent the characteristics of an instance affects the performance of a MIP solver when attempting to solve instances of the two formulations. We focus on two instance characteristics in this analysis: the number of shipments, $|\mathcal{K}|$, and the capacity division underlying the cost structure. The third is the effectiveness of the consolidation pruning speed-up technique proposed in Sect. 3. The fourth and final issue we consider is a comparison of the difficulty of solving instances of $CONS(\mathcal{D})$ when there are piecewise linear costs, as considered in this paper, as opposed to the traditional cost structure considered in the SSNDP.

4.4 Performance of Formulations

To compare the performance of the different formulations and speed-up techniques we focus on three statistics: (1) the number of instances solved, (2) the average time to solve an instance, and, (3) the average optimality gap of instances not solved. We report these statistics aggregated over instances with same number of shipments, $|\mathcal{K}|$. Table 1a, b, and c report these statistics for each of the three capacity divisions.

We see that the MIP solver is capable of solving all but one instance of the consolidation-based formulation and only two of the time expanded network-based formulation. Focusing on the consolidation-based formulation, $DVCONS$, we see that instances become harder to solve as the number of shipments in the instance increases. The correlation between the number of segments in the cost function and the difficulty of an instance is not clear. Theoretically speaking, the number of segments in the cost function has no impact on the number of variables or constraints

Table 1 Performance statistics, 15 instances per group

	DVCONS			*TEN*		
$\|\mathcal{K}\|$	% Solved	Time	Gap unsolved	% Solved	Time	Gap unsolved
100	5	8.45	N/A	0	N/A	1.69%
150	5	929.46	N/A	0	N/A	5.82%
200	4	2683.19	5.37%	0	N/A	6.92%
Summary	14	1207.04	5.37%	0	N/A	4.81%

(a) Capacity division into 3 segments

	DVCONS			*TEN*		
$\|\mathcal{K}\|$	% Solved	Time	Gap unsolved	% Solved	Time	Gap unsolved
100	5	5.31	N/A	2	5905.48	5.32%
150	5	35.54	N/A	0	N/A	12.80%
200	5	392.66	N/A	0	N/A	33.16%
Summary	15	144.50	N/A	2	5905.48	17.09%

(b) Capacity division into 4 segments

	DVCONS			*TEN*		
$\|\mathcal{K}\|$	# Solved	Time	Gap unsolved	# Solved	Time	Gap unsolved
100	5	6.06	N/A	0	N/A	5.10%
150	5	37.75	N/A	0	N/A	15.29%
200	5	538.36	N/A	0	N/A	35.93%
Summary	15	194.06	N/A	0	N/A	18.77%

(c) Capacity division into 5 segments

in this model. But we see that the three segments instances take longer to solve. This is in contrast to the time expanded network-based formulation in which the number of variables and constraints increases in the number of segments. The impact of this can be seen by noting that the average gap of unsolved instances of the TEN formulation increases in the number of segments. Finally, we also solved instances of TEN strengthened with the valid inequalities (12) added in an *a priori* manner. We note that with those valid inequalities, CPLEX still solved only two instances of the TEN and the average gap of those unsolved was larger. Adding such inequalities in a cutting-plane type manner may yield better results. However, we still conclude that the consolidation-based formulation is easier to solve.

4.5 Impact of Consoldation Pruning

We next consider in Table 2 the impact of consolidation pruning. We report the same summary statistics as above, again aggregated over instances with the same number of shipments. We report results separately by the number of segments in the piecewise linear cost function.

Table 2 *DVCONS* with consolidation pruning

	3 segments			4 segments			5 segments		
	# Solved	Time	Gap	# Solved	Time	Gap	# Solved	Time	Gap
$\|\mathcal{K}\|$			Unsolved			Unsolved			Unsolved
100	5	5.78	N/A	5	2.70	N/A	5	2.82	N/A
150	5	480.90	N/A	5	13.72	N/A	5	8.92	N/A
200	5	2006.51	N/A	5	364.62	N/A	5	773.96	N/A
Summary	15	831.06	N/A	15	127.01	N/A	15	261.90	N/A

Table 3 SSNDP with per vehicle transportation costs

	$CONS(\mathcal{D})$			$CONS(\mathcal{D})$ with pruning		
$\|\mathcal{K}\|$	# Solved	Time	Gap unsolved	# Solved	Time	Gap unsolved
100	5	108.22	N/A	5	81.16	N/A
150	1	1404.58	2.42%	1	654.21	2.08%
200	0	N/A	6.43%	0	N/A	3.59%
Summary	6	756.40	4.43%	6	367.68	2.84%

We see that the consolidation pruning technique enables the solver to solve every instance and often in less time, on average.

4.6 Comparison to the Classical Model

Finally, we seek to assess whether the SSNDP with Piecewise Linear costs is easier or harder to solve than the variant of the SSNDP that is traditionally studied wherein transportation costs are paid on a per-vehicle basis. To understand this, we formulate and solve the consolidation-based SSNDP formulation of the instances considered in this experimental study. We note that these instances were based on the same per-vehicle transportation costs, C_{ij}^{FTL}, that were used to define the piecewise linear cost functions. We report performance statistics for these instances in Table 3.

We see that the variant with piecewise linear costs is easier to solve than the per-vehicle cost structure traditionally modeled and solved. We hypothesize that this may be due to the traditional cost structure requiring a set of integer decision variables that model the number of vehicles used on arc $(i, j) \in \mathcal{A}$.

5 Conclusions and Future Work

Piecewise linear cost structures often appear in transportation and logistics contexts wherein the per unit price of a good or service exhibits economies of scale. Such a structure is often seen in the context of Less-than-truckload freight transportation wherein a carrier will charge a shipper a per unit of volume price that decreases as the total volume transported increases. The planning of transportation operations for such a carrier can be assisted with the solution of the Scheduled Service Network Design Problem. In this chapter, we considered such a context and cost structure and studied the Scheduled Service Network Design Problem with Piecewise Linear Costs. One practical setting in which the solution of this problem is relevant is when a shipper outsources its freight transportation to a dedicated carrier and the goods the shipper seeks to have transported are small relative to vehicle capacity. While such cost structures are prevalent in practice and modeling them is fairly straightforward, the resulting mathematical programs are notoriously difficult to solve.

Recently, a new formulation was proposed for the Scheduled Service Network Design Problem (SSNDP) in which transportation costs are paid on a per-vehicle basis. This new formulation is based on enumerations of shipments that will be transported together and are referred to as "consolidations." It has been shown that this new formulation is much easier to solve, computationally-speaking, than the classical formulation of the SSNDP that is based on a time expanded network. Thus, in this chapter we studied whether adapting the ideas underlying that formulation to the case wherein costs are piecewise linear yields a formulation that is similarly easier to solve than a formulation based on a time-expanded network. We proposed such a formulation and with a computational study demonstrated that the consolidation-based formulation was easier to solve. We also adapted a speed-up technique for solving a consolidation-based formulation of the SSNDP with per-vehicle transportation costs to the case of piecewise linear costs.

Regarding future work, we see avenues for both algorithmic and modeling work. While we observed that instances of the proposed consolidation-based formulation are easier to solve than those of the time expanded network-based formulation, its enumerative nature can require a large number of consolidation-based variables. Thus, solving instances of this formulation that involve large numbers of commodities (and hence consolidations) likely necessitates a solution method that generates consolidations dynamically. The piecewise linear cost structure will likely present algorithmic challenges not present in a comparable method for a consolidation-based formulation of the SSNDP with the traditional cost structure. From a modeling perspective, we considered a shipper-oriented model wherein the transportation costs on each arc have a piecewise linear structure. However, taking the perspective of a carrier will likely yield a different model as a carrier with its own fleet may rely on both owned and outsourced transportation. In this case, the cost structure would involve a combination of paying for the use of owned vehicles, as in the classical SSNDP, and outsourcing transportation to a third party, as in the model considered in this paper. More generally, one could consider extending the consolidation based formulation to integrate the variety of features that have been proposed in the SSNDP literature, such as holding costs, time-dependent travel times, merge-in-transit or resource constraints,. . .

Acknowledgments *We would like do dedicate this work to Bernard Gendron for his numerous contributions to the topic and his extreme kindness.*

References

Arslan, O., Archetti, C., Jabali, O., Laporte, G., & Speranza, M. G. (2020). Minimum cost network design in strategic alliances. *Omega, 96*, 102079.

Bakir, I., Erera, A., & Savelsbergh, M. (2021). Motor carrier service network design. In: T. G. Crainic, M. Gendreau, & B. Gendron (Eds.), *Network design with applications to transportation and logistics* (pp. 427–467). Springer.

Balakrishnan, A., & Graves, S. C. (1989). A composite algorithm for a concave-cost network flow problem. *Networks, 19*(2), 175–202.

Boland, N., Hewitt, M., Marshall, L., & Savelsbergh, M. (2017). The continuous-time service network design problem. *Operations Research, 65*(5), 1303–1321.

Chabot, T., Bouchard, F., Legault-Michaud, A., Renaud, J., & Coelho, L. C. (2018). Service level, cost and environmental optimization of collaborative transportation. *Transportation Research Part E: Logistics and Transportation Review, 110*, 1–14.

Crainic, T. G., & Hewitt, M. (2021). Service network design. In: T. G. Crainic, M. Gendreau, & B. Gendron (Eds.), *Network design with applications to transportation and logistics* (pp. 347–382). Springer.

Crainic, T. G., Gendreau, M., & Gendron, B. (2021). *Network design with applications to transportation and logistics*. Springer Nature.

Croxton, K. L., Gendron, B., & Magnanti, T. L. (2003). Models and methods for merge-in-transit operations. *Transportation Science, 37*(1), 1–22.

Croxton, K. L., Gendron, B., & Magnanti, T. L. (2007). Variable disaggregation in network flow problems with piecewise linear costs. *Operations Research 55*(1), 146–157.

Ford, L. R., & Fulkerson, D. R. (1958). Constructing maximal dynamic flows from static flows. *Operations Research, 6*(3), 419–433.

Ford, L. R., & Fulkerson, D. R. (1962). *Flows in networks*. Princeton University Press.

Fortz, B., Gouveia, L., & Joyce-Moniz, M. (2017). Models for the piecewise linear unsplittable multicommodity flow problems. *European Journal of Operational Research, 261*(1), 30–42.

Frangioni, A., & Gendron, B. (2009). 0–1 reformulations of the multicommodity capacitated network design problem. *Discrete Applied Mathematics 157*(6), 1229–1241.

Frangioni, A., & Gendron, B. (2021) Piecewise linear cost network design. In: T. G. Crainic, M. Gendreau, & B. Gendron (Eds.), *Network design with applications to transportation and logistics* (pp. 167–185). Springer.

Gendron, B., & Gouveia, L. (2017). Reformulations by discretization for piecewise linear integer multicommodity network flow problems. *Transportation Science, 51*(2), 629–649.

Hewitt, M. (2019). Enhanced dynamic discretization discovery for the continuous time load plan design problem. *Transportation Science, 53*(6), 1731–1750.

Hewitt, M., & Lehuédé, F. (2022). The service network scheduling problem. Technical Report, Quinlan School of Business, Loyola University Chicago. https://hal.archives-ouvertes.fr/hal-03598983/.

Hewitt, M., & Lehuédé, F. (2023). New formulations for the scheduled service network design problem. *Transportation Research Part B: Methodological, 172*, 117–133.

Lai, M., Cai, X., & Hall, N. G. (2022). Cost allocation for less-than-truckload collaboration via shipper consortium. *Transportation Science, 56*(3), 585–611.

Studio-CPLEX IICO (2022). Users manual-version 22 release 1.0.

Tang, X., Lehuédé, F., Péton, O., & Pan, L. (2019). Network design of a multi-period collaborative distribution system. *International Journal of Machine Learning and Cybernetics, 10*, 279–290.

Van Rossum, G., & Drake, F. L. (2010). *The python language reference. Python Software Foundation Amsterdam.*

Vielma, J. P., Ahmed, S., & Nemhauser, G. (2010). Mixed-integer models for nonseparable piecewise-linear optimization: unifying framework and extensions. *Operations Research, 58*(2), 303–315.

Multi-Layer Network Design for Consolidation-Based Transportation Planning

Teodor Gabriel Crainic

1 Introduction

Multilayered networks are used to represent complex natural, social, biological, and technological systems, where each layer stands for a particular component of the system, the components interacting in multiple types of relationships (Kivelä et al., 2014; Areta & Moreno, 2019). One finds a number of network types in the Operations Research literature corresponding to this broad definition, e.g., the *multi-echelon* (Cordeau et al., 2006), *multi-tier* (Crainic et al., 2004, 2009, 2021c), *multilevel* (Balakrishnan et al., 1994; Costa et al., 2011), and *hierarchical* (Obreque et al., 2010; Lin, 2010) networks. The network components within any one of these problem contexts and definitions correspond to particular networks making up a transportation, logistics, or telecommunication system, and interact through transfer links or nodes, providing the means to move flows over multi-component paths. Thus, for example, each component may correspond to the routes of a particular container-shipping navigation company in a regional or global intermodal study, transfers taking place in ports, or to the network of a particular public-transport mode in an urban transportation study, passengers transferring from one mode to another at common stations. It is noteworthy however that, in all these cases, the definitions of the links of a component do not depend on the definitions of the links of another component. A tramway link is not defined in terms of a path of bus or bicycle links, for example. Similarly, the routes between regional warehouses and city fulfillment terminals (first echelon) are not defined in terms of the route segments of delivery tours from the latter to the stores (second echelon).

T. G. Crainic (✉)
CIRRELT and Département d'analytique, opérations et technologies de l'information, École des sciences de la gestion, Université du Québec à Montréal, Montréal, QC, Canada
e-mail: TeodorGabriel.Crainic@cirrelt.net; crainic.teodor@uqam.ca

T. G. Crainic et al. (eds.), *Combinatorial Optimization and Applications*,
International Series in Operations Research & Management Science 358,
https://doi.org/10.1007/978-3-031-57603-4_9

The Operations Research *multi-layer network* term is generally associated to more complex problem settings, encountered in planning transportation and telecommunication systems, in which an arc in a given layer is defined with respect to a set of arcs in another layer, the arcs in the set often making up a path or a cycle. In freight railway tactical planning, for example, a block (group of cars handled together as a unit) is defined, for possible selection in a block layer, in terms of the path of train-service arcs which will transport it, if selected in the service layer (Chouman & Crainic, 2021; Zhu et al., 2014). Such interwoven definitions imply several connectivity relations and requirements in terms of both design (arc or node selection) and flow-distribution decisions, raising challenging network-design modelling and algorithmic issues.

Crainic et al. (2022) reviewed and synthesized the *Multi-layer Network Design* (*MLND*) literature, presenting a first version of a formal framework and taxonomy of the field. The authors focused on what may be described as the "basic" aspects of the multi-layer problem class, i.e., two layers and one-to-one connectivity requirements between two layers (even when more than two layers are considered), these requirements involving either the design or the flow decisions, but not both.

One continually witnesses, however, the emergence of studies addressing problems with more than two layers interacting in both design and flow decisions. The field of medium to long-time planning of consolidation-based freight transportation is particularly active in this respect, following the increasing interest in integrating the various aspects of the planning process within a comprehensive decision-support methodology. The simultaneous optimization of the service network, the demand-flow itineraries using this network, and the high-level management of resources supporting those services is a case in point.

We focus on these issues in this chapter. The goals are first, to recall and enhance the basic multi-layer definitions and formulations, second, to introduce the definitions and formulations for richer multi-layer networks, with more than two layers and connectivity relations involving several layers simultaneously, and, third, to illustrate these concepts in the context of multi-layer service network design models of consolidation-based freight transportation planning. Following a brief summary of the basic notation, definitions, and formulation of multi-layer network design making up Sect. 2, Sects. 3, 4, and 5 address these three goals, respectively. We conclude in Sect. 6.

2 Multi-Layer Network Design

We recall the basic notation, definitions, and formulation of multi-layer network design problems as formalized in Crainic et al. (2022).

A multi-layer network consists of several networks, one on each layer, coupled through *arc-definition* relations and *connectivity* settings and requirements. Inter-layer arcs complete the network, providing the means for the circulation of flows among the appropriate layers.

Inter-layer connectivity may involve two layers only or several. Most of the literature concerns the former case, as reviewed in Crainic et al. (2022). Hence, while the definitions in this section are general, we illustrate them using two layers only.

Let $\mathcal{L}$ be the set of layers of multi-layer network $\mathcal{G} = (\mathcal{N}_l, \mathcal{A}_l) = \bigcup_{l \in \mathcal{L}} \{\mathcal{G}_l = (\mathcal{N}_l, \mathcal{A}_l)\}$, where $\mathcal{G}_l$ is the network on layer $l \in \mathcal{L}$, with $\mathcal{N}_l$ and $\mathcal{A}_l$ the corresponding sets of nodes and arcs. Some of the nodes in $\mathcal{N}_l, l \in \mathcal{L}$, may belong exclusively to layer l, while others may be shared with the layers involved in the connectivity requirements. For simplicity of presentation, but without loss of generality, let's assume that all the arcs in a set $\mathcal{A}_l, l \in \mathcal{L}$, are design arcs, i.e., that they must be selected in order to be included in the final network.

Let $l, l' \in \mathcal{L}$, be two layers of $\mathcal{G}$ coupled by an arc-definition specifying how an arc in layer l' is related to a subset of arcs in layer l. According to Crainic et al. (2022), we then say that l' is *supported* by l and that l is *supporting* l'. To illustrate, consider the 2-layer network in Fig. 1, where arcs a, b, and c of the supported layer l' are defined by the sets of arcs (paths, actually) in the supporting layer l $(\beta, \delta, \epsilon)$, (α, ϵ), and $(\gamma, \delta, \epsilon)$, respectively.

The multi-layer network $\mathcal{G}$ is to be designed to satisfy the multicommodity, origin-destination (OD), demand $\mathcal{K} = \bigcup_{l \in \mathcal{L}} \mathcal{K}_l$. Notice that, most cases of interest either involve a single set of OD demands only, or one may easily transform the problem to a single-demand-set setting (e.g., when several demands of different product classes are defined between the same pair of origin-destination nodes, one duplicates these nodes and thus creates "new" OD commodities on the same layer Crainic et al., 2021a). Consequently, for simplicity of presentation but without loss of generality, we assume in the following that only one set of OD demands $\mathcal{K}$ is defined on a given layer $\varkappa \in \mathcal{L}$, each $k \in \mathcal{K}$ requiring the transportation of d^k units from its origin $O(k) \in \mathcal{N}_\varkappa$ to its destination $D(k) \in \mathcal{N}_\varkappa$ primarily through the arcs of $\mathcal{A}_\varkappa$. Notice that, the flows on layer $\varkappa$ may be projected on the layers associated with it through arc definitions. In the railway planning case invoked above, for example, flows defined and circulating on the block layer may be summed up to compute the total railcar flow on the train service layer. Sections 3 and 4 detail such connectivity settings and the associated constraints.

The *arc-definition relations* specify that an arc in a given layer is tightly related to (defined by) a set of arcs in at least another layer. The arcs on a defining layer

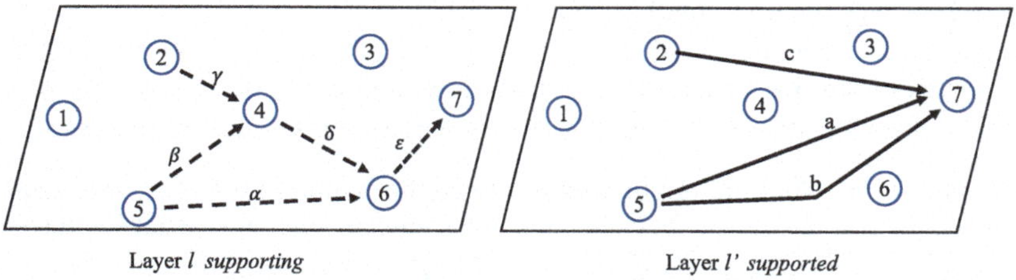

Fig. 1 Illustration of arc-definition connectivity

are often, but not always, making up a path or a cycle. Turning again to the railway tactical planning example, a block arc is defined by a path of train-service arcs making up its route from origin to destination.

The *connectivity requirements* specify the degree and type of relations between the arc decision variables and attribute values of coupled (*supporting*, *supported*) layers, yielding constraints in the related MLND formulations (Sects. 3 and 4).

The connectivity *degree* indicates whether two or more layers are involved and the direction of involvement:

One-to-one: Two layers, one supporting and the other being supported (Fig. 1); A pair of such (supporting, supported) layers in $\mathcal{G}$ is denoted (l, l'), $l, l' \in \mathcal{L}$;

Many-to-one: Several, $m + 1$, layers involved; Represented by the couple $(\mathcal{L}(l'), l')$, it refers to the relations between $m > 1$ supporting layers, gathered in set $\mathcal{L}(l') \subset \mathcal{L}$, and a supported layer $l' \in \mathcal{L}$; The design of a train service (on a supported layer) which may be operated by engines of three different types, the activity of each type being represented on a different supporting layer, illustrates the case;

One-to-many: Several, $m+1$, layers involved; Represented by the couple $(l, \mathcal{L}(l))$, it refers to the relations between a supporting layer $l \in \mathcal{L}$ and $m > 1$ supported layers, gathered in set $\mathcal{L}(l) \subset \mathcal{L}$; The simultaneous design, on the same infrastructure, of a passenger-train service network and a freight-train service network (on two supported layers), which are to be operated by engines of a unique fleet, managed on a supporting layer, illustrates the case.

Let $C = \{(l, l') \cup (\mathcal{L}(l'), l') \cup (l, \mathcal{L}(l)),\ l, l' \in \mathcal{L}\}$ be the *connectivity-requirement* set of groups of *supporting* and *supported* layers in $\mathcal{G}$. We define three types of layer connectivity relations among the layer groups in C (Crainic et al., 2022, introduced the first two):

Design The selection of an arc in a layer requires the selection of all or some of its supporting arcs in another layer (at least one, for the many-to-one case); Thus, to open arc c in the supported layer l' of Fig. 1, one must open all the arcs defining it, i.e., γ, δ, ϵ, in the supporting layer l. Design connectivity is fundamental in the definition of multi-layer networks.

Flow The flow on an arc in a layer is a function of the flows on a set of arcs of at least another layer; Thus, for example, the flow on arc ϵ in layer l (Fig. 1), supporting several arcs, a, b, and c, in the supported layer l', could be the sum of the flows on these three arcs.

Attribute Generalizing the flow type of connectivity, it concerns the relations between the attributes, e.g., cost, time, and capacity, of the links of the layers involved; Sect. 4.1 further investigates this connectivity type.

Let arcs $a \in \mathcal{A}_l, l \in \mathcal{L}$, be characterized by a fixed cost f_{al}, commodity-specific unit flow costs c^k_{al}, $k \in \mathcal{K}$, and flow capacity u_{al} (as usual in network design, commodity-specific capacities u^k_{al}, $k \in \mathcal{K}$, and associated constraints, may be defined; for simplicity of presentation but without loss of generality, we do not include those in the following).

A generic multicommodity, fixed-cost, capacitated multi-layer network design formulation may then be introduced (Crainic et al., 2022) with the following decision variable vectors:

Design $\mathbf{y} = [y_{al}] \in \mathbb{Y}$, where $y_{al} = 1$ if arc $a \in \mathcal{A}_l$ of layer l is selected, 0, otherwise; Alternatively, $y_{al} \in \mathbb{N}$ when the arc may be selected more than once (e.g., the departure frequency of a selected transportation service);

Flow $\mathbf{x} = [x^k_{al}] \in \mathbb{X}$ indicating the quantity of demand $k \in \mathcal{K}$ assigned to arc a of layer l; Depending upon the application, the flow variable may be continuous or integer, but always non-negative.

Let $\mathcal{A}^+_l(i)$ and $\mathcal{A}^-_l(i)$ represent the sets of outgoing and incoming arcs of node $i \in \mathcal{N}_l$, and w^k_i equal 1 if $i = O(k)$, -1 if $i = D(k)$, and 0, otherwise, for all $i \in \mathcal{N}_l, k \in \mathcal{K}, l \in \mathcal{L}$. The formulation may then be written as

$$\min \sum_{l \in \mathcal{L}} \left\{ \sum_{a \in \mathcal{A}_l} f_{al} y_{al} + \sum_{k \in \mathcal{K}} \sum_{a \in \mathcal{A}_l} c^k_{al} x^k_{al} \right\} \tag{1}$$

$$\text{s.t.} \sum_{a \in \mathcal{A}^+_l(i)} x^k_{al} - \sum_{a \in \mathcal{A}^-_l(i)} x^k_{al} = w^k_i, \qquad i \in \mathcal{N}_l, k \in \mathcal{K}, l \in \mathcal{L}, \tag{2}$$

$$\sum_{k \in \mathcal{K}} x^k_{al} \le u_{al} y_{al}, \qquad a \in \mathcal{A}_l, \tag{3}$$

$$(\mathbf{y}, \mathbf{x}) \in (\mathcal{Y}, \mathcal{X})_{ll'}, \qquad (l, l') \in C, l \in \mathcal{L}, \tag{4}$$

$$\mathbf{y} \in \mathbb{Y}, \mathbf{x} \in \mathbb{X}. \tag{5}$$

The objective function (1) minimizes the total cost of selecting and using arcs on all the layers of the network. Constraints (2) are the classical *flow-conservation* equations ensuring that the demand flows are routed from their origins to their destinations in each layer, without gains or losses at intermediary nodes. Notice that, flow conservation has to be enforced on a layer only, usually the layer of definition, when the projections on the other layers are simple aggregation (Sect. 3). For example, when demand is defined on the supported layer (blocks in the railway planning case), flows on the supporting layer arcs (train services) equal the sum of the flows on the corresponding supported arcs (the blocks making up the train).

The aggregated *linking capacity* constraints (3) ensure that only selected, and paid for, arcs are used, enforcing the associated capacity restriction with respect to the total flow assigned to each one of them. Relations (5) define the domains of the decision variables. (Particular so-called network design side constraints, e.g., limiting the total budget or imposing topological design conditions on particular layers, such as the equality between the number of entering and exiting design arcs at nodes, are not discussed in this chapter as they are not related to the multi-layer nature of the problem; we refer the interested reader to Crainic et al., 2021a).

Relations (4) stand for the sets of constraints corresponding to the design, flow, or attribute connectivity requirements proper to the multi-layer network design application at hand. We discuss general constraint classes in Sects. 3 and 4.

3 MLND Basic Connectivity Requirements and Constraints

The so-called basic connectivity definitions and requirements are found in 2-layer network design problem settings, which were the object of most of the early contributions. We focus in this section on design and flow-connectivity constraints (see Crainic et al., 2022, for the initial definitions and particular applications).

The set C then reduces to $\{(l, l'),\ l, l' \in \mathcal{L}\}$, that is, the set of (supporting, supported) pairs of layers in $\mathcal{G}$ involved in one-to-one connectivity requirements (the only type possible in this context). We use the notation α, β, etc. for arcs on a supporting layer $l \in C$ and a, b, etc. for arcs on a supported layer $l' \in C$. We then define the following sets:

$\mathcal{A}_{l'}(\alpha l)$ Set of *supported arcs* in layer l' by the supporting arc α in layer l, i.e., $\mathcal{A}_{l'}(\alpha l) = \{a \in \mathcal{A}_{l'} \mid a \text{ is supported by } \alpha\}$, $\mathcal{A}_{l'}(\alpha l) \subseteq \mathcal{A}_{l'}$;

$\mathcal{A}_l(al')$ Set of *supporting arcs* in layer l of the supported arc a in layer l', i.e., $\mathcal{A}_l(al') = \{\alpha \in \mathcal{A}_l \mid \alpha \text{ supports } a\}$, $\mathcal{A}_l(al') \subset \mathcal{A}_l$; This corresponds to the arc definition (without its topology) of a.

Design-connectivity constraints enforce existence relations between supported and supporting arcs.

Also called *all-design* linking constraints for supported arcs, the relation (6) states that all the supporting arcs (e.g., α and ϵ of layer l of Fig. 1) must be selected in order for a supported arc (b of layer l') to be eligible for selection,

$$y_{al'} \leq y_{\alpha l},\quad \alpha \in \mathcal{A}_l(al'), a \in \mathcal{A}_{l'}, (l, l') \in C, \tag{6}$$

while *min-design* linking constraints (7) are introduced when a single supporting arc has to be selected only, in order for the supported arc to be eligible for selection,

$$y_{al'} \leq \sum_{\alpha \in \mathcal{A}_l(al')} y_{\alpha l},\quad a \in \mathcal{A}_{l'}, (l, l') \in C. \tag{7}$$

Figure 2 illustrates fundamental occurrences of design-connectivity relations and the utilization of constraints (6) and (7). The lower-left side of the figure illustrates the case where arc a of supported layer l' is defined by a link, in other words, a single-link path $\{\alpha\}$, and a path, $\{\delta, \beta\}$, in the supporting layer l. Assuming link a may exist if either one of the two paths is selected, the design-connectivity is enforced through a type (7) constraint for the alternative definitions

$$y_{al'} \leq y_{\alpha l} + y_{\delta l} + y_{\beta l}, \tag{8}$$

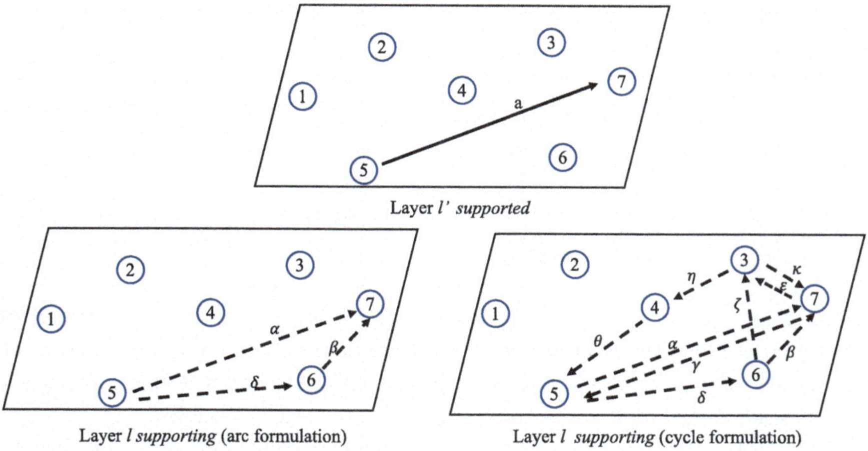

Fig. 2 Illustration min-design linking constraints

and two type-(6) constraints to enforce the two-link path definition

$$\begin{aligned} y_{al'} &\leq y_{\delta l}, \\ y_{al'} &\leq y_{\beta l}. \end{aligned} \tag{9}$$

Obviously, the last two constraints may be compressed into a single one when design variables are defined for path selection in the supporting layer.

The lower-right side of Fig. 2 generalizes the previous case and illustrates the situation where a supporting arc or path defining an arc in a supported layer is part of one or several larger network structures, e.g., cycles. Thus, we see arc α belonging to cycle $\{\alpha, \varepsilon, \eta, \varphi\}$, arc β being part of $\{\delta, \beta, \gamma\}$, while arcs δ and γ belong not only to the latter cycle but also to $\{\delta, \zeta, \kappa, \gamma\}$. Such a case is often encountered in service network design applications when any of a number of possible resources (generally operating cycling sequences of services) may be used to operate a service (Crainic & Hewitt, 2021). Notice that the constraints (8) and (9) are sufficient in this case as well, the cycle structures being ensured in the supporting layer, usually through cycle-selection (design) decision variables and constraints imposing that each arc belongs at most to a selected cycle only.

When a supported arc requires the support of a given number $n_{al'}$ of supporting arcs, the min-design linking constraints take the form

$$n_{al'} y_{al'} \leq \sum_{\alpha \in \mathcal{A}_l(al')} y_{\alpha l}, \quad a \in \mathcal{A}_{l'}, (l, l') \in \mathcal{C}. \tag{10}$$

Design-connectivity issues from the point of view of the supporting layers impose particular constraints as well. Thus, the *min-design* linking constraints (11), or (12), enforce the requirement that selecting an arc supporting one or several arcs in a different layer implies that at least one, or $n_{\alpha l} \leq |\mathcal{A}_{l'}(\alpha(l))|$, respectively, of those must be selected as well:

$$y_{\alpha l} \leq \sum_{a \in \mathcal{A}_{l'}(\alpha l)} y_{al'}, \quad \alpha \in \mathcal{A}_l, (l, l') \in C, \tag{11}$$

$$n_{\alpha l} y_{\alpha l} \leq \sum_{a \in \mathcal{A}_{l'}(\alpha l)} y_{al'}, \quad \alpha \in \mathcal{A}_l, (l, l') \in C. \tag{12}$$

Note that, min-design linking constraints are not always present, at least, not in transportation-planning applications. Thus, for example, a link on a resource layer could not be supporting any link in a supported service network design layer when it corresponds to an empty move to reposition the resource as needed.

Finally, it is noteworthy that design requirements between two layers may imply not only the existence of one given the other, but also the amplitude of the relation. This characteristic, proper to multi-layer networks, is captured through the *design capacity* $v_{\alpha l}$ of a supporting arc $\alpha \in \mathcal{A}_l$, limiting the design-related measures of its supported arcs. In its simpler expression, it limits the number of supported arcs, which may be selected simultaneously, by imposing the *design-capacity* constraints

$$\sum_{a \in \mathcal{A}_{l'}(\alpha l)} y_{al'} \leq v_{\alpha l} y_{\alpha l}, \quad \alpha \in \mathcal{A}_l, (l, l') \in C. \tag{13}$$

Thus, setting $v_{\alpha l} = 2$ to the arcs in layer l of Fig. 1 and imposing constraints (13) means that at maximum two of the three arcs a, b, and c on layer l' may be selected. Such a design-capacity constraint on the attributes of a given train service (supporting layer) in a freight railway tactical planning application could thus limit the total number of different blocks (supported layers) assigned to it.

Flow-Connectivity refers to relations between the flows on the arcs of a (supporting, supported) layer pair.

Flow-accumulation constraints address the case when the demand $\mathcal{K}$ is defined on the supported layer, i.e., $\varkappa = l'$, and state that the commodity flow on a supporting arc equals the sum of that commodity flows on all its supported arcs:

$$x^k_{\alpha l} = \sum_{a \in \mathcal{A}_{l'}(\alpha l)} x^k_{al'}, \quad \alpha \in \mathcal{A}_l, \ k \in \mathcal{K}, (l, l') \in C. \tag{14}$$

Recall that, according to the particular application, demand flows may or may not be split ("bifurcated" in the telecommunication literature), that is, each commodity may either flow on several paths between its origin and destination, or must move through a single path, respectively. When flows cannot be split on the supported layer, the supporting layer inherits the property through the flow-accumulation constraints.

Notice that, when demand is measured in the same units as the capacity of the supporting arc, e.g., number of vehicles or tons, combining the flow-accumulation relations (14) and the linking-capacity constraints (3) enforces the requirement that the sum of the flows on all supported arcs does not violate the supporting-arc capacity.

When this is not the case, that is, when the demand unit of measure, e.g., the container or railcar, is different from the capacity measure of the supporting arc, e.g., length (very frequent in railway planning), explicit *flow-attribute linking-capacity* constraints must be added (complementary to the linking-capacity constraints (3)). Let φ_k be the attribute of commodity $k \in \mathcal{K}$ corresponding to the measure of capacity $u_{\alpha l}$ of arcs in a supporting layer $l \in \mathcal{L}$. The flow-attribute linking-capacity constraints may then be written as:

$$\sum_{a \in \mathcal{A}_{l'}(\alpha l)} \sum_{k \in \mathcal{K}} \varphi_k x^k_{al'} \leq u_{\alpha l} y_{\alpha l}, \quad \alpha \in \mathcal{A}_l, (l, l') \in C. \tag{15}$$

4 Designing Richer Multilayer Networks

Rich MLND problems involve different, more complex arc-definition relations and connectivity requirements compared to the basic problem setting discussed in the previous section. Connectivity requirements among the attributes of the arcs involved in a (supporting, supported) layer relation are discussed in Sect. 4.1. Arc-definition relations involving more than two layers, and the associated connectivity requirements, are the topic of Sect. 4.2. It is noteworthy that rich MLND applications often include several arc definitions and connectivity requirements in the same formulation as discussed in Sect. 5.

4.1 Attribute Connectivity

Network links may display a wide variety of attributes. Commodity flows together with fixed and unit commodity transportation costs are among the most usual arc attributes, where cost may indeed be a monetary value, but may also be distance or time. It is not unusual to have these three cost attributes simultaneously within an application, while other "cost" measures may also be found, somewhat less frequently (e.g., the touristic interest of the urban zone represented by the arc), however. *Capacity* is another widely encountered link attribute limiting the flow of commodities, total, commodity-specific, or both, one may assign to the arc.

Section 3 addressed inter-layer relations in terms of flows. What about relations among the other attributes, including costs and the "physical" attributes mentioned

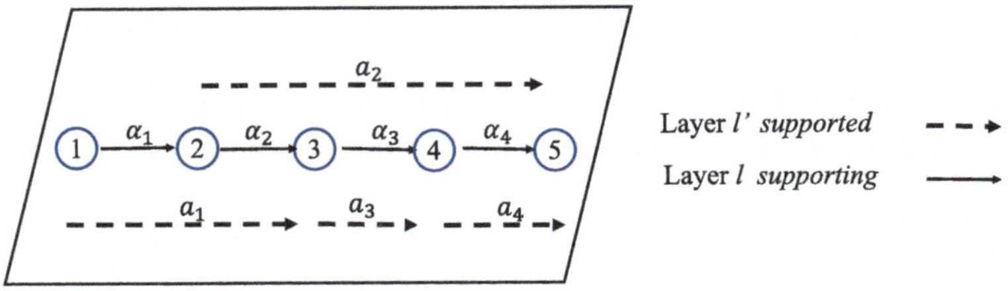

Fig. 3 Attribute connectivity illustration

above? We aim to start answering this question in this section. The (supporting, supported) arcs of Fig. 3 serve as illustration. The supporting layer l has four arcs, which support four arcs in layer l', such that $\mathcal{A}_l(a_1) = \{\alpha_1, \alpha_2\}$, $\mathcal{A}_l(a_2) = \{\alpha_2, \alpha_3, \alpha_4\}$, $\mathcal{A}_l(a_3) = \{\alpha_3\}$, and $\mathcal{A}_l(a_4) = \{\alpha_4\}$.

We first notice that not all attributes of interacting-layer arcs are always related. Thus, for example, the fixed costs of supported arcs are often defined independently of those of the supporting ones. Second, many such relations are of a definitional and additive nature. *Distance* generally belongs to this class of attributes, as do unit commodity costs, *time*-related measures, e.g., duration, transport, delay, and waiting times, and most node measures.

We thus introduce the *supported arc additive-attribute* definition stating that the value $F_\varrho(\cdot)$ of any additive attribute ϱ of a supported-layer arc a is given by the sum of the values of the corresponding attributes of the supporting arcs, i.e.,

$$F_\varrho(al') = \sum_{\alpha \in \mathcal{A}_l(al')} F_\varrho(\alpha l), \;\; \varrho, \; a \in \mathcal{A}_{l'}. \tag{16}$$

Definition (16) must be verified for the relevant attributes when all the potential arcs on the supporting and supported layers are given as input in the problem setting, e.g., when a potential block in a railway planning application is defined *a priori* as moving on a given sequence of potential train services. The situation is different when the arcs of the supported layer are to be dynamically generated during problem solving, in which case, constraints of the (16) type have to be included in the variable-generation model or procedure.

A different case is observed when addressing arc capacity, as feasibility issues have to be addressed. Similar to the discussion above, when the potential supported arcs are pre-defined, verifying that their capacities are not higher than the lowest capacity among the respective supporting arcs, i.e.,

$$u_{al'} \leq \min_{\alpha \in \mathcal{A}_l(al')} \{u_{\alpha l}\}, \;\; a \in \mathcal{A}_{l'}, \tag{17}$$

guarantees feasibility, together with intra-layer linking (3), the (supporting, supported) all-design design-connectivity (6), and the flow-connectivity, e.g., (14), constraints of the MLND model.

The situation is less straightforward when the arc capacities in the supported layer are not known *a priori*, rather belonging to the set of decisions characterizing the problem setting. In such cases, the capacity of a supported arc a in layer l' becomes a decision variable, its "optimal" value to be determined by the interplay among the design decisions in both layers, the capacity of each supporting arc $\alpha \in \mathcal{A}_l(al')$, $(l, l') \in C$, and the allocation of the latter to all the arcs it supports, that is, $a \in \mathcal{A}_{l'}(\alpha l)$. The *multi-layer network design problem with capacity decisions* then aims to determine simultaneously the selection of the design arcs on all layers, the arc capacities on the supported layer, and the distribution of demand flows over the resulting multi-layer network, to minimize the total generalized cost of the system.

To illustrate, consider the case of railway tactical planning, where a train service of known capacity (length or number of cars, usually based on demand as well as physical and operational policies) may haul one or several of a given set of blocks. While each of the latter has a maximum capacity (based, e.g., on the physical characteristics of the terminal where it is built), its actual capacity cannot be higher than the residual capacity of the service once the capacities of the other blocks selected to be moved by the service have been determined. The tactical plan should therefore provide not only the selected service network and the composition of each service in terms of the blocks it hauls, but also indications regarding the maximum volume of demand which may be assigned to each block. The latter information is important during operations when the actual blocks and trains are built, particularly when the blocks are not built within the railway's own terminals (e.g., in the port terminal by port crews working with guidelines provided by the railway). One has, therefore, to allocate the capacity of the service, the supporting arc, to blocks it hauls, the supported arcs, and this has to be done for all supporting and supported arcs simultaneously with their own selection.

Define the decision variables $\vartheta_{al'}$, $a \in \mathcal{A}'$, the capacity of arc a in the supported layer l', and let $c_{al'}$ be the unit cost of assigning capacity to arc a. The range of $\vartheta_{al'}$:

$$0 \leq \vartheta_{al'} \leq u_{al'} = \min_{\alpha \in \mathcal{A}_l(al')} \{u_{\alpha l}\}. \tag{18}$$

When $|\mathcal{A}_{l'}(\alpha l)| = 1$, for $\alpha \in \mathcal{A}_l(al')$, i.e., when there are no other arcs in layer l' competing for the capacity of the supporting arc α, constraints (6) and the feasibility range (18) are sufficient.

Shared-capacity requirements and constraints are involved when, on the contrary, the supporting-arc capacity must be allocated among several supported arcs, i.e., when $|\mathcal{A}_{l'}(\alpha l)| > 1$, each allocation being bounded from above by the *residual* capacity of the supporting arc given the allocations to the other supported arcs.

From the point of view of the supported arcs, shared-capacity requirements may be represented by:

$$\vartheta_{al'} \leq \min_{\alpha \in \mathcal{A}_l(al')} \left\{ u_{\alpha l} y_{\alpha l} - \sum\nolimits_{b \in \mathcal{A}_{al'}(\alpha l) \setminus \{a\}} \vartheta_{bl'} y_{bl'} \right\}, \quad a \in \mathcal{A}_{l'}, (l, l') \in C, \tag{19}$$

where $\mathcal{A}_{al'}(\alpha l) \subseteq \mathcal{A}_{l'}(\alpha l)$ stands for the set of arcs supported by α which includes a. Recalling that, the (supporting, supported) all-design design-connectivity (6) constraints also apply in this case, the previous expression becomes

$$\vartheta_{al'} \leq \min_{\alpha \in \mathcal{A}_l(al')} \left\{ u_{\alpha l} - \sum\nolimits_{b \in \mathcal{A}_{al'}(\alpha l) \setminus \{a\}} \vartheta_{bl'} y_{bl'} \right\}, \quad a \in \mathcal{A}_{l'}, (l, l') \in C. \tag{20}$$

Indeed, when the supporting arc is not selected, i.e., $y_{\alpha l} = 0$, all the supported arcs are not selected either, i.e., $y_{al'} = y_{bl'} = 0$, and the corresponding arc flows are also zero. Any capacity distribution satisfying relations (18) would do in this case, in particular that of an initial solution setting $\vartheta_{al'} = 0$. Moreover, when $c_{al'} > 0$, any solution with a positive capacity on such an arc will have a higher cost, hence not optimal, compared to a solution with zero capacity. Hence, to illustrate, the capacities of the four supported arcs in Fig. 3 would have to satisfy the system of non-linear relations:

$$\begin{aligned}
\vartheta_{a_1} &\leq u_{\alpha_1} \\
\vartheta_{a_1} &\leq u_{\alpha_2} - \vartheta_{a_2} y_{a_2} \\
\vartheta_{a_2} &\leq u_{\alpha_2} - \vartheta_{a_1} y_{a_1} \\
\vartheta_{a_2} &\leq u_{\alpha_3} - \vartheta_{a_3} y_{a_3} \\
\vartheta_{a_2} &\leq u_{\alpha_4} - \vartheta_{a_4} y_{a_4} \\
\vartheta_{a_3} &\leq u_{\alpha_3} - \vartheta_{a_2} y_{a_2} \\
\vartheta_{a_4} &\leq u_{\alpha_4} - \vartheta_{a_2} y_{a_2}.
\end{aligned}$$

One can also examine the shared-capacity requirements from the point of view of the supporting-layer arcs, which yields relations:

$$\sum\nolimits_{a \in \mathcal{A}_{l'}(\alpha l)} \vartheta_{al'} y_{al'} \leq u_{\alpha l}, \quad \alpha \in \mathcal{A}_l, \ (l, l') \in C, \tag{21}$$

where the $y_{\alpha l}$ design variables on the supporting layer are dropped given the all-design design-connectivity relations. A two-layer network design with capacity decisions formulation may then be written (recall that, $\mathcal{L} = \{l, l'\}$ with $C = \{(l, l')\}$ being the couple of (supporting, supported) layers):

$$\min \sum_{\alpha \in \mathcal{A}_l} f_{\alpha l} y_{\alpha l} + \sum_{a \in \mathcal{A}_{l'}} (f_{al'} y_{al'} + c_{al'} \vartheta_{al'}) + \sum_{k \in \mathcal{K}} \sum_{\varkappa \in \{l, l'\}} \sum_{\alpha \in \mathcal{A}_\varkappa} c^k_{a\varkappa} x^k_{a\varkappa} \tag{22}$$

$$\text{s.t.} \sum_{a \in \mathcal{A}^+_{l'}(i)} x^k_{al'} - \sum_{a \in \mathcal{A}^-_{l'}(i)} x^k_{al'} = w^k_i, \qquad i \in \mathcal{N}_{l'}, k \in \mathcal{K}, \tag{23}$$

$$\sum_{k \in \mathcal{K}} x^k_{al'} \le \vartheta_{al'} y_{al'}, \qquad a \in \mathcal{A}_{l'}, \tag{24}$$

$$\sum_{k \in \mathcal{K}} x^k_{\alpha l} = \sum_{k \in \mathcal{K}} \sum_{a \in \mathcal{A}_{l'}(\alpha l)} x^k_{al'} \le u_{\alpha l} y_{\alpha l}, \qquad \alpha \in \mathcal{A}_l, \tag{25}$$

$$\sum_{a \in \mathcal{A}_{l'}(\alpha l)} \vartheta_{al'} y_{al'} \le u_{\alpha l} y_{\alpha l}, \qquad \alpha \in \mathcal{A}_l, \tag{26}$$

$$y_{al'} \le y_{\alpha l}, \qquad a \in \mathcal{A}_{l'}, \alpha \in \mathcal{A}_l(al'), \tag{27}$$

$$0 \le \vartheta_{al'} \le u_{al'}, \qquad a \in \mathcal{A}_{l'}, \tag{28}$$

$$\mathbf{y} \in \mathbb{Y}, \mathbf{x} \in \mathbb{X} \qquad . \tag{29}$$

A linear version of this formulation is obtained by, first, using (24) to replace (26) with

$$\sum_{a \in \mathcal{A}_{l'}(\alpha l)} \sum_{k \in \mathcal{K}} x^k_{al'} \le \sum_{a \in \mathcal{A}_{l'}(\alpha l)} \vartheta_{al'} y_{al'} \le u_{\alpha l} y_{\alpha l}, \quad \alpha \in \mathcal{A}_l, \tag{30}$$

and, second, by replacing constraints (24) and (28) with the sets of constraints

$$\sum_{k \in \mathcal{K}} x^k_{al'} \le \vartheta_{al'}, \qquad a \in \mathcal{A}_{l'}, \tag{31}$$

$$\vartheta_{al'} \le u_{al'} y_{al'}, \qquad a \in \mathcal{A}_{l'}, \tag{32}$$

$$\vartheta_{al'} \ge 0, \qquad a \in \mathcal{A}_{l'}. \tag{33}$$

Rouhani et al. (2024) further detail these issues and present Benders decomposition solution methods.

4.2 *Multi-Layer Connectivity*

As briefly mentioned in Crainic et al. (2022), multi-layer network-design models with more than two design layers are increasingly proposed when addressing the planning and management of complex systems in as a comprehensive way as possible. This trend is particularly observed in consolidation-based freight transportation planning (Sect. 5).

The concepts, notation, and definitions presented earlier in this section certainly also apply to L-layer networks with $L > 2$, in particular the one-to-one connectivity discussed in Sects. 3 and 4.1. On the other hand, inter-layer relations may involve more than two layers as, for example, when a transportation service (supported layer) may be operated by one of two (or more) different resource types (the supporting layers) or even a combination of those, as discussed in more detail in Sect. 5. We therefore generalize the previous definitions and formulations to many-to-one and one-to-many connectivity relations in L-layer networks with $|\mathcal{L}| > 2$.

Three general *many-to-one design-connectivity* classes may be encountered when the arcs of a supported layer are defined in terms of several supporting layers: exclusive, required, and complementary.

The arc-definitions of an arc on a supported layer, $l' \in \mathcal{L}$, in terms of the arcs of the supporting layers $l \in \mathcal{L}(l')$, together with the design-connectivity requirements, are said to be *exclusive* when at most one of these definitions may be selected. Hence, for example, only one of a given number of ship types (supporting layers) may be assigned to a particular navigation line/service (supported layer).

Additional decision variables must be added to the MLND formulations to capture such exclusive design-assignment decisions. Let $y_{al'l} = 1$ if the $\mathcal{A}_l(al'), l \in \mathcal{L}(l')$, definition of arc $a \in \mathcal{A}_{l'}$ is selected, 0, otherwise. The *design-exclusive-assignment* arc-definition Eq. (34), together with constraints (6), (7), or (10) written for the new $y_{al'l}$ variables, enforce the exclusivity of the connectivity relations,

$$y_{al'} = \sum_{l \in \mathcal{L}(l')} y_{al'l}, \quad a \in \mathcal{A}_{l'}, (\mathcal{L}(l'), l') \in C. \tag{34}$$

The supporting-supported layer relation is *required* when at least an arc must be selected on each of the supporting layers in $\mathcal{L}(l')$ for the supported arc l' to be selected. Thus, for example, both traction-power units, e.g., locomotives or tractors, and crews are required to operate rail or road freight transportation services, respectively. This translates into

$$n_{al'l} y_{al'} = y_{\alpha l}, \quad \alpha \in \mathcal{A}_l(al'), (\mathcal{L}(l'), l') \in C, \tag{35}$$

where, $n_{al'l}$ generalizes the definition of Sect. 3, to indicate the required number of arcs in the supporting layer l, assuming of course that $|\mathcal{A}_l(al')| > 1$.

Notice that, this formulation assumes binary arc-selection decision variables in layer l, which, when $n_{al'l} > 1$, implies parallel supporting arcs in layer l, as illustrated in Fig. 2. Alternatively, non-negative integer values may be allowed for the arc-design variables in the supporting layer l, modelling, for example, the number of resource units moving on the link. The required design-connectivity relations may then be modelled as constraints (36), together with the *design-utilization* constraints (37) indicating how the intensity on the supporting arc is shared among the supported ones (design-capacity constraints (13) may complete the model):

$$n_{al'l} y_{al'} \leq y_{\alpha l}, \quad \alpha \in \mathcal{A}_l(al'), (\mathcal{L}(l'), l') \in C, \tag{36}$$

$$y_{\alpha l} = \sum_{l' \in \mathcal{L}(l)} \sum_{a \in \mathcal{A}_{l'}(\alpha l)} n_{al'l} y_{al'}, \quad \alpha \in \mathcal{A}_l, l \in C. \tag{37}$$

Complementary, or additive (Crainic et al., 2022), arc-definitions and design-connectivity requirements generally stand for the possibility to select more than one definition for an arc on a supported layer $l' \in \mathcal{L}$, among its supporting layers $l \in \mathcal{L}(l')$. Hence, for example, a freight train service may be defined in terms of power units of, say, three types of locomotives, yielding three layers similar to the one illustrated in Fig. 2. This translates into *design-complement-assignment* arc-definition Eq. (38):

$$y_{al'} \leq \sum_{l \in \mathcal{L}(l')} y_{al'l}, \quad a \in \mathcal{A}_{l'}, (\mathcal{L}(l'), l') \in C. \tag{38}$$

It seems clear, however, that these constraints would not capture the actual setting in transportation applications, as the same service cannot be operated simultaneously with different definitions. In such cases, complementary design connectivity also models the requirements in terms of how to combine the selected definitions to achieve a certain level of particular attributes of the supported arc, e.g., a service requiring a certain minimal power, which can be obtained through different combinations of resource types.

Let $\varphi_{al'}$ be the intensity of some specific attribute (power in the previous example) the arc a on layer $l' \in \mathcal{L}$ must achieve through a combination of arcs on a subset of its supporting layers $l \in \mathcal{L}(l')$, and let $\varphi_{ll'}$ be the value of the corresponding attribute on an arc $\alpha \in \mathcal{A}_l(al')$ (these values may be different for different (l, l') pairs). The *design-complement* constraints then become

$$\varphi_{al'} y_{al'} \leq \sum_{l \in \mathcal{L}(l')} \varphi_{ll'} y_{al'l}, \quad a \in \mathcal{A}_{l'}, (\mathcal{L}(l'), l') \in C. \tag{39}$$

The $y_{al'l}$ assignment decision variables are defined as non-negative integers when several "units" of supporting arcs have to be assigned to a supported arc in order to achieve the required attribute value.

The supporting-layer-specific design-connectivity requirements and constraints (11)–(13), defined for the one-to-one connectivity degree class (Sect. 3), apply straightforwardly to both many-to-one and one-to-many classes. Moreover, in the latter case, one can generalize them to (1) force that at least a certain number of supported links be selected in order to select the supporting link, constraints (40) or, (2), limit the number of supported links one may select, constraints (41):

$$n_{\alpha l} y_{\alpha l} \leq \sum_{l' \in \mathcal{L}(l)} \sum_{a \in \mathcal{A}_{l'}(\alpha l)} y_{al'}, \quad \alpha \in \mathcal{A}_l, (l, \mathcal{L}(l)) \in C, \tag{40}$$

$$\sum_{l' \in \mathcal{L}(l)} \sum_{a \in \mathcal{A}_{l'}(\alpha l)} y_{al'} \leq v_{\alpha l} y_{\alpha l}, \quad \alpha \in \mathcal{A}_l, (l, \mathcal{L}(l)) \in C. \tag{41}$$

Similar generalizations apply to the *flow-connectivity* relations and constraints (14), yielding the *total-flow-accumulation* constraints (42). One may also define, when relevant, layer-specific flow-accumulation decision variables, $x_{\alpha l}^{kl'} \geq 0, \alpha \in \mathcal{A}_l, k \in \mathcal{K}, (l, \mathcal{L}(l)) \in C$, with $x_{\alpha l}^{k} = \sum_{l' \in \mathcal{L}(l)} x_{\alpha l}^{kl'}$. The *layer-specific flow-accumulation* constraints then become Eq. (43). Finally, flow-attribute capacity connectivity may also be generalized, yielding constraints (44).

$$x_{\alpha l}^{k} = \sum_{l' \in \mathcal{L}(l)} \sum_{a \in \mathcal{A}_{l'}(\alpha l)} x_{al'}^{k}, \quad \alpha \in \mathcal{A}_l, k \in \mathcal{K}, (l, \mathcal{L}(l)) \in C, \tag{42}$$

$$x_{\alpha l}^{kl'} = \sum_{a \in \mathcal{A}_{l'}(\alpha l)} x_{al'}^{k}, \quad \alpha \in \mathcal{A}_l, k \in \mathcal{K}, (l, \mathcal{L}(l)) \in C, \tag{43}$$

$$\sum_{l' \in \mathcal{L}(l)} \sum_{a \in \mathcal{A}_{l'}(\alpha l)} \sum_{k \in \mathcal{K}} \varphi_k x_{al'}^{k} \leq u_{\alpha l} y_{\alpha l}, \quad \alpha \in \mathcal{A}_l, (l, \mathcal{L}(l)) \in C. \tag{44}$$

5 Multi-Layer Network Design for Consolidation-based Freight Transportation Planning

From its first appearance in Western vocabulary in the fifteenth century, *consolidation*, in the broadest sense of the term, means to join together several diverse things into one whole and, thus, to unite and strengthen that whole (Merriam-Webster, 2023). The concept and process of consolidation is used extensively in a large gamut of human activity fields, particularly in transportation where it refers to grouping together freight loads, vehicles, or people for economically and operationally efficient transport for the entire, or part of, the journey between their respective origins and destinations. Consolidation is performed by passenger or freight transport companies and systems, commonly identified as *carriers* (Crainic & Laporte, 1997). Public-transport carriers, operating urban (bus, trolleybus, light rail, collective taxi, etc.) or inter-urban (coach, rail, airplane, ship, etc.) networks, consolidate passengers who do not desire or cannot move by a dedicated vehicle (Mauttone et al., 2021).

Less-than-truckload (LTL) motor carriers, railroads, ocean/maritime liner navigation companies, land- and water (coastal, river, etc)-based intermodal carriers, postal and small-package transportation companies perform consolidation-based services for freight (Crainic, 2003; Crainic & Kim, 2007; Crainic et al., 2021b; Crainic & Hewitt, 2021). Consolidation is also at the heart of innovative transportation-system business and operation models introduced for urban, e.g., City Logistics (Crainic et al., 2023; Crainic & Montreuil, 2023) and interurban, e.g., Physical Internet (Ballot et al., 2014), Synchromodality (Ambra et al., 2021; Giusti et al., 2019), and multi-stakeholder (Taherkhani et al., 2022) settings.

Consolidation offers the means to lower unit-transportation costs and raise service quality, through economies-of-scale of loading large groups of loads on large and efficient vehicles (or convoys) travelling frequently. For shippers with relatively small loads (or value), it means avoiding paying high fees for small vehicles or excess waiting for a sufficiently large volume of freight to warrant a larger vehicle. The downside of these benefits is that carrier consolidated services cannot be tailored-made for each individual customer request for transportation. Consolidation-based services must rather be planned to address the potential demand of as many customers as possible, in a manner to satisfy the closest possible their requests, preferences, and expectations. The goal for consolidation-based freight carriers is to be profitable while achieving these demand-service objectives and, thus, to best use their material and human resources, and operate at the lowest possible cost.

Setting up such a service network is generally part of tactical, medium-term, planning, addressed in most cases through *Service Network Design, SND*, methodology. Crainic et al. (2021a) presents state-of-the-art literature and methodology syntheses and reviews for the general SND problem (Crainic & Hewitt, 2021) and applications, including rail (Chouman & Crainic, 2021), motor-carrier interurban road transport (Bakir et al., 2021), liner-shipping navigation (Christiansen et al., 2021), and City Logistics (Crainic et al., 2021c).

Closely related to network design (Magnanti & Wong, 1984; Crainic et al., 2021a), SND presents particular characteristics, stemming from the various applications and the increasingly major trend of integrating into the same model and decision process several system aspects and related decisions. The simultaneous optimization of the service network, with the corresponding demand itineraries, and the high-level management of resources supporting those services is a case in point. Such trends yield multi-layer network design problem settings and formulations, which inspired quite a number of the generic developments of the previous sections.

We discuss the main classes of SND problem settings (following, for the most part, the vocabulary and notation of Crainic and Hewitt, 2021) and multi-layer network design formulations. We identify common characteristics, review briefly the literature, and discuss challenges and possible future developments. The fundamental concepts and definitions are briefly recalled in Sect. 5.1. Two and L-layer SND problem settings and formulations are addressed in Sects. 5.2 and 5.3, respectively.

5.1 Single-Layer Service Network Design

Consolidation-based carriers operate on physical networks made up of terminals (nodes) connected by physical, e.g., roads and rail tracks, or conceptual, e.g., sea and air links, arcs. Terminals are facilities where most of the demand originating in nearby regions is brought in to be consolidated (sorted, grouped, loaded into vehicles) before transport, and where the demand flows terminating their journeys are processed (unloaded, deconsolidated) before distribution to the final destinations. Rail stations and yards, LTL terminals, deep-sea and river/canal ports, in-land intermodal platforms, and airports are examples of terminals in consolidation-based networks. Some of these terminals, the so-called *hubs*, play a particularly important role in such systems, by structuring the flows for long-haul transport to take full advantage of the economies of scale of consolidation. Most terminals in a given region are connected to such a large terminal. The loads originating herein, most of which cannot be economically shipped directly, are then sent to that hub to be classified (sorted) and consolidated into larger flows, which are routed to other hubs by high-frequency, high-capacity services. Loads may thus go through more than one intermediary hub before reaching their regional destination terminal, being either transferred from one service to another or undergoing re-classification and re-consolidation.

For network-planning purposes, the multicommodity *demand* for transportation is defined between pairs of terminals in the physical network. Besides the origin and destination terminals, each commodity is also characterized by its amplitude (e.g., quantity, weight, or volume), unit revenue or transportation cost (or both), time-related requirements, i.e., availability date (and time) at the origin terminal and due date or time interval of delivery at the destination, as well as, possibly, various service requirements, e.g., vehicle type, handling conditions, etc.

Carriers respond to demand by offering a network of more or less scheduled services between their terminals. A *service* follows a route through the physical network moving either directly between its origin and destination, or stopping at intermediary terminals to drop and pick up loads and, eventually, vehicles when convoys are being operated (e.g., car and blocks for railroads and trailers for LTL motor carriers operating multi-trailer road trains). The set of services the carrier selects to operate makes up the *service network*. *Itineraries* are then defined for each individual origin-destination (OD) demand, specifying how it is moved through the service network, i.e., the sequence of services and the operations at inter-service terminals, transfer or classification and consolidation.

Tactical planning addresses the issue of building an operationally and economically efficient transportation plan, that is, the service network and schedule to satisfy the contemplated demand. The plan is generally built for a medium-term planning horizon, the so-called season. The carrier determines the season's duration, from 1 to 6 months, more rarely a year, depending on the transportation mode(s) involved. The goal is a "stable" operation context under the tactical plan, in terms of environment (regulatory, economic, seasonal, etc.), available resources, and *regular* demand, that is, demand requiring service at regular intervals (e.g., daily or weekly), at some expected (or contracted) time, with a certain predicted (or contracted) amplitude. This regularity is inherited by the transportation plan, which includes regular services for a given *schedule length*, e.g., a day or a week, defined relative to the regularity of demand. The plan is then to be repetitively applied. schedule-length period after schedule-length period, for the duration of the season.

The basic, *static*, single-layer Service Network Design problem setting involves selecting services, out of a set of potential services, to satisfy a set of OD demands, assuming the regularity and stability mentioned previously. Each individual demand is defined by its origin and destination nodes in the physical network, as well as the quantity of freight to move between these. The nodes of the SND network correspond to the nodes of the physical network, while the arcs correspond to the considered services among which the selection has to be made. When services include intermediary stops, each service leg, that is, the path in the physical network connecting two consecutive terminals on the route of the service, yields a particular arc. The selection, and the cost it incurs, are still per service, however, not per leg. Arcs (services) are characterized by the fixed selection (activation) cost, a unit flow transportation cost, as well as, potentially, commodity-specific ones, and a capacity, possibly leg specific. The SND network may include parallel arcs representing different services or service legs which could run between the same pair of terminals.

Two sets of decision variables are defined, design and flow. *Design*, or *service-selection*, variables are binary in most papers in the literature, but may also take non-negative integer values to decide the service frequency, that is, the number of times it will be repeated within the schedule length. The continuous or integer-valued *flow* variables determine the quantity of each demand moved on each arc of the network. They become binary, selecting whether the arc belongs to the itinerary of the demand or not, when demand cannot be split. The objective function of the SND formulation minimizes the total system cost, computed as the sum of the total design cost for the services selected and the total flow-distribution cost. The formulation includes flow conservation constraints for each demand, design-flow linking and capacity constraints for each arc and demand, plus possible constraints modelling particular system requirements (e.g., terminal capacity).

The notation and formulation of this single-layer problem are generally very similar to that of the classic network design problem and are the same, without the layer-specific notation (indexing) and constraints, to that of Sect. 2 and formulation (1)–(3), plus constraints (5) standing for system requirements and the decision-variable types.

Single-layer *time-dependent* problem settings and formulations arise when time-requirements, often availability at origin and due-date at destination, characterize demand. Services are then also defined by time characteristics, generally taking the form of a *schedule* specifying the departure and arrival times at each terminal on their routes (the degree of precision of the schedule may vary with the particular application). A so-called *time-space network* is the widely-used modelling device in such cases, being obtained by, implicitly or explicitly, extending the physical network along the time dimension with a given granularity partitioning the schedule length into periods. Demands and potential services are then represented on the time-space network, their respective origins, destinations, and intermediate service stops being defined as the copy of the respective physical node at the appropriate time moment. The moving arcs of the time-space network are the service legs operating in space and time, while holding arcs connect two consecutive copies of the physical nodes.

The single-layer SND defined on such a time-space network is generally named *Scheduled Service Network Design, SSND*, problem setting and formulation. It is noticeable that, once the corresponding definitions of the system parameters and decision variables are adjusted for the time dimension, the model is the same, mutatis mutandis, as that mentioned above for the SND (with a much higher computational complexity and difficulty). See Crainic and Hewitt (2021) for richer SSND settings, which we also discuss in the following.

5.2 Two-Layer Service Network Design

The previous SND formulations are single-layer. Two-layer formulations are found when the service involves convoys, that is, more than a single vehicle and particular vehicle organization structures. They are also found when resource management concerns are explicitly integrated into the tactical planning process.

The first case is typical of railway freight transportation where railcars are consolidated into *blocks*, which are then assembled into and moved between their respective origins and destinations by *train services* (Chouman & Crainic, 2021). This *classification* process aims to reduce the cost of handling individual railcars at intermediate terminals, and is particularly important when trains are long and travel long distances. In such a setting, railcars with possibly different origins and destinations are classified (sorted) at some terminal (origin or intermediary) and consolidated into a block, that is, into a group of cars which travel together for part of their journeys (until the destination of the block, where it is disbanded and the cars not at their destination terminal undergo classification again), being handled as a single unit when transferred from one service to another. Similar issues may be observed for other modes when convoys are involved, such as road-trains and platoons for road transport, and multi-barge assemblies for river/canal navigation.

The tactical planning problem then involves two related sets of design decisions, first, the services and, second, the "blocks", the latter being defined as sequences of the former. As also discussed in Crainic et al. (2022), SSND models in such cases are defined on a two or three-layer time-space network. The first design layer targets the train-service selection and supports the second design layer targeting the block selection, each supported potential block being defined as a path of arcs in the train-service layer. The OD demand and the flow-decision variables are defined on the supported block layer in a two-layer formulation. A separate layer may be defined for the handling of demand at terminals, availability at origin, classification, arrival at destination for delivery, etc. Such a three-layer formulation simplifies the modelling of complex terminal operations, classification in particular, as shown in Zhu et al. (2014), where the time-space network on each layer is built on two time-nodes for each physical one. This representation provides the means to capture the incoming and outgoing flow at the node at each particular time period, the link connecting the two nodes modelling the terminal activities, classification in particular.

Such two-layer formulations include one-to-one design and flow connectivity requirements and constraints, and thus belong to the Design-Flow Connect class of the MLND taxonomy proposed by Crainic et al. (2022). The former are of the all-design connectivity type, imposing constraints (6), while flow-accumulation constraints (14) belong to the latter group.

A different class of two-layer MLND problems emerges when integrating SND and the issue of providing the resources required to support the selected services. The initial contributions (e.g. Kim et al., 1999; Andersen et al., 2009b; Pedersen et al., 2009) focused on the basic case of a single resource type, a single resource unit required to operate a service, and given levels of available resources at each terminal. The resource movements in space and time are not explicitly addressed, however. The formulations require only the service network to be *balanced*, i.e., the same number of services, hence, resources, to flow into and out of every node (topological constraints on the node in and out degrees). Consequently, design-balanced SND models, without explicit resource management, do not require multi-layer formulations.

Most human (crews) and material (power units, loading units, vehicles, etc.) mobile resources in transportation are operated according to cycles in space and time, generally starting and ending at the same terminal. At the operational planning level, the management of mobile resources yields rather complex problems due to the various requirements of work agreements, periodic inspections, and maintenance, notably. The integration of resource-management concerns into tactical service-network requires a more macro view of the resource cycles to capture the essence of operations without unnecessary detail. This yields a two-layer MLND problem, with services being selected on the supported layer and resource cycles being selected on the supporting one. The demand is then moved on the service layer as in Andersen et al. (2009a), which shows the superiority of cycle-based formulations compared to the design-balanced ones, and Crainic et al. (2014), where explicit service and resource-cycle selection decisions are introduced together with a number of resource management rules (e.g., duration limit and number of returns to the origin terminal within a cycle). All-design connectivity constraints (6) link the two design layers only (no flow connectivity is considered), placing this MLND problem class in the Design Connect category of the Crainic et al. (2022) taxonomy.

5.3 Multi-Layer Service Network Design

Richer L-layer SSND models are increasingly proposed, mainly due to the growing interest in adequately addressing problem settings with several service classes or providers and several resource types connected through various design and flow connectivity requirements. They are generally known as SSND with resource management, *SSND-RM*, and SSND with revenue and resource management *SSND-RRM*, and belong to the Design-Flow Connect class of the MLND taxonomy

(Crainic et al., 2022). They involve several, but not all, connectivity relations discussed previously.

A first observation is the worth of explicitly including a *commodity layer* to manage the entry and exit of OD demand into and out of the system, as well as the various activities involving demand flows in terminals, e.g., railcar consolidation and waiting for departure (Zhu et al., 2014) or container waiting for assignment and loading on selected intermodal railcars (Kienzle et al., 2024).

The commodity layer is then part of multi-layer network design formulations with several service and resource layers. Crainic et al. (2018) extends the model and solution method proposed in Crainic et al. (2014) to include service-outsourcing possibilities in some parts of the network, and several resource layers representing different resources with particular management rules. The outsourcing possibilities are modelled as a particular resource type, each arc supporting a particular service that may be outsourced. Thus, while own resources move on cycles, outsourcing ones do not. The authors also explicitly include the possibility to acquire or rent new resources and the decisions to (re-)position resources to terminals before the tactical plan starts to be applied. Services require one unit of one of the resource types available, yielding min-design connectivity constraints (7). Hewitt et al. (2019) extends these concepts and propose a L-layer SSND-RM model explicitly integrating demand uncertainty. This is one of the very few contributions addressing uncertainty issues in multi-layer network design.

Intermodal rail transportation planning raises additional challenges compared to the general rail case. Indeed, containers come in different types (dimensions and particular requirements, e.g., power needs to for refrigeration), as do railcars, which also come in different settings in terms of the type and number of containers they may carry. The problem becomes even more complex when, as in North America, containers may be stacked, one on top of another, with particular feasibility and security stacking rules. The proposed L-layer SSND-RM (Kienzle et al., 2024) thus includes train-service and block design layers, a commodity (several types of containers) layer, and a railcar layer corresponding to as many resource layers as the different railcar fleets considered in the problem. The movement of resources in this model is represented on arcs, the management of the various railcar fleets involving the assignment and loading of containers, the management of inventories at terminals at appropriate time moments, the possible empty movements to fulfill the needs and ensure inventory conservation in space and time, and the initial allocation to terminals. The model introduces constraints describing the container-to-railcar assignment and loading activity. It also includes one-to-one all-design connectivity constraints (6) between the service and block layers, flow-accumulation connectivity constraints similar to (42) between the railcar and block layers, and flow-attribute capacity constraints (26) relative to the total length of train services.

Bilegan et al. (2022) proposes what appears to be the first SSND-RRM contribution, integrating revenue and resource management considerations and the SSND modelling framework to address the tactical planning problem of an intermodal barge transportation system. The revenue management context is represented by the segmentation of demand and service types. Three types of customers are considered, regular with established contracts or understandings for the planning period considered, which must be satisfied; partial-spot, which may be satisfied, if selected, for part of their forecast demand; and full-spot, which must be fully satisfied, if selected. Two service levels are also defined, standard and express, with corresponding fare classes being defined for each demand type. The OD demand is associated to the service layer. The problem setting considers two resource types, that is, two vessel types with particular characteristics in terms of capacity and speed. Vessels operate cyclic routes made up of service legs, the corresponding inter-service leg, holding arcs at intermediary terminals, as well as waiting-at-terminal arcs when idly waiting for the next service to start. The goal is to maximize profit by selecting the appropriate services, and the vessels to support them, to satisfy the regular and selected spot demand. The authors wrote a "flat" model, projecting the two resource layers onto the service layer, by *a priori* associating a vessel to each potential service (which implies that parallel service arcs are possible), and enforcing vessel-type specific fleet size and design-balancing constraints at each port (node) at all time periods.

6 Conclusions and Perspectives

Multi-layer network design emerges as an important problem class within the network-design combinatorial optimization field. On the one hand, it raises interesting methodological challenges stemming from the problem structure and the relations among the design and flow variables and attributes defined on the different layers. On the other hand, it provides the means to adequately represent and address complex planning problems in major application domains, transportation, logistics, and telecommunications, in particular.

Following the structuring of the field proposed in Crainic et al. (2022), this chapter updated the basic concepts and properties of MLND, and developed the connectivity-requirement definitions and modelling to address richer problem settings. Attribute-based connectivity was thus defined, in particular related to non-additive attributes, such as arc capacities. The chapter also significantly extended the study of connectivity relations involving the design and flow-distribution decisions on more than two related layers. These relations and the corresponding constraint representation within MLND formulations were then discussed from the point of view of tactical planning in consolidation-based freight transportation and the service network design methodology used to address it.

The field is still emerging, however, particularly when more than two layers are involved. Several relations described in this chapter are only defined in mathematical terms, applications still waiting to be proposed, as the brief SSND literature analysis illustrates. Moreover, other problem characteristics, important methodologically and in practice alike, e.g., uncertainty, robustness, and resilience, have still to be studied. Continuing to address the MLND challenges appears thus important, particularly given the strong mutually beneficial relations between methodological development and applications proper to Operations Research.

Solution-method developments are particularly necessary. As already observed in Crainic et al. (2022), this development may follow two related directions. First, extend the methods proposed for the general network design problems (see, e.g., Crainic and Gendron, 2021; Crainic and Gendreau, 2021, for recent surveys on exact and heuristic methods, respectively). Second, take advantage of the MLND structure to strengthen formulations and develop exact, meta-heuristic, and matheuristic solution methods. Decomposition appears particularly relevant given the multi-layer structure of the problem class, Lagrangian, Benders, and Integrative Cooperative Search offering rich research perspectives.

Acknowledgments As the entire book, this chapter is dedicated to the loving memory of Bernard. We miss him so much.

While working on this paper, the author was Adjunct Professor, Department of Computer Science and Operations Research, Université de Montréal, Canada. We gratefully acknowledge the financial support provided by the Canadian Natural Sciences and Engineering Research Council (NSERC) through its Discovery and Collaborative Research and Development grant programs, as well as by the Fonds de recherche du Québec through their Teams and CIRRELT infrastructure grants. The author wishes to express his gratitude toward the unknown referees and the colleagues and friends who offered many very constructive comments. He assumes all errors still in the chapter.

Appendix: Flow Accumulation and Conservation Conformity

Crainic et al. (2022) discuss the issue of so-called reachable arcs and the apparent contradiction between flow-connectivity and flow-conservation constraints. Two arcs (x_1, y_1) and (x_2, y_2) are defined as reachable when there is a path from x_1 to y_2, or from x_2 to y_1. Thus arcs a and b of the supported layer l' of Fig. 4 are reachable, assuming there is a path in l' from node 3 to node 4. Define arcs a and b as the paths $\{\alpha, \beta\}$ and $\{\epsilon, \beta, \zeta\}$ is the supporting layer l. Assume d units of flow are to be moved from node 1 to node 6 on the supporting and supported arcs. The arc labels in the figure display the arc identifier and the quantity of flow assigned to it in this chapter.

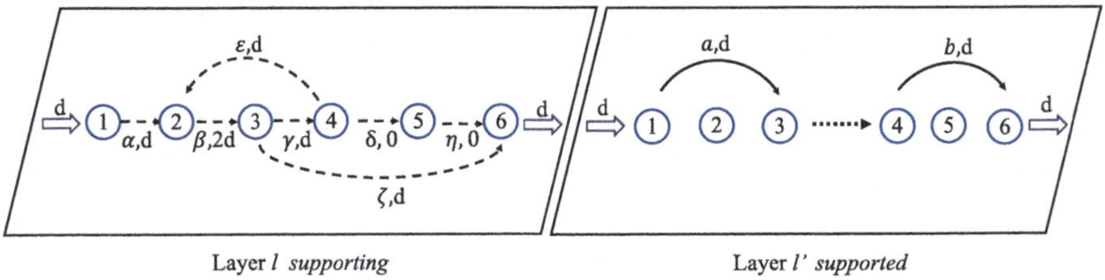

Fig. 4 Flow accumulation and conservation illustration

Crainic et al. (2022) claim that the flow-accumulation constraints force a flow of $2d$ units on arc β, which contradicts the flow-conservation constraints at node 3 putting d units on arc ζ and 0 units on arc γ. Yet, the equality between the flows on the two layers imposed by the flow-accumulation constraints operates in both directions, impacting also the flow distribution on the supporting layer. Hence, a solution sending d units on arcs γ and ζ, and 0 units on arcs δ and ν (see Fig. 4,) respects the two types of constraints. No contradiction is observed in this case.

One may ask when such cases may arise. Clearly, the "cost" attributes on the supporting and supported layers would have an impact and should therefore reflect such flow distributions. More generally, the problem setting must require it. We are not aware of general settings with network configurations such at those of Fig. 4. The only situation we are aware of concerns a particular setting of the Scheduled Service Network Design (SSND) problem (Crainic et al., 2021b; Kienzle et al., 2024). Recall that SSND models are often formulated on time-space networks covering activities for a given schedule length. Wrap-around arcs are then defined to represent activities, e.g., holding and services, starting during the schedule length but terminating in the next occurrence of the plan (schedule). As resources move according to (explicit or implicit, depending on the problem setting) cycles and all arcs point in the same time direction, the situation discussed above does not happen. The case may be observed, however, when resources are injected into the system at the first occurrence of the plan, and representations similar to that around node 2 in the figure are built. (Notice, however, that Fig. 4 is not an illustration of a time-space network with wrap-around arcs as some flows, on arc ζ, do not return.)

References

Ambra, TK., Mommens, C., & Macharis, C. (2021). Intermodal and synchromodal freight transport. In: R. Vickerman, et al. (Eds.), *International encyclopedia of transportation* (Vol. 3, pp. 456–462). Elsevier.

Andersen, J., Crainic, TG., & Christiansen, M. (2009a). Service network design with asset management: formulations and comparative analyzes. *Transportation Research Part C: Emerging Technologies, 17*(2), 197–207.

Andersen, J., Crainic, TG., & Christiansen, M. (2009b). Service network design with management and coordination of multiple fleets. *European Journal of Operational Research, 193*(2), 377–389.

Areta, A., & Moreno, Y. (2019). Multilayer networks in a nutshell. *Annual Review of Condensed Matter Physics, 10*, 45–62.

Bakir, I., Erera, A., & Savelsbergh, M. (2021). Motor carrier service network design. In: TG. Crainic, M. Gendreau, B. Gendron (Eds.), *Network design with applications in transportation and logistics* (Chap. 14, pp. 425–467). Springer.

Balakrishnan, A., Magnanti, T. L., Mirchandani, P. (1994). A dual-based algorithm for multi-level network design. *Management Science, 40*(5), 567–581.

Ballot, E., Montreuil, B., & Meller, R. D. (2014). *The physical internet: The network of logistics networks*. La Documentation Française.

Bilegan, IC., Crainic, TG., & Wang, Y. (2022). Scheduled service network design with revenue management considerations and an intermodal barge transportation illustration. *European Journal of Operational Research, 300*(1), 164–177.
Chouman, M., & Crainic, TG. (2021). Freight railroad service network design. In: TG. Crainic, M. Gendreau, & B. Gendron (Eds.), *Network design with applications in transportation and logistics* (Chap 13, pp. 383–426). Springer.
Christiansen, M., Hellsten, E., Pisinger, D., Sacramento, D., & Vilhelmsen, C. (2021). Liner shipping service network design. In: TG. Crainic, M. Gendreau, B. Gendron (Eds.), *Network design with applications in transportation and logistics* (Chap. 14, pp. 469–505). Springer.
Cordeau, J. F., Pasin, F., & Solomon, M. M. (2006). An integrated model for logistics network design. *Annals of Operations Research, 144*(1), 59–82.
Costa, A. M., França, P. M., & Lyra Filho, C. (2011). Two-level network design with intermediate facilities: An application to electrical distribution systems. *Omega, 39*(1), 3–13.
Crainic, TG. (2003). Long-haul freight transportation. In: R. W. Hall (Ed.), *Handbook of Transportation Science* (2nd ed., pp 451–516). Kluwer Academic Publishers.
Crainic, TG., & Gendreau, M. (2021). Heuristics and metaheuristics for fixed-charge network design. In: TG. Crainic, M. Gendreau, B. Gendron (Eds.), *Network design with applications in transportation and logistics* (Chap 4, pp. 91–138). Springer.
Crainic, TG., & Gendron, B. (2021). Exact methods for fixed-charge network design. In: TG. Crainic, M. Gendreau, & B. Gendron (Eds.), *Network design with applications in transportation and logistics* (Chap 3, pp. 29–89). Springer.
Crainic, TG., & Hewitt, M. (2021). Service network design. In: TG. Crainic, M. Gendreau, & B. Gendron (Eds.), *Network design with applications in transportation and logistics* (Chap 12, pp. 347–382). Springer.
Crainic, TG., & Kim, KH. (2007). Intermodal transportation. In: C. Barnhart, G. Laporte (Eds.), *Transportation, handbooks in operations research and management science* (Vol. 14, Chap. 8, pp. 467–537). North-Holland.
Crainic, TG., & Laporte, G. (1997). Planning models for freight transportation. *European Journal of Operational Research, 97*(3), 409–438.
Crainic, TG., Klibi, W., & Montreauil, B. (2023). Hyperconnected city logistics: A conceptual framework. In: E. Marcucci, V. Gatta, M. Le Pira (Eds.), *Handbook on city logistics and urban freight* (Chap. 20, pp. 398–421). Edward Elgar.
Crainic, TG., Ricciardi, N., & Storchi, G. (2004). Advanced freight transportation systems for congested urban areas. *Transportation Research Part C: Emerging Technologies, 12*(2), 119–137.
Crainic, TG., Ricciardi, N., & Storchi, G. (2009). Models for evaluating and planning city logistics transportation systems. *Transportation Science, 43*(4), 432–454.
Crainic, TG., Hewitt, M., Toulouse, M., & Vu, DM. (2014). Service network design with resource constraints. *Transportation Science, 50*(4), 1380–1393.
Crainic, TG., Hewitt, M., Toulouse, M., & Vu, DM. (2018). Scheduled service network design with resource acquisition and management. *EURO Journal on Transportation and Logistics, 7*(3), 277–309.
Crainic, TG., Gendreau, M., & Gendron, B. (2021a). Fixed-charge network design problems. In: TG. Crainic, M. Gendreau, & B. Gendron (Eds.), Network design with applications in transportation and logistics (Chap. 2, pp. 15–28). Springer.
Crainic, TG., Gendreau, M., & Gendron, B. (Eds.) (2021b). *Network design with applications in transportation and logistics*. Springer.
Crainic, TG., Perboli, G., & Ricciardi, N. (2021c). City logistics. In: TG. Crainic, M. Gendreau, B. Gendron (Eds.), *Network design with applications in transportation and logistics* (Chap. 16, pp. 507–537). Springer.
Crainic, TG., Gendron, B., & Kazemzadeh, M. (2022). A taxonomy of multilayer network design and a survey of transportation and telecommunication applications. *European Journal of Operational Research, 303*(1), 161–179.

Crainic, TG., Gonzalez-Feliu, J., Ricciardi, N., Semet, F., & Van Woensel, T. (2023). Operations research for planning and managing city logistics systems. In: E. Marcucci, V. Gatta, M. Le Pira (Eds.), *Handbook on city logistics and urban freight* (Chap. 10, pp. 190–223). Edward Elgar.

Giusti, R., Manerba, D., & Tadei, R. (2019). Synchromodal logistics: An overview of critical success factors, enabling technologies, and open research issues. *Transportation Research Part E: Logistics and Transportation Review, 129*, 92–110.

Hewitt, M., Crainic, T. G., Nowak, M., & Rei, W. (2019). Scheduled service network design with resource acquisition and management under uncertainty. *Transportation Research Part B: Methodological, 128*, 324–343.

Kienzle, J, Crainic, TG., Frejinger, E., & Bisaillon, S. (2024). The intermodal railroad blocking and railcar fleet management planning problem. Technical Report, CIRRELT-2024, Centre interuniversitaire de recherche sur les réseaux d'entreprise, la logistique et les transports, Université de Montréal, Montréal, QC, Canada

Kim, D., Barnhart, C., Ware, K., & Reinhardt, G. (1999). Multimodal express package delivery: A service network design application. *Transportation Science, 33*(4), 391–407.

Kivelä, M., Arenas, A., Barthelemy, M., Gleeson, JP., Moreno, Y., & Porter, MA. (2014). Multilayer networks. *Journal of Complex Networks, 2*, 203–271.

Lin, C. C. (2010). The integrated secondary route network design model in the hierarchical hub-and-spoke network for dual express services. *International Journal of Production Economics, 123*(1), 20–30.

Magnanti, TL., & Wong, RT. (1984). Network design and transportation planning: models and algorithms. *Transportation Science, 18*(1), 1–55.

Mauttone, A., Héctor, C., Urquhart, & ME. (2021). Public transportation. In: TG. Crainic, M. Gendreau, & B. Gendron (Eds.), *Network design with applications in transportation and logistics* (Chap. 17, pp. 539–565). Springer.

Merriam-Webster. (2023). Dictionary & thesaurus. https://www.merriam-webster.com/, accessed 1st June 2023.

Obreque, C., Donoso, M., Gutiérrez, G., & Marianov, V. (2010). A bnranch and cut algorithm for the hierarchical network design problem. *European Journal of Operational Research, 200*(1), 28–35.

Pedersen, MB., Crainic, TG., & Madsen, OBG. (2009). Models and tabu search meta-heuristics for service network design with asset-balance requirements. *Transportation Science, 43*(2), 158–177.

Rouhani, A., Crainic, TG., Rei, W., & Klibi, W. (2024). Benders decomposition for two-layer network design with capacity decisions. Technical Report, CIRRELT-2024, Centre interuniversitaire de recherche sur les réseaux d'entreprise, la logistique et les transports, Université de Montréal, Montréal, QC, Canada.

Taherkhani, G., Bilegan, I., Crainic, TG., Gendreau, R., & Rei, W. (2022). Tactical capacity planning in an integrated multi-stakeholder freight transportation system. *OMEGA, 110*, article 102628.

Zhu, E., Crainic, TG., & Gendreau, M. (2014). Scheduled service network design for freight rail transportations. *Operations Research, 62*(2), 383–400.

Separable Lagrangian Decomposition for Quasi-Separable Problems

Antonio Frangioni, Bernard Gendron, and Enrico Gorgone

1 Introduction

The Lagrangian relaxation approach in (Linear) Mixed-Integer Programming is described in general terms as being applied to Mixed-Integer (Linear) Programs of the form

$$(P) \qquad \min\{\, cx \;:\; Cx = b \;,\; x \in X \,\},$$

where X is an "easy" set, in the sense that (P) would be "much easier" to solve if the "complicating constraints" $Cx = b$ could be removed. In general, removing the constraints outright is not a feasible option; an alternative is to form the *Lagrangian relaxation* of (P) with respect to the complicating constraints, i.e., the (sub)problem

$$(P_\lambda) \qquad \phi(\lambda) = \lambda b + \min\left\{\, cx + \lambda(b - Cx) \;:\; x \in X \,\right\}$$

for a fixed vector λ of *Lagrangian multipliers*. (P_λ) can then be "easily" solved providing a valid lower bound $\phi(\lambda) \;\leq\; v(P)$, with $v(\cdot)$ indicating the optimal

A. Frangioni (✉)
Dipartimento di Informatica, Università di Pisa, Pisa, Italy
e-mail: frangio@di.unipi.it

B. Gendron
CIRRELT and Department of Computer Science and Operations Research, Université de Montréal, Montréal, QC, Canada
e-mail: Bernard.Gendron@cirrelt.ca

E. Gorgone
Dipartimento di Matematica e Informatica, Università degli Studi di Cagliari, Cagliari, Italy
e-mail: egorgone@unica.it

T. G. Crainic et al. (eds.), *Combinatorial Optimization and Applications*,
International Series in Operations Research & Management Science 358,
https://doi.org/10.1007/978-3-031-57603-4_10

value of an optimization problem. One can then consider finding the *best possible* Lagrangian relaxation, i.e., the solution of the *Lagrangian dual*

$$(D) \qquad \max\{\, \phi(\lambda) = v(P_\lambda) \;:\; \lambda \in \mathbb{R}^n \,\}$$

of (P) with respect to the complicating constraints: clearly, $v(D) \leq v(P)$, i.e., (D) is still a relaxation of (P), but possibly providing better bounds than the standard continuous one. Solving the Lagrangian dual of (P), possibly approximately, yields information that can be used to drive solution methods to (P), be them exact (Holmberg & Yuan 2000; Gendron & Larose 2014) or heuristic (Borghetti et al. 2003; Crainic et al. 2004). Crucially, (D) can be "easily" solved since it is a concave maximization problem. On first reading, with $X = \{\, \bar{x}_1 \,,\, \bar{x}_2 \,,\, \dots \}$ finite but possibly very large set, one sees that

$$(D) \qquad \max\{\, v \;:\; v \leq c\bar{x} + \lambda(b - C\bar{x}) \;\; \bar{x} \in X \,\}, \tag{1}$$

i.e., a (possibly, semi-infinite) linear program. While typically (D) cannot be written down explicitly, it can be solved by *row generation*; that is, one takes a "small" $\mathcal{B} \subset X$ to define the (Dual) *Master Problem*

$$(D_{\mathcal{B}}) \qquad \max\{\, v \;:\; v \leq c\bar{x} + \lambda(b - C\bar{x}) \;\; \bar{x} \in \mathcal{B} \},\tag{2}$$

clearly a relaxation to (D).

W.l.o.g. it is assumed that an optimal solution $(\, \tilde{v} \,,\, \tilde{\lambda} \,)$ of $(D_{\mathcal{B}})$ exists. In fact, there are appropriate tricks available to ensure it happens (Frangioni 2005). The solution $(\, \tilde{v} \,,\, \tilde{\lambda} \,)$ can be easily checked for optimality of the whole (D) by just solving $(P_{\tilde{\lambda}})$, i.e., computing $\phi(\tilde{\lambda})$. Indeed, $\tilde{v} \leq \phi(\tilde{\lambda})$ clearly implies $\tilde{v} = \phi(\tilde{\lambda}) = v(D)$, i.e., that the problem is solved. If the condition does not hold, any optimal $\bar{x}$ to $(P_{\tilde{\lambda}})$ corresponds to a violated constraint of (1) that can be added to $\mathcal{B}$, and the process iterated. The corresponding *Cutting Plane algorithm* (CP) is globally convergent and, while not efficient in its by-the-book incarnations, it is the basis of the most efficient algorithms for solving (D) with high degree of accuracy. These algorithms are known as variants of the *bundle method* (Frangioni 2020), although simpler alternatives exist if only coarse solutions are desired (Frangioni et al. 2017).

The primal viewpoint of Lagrangian duality is also well-known, and can be easily expressed by just noting that $(D_{\mathcal{B}})$ is a simple LP and has dual

$$(P_{\mathcal{B}}) \qquad \min\left\{\, \textstyle\sum_{\bar{x}\in\mathcal{B}} c\bar{x}\theta_{\bar{x}} \;:\; \textstyle\sum_{\bar{x}\in\mathcal{B}}(b - C\bar{x})\theta_{\bar{x}} = 0 \;\;,\;\; \theta \in \Theta \right\}, \tag{3}$$

where $\Theta = \{\, \theta \geq 0 \;:\; \sum_{\bar{x}\in\mathcal{B}} \theta_{\bar{x}} = 1 \,\}$ (the unitary simplex) indicates that a convex hull operation is taking place. Indeed, linearity of the constraints easily allows to recast (3) as

$$(P_{\mathcal{B}}) \qquad \min\{\, cx \;:\; Cx = b \;,\; x \in X_{\mathcal{B}} = conv(\mathcal{B}) \,\}, \tag{4}$$

where $conv(\cdot)$ denotes the convex hull. That is, the (primal) master problem is optimising the original objective on the intersection between the feasible region defined by the complicating constraints and the convex hull $X_{\mathcal{B}}$ generated by the known set of points in $\mathcal{B}$. With $\mathcal{B} = X$ (and ignoring formal complications for brevity), (4) is the *Dantzig-Wolfe reformulation* of the *convexified relaxation* of (P)

$$(\tilde{P}) \qquad \min\{\, cx \;:\; Cx = b \;,\; x \in conv(X) \,\} \tag{5}$$

to which then (D) is equivalent. This immediately implies that, under appropriate assumptions (D) can provide tighter bounds than the ordinary continuous relaxation. This happens when X is not "too easy", i.e., its linear formulation, say $Ex \leq d$, does not have the integrality property (Frangioni 2005). This fact justifies the interest in the technique.

This succinct description is however neglecting a number of details that are crucial computationally. Perhaps the most important one is that often (but not always) the relaxed constraints are "complicating" because they are "linking"; that is, $X = X^1 \times X^2 \times \ldots \times X^k$ is a Cartesian product of sets that have no joint constraints other than these. This means that the problem can be written as

$$(P) \qquad \min \left\{ \sum_{h \in K} c^h x^h \;:\; \sum_{h \in K} C^h x^h = b \;,\; x^h \in X^h \;\; h \in K \right\} \tag{6}$$

with $K = \{1, \ldots, k\}$. Hence, the Lagrangian relaxation is separable in k independent subproblems:

$$(P_\lambda) \qquad \phi(\lambda) = \lambda b + \sum_{h \in K} \min \left\{ (c^h - \lambda A^h) x^h \;:\; x^h \in X^h \right\}.$$

One advantage is of potentially making it faster to solve (P_λ) by means of parallel processing techniques (Cappanera & Frangioni 2003), but this is likely not the most important aspect. Rather, having independent subproblems allows to collect independent subsets $\mathcal{B}^h \subset X^h$ and use them to construct *disaggregated cuts* in the Dual Master Problem (Bacaud et al. 2001; Borghetti et al. 2003). The disaggregation makes it roughly a factor of k larger in terms of constraints, and therefore more costly to solve, after the same number of iterations and save *selection* and *compression* techniques that can be used to reduce its size. Nonetheless, this typically results in a much faster convergence and therefore ultimately better performances (Jones et al. 1993).

The reason is easily understood by looking at the *disaggregated Primal Master Problem*

$$(P_{\mathcal{B}}) \quad \min \sum_{h \in K} c^h \sum_{\bar{x}^h \in \mathcal{B}^h} \bar{x}^h \theta_{\bar{x}^h} \tag{7a}$$

$$\text{s.t.} \sum_{h \in K} A^h \sum_{\bar{x}^h \in \mathcal{B}^h} \bar{x}^h \theta_{\bar{x}^h} = b \tag{7b}$$

$$\sum_{\bar{x}^h \in \mathcal{B}^h} \theta_{\bar{x}^h} = 1 \qquad h \in K. \tag{7c}$$

It is plain to see that (3) is the restriction of (7) where all the components $\bar{x}^h$ corresponding to the solution $\bar{x}$ obtained at some iteration must have the same convex multiplier. Hence, for the same $\mathcal{B} \subseteq X$ (7) has more solutions than (3). Intuitively, forcing each component of $\bar{x}$ to have the same multiplier does not make sense: in other words, the Cartesian product of the $conv(\mathcal{B}^h)$ is a better approximation of $conv(X)$ than $conv(\mathcal{B})$, usually so much so as to amply counterbalance the (significant) increase in the number of variables. This is largely because the process terminates after much fewer iterations, hence the average size of the solved master problems (often, but not always the computational bottleneck of the process) may still be significantly lower.

The underlying intuition is that the structure of the original problem (P) should be exploited as much as possible when constructing its Dantzig-Wolfe reformulation. Another instance of this concept is the "easy components" variant of the approach (Frangioni & Gorgone 2014), which exploits the fact that the different Lagrangian subproblems need not be all of the same type. For illustration, consider an original problem (P) of the form

$$(P) \quad \min \left\{ c^1 x^1 + c^2 x^2 \;:\; C^1 x^1 + C^2 x^2 = b \;,\; x \in X^1 \;,\; E x^2 \leq d \right\}, \tag{8}$$

i.e., where the feasible region X is the Cartesian product of a generic set X^1 (which we only know we can efficiently optimise upon) and of a set X^2 that is representable by a "simple, small" (linear) formulation. Clearly one could apply the disggregated approach (7), as optimising over X^2 is "easy". However, this is not the best approach: representing of X^2 via its vertices yields an exponential number of points, while a compact linear formulation is available. An arguably better solution is that only the first group of variables is Dantzig-Wolfe reformulated, while the second is just kept as it is, i.e.,

$$(P_{\mathcal{B}}) \quad \min c^1 \left(\textstyle\sum_{\bar{x}^1 \in \mathcal{B}^1} \bar{x}^1 \theta_{\bar{x}^1} \right) + c^2 x^2 \tag{9a}$$

$$\text{s.t.} \;\; C^1 \left(\textstyle\sum_{\bar{x}^1 \in \mathcal{B}^1} \bar{x}^1 \theta_{\bar{x}^1} \right) + C^2 x^2 = b \tag{9b}$$

$$\textstyle\sum_{\bar{x}^1 \in \mathcal{B}} \theta_{\bar{x}^1} = 1 \;,\; \theta \geq 0 \;,\; E x^2 \leq d. \tag{9c}$$

While such a master problem can initially be larger than the standard one, it contains "perfect" information on the feasible set of the x_2 variables, which therefore the solution process does not need to painstakingly accrue iteratively. This clearly holds for any number of "easy" and "difficult" components. This approach usually decreases the total number of iterations required to solve the problem and provides a positive trade-off on the total solution time (Scuzziato et al. 2018).

Hence, the Dantzig-Wolfe reformulation approach underlying Lagrangian duality is not necessarily a rigid scheme; rather, it possesses degrees of flexibility that can be exploited to make maximal use of the existing structure of the underlying problem. In this paper we consider one structure that, to the best of our knowledge, has not been analysed in this sense despite being relatively common. This is the case of a problem having several groups of semi-continuos variables x_i, each one "governed" by a single binary variables y_i, as in

$$A_i x_i \leq b_i y_i \qquad i \in N \tag{10}$$

$$x_i \in X_i \cup \{0\} \qquad i \in N \tag{11}$$

$$y \in Y, \tag{12}$$

where $X_i \subseteq \{ x_i \in \mathbb{R}^{n_i} : A_i x_i \leq b_i \}$. That is, the semi-continuos variables have a (potentially) disjoint feasible region, which we denote by $X'_i = X_i \cup \{0\}$. When $y_i = 0$ the condition $x_i = 0$ is enforced, while when $y_i = 1$ the condition $x_i \in X_i$ rather is. The textbook example is that of a single variable $x_i \in \{0\} \cup [l_i, u_i]$, with $0 < l_i < u_i < \infty$, describing a machinery or an activity that is either off, and therefore produces nothing, or on and therefore has to produce at least a minimum positive quantity (l_i). Although $0 \in X_i$ can also happen, a cost may be incurred to "activate the process", i.e., to have $y_i = 1$ (that is, the objective contains a term $c_i y_i$ with $c_i \neq 0$). Thus, the case where $0 \in X_i$ is typically not easier than $0 \notin X_i$ and we consider the latter for full generality.

Removing a set of complicating constraints, say $Dy + \sum_{i \in N} C_i x_i = b$, the problem does not decompose as a sum of independent subproblems, but can still be solved by applying two separate solution approaches "in cascade"—one of which exploiting a degree of decomposability—because the variables can be partitioned into k groups of semi-continuous ones. This is due to the fact that there are no other constraints—apart for the complicating ones—linking variables in different blocks, save for those linking each group to the binary variable "governing" it, and (possibly) the ones linking the binary variables among themselves. In such a case, the Lagrangian subproblem is formally undivided, and therefore the use of disaggregated master problems a-la (7) is not possible, which may have a significant detrimental effect on the ability to efficiently solve the Lagrangian dual. This is counter-intuitive, in that there clearly is a significant element of separability in these *quasi-separable problems*. We extend and computationally validate the results of (Frangioni et al. 2020) by showing that a relatively minor modification of the master problem formulation allows to efficiently inject a full separable structure into the approach, at the cost of a (likely) relatively minor increase of the number of Lagrangian multipliers. We contrast our result with the "brute force" approach in which all the constraints linking the semi-continuous variables and the corresponding integer ones are relaxed, which yields an analogous final effect with a possibly weaker bound, but at the cost of a (likely) much greater number of Lagrangian multipliers. After having introduced the idea in the master problem space we link it to an application of the *Lagrangian decomposition* approach which gives it a different perspective.

The structure of the Chapter is the following. In Sect. 2 we formally introduce the family of quasi-separable problems and discuss different possible master problem (Dantzig-Wolfe) formulations for them. In Sect. 3 we describe our target application, the fixed-charge multicommodity capacitated network design problem, and how the results of the previous sections apply to it. Then, in Sect. 4 we describe our computational experience aimed at verifying the practical efficiency of the newly proposed approach, and Sect. 5 provides some closing discussion.

2 Master Problem Reformulations for Quasi-Separable Problems

We now formally present the relevant formulations. For a finite index set $N = \{1, \dots, n, \}$, a set $Y \subseteq \{0, 1\}^n$ is given, represented by a vector of binary variables $y = [y_i]_{i \in N}$, together with n sets $X_i \subseteq \mathbb{R}^{n_i}$ and the corresponding partitioned vector(s) of variables $x = [x_i]_{i \in N}$ as in (10)–(12). That is, each x_i is a semi-continuous variable "governed" by the corresponding y_i, in the sense that $y_i = 0 \Longrightarrow x_i = 0$, while $y_i = 1 \Longrightarrow x_i \in X_i$. We assume that (10) define a compact polyhedron when $y_i = 1$, i.e., such that $\{ x_i \in \mathbb{R}^{n_i} \ : \ A_i x_i \leq 0 \} = \{ 0 \}$. Clearly, these constraints also influence the shape of the feasible set for x_i; that is, without loss of generality we can assume that $X_i \subseteq \{ x_i \in \mathbb{R}^{n_i} \ : \ A_i x_i \leq b_i \}$ (which immediately implies that each X_i is compact). The sets X_i may entail integrality restrictions, but since we are concerned with relaxations (of the Lagrangian type) we will actually only work with the convex hull of X_i; hence, to ease on notation we will consider X_i to be convex, and compact. The relevant object is then the set $\bar{X}_i$ of the extreme points of X_i, so that $X_i = conv(\bar{X}_i)$. It is intuitive, and it will be confirmed later on, that a slight difference exists between the case where $0 \in X_i$ and that where $0 \notin X_i$. It will therefore be expedient to introduce $X_i' = X_i \cup \{0\}$, which may or may not coincide with X_i, and its set of extreme points $\bar{X}_i'$. The set Y is decidedly integer, but as Lagrangian relaxations cannot distinguish a set from its convex hull, we will basically only work with the latter; yet, because of its binary nature, $Y = ext(Y)$ and therefore we don't need to explicitly define the set of extreme points. An "abstract" formulation of our problem (neglecting the details of how X_i' and Y are formulated) is

$$\max dy + \textstyle\sum_{i \in N} c_i x_i \tag{13}$$

$$\text{s.t. } Dy + \textstyle\sum_{i \in N} C_i x_i = b \tag{14}$$

$$A_i x_i \leq b_i y_i \qquad i \in N \tag{15}$$

$$x_i \in X_i' \qquad i \in N \tag{16}$$

$$y \in Y. \tag{17}$$

Note that in the problem we have to impose $x_i \in X_i'$, rather than $x_i \in X_i$ (although the two may indeed be equal), to make it possible to set $x_i = 0$ when $y_i = 1$. The m constraints (14) are the "linking" ones, whose removal makes the problem (much) easier to solve. Hence, it may make sense to form the standard

Lagrangian relaxation w.r.t. (14)

$$\phi(\lambda) = \lambda b + \max\left\{(d - \lambda D)y + \sum_{i\in N}(c_i - \lambda C_i)x_i \;:\; (15),\ (16),\ (17)\right\} \tag{18}$$

and aim at (efficiently) solving the corresponding Lagrangian dual. This requires being able to compute $\phi(\lambda)$ efficiently. Contrarily to what often happens with Lagrangian relaxation, however, $\phi(\lambda)$ is not easy to solve because it decomposes in the sum of independent problems. Rather, due to the semi-continuous nature of the x_i one can first solve independent subproblems in those variables, and then a unique problem in the y ones using the results. More formally, one can define

$$\phi_i(\lambda) = \max\left\{ (c_i - \lambda C_i)x_i \;:\; x_i \in X_i \right\} \tag{19}$$

(where note that, unlike in (16), the original X_i is used instead of the "extended" X_i', although of course the two may coincide). Clearly, $\phi_i(\lambda)$ is the best value one can ever hope for, for the fixed λ, to be provided by the x_i variables; we will denote by x_i^* any one of the corresponding optimal solutions, eschewing the dependency on λ for notational simplicity. It is then clear that

$$\phi(\lambda) = \lambda b + \max\left\{ \sum_{i\in N}(d_i - \lambda D^i + \phi_i(\lambda))y_i \;:\; y \in Y \right\}. \tag{20}$$

Provided that (19) (n independent problems) and (20) (one single problem concerning the y variables only) can be efficiently solved, the Lagrangian function can be easily computed. For any y^* optimal for (20), clearly $(\, y^*\,,\, \tilde{x}\,)$ is optimal for (18) with $\tilde{x}_i = 0$ if $y_i^* = 0$, and $\tilde{x}_i = x_i^*$ if $y_i^* = 1$. Hence, $(b - Dy^* + \sum_{i\in N} C_i\tilde{x}_i)$ is a valid subgradient of ϕ in λ—providing the linear inequality used in the *aggregated* dual master problem (1)—which can be used in any of the known approaches to seek for the optimal λ^*.

While all this is perfectly sensible, we can argue that the process somewhat "hides" some of the precious structure of the problem. In particular, computing the Lagrangian function actually boils down to solving $n+1$ separate subproblems, but *not independent* ones. Hence the Lagrangian function is not a simple sum-function, and therefore *disaggregated master problems* a-la (7) cannot be used, losing one major opportunity at improving convergence speed. There is of course a simple way to make the function separable: also relax (in Lagrangian fashion) the constraint (15) with multipliers $\mu = [\mu_i]_{i\in N} \geq 0$, which yields the Lagrangian function

$$\phi(\lambda, \mu) = \lambda b + \psi(\lambda, \mu) + \sum_{i\in N} \psi_i(\lambda, \mu_i) \tag{21}$$

with

$$\psi_i(\lambda, \mu_i) = \max\left\{ (c_i - \lambda C_i - \mu_i A_i)x_i \;:\; x_i \in X_i \right\}, \tag{22}$$

$$\psi(\lambda, \mu) = \max\left\{ \sum_{i\in N}(d_i - \lambda D^i - \mu_i b_i)y_i \;:\; y \in Y \right\}. \tag{23}$$

Clearly, (22) and (23) can be expected to be efficiently solvable if (19) and (20) are; note that, actually, in (19) one may use X_i *without* the constraints $A_i x_i \leq b_i$, should this help in any way. Then, $\phi(\lambda, \mu)$ is a classical sum-function for which the corresponding approaches can be used. The dual space is extended when relaxing more constraints, which is necessary to obtain a better decomposition; hence, the Lagrangian dual corresponding to maximising (18) can provide a strictly better bound than that corresponding to maximising (21). However, the results of (Frangioni et al. 2020, §3) show that this is not happening if Y has the integrality property, as in our application (cf. § 3); we will return to this equivalence later on. Thus, in this case the two Lagrangian approaches provide alternative ways to compute the same (strong) bound. However, the drawback of the disaggregated approach is that (15) can be "many", which means that the (λ, μ) space might be much larger than the λ space; this increase in size may well negate the advantage of having a disaggregated function, as our computational results will confirm. One would then therefore have a way to reap the algorithmic benefits of disaggregation without the significant extra cost incurred by relaxing all the constraints (15).

2.1 The Master Problem Reformulation

Our proposal hinges on a reformulation of the two classical Dantzig-Wolfe reformulations of the problem corresponding to the two Lagrangian approaches described above. For the first, we define by YX the set of extreme points $(\bar{y}, \bar{x} = [\bar{x}_i]_{i \in N})$ of (18), that can be obtained (for any given value of λ) with the procedure illustrated above; then, the standard the Dantzig-Wolfe relaxation of (13)–(17) is

$$\max \sum_{(\bar{y},\bar{x}) \in YX} \big(d\bar{y} + \sum_{i \in N} c_i \bar{x}_i \big) \theta_{(\bar{y},\bar{x})} \tag{24}$$

$$\text{s.t.} \sum_{(\bar{y},\bar{x}) \in YX} \big(Dy + \sum_{i \in N} C_i x_i \big) \theta_{(\bar{y},\bar{x})} = b \tag{25}$$

$$\sum_{(\bar{y},\bar{x}) \in YX} \theta_{(\bar{y},\bar{x})} = 1 \tag{26}$$

$$\theta_{(\bar{y},\bar{x})} \geq 0 \qquad (\bar{y}, \bar{x}) \in YX. \tag{27}$$

For the decomposable version (21), the Dantzig-Wolfe relaxation rather is

$$\max \sum_{\bar{y} \in Y} (d\bar{y}) \theta_{\bar{y}} + \sum_{i \in N} \sum_{\bar{x}_i \in \bar{X}_i} (c_i \bar{x}_i) \theta_{\bar{x}_i} \tag{28}$$

$$\text{s.t.} \sum_{\bar{y} \in Y} (D\bar{y}) \theta_{\bar{y}} + \sum_{i \in N} \sum_{\bar{x}_i \in \bar{X}_i} (C_i \bar{x}_i) \theta_{\bar{x}_i} = b \tag{29}$$

$$\sum_{\bar{x}_i \in \bar{X}_i} (A_i \bar{x}_i) \theta_{\bar{x}_i} \leq \sum_{\bar{y} \in Y} (b_i \bar{y}_i) \theta_{\bar{y}}, \qquad i \in N \tag{30}$$

$$\sum_{\bar{y} \in Y} \theta_{\bar{y}} = 1 \tag{31}$$

$$\sum_{\bar{x}_i \in \bar{X}_i} \theta_{\bar{x}_i} \leq 1, \qquad i \in N \tag{32}$$

$$\theta_{\bar{y}} \geq 0, \qquad \bar{y} \in Y \tag{33}$$

$$\theta_{\bar{x}_i} \geq 0 \qquad \bar{x}_i \in \bar{X}_i . \tag{34}$$

We remark that the constraints (32) would ordinarily be equalities. In fact, they are so if the sum is extended to $\bar{X}'_i$, i.e., if 0 is added to the extreme points: then, the variable θ_0 acts as a slack—it has zero coefficients everywhere else except (32)—so we can rather use $\bar{X}_i$ as the index set and consider the constraint an inequality (note, again, that $0 \in X_i$ might happen anyway). What we propose here is a modification of (28)–(34) that directly exploits the specific quasi-separable structure, i.e., the constraints (15) defining the semi-continuous variables:

$$\max \; \textstyle\sum_{\bar{y} \in Y} (d\bar{y})\theta_{\bar{y}} + \sum_{i \in N} \sum_{\bar{x}_i \in \bar{X}_i} (c_i \bar{x}_i)\theta_{\bar{x}_i} \tag{28}$$

$$\text{s.t.} \; \textstyle\sum_{\bar{y} \in Y} (D\bar{y})\theta_{\bar{y}} + \sum_{i \in N} \sum_{\bar{x}_i \in \bar{X}_i} (C_i \bar{x}_i)\theta_{\bar{x}_i} = b \tag{29}$$

$$\textstyle\sum_{\bar{x}_i \in \bar{X}_i} \theta_{\bar{x}_i} = \sum_{\bar{y} \in Y} \theta_{\bar{y}}, \qquad i \in N \tag{35}$$

$$(31) \; , \quad (32) \; , \quad (33) \; , \quad (34). \tag{36}$$

Basically, the idea is just to replace (30) with the simpler, "logically equivalent" constraint (35). If Y is an "easy set" (which implies the integrality property and therefore the result of (Frangioni et al. 2020, §3)) then (28)–(34) can be given the "easy component" treatment (9), which just corresponds to the *partial* DW reformulation whereby the X_i are expressed as convex combination of their extreme points, but Y is not:

$$\max \; dy + \textstyle\sum_{i \in N} \sum_{\bar{x}_i \in \bar{X}_i} (c_i \bar{x}_i)\theta_{\bar{x}_i} \tag{37}$$

$$\text{s.t.} \; Dy + \textstyle\sum_{i \in N} \sum_{\bar{x}_i \in \bar{X}_i} (C_i \bar{x}_i)\theta_{\bar{x}_i} = b \tag{38}$$

$$\textstyle\sum_{\bar{x}_i \in \bar{X}_i} (A_i \bar{x}_i)\theta_{\bar{x}_i} \leq b_i y_i \qquad i \in N \tag{39}$$

$$(17) \quad (32) \; , \quad (34). \tag{40}$$

Our proposal may actually look more striking if applied to this "easy component" version, where the constraint to be replaced is rather (39):

$$\max \; dy + \textstyle\sum_{i \in N} \sum_{\bar{x}_i \in \bar{X}_i} (c_i \bar{x}_i)\theta_{\bar{x}_i} \tag{37}$$

$$\text{s.t.} \; Dy + \textstyle\sum_{i \in N} \sum_{\bar{x}_i \in \bar{X}_i} (C_i \bar{x}_i)\theta_{\bar{x}_i} = b \tag{38}$$

$$\textstyle\sum_{\bar{x}_i \in \bar{X}_i} \theta_{\bar{x}_i} = y_i, \qquad i \in N \tag{41}$$

$$(17) \; , \quad (32) \; , \quad (34). \tag{42}$$

That is, the only required modification is that the "many" extra linking constraints (15)/(30)/(39), that one has to relax if separability has to be attained, are replaced with "much fewer" linking constraints (35)/(41): exactly one for each $i \in N$. According to (Frangioni et al. 2020, Corollary 1), (28)–(36) provides the same bound as (24)–(27). If Y has the integrality property then (37)–(40) yields the same bound and therefore (37)–(42) also does; hence, due to the cited result, every version is equivalent to (28)–(34). We will comment further on these equivalences in the next subsection. Here we focus on the fact that we can efficiently solve (37)–(42) by pricing procedures that are obvious modifications to (22)/(23). We start by introducing the extended set W_i of vectors of the form $[\, x_i \,,\, v \,]$, with $v \in \{0\,,\,1\}$, where $x_i \in X_i$ and $v = 1$ if $x_i \neq 0$, while $v = 0$ if $x_i = 0$. Then, one is solving

$$\psi_i(\lambda\,,\,\mu_i) = \max\left\{ (c_i - \lambda C_i)x_i + \mu_i v \;:\; [\,x_i\,,\,v\,] \in W_i \right\}, \tag{43}$$

which is clearly not significantly more expensive than (22). In fact, if $\psi_i(\lambda) + \mu_i > 0$, which clearly implies $x_i^* \neq 0$, then this is kept as the solution, and therefore $v^* = 1$; otherwise, the solution $[\,x_i^*\,,\,v^*\,] = [\,0\,,\,0\,]$ is preferable (it has 0 value, which is better than negative in a maximization problem) and it is taken instead. The required variant of (23) is rather just

$$\psi(\lambda\,,\,\mu) = \max\left\{ \textstyle\sum_{i \in N}(d_i - \lambda D^i - \mu_i)y_i \;:\; y \in Y \right\}. \tag{44}$$

Given the optimal solutions $[\,x_i^*\,,\,v^*\,]$ and y^* of (43) and (44), respectively,

$$\begin{bmatrix} -C_i x_i^* \\ e_i v^* \end{bmatrix} \in \partial\psi_i(\lambda\,,\,\mu_i) \qquad \text{and} \qquad \begin{bmatrix} -D_i y_i^* \\ -y^* \end{bmatrix} \in \partial\psi(\lambda\,,\,\mu)$$

(where e_i is the i-th vector of the canonical base of $\mathbb{R}^n$) are clearly the right new columns to be added to (28)–(36). In the "easy component" version, only (43) need be solved. Note that (35)/(41) together with the "true" form of (32), i.e.,

$$\textstyle\sum_{\bar{x}_i \in \bar{X}_i} \theta_{\bar{x}_i} = 1$$

would imply $y_i = 1$ in each feasible solution to the master problem. Yet, our solution avoids the issue since the multiplier corresponding to the all-0 x_i solution do not participate to (35)/(41), as the column inserted in the master problem is $[\,0\,,\,0\,]$ when $x_i^* = 0$.

The master problems (28)–(36) and (37)–(42) have typically far fewer "extra" multipliers μ than their original counterparts (28)–(34) and (37)–(40); the reduced dual space should make the solution of the corresponding Lagrangian duals easier, while retaining the separability advantage over the aggregated (24)–(27). This will be confirmed in the computational section.

2.2 Relationship with Lagrangian Decomposition

We now discuss an alternative derivation of the results of the previous paragraph in terms of the well-known *Lagrangian decomposition* approach (Guignard & Kim 1987). The idea is to reformulate (13)–(17) as

$$\max\ dy + \textstyle\sum_{i \in N} c_i x_i \tag{13}$$

$$\text{s.t.}\ \ Dy + \textstyle\sum_{i \in N} C_i x_i = b \tag{14}$$

$$A_i x_i \leq b_i v_i \qquad i \in N \tag{45}$$

$$v_i = y_i \qquad i \in N \tag{46}$$

$$v_i \in \{0,\ 1\} \qquad i \in N \tag{47}$$

$$x_i \in X_i' \qquad i \in N \tag{16}$$

$$y \in Y. \tag{17}$$

That is, we create an explicit copy v of the design variables y and we use it in the semi-continuous constraints (45), while keeping the original y in whatever constraints are needed to express (17): of course we need also add the corresponding "copy constraints" (46), that are then treated as "complicating" ones. In so doing the Lagrangian relaxation decomposes in the problems (43) and (44) as in the previous section, and one readily recovers the corresponding master problems (28)–(36) and (37)–(42). Looked at in reverse, this provides an interesting interpretation of the master problem of the Lagrangian Decomposition in our model, justifying why the approach should be expected to provide better results than the direct relaxation of the semi-continuous constraints: basically, one is using a more compact form *in the master problem* of the logical constraints "$y_i = 0 \implies x_i = 0$".

In turn, this interpretation also provides a different way to prove the equivalence results of (Frangioni et al. 2020, §2). In fact, once the complicating constraints (14) are relaxed, the formulation (13)–(17) is a *star-shaped MI(N)LP* (Bacci et al. 2023): its variables can be partitioned into $n + 1$ subsets, the "core" one that contain all linking variables and n "ray" ones, each of which only shares one single binary variable with the "core". (Bacci et al. 2023, Lemma 1) (applied repeatedly as in (Bacci et al. 2023, Theorem 1)) then proves that a formulation of the convex hull of the whole problem can be simply obtained by concatenating the formulations of the convex hulls of the $n + 1$ subsets. Applied in our DW reformulation setting, this easily proves that the relaxation of (45) and (46) provides a bound that is as strong as that of the aggregated relaxation (19)–(20). The bound can actually be stronger than the one obtained by the natural disaggregated relaxation in case Y does not have the integrality property; however, we are more interested in the impact on the computational performances of the solution to the Lagrangian dual. For this reason in the next Section we will concentrate on an application where Y does have the integrality property, so that the result of (Frangioni et al. 2020, §3) applies negating any bound advantage to the new approach and thereby being a "worst case scenario" for the technique.

3 Application to Multicommodity Network Design

The Fixed-Charge Multicommodity Capacitated Network Design problem (FC-MCND) is a general network design problem with many applications (see, e.g., Crainic et al. 1999; Chouman et al. 2016; Frangioni & Gorgone 2014; Frangioni et al. 2017 and the references therein). Efficiently computing tight lower bounds on its optimal value is crucial for solution approaches, and Lagrangian techniques have been repeatedly shown to be competitive (Balakrishnan & Magnanti 1989; Crainic et al. 2004, 2001; Frangioni & Gorgone 2014; Frangioni et al. 2017, 2018; Gendron 2019; Holmberg & Yuan 2000; Crainic & Gendron 2021).

The data of FC-MCND is a directed graph $G = (N, A)$, where F_i and B_i respectively denote the set of outbound and inbound arcs of node $i \in N$, and a set of commodities K. Each $k \in K$ has a *deficit vector* $b^k = [b_i^k]_{i \in N}$ that denotes the net amount of flow asked at each node. Each arc $(a_+, a_-) = a \in A$ can only be used if the corresponding *fixed cost* $f_a > 0$ is paid, in which case the *mutual capacity* $u_a > 0$ bounds the total amount of flow on a, while *individual capacities* u_a^k bound the flow of commodity k. The *routing cost* c_a^k has to be paid for each unit of commodity k moving through a. The flow variables are denoted by x_a^k, for each $a \in A$ and $k \in K$, while the design variables are denoted by y_a, $a \in A$. While the most prevalent form of the problem in the literature does not have constraints directly linking design variables among themselves, making the design set Y just a binary hypercube, we assume the presence of a simple *budget constraint* that limits the maximum number of arcs to be opened, irrespectively of their cost. The formulation is then

$$\min \textstyle\sum_{k \in K} \sum_{a \in A} c_a^k x_a^k + \sum_{a \in A} f_a y_a \tag{48}$$

$$\text{s.t. } \textstyle\sum_{a \in F_i} x_a^k - \sum_{a \in B_i} x_a^k = b_i^k \qquad i \in N\,,\ k \in K \tag{49}$$

$$\textstyle\sum_{k \in K} x_a^k \leq u_a \qquad a \in A \tag{50}$$

$$x_a^k \leq u_a^k y_a \qquad a \in A\,,\ k \in K \tag{51}$$

$$0 \leq x_a^k \leq u_a^k \qquad a \in A\,,\ k \in K \tag{52}$$

$$\textstyle\sum_{a \in A} y_a \leq K_A \tag{53}$$

$$y_a \in \{0, 1\} \qquad a \in A \tag{54}$$

A classical approach to FC-MCND for deriving lower bounds on its optimal value is the *knapsack relaxation* (KR) (Crainic et al. 2001; Frangioni & Gorgone 2014; Frangioni et al. 2018), whereby one dualizes the flow conservation constraints (49) with multipliers $\lambda = [\lambda_i^k]_{i \in N, k \in K}$, yielding the objective function

$$\min \sum_{a \in A} \big[\sum_{k \in K} (c_a^k + \lambda_{a_+}^k - \lambda_{a_-}^k) x_a^k + f_a y_a \big] + \sum_{i \in N} \sum_{k \in K} \lambda_i^k b_i^k$$

If the budget constraint (53) were not present, the Lagrangian relaxation would reduce to $|A|$ continuous knapsack problems (KP) of the type

$$g^a(\lambda) = \min \big\{ \sum_{k \in K} [\bar{c}_a^k x_a^k] \,:\, \sum_{k \in K} x_a^k \leq u_a \,,\; (52) \big\} \tag{55}$$

with $\bar{c}_a^k = c_a^k + \lambda_{a_+}^k - \lambda_{a_-}^k$; once $g^a(\lambda)$ is computed, the trivial one-dimensional problem

$$\min \big\{ (g^a(\lambda) + f_a) y_a \,:\, (54) \big\}$$

can be solved independently for all arcs, retaining full separability. The presence of (53) immediately changes this to a quasi-separable setting, in which the (simple) knapsack problem on the y variables

$$\min \big\{ \sum_{a \in A} (g^a(\lambda) + f_a) y_a \,:\, (53) \;,\; (54) \big\}$$

has to be solved, linking the $|A|$ KP together and killing separability in the process. The "obvious" way to retain separability is dualizing the forcing capacity constraints (51) too, i.e., working in the (much) extended multipliers space $\lambda = [\gamma \,,\, \beta] = [\, \gamma_i^k \,,\, \beta_a^k \geq 0 \,]_{i \in N \,,\, a \in A \,,\, k \in K}$. This yields the objective function

$$\min \sum_{k \in K} \sum_{a \in A} (c_a^k + \gamma_{a_+}^k - \gamma_{a_-}^k + \beta_a^k) x_a^k + \sum_{a \in A} (f_a - \sum_{k \in K} \beta_a^k u_a^k) y_a$$

(besides the constant term $f^0(\lambda) = \sum_{i \in N} \sum_{k \in K} \gamma_i^k b_i^k$), whose minimization subject to the remaining (50), (52)–(54) reduce to $|A|$ KP of the form (55) except with $\bar{c}_a^k = c_a^k + \gamma_{a_+}^k - \gamma_{a_-}^k + \beta_a^k$. Independently from these one solves

$$f^1(\lambda) = \min \big\{ \sum_{a \in A} (f_a - \sum_{k \in K} \beta_a^k u_a^k) y_a \,:\, (53) \big\} .$$

This leads to a decomposable objective function of the form $f^0(\lambda) + f^1(\lambda) + \sum_{a \in A} g^a(\lambda)$, which we call the Fully decomposed Relaxation (FR).

According to what has been discussed in Sect. 2.2, the new form of the master problem can be obtained, following (45)–(47), by the simple approach of reformulating FC-MCND as

$$\min \sum_{k \in K} \sum_{a \in A} c_a^k x_a^k + \sum_{a \in A} f_a y_a \tag{48}$$

$$\text{s.t.} \sum_{a \in F_i} x_a^k - \sum_{a \in B_i} x_a^k = b_i^k \qquad i \in N \,,\; k \in K \tag{49}$$

$$\sum_{k \in K} x_a^k \leq u_a v_a \qquad a \in A \tag{56}$$

$$x_a^k \le u_a^k v_a \qquad a \in A\,,\ k \in K \qquad (57)$$

$$v_a = y_a \qquad a \in A \qquad (58)$$

$$0 \le x_a^k \le u_a^k \qquad a \in A\,,\ k \in K \qquad (52)$$

$$\textstyle\sum_{a \in A} y_a \le K_A \qquad (53)$$

$$y_a \in \{0, 1\} \qquad a \in A \qquad (54)$$

$$v_a \in \{0, 1\} \qquad a \in A. \qquad (59)$$

Then, one relaxes (58) in addition to (49) with multipliers $\lambda = [\mu_a, \lambda_i^k]_{a \in A, i \in N, k \in K}$; this yields the objective function:

$$\min \textstyle\sum_{a \in A} \big[\ \sum_{k \in K} (c_a^k + \lambda_{a_+}^k - \lambda_{a_-}^k) x_a^k + f_a y_a + \mu_a (v_a - y_a)\ \big]$$

(besides the usual constant term), whose minimization subject to (56)–(59) reduces to $|A|$ independent KP and a simple problem defined on the y variables. Indeed, for each arc $a \in A$, one has to solve the usual (55) with $\bar{c}_a^k = c_a^k + \lambda_{a_+}^k - \lambda_{a_-}^k$, and then the trivial

$$f^a(\lambda) = \min \{\ (\ g^a(\lambda) + \mu_a\ ) v_a\ :\ (59)\ \}\ .$$

Independently from the evaluation of $f^a(\lambda)$, one solves

$$f^1(\lambda) = \min \big\{ \textstyle\sum_{a \in A} (\ f_a - \mu_a\ ) y_a\ :\ (53) \big\}\ ,$$

overall leading again to a decomposable objective function of the form $f^0(\lambda) + f^1(\lambda) + \sum_{a \in A} f^a(\lambda)$, except with the terms defined differently from the (FR) case. We call this model the Quasi-separable Relaxation (QR).

4 Numerical Experiments

We have solved all the Lagrangian duals by means of a *bundle method*, i.e., a stabilised version of the Cutting Plane algorithm that corresponds to the naïve version of the Dantzig-Wolfe decomposition. Proper stabilisation is well-known to be crucial for performances and typically corresponds to modifications to the master problem: luckily, most general forms of stabilisation change the master problem in a simple way and are therefore independent from other modifications aimed at ensuring better performances, see, e.g., (Frangioni & Gendron 2013; Frangioni & Gorgone 2014). Thus we have so far avoided to explicitly describe stabilisation and we will continue to do so; the interested reader can consult, e.g., (Frangioni 2020) for details.

The implementation of the bundle method we have been using is in C++, within the `NDOSolver/FiOracle` open-source project available at https://gitlab.com/frangio68/ndosolver_fioracle_project that have already been successfully used a number of times in this context (Frangioni & Gorgone 2014; Frangioni et al. 2017, 2018). The code was compiled with GNU `g++ 4.4.5` (with `-O3` optimization option) and ran on an Opteron 6174 processor (2.2 GHz) with 32 GB of RAM, under a i686 GNU/Linux operating system.

The numerical experiments have been performed on 80 randomly generated instances already used in several papers (Frangioni & Gorgone 2014; Frangioni et al. 2017), and available at https://commalab.di.unipi.it/datasets/mmcf/#Canad.

These instances are of problems without constraints on the design variables, so we had to figure out a significant value for the constant K_A of the arc budget constraint (53). This has been chosen in the following way: first, we solved the continuous relaxation of the model excluding (53) by a general-purpose solver, then we computed the number aa of active arcs (with $y_a > 0$) in the optimal continuous solution and we set $K_A = 0.8 * aa$. The factor 0.8 was the smallest one that consistently produced feasible solutions: any K_A chosen with a smaller factor produces some unfeasible instances.

Since KR results in a single Lagrangian relaxation and therefore a unique function, we have used an aggregated bundle method (BKR) to solve it. The advantage is that we have therefore been able to solve the (aggregated) master problem with (quadratic) stabilization by using an efficient specialized quadratic solver (Frangioni 1996). Despite this the master problem time is still by far the bottleneck of the approach, and therefore we have experienced with using different maximum sizes of the bundle (set of collected subgradients/columns): this is possible with quadratic stabilisation due to *selection* and *aggregation* techniques (Frangioni 2020). In particular we report the results with a "small" maximum bundle size and a "large" one, set to 10 and 4000, respectively; in Table 1 we these are the columns labeled *BKR-10* and *BKR-4k*, respectively. We remark with bundle methods with "small" ("mininal") maximum bundle size behave in practice somehow like subgradient-type methods (Bahiense et al. 2002), and therefore BKR-10 can be considered a proxy of what a subgradient approach may provide, except subgradient methods hardly ever provide solutions with accuracy greater than `1e-4` (Frangioni et al. 2017), while bundle methods can, as our results show.

To solve FR and QR we have rather used a disaggregated bundle method where each subproblem produces separate subgradients/columns in the master problem, a-la (7). Since a specialized solver for the disaggregated master problem is not available yet, the master problem has been solved by IBM-ILOG `CPLEX 20.1`. We have therefore performed a significant tuning phase to find the stabilising term and the solution algorithm that work best in this case. Unlike in other cases (Frangioni & Gendron 2013; Frangioni & Gorgone 2014) quadratic stabilisation performed better than the linear one and it has been selected; among the applicable CPLEX algorithms, the *sifting* one turned out to be the better choice. For QR we have also adopted the "easy components" approach discussed above, while for FR we did not as it did not bring any significant performance improvement. The results

Table 1 Comparison of the different approaches

	Dimension			BKR-10	BKR-4k		BFR	BQR	
#	$\|N\|$	$\|A\|$	$\|K\|$	Gap	Time	Gap	Gap	Time	Gap
p33	20	226	40	5.7e–6	227.68	6.6e–7	4.6e–2	5.27	1.3e–07
p34	20	226	40	8.2e–6	233.14	3.5e–7	5.4e–2	5.36	3.3e–07
p35	20	226	40	7.3e–6	260.01	8.6e–7	8.9e–2	5.83	3.3e–09
p36	20	226	40	9.6e–6	57.02	8.5e–7	9.3e–2	4.59	3.8e–07
p37	20	230	200	5.1e–4	—	3.2e–4	9.2e–2	3954.59	1.4e–07
p38	20	230	200	4.8e–4	—	3.2e–4	5.7e–2	3724.92	2.6e–07
p39	20	230	200	4.5e–6	—	2.5e–5	4.5e–2	964.00	1.3e–09
p40	20	230	200	5.0e–6	—	1.5e–5	5.1e–2	838.73	4.7e–09
p41	20	292	40	3.2e–6	212.67	3.1e–8	4.9e–2	6.75	2.5e–08
p42	20	292	40	3.3e–6	130.07	2.6e–8	7.3e–2	6.66	2.8e–10
p43	20	292	40	9.9e–6	193.61	3.0e–8	9.0e–2	5.25	5.9e–10
p44	20	292	40	5.2e–6	134.04	1.3e–6	1.3e–1	6.56	2.3e–07
p45	20	292	200	6.4e–5	—	3.4e–5	2.7e–1	1375.14	7.7e–09
p46	20	292	200	5.5e–6	—	9.5e–5	9.9e–2	775.76	9.8e–07
p47	20	292	200	6.4e–6	—	7.0e–6	2.0e–2	552.91	7.8e–07
p48	20	292	200	6.2e–6	—	5.2e–5	8.4e–2	897.78	1.1e–08
p49	30	519	100	6.9e–6	—	6.4e–5	5.5e–2	261.61	2.8e–09
p50	30	519	100	1.1e–5	—	1.7e–4	7.0e–2	379.68	2.2e–09
p51	30	519	100	8.4e–6	—	2.5e–5	6.0e–2	122.46	3.2e–10
p52	30	519	100	9.8e–6	—	2.1e–5	9.4e–2	163.40	8.4e–09
p53	30	519	400	7.9e–6	—	7.1e–6	7.6e–2	11624.50	1.5e–10
p54	30	519	400	3.1e–5	—	5.3e–5	1.8e–1	14904.24	2.9e–09
p55	30	519	400	1.4e–5	—	5.6e–5	1.0e–1	7890.78	2.9e–10
p56	30	519	400	1.2e–4	—	1.8e–4	1.7e–1	15731.90	3.5e–07
p57	30	684	100	7.0e–6	—	2.9e–5	7.4e–2	155.03	1.5e–09
p58	30	684	100	6.1e–6	—	3.7e–5	7.6e–2	169.55	2.3e–08
p59	30	684	100	1.9e–5	—	8.2e–5	5.3e–2	164.25	3.2e–09
p60	30	684	100	7.2e–6	—	3.9e–5	9.7e–2	124.95	5.3e–07
p61	30	692	400	1.5e–5	—	1.2e–4	7.0e–2	11960.57	3.8e–09
p62	30	692	400	1.1e–5	—	1.4e–4	1.2e–1	16747.33	7.4e–07
p63	30	692	400	6.6e–6	—	9.5e–5	1.0e–1	9327.00	2.6e–07
p64	30	692	400	1.3e–6	—	2.6e–4	1.1e–1	8720.21	7.0e–08

of the bundle method applied to the Lagrangian dual of FR and QR are reported in Table 1 under the columns *BFR* and *BQR*, respectively. All codes have been ran with a time limit equal to 18,000 s, no limit on the number of iterations and a required relative gap tolerance of `1e-6`.

Table 1 reports the comparison among all the three approaches. The first column indicates the name of the instance, after which we reported the dimensions in terms of number of nodes, arcs and commodities, respectively. For each method we have reported the total execution time at termination and the relative gap between the optimal bound attainable (the same for all approaches, as discussed in the previous sections) and that provided by the Lagrangian approach in the columns called *time* and *gap*, respectively. If the time column is not present, this means that the time limit has been reached for all instances; this happens for BKR-10 and BFR. Otherwise, time limit being reached is indicated by "—".

The results show that BQR significantly outperforms all the other approaches for all the instances. BKR-4k at least manages to terminate in the smallest instances but not for medium and large ones; on the others it provides between about four or five significant digits of accuracy. By dint of being able to perform more iterations, due to the much leaner master problem, BKR-10 sometimes achieves better accuracy than BKR-4k in the same time, at times close to the required 6 significant digits; however, it never really terminates within the time limit. The naïve fully disaggregated approach, BFR, basically does not work at all: the much larger master problem size allows for performing far fewer iterations in the allotted time, and the much larger dual space makes it way more difficult to find a good dual solution with that effort. On the contrary, the new BQR approach always manages to find solutions with the required accuracy, albeit possibly at a significant computational cost for the largest instances (that are, however, notoriously hard to solve). Note that, by the results of (Frangioni et al. 2017), the alternative provided by subgradient-type methods would ever be able to achieve such accuracies, except possibly using smoothing techniques (Frangioni et al. 2018) but at the cost of a disproportionate number of iterations. All in all the results confirm that the proposed modification of master problem can significantly improve performances of a Lagrangian-type approach.

5 Conclusions

While the Dantzig-Wolfe/Lagrangian approach to compute tight bounds for large structured combinatorial optimization problems is well-known, its by-the-book application hardly directly provides compelling results. One of the main ways to improve its performances is to tinker with the master problem to better reflect the structure of the underlying original model. This has been already done a number of times, with Multicommodity flow and Network Design problems being a prime application, with good success (Frangioni & Gendron 2013; Frangioni & Gorgone 2014); usually general stabilisation approaches (Frangioni 2020) have also to be used at the same time, but this is in general possible and relatively easy to do (with the appropriate software). This work provides yet another step in this direction by developing a special modification for models where the crucial separability structure is not naturally present, but can be revealed by appropriate (master problem) reformulation techniques. Future work in this direction comprise the development of full-fledged approaches to the underlying problem, be them exact or heuristic, that exploit the Lagrangian bound and the corresponding primal information.

Acknowledgments The first and third authors want to express their profound gratitude to the second author for the hugely positive role that he has had in their scientific and human lives, only a tiny bit of which has been initiating the research leading to this Chapter.

References

Bacaud, L., Lemaréchal, C., Renaud, A., & Sagastizábal, C. (2001). Bundle methods in stochastic optimal power management: A disaggregated approach using preconditioners. *Computational Optimization and Applications, 20*, 227–244.

Bacci, T., Frangioni, A., Gentile, C., Tavlaridis-Gyparakis, K. (2023). New minlp formulations for the unit commitment problem with ramping constraints. Operations Research. (Published Online:9 May 2023). https://doi.org/10.1287/opre.2023.2435

Bahiense, L., Maculan, N., & Sagastizábal, C. (2002). The volume algorithm revisited: Relation with bundle methods. *Mathematical Programming, 94*(1), 41–70.

Balakrishnan, A., & Magnanti, T. (1989). A dual-ascent procedure for large-scale uncapacitated network design. *Operations Research, 37*(5), 716–740.

Borghetti, A., Frangioni, A., Lacalandra, F., & Nucci, C. (2003). Lagrangian heuristics based on disaggregated bundle methods for hydrothermal unit commitment. *IEEE Transactions on Power Systems, 18*(1), 313–323.

Cappanera, P., & Frangioni, A. (2003). Symmetric and asymmetric parallelization of a cost-decomposition algorithm for multi-commodity flow problems. *INFORMS Journal on Computing, 15*(4), 369–384.

Chouman, M., Crainic, T., & Gendron, B. (2016). Commodity representations and cut-set-based inequalities for multicommodity capacitated fixed-charge network design. *Transportation Science, 51*(2), 395–789.

Crainic, T., Frangioni, A., & Gendron, B. (1999). Multicommodity capacitated network design. In P. Soriano, & B. Sanso, (Eds.), *Telecommunications network planning* (pp 1–19). Kluwer Academics.

Crainic, T., Frangioni, A., & Gendron, B. (2001). Bundle-based relaxation methods for multicommodity capacitated fixed charge network design problems. *Discrete Applied Mathematics, 112*, 73–99.

Crainic, T., & Gendron, B. (2021). Exact methods for fixed-charge network design. In *Network Design with Applications to Transportation and Logistics* (pp 29–88). Springer.

Crainic, T., Gendron, B., & Hernu, G. (2004). A slope scaling/lagrangean perturbation heuristic with long-term memory for multicommodity capacitated fixed-charge network design. *Journal of Heuristics, 10*(5), 525–545.

Frangioni, A. (1996). Solving semidefinite quadratic problems within nonsmooth optimization algorithms. *Computers & Operations Research, 21*, 1099–1118.

Frangioni, A. (2005). About Lagrangian methods in integer optimization. *Annals of Operations Research, 139*, 163–193.

Frangioni, A. (2020). Standard bundle methods: Untrusted models and duality. In *Numerical nonsmooth optimization: state of the art algorithms* (pp. 61–116). Springer.

Frangioni, A., & Gendron, B. (2013). A stabilized structured Dantzig-Wolfe decomposition method. *Mathematical Programming, 140*, 45–76.

Frangioni, A., & Gorgone, E. (2014). Generalized bundle methods for sum-functions with "easy" components: Applications to multicommodity network design. *Mathematical Programming, 145*(1), 133–161.

Frangioni, A., Gorgone, E., & Gendron, B. (2017). On the computational efficiency of subgradient methods: A case study in combinatorial optimization. *Mathematical Programming Computation, 9*, 573–604.

Frangioni, A., Gendron, B., & Gorgone, E. (2018). Dynamic smoothness parameter for fast gradient methods. *Optimization Letters, 12*(1), 43–53.

Frangioni, A., Gendron, B., & Gorgone, E. (2020), Quasi-separable dantzig-wolfe reformulations for network design. In: M. Baiou, B. Gendron, O. Gunluk, & A. Mahjoub (Eds.), *Proceedings of the International Symposium on Combinatorial Optimization ISCO 2020. Lecture notes in computer science* (vol. 12176, pp. 227–236). Springer.

Gendron, B. (2019). Revisiting Lagrangian relaxation for network design. *Discrete Applied Mathematics, 261*, 203–218.

Gendron, B., & Larose, M. (2014). Branch-and-price-and-cut for large-scale multicommodity capacitated fixed-charge network design. *EURO Journal on Computational Optimization, 2*, 55–75.

Guignard, M., & Kim, S. (1987). Lagrangian decomposition: A model yelding stronger lagrangian bounds. *Mathematical Programming, 39*, 215–228.

Holmberg, K., & Yuan, D. (2000). A lagrangian heuristic based branch-and-bound approach for the capacitated network design problem. *Operations Research, 48*(3), 461–481.

Jones, K., Lustig, I., Farwolden, J., & Powell, W. (1993). Multicommodity network flows: The impact of formulation on decomposition. *Mathematical Programming, 62*, 95–117.

Scuzziato, M., Finardi, E., & Frangioni, A. (2018). Comparing spatial and scenario decomposition for stochastic hydrothermal Unit Commitment problems. *IEEE Transactions on Sustainable Energy, 9*(3), 1307–1317.

Decomposition-Based Algorithms for Mixed-Integer Linear Programs with Integer Subproblems

María-I. Restrepo, Bernard Gendron, and Louis-Martin Rousseau

1 Introduction

Mathematical Optimization (MO) has been a method of choice for more than half a century to tackle hard combinatorial problems emerging from applications in supply chain logistics, services, healthcare, etc. When MO models become too large to be solved directly, most researchers and practitioners resort to mathematical decomposition methods (see Wolsey, 2020) that can address formulations with an exponential number of variables, constraints, or both.

This chapter considers mixed-integer linear programs (MILP) solved via Benders Decomposition (BD). In its classic form, BD is meant to solve MILP in which only the master problem decision variables are integer or discrete, while the subproblem variables are continuous. The main objective of BD is to solve problems with complicating variables, which, when temporarily fixed to a value, yield a continuous subproblem significantly easier to handle. We refer the reader to Rahmaniani et al. (2017) for a comprehensive survey of this powerful technique.

However, when Benders subproblems involve discrete decisions, Benders decomposition in its original form is no longer applicable, and a more complex algorithmic strategy must be adopted. In this case, the problem complexity increases significantly since the generation of valid integer Benders cuts becomes non-trivial

M.-I. Restrepo (✉)
IMT Atlantique, Nantes, France
e-mail: maria-isabel.restrepo-ruiz@imt-atlantique.fr

B. Gendron
CIRRELT and DIRO, Université de Montréal, Montréal, QC, Canada

L.-M. Rousseau
CIRRELT and MAGI, Polytechnique Montréal, Montréal, QC, Canada
e-mail: louis-martin.rousseau@cirrelt.net

T. G. Crainic et al. (eds.), *Combinatorial Optimization and Applications*,
International Series in Operations Research & Management Science 358,
https://doi.org/10.1007/978-3-031-57603-4_11

due to the duality gap. Moreover, the optimal value of the objective function becomes non-convex and discontinuous in general. In the next section, we detail how researchers have tried to address this issue in several ways.

In this chapter, we study a special class of problems that, once reformulated through Benders decomposition, exhibit mixed-integer variables in the master problem and subproblems. In these problems, some master problem variables are bounded, and cardinality constraints restrict their use. These cardinality constraints are present in many practical optimization problems that involve the selection of subsets of specified size from a collection of items. We focus on improving the master problem's approximation of the impact of the subproblem on the objective cost. To achieve this, we propose two Benders decomposition approaches. The first one presents an algorithmic strategy alternating between linear and mixed-integer subproblems. This approach is based on the ideas presented in Angulo et al. (2016) and Restrepo et al. (2018), where valid classical optimality cuts are added to the master problem to avoid time-consuming exact evaluations of the subproblem cost function. These valid cuts are obtained from solving the subproblem's linear programming (LP) relaxation. When no more of these classical Benders cuts can be added to the problem, a new type of integer Benders optimality cut is added to the master problem. In this cut we exploit the fact that exactly r binary variables take the value one, and this yields a strong cut than the one presented in Laporte and Louveaux (1993) and Angulo et al. (2016). As a second approach, we propose to add the relaxation of the Benders subproblem variables in the master problem and to include these variables in the composition of integer Benders optimality cuts. As in Gendron et al. (2016), we show that this provides an efficient way to consider the subproblem costs in the master problem to ensure that effective lower bounds are computed. We compare the performance of the two methods on two different problems: the multi-commodity facility location problem and the multi-activity shift scheduling problem. The computational results show that both algorithms usually outperform a non-decomposition approach in terms of computational effort and scalability and that the performance of the two algorithms, when compared to each other, depends on the characteristics and size of the instances tested.

The chapter is organized as follows. Section 2 gives a brief overview of related research on decomposition methods for mixed-integer programming and stochastic mixed-integer programming. Section 3 presents a formal definition of the family of problems under study. In Sect. 4, we describe the two algorithmic approaches. The descriptions of the multi-commodity facility location problem and the multi-activity shift scheduling problem are presented in Sects. 5 and 6, respectively. Computational results from experiments on real-world-based and randomly generated instances are reported in Sect. 7. Section 8 concludes the chapter and outlines future research directions.

2 Related Work

Decomposition methods for problems with integer subproblems have been used successfully in deterministic mixed-integer and stochastic mixed-integer problems (SMIPs). SMIPs usually involve discrete decisions in both the master problem (*first-stage problem*) and subproblems (*second-stage problems*), in addition to the presence of a large number of scenarios that originate from the realization of the uncertain parameters. Laporte and Louveaux (1993) present one of the earliest finite exact branch-and-cut (B&C) algorithms for SMIPs with complete recourse (i.e., the second-stage problems are feasible for any realization of the stochastic parameters) and with binary first-stage variables and continuous and discrete second-stage variables. This approach solves each second-stage MIP to optimality at each iteration and constructs first-stage cuts using the objective function values of the second-stage problems. The potential drawback of this approach is the lack of scalability since each iteration requires the solution of multiple NP-hard MIPs in the second stage. Generalizing the work of Laporte and Louveaux (1993), Carøe and Tind (1998) present a method based on classical Benders decomposition and general duality theory to solve SMIPs. The authors propose to solve the second-stage problems to optimality using a pure cutting plane algorithm with Gomory cuts. The method can solve problems that include continuous, binary, and discrete first-stage decisions and binary and discrete second-stage decisions. In addition to solving problems where first-stage decisions give rise to infeasible second-stage problems (incomplete recourse). A generalization of the linear programming dual used in classical Benders decomposition to an "inference dual" is proposed in Hooker and Ottosson (2003). Since such inference dual must be definable for any type of subproblem, not just linear ones, and must provide an appropriate bound on the optimal value, its solution takes the form of a logical deduction that yields valid Benders cuts.

The ability to reuse cutting planes from one subproblem to the next is useful from the viewpoint of potentially reducing the computational effort required to solve integer subproblems while providing valid Benders cuts that improve the lower bound of the master problem. In that vein, Sherali and Fraticelli (2002) develop a Benders approach using concepts from reformulation-linearization techniques and lift-and-project cuts to solve discrete optimization problems that yield 0–1 mixed-integer subproblems, such as those encountered in SMIPs. The cutting planes are derived in terms of functions of the first-stage variables, enabling them to be re-used in subsequent subproblems simply by revising them according to the corresponding first-stage solutions. Globally valid Benders cuts are also obtained by recognizing these cuts as functions of the first-stage variables. Sen and Higle (2005) consider SMIPs with binary first-stage variables and binary and continuous second-stage variables. Since the authors assume that the recourse matrix is fixed, they develop disjunctive cuts for the second stage that can be valid across all scenarios by calculating an appropriate right-hand side function.

Alternative decomposition methods in which the second stage integer problems are solved using B&C methods are presented in Sen and Sherali (2006). The authors propose two approaches. The first approach uses second-stage variables as continuous decisions in the first stage. The second approach creates a master problem using only first-stage decisions. Numerical experiments on both approaches show that the former is convenient in a deterministic setting or for SMIPs with only a few second-stage scenarios. On the contrary, if the number of outcomes in the second stage is very large, the second approach allows a more direct facility to create a master program that uses only the first-stage variables. Sherali and Zhu (2006) proposes a decomposition-based branch-and-bound (B&B) algorithm for solving two-stage stochastic programs with mixed-integer first- and second-stage variables. Lower bounds for the nodes in the B&B tree are computed by applying a modified Benders approach that coordinates a master program with lower-bounding scenario-based second-stage subproblems. These subproblems are derived by sequentially constructing a certain partial convex hull representation of the two-stage solution space. Gade et al. (2014) consider a class of two-stage SMIPs with binary variables in the first stage and general integer variables in the second stage. The authors develop decomposition algorithms by utilizing Gomory cuts to obtain iteratively tighter approximations of the second-stage integer programs. These Gomory cuts are parametrized by the first-stage decisions, thus allowing the adoption of a decomposition framework. Unlike standard Benders decomposition, the second-stage LP relaxation is updated from one iteration to the next, and moreover, these approximations include affine functions of first-stage variables. Zhang and Küçükyavuz (2014) study two-stage SMIPs with general integer variables in both stages. Based on Benders decomposition, the authors propose an algorithm that utilizes Gomory cuts in both stages. The first-stage decision variables parameterize the Gomory cuts for the second-stage scenario subproblems.

More recent approaches to solve SMIPs include a Benders decomposition method that alternates between linear and mixed-integer subproblems (Angulo et al., 2016), a B&C algorithm with its nodes inheriting Benders cuts that are valid for their ancestor nodes (Qi & Sen, 2017), and a generalized Benders decomposition-based B&C algorithm (Li & Grossmann, 2019). Angulo et al. (2016) consider problems with first-stage binary variables and second-stage integer variables and propose to use the linear relaxation of the recourse function as the unique intermediate approximation for the second-stage function. The authors also propose to use a cut-generating linear program that considers information from all solutions found up to any given stage of the method. The alternating cuts strategy works better in problems where the computational bottleneck of the problem is in evaluating the recourse function. Qi and Sen (2017) consider problems with bounded integer decision variables in both stages. The authors develop two closely related convexification schemes, which use multi-term disjunctive cuts to obtain approximations of the second-stage mixed-integer programs. One of the main advantages of the decomposition scheme is that the authors use a Benders-based B&C approach in which linear programming approximations are strengthened sequentially. Li and Grossmann (2019) develop a decomposition algorithm for SMIPs that performs spatial B&B search on the first stage variables. Each node in the B&B search is solved with a Benders-like decomposition algorithm where both Lagrangean cuts and Benders cuts are included in the Benders master problem. The Lagrangean cuts are derived from Lagrangean decomposition. The Benders cuts are derived from the Benders subproblems, which are convexified by cutting planes.

Cordeau et al. (2001), Kim and Mehrotra (2015), and Elçi and Hooker (2022) present alternative approaches inspired by practical applications to solve problems that include integer variables in the master problem and subproblems. Cordeau et al. (2001) introduce a model and a heuristic 3-phase solution methodology for the simultaneous routing of aircraft and scheduling of crews. The idea is to first solve the LP relaxation of the master problem and subproblems. In the second phase, the problem is solved by imposing integrality constraints on the variables of the master problem. The third and last phase consists in adding the integrality constraints in the subproblem variables, and solving the integer subproblem once the values of the master problem variables are held fixed. Kim and Mehrotra (2015) address an integrated staffing and scheduling problem under demand uncertainty that is formulated as a two-stage SMIP. The authors propose new mixed-integer rounding inequalities for the second-stage problem to convexify the recourse function and to obtain a tight formulation that describes the convex hull of feasible solutions in the second stage. Elçi and Hooker (2022) apply logic-based Benders decomposition to two-stage stochastic planning and scheduling problems in which the second stage corresponds to a scheduling task. The authors solve the master problem with mixed integer linear programming and the subproblem with constraint programming. They use no-good cuts as well as analytic logic-based cuts as Benders cuts.

The Benders decomposition methods presented in this work have some similarities to the approach presented in Angulo et al. (2016), since our algorithmic strategy also alternates between linear and mixed-integer subproblems. However, we exploit the fact that exactly r binary master problem variables take the value one and this yields a strong cut than the one presented in Angulo et al. (2016). In addition, we propose to add the relaxation of the Benders subproblem variables to the master problem and to include these variables in the composition of integer Benders optimality cuts. As in Gendron et al. (2016), we show that this provides an efficient way to consider the subproblem costs in the master problem to ensure that effective lower bounds are computed.

3 Problem Description

In this section, we formally define the general problem addressed in our work. We also present the assumptions needed to apply the proposed Benders decomposition approaches. We consider MILPs of the form

$$\text{(MILP)} \min f^T x + d^T v + c^T y \tag{1}$$

$$\mathbb{1}^T x = r \tag{2}$$

$$\mathbf{A}x + \mathbf{D}v \leq b \tag{3}$$

$$\mathbf{T}x + \mathbf{W}y = h \tag{4}$$

$$x \in X,\ x_i \in \{0, 1\},\ i \in I \tag{5}$$

$$v \in V \tag{6}$$

$$y \in Y, \tag{7}$$

where the set X is described by linear constraints. Similarly, we assume that V and Y are bounded and represented as follows:

$$V = \{v \mid \mathbf{M}v \geq a, v \geq 0, v_j \text{ integer for } j \in J_2\} \tag{8}$$

$$Y = \{y \mid \mathbf{N}y \geq e, y \geq 0, y_k \text{ integer for } k \in K_2\}, \tag{9}$$

where index sets J_2, K_2 provide the subsets of variables that are restricted to be integer/binary. Observe that we address a special kind of problem where:

1. The utilization of variables x is restricted by a cardinality constraint (constraint (2)).
2. For any value of x, constraints (4) are always feasible.
3. The problem (1)–(7) can be decomposed into a master problem with x and v representing the master problem variables and a subproblem where y represents the subproblem variables.

MILP problems such as (1)–(7) in which all y variables are continuous (e.g., the set K_2 is empty) can be solved by using the classical version of Benders decomposition. The main objective of this algorithm is to solve problems with complicating variables (e.g., variables x), which, when temporarily fixed to a value $\bar{x}$, yield a problem significantly easier to handle. Therefore, assuming that integrality constraints on y variables are relaxed, problem (1)–(7) can be decomposed into a *Benders master problem* and a *Relaxed Benders subproblem*, denoted as (MP) and (RSP), respectively. The formulation of (RSP) is as follows:

$$\text{(RSP)}\ \min c^T y \tag{10}$$

$$\mathbf{W}y = h - \mathbf{T}\bar{x} \tag{11}$$

$$\mathbf{N}y \geq e \tag{12}$$

$$y \geq 0 \tag{13}$$

Since we assume that for any value of $\bar{x}$ (RSP) is always feasible, when (RSP) is solved to obtain classical Benders cuts, no *feasibility cuts* will be generated. We introduce the following notation to define the *optimality cuts*. Let α and π be the dual variables associated with constraints (11) and (12), respectively. Let $\boldsymbol{\Delta}$ be the polyhedron defined by the constraints of the dual of model (10)–(12). Let $E_{\mathbf{\Delta}}$ be the set of extreme points of $\mathbf{\Delta}$, and let θ be a variable that represents the value of the objective of (RSP). The Benders optimality cuts are then defined as follows

$$\theta \geq \alpha^T e + \pi^T h - \pi^T \mathbf{T}x,\ \alpha, \pi \in E_{\mathbf{\Delta}} \tag{14}$$

Optimality cuts ensure that the value of θ is greater than or equal to the value of the Benders subproblem. We have assumed that Benders subproblems are linear programs to derive these cuts, i.e., the integrality of variables y is relaxed. The relaxation of the model (MILP) obtained by relaxing these integrality constraints can thus be reformulated as the following master problem

$$\text{(MP)}\ \min f^T x + d^T v + \theta \tag{15}$$

$$\mathbb{1}^T x = r \tag{16}$$

$$\theta \geq \alpha^T e + \pi^T h - \pi^T \mathbf{T}x,\quad \alpha, \pi \in E_{\mathbf{\Delta}} \tag{17}$$

$$x \in X, x_i \in \{0, 1\},\ i \in I \tag{18}$$

$$v \in V \tag{19}$$

Optimality cuts (17) do not need to be exhaustively generated since only a subset of them will be active in an optimal solution. Hence, an iterative cutting plane algorithm can generate only the subset of cuts that will yield an optimal solution of (MP).

As mentioned above, when Benders subproblems involve discrete decisions, Benders decomposition in its original form is no longer applicable, and a more complex algorithmic strategy must be adopted. To address this problem, we propose two approaches. The first one presents an algorithmic strategy that alternates between linear and mixed-integer subproblems. In this approach, valid classical optimality cuts are added to the master problem to avoid time-consuming exact evaluations of the subproblem cost function. These valid cuts are obtained from solving problem (RSP). When no more classical Benders cuts can be added to the problem, a new type of integer Benders optimality cut is added to the master problem. This optimality cut exploits the fact that exactly r binary variables take the value one (constraints (2)). As a second approach, we propose to add the relaxation of the Benders subproblem variables y in the master problem and to include these y variables in the composition of integer Benders optimality cuts. The next section presents these two decomposition approaches.

4 Methodology

This section presents two decomposition-based approaches to solve problems including integrality restrictions in the subproblem variables. Section 4.1 presents the first algorithmic strategy that alternates between linear and mixed-integer subproblems. Section 4.2 presents the second algorithmic strategy that adds the relaxation of the Benders subproblem variables to the master problem and that includes these variables in the composition of integer Benders optimality cuts.

4.1 Algorithm 1: Alternating Between Classical and Integer Benders Cuts

Let $t \geq 0$ be the index of each iteration. Let $\bar{\theta}(t), \bar{x}(t), \bar{v}(t)$ denote the optimal value of variables θ, x and v at iteration t, respectively. Note that $\zeta^l(t) = f^T\bar{x}(t) + d^T\bar{v}(t) + \bar{\theta}(t)$ is a lower bound on (1)–(7) at iteration t. Let $\bar{Z}(t)$ be the optimal value of the Benders subproblem (10)–(12) at iteration t (the integrality constraints on variables y are relaxed). Let $Z(t)$ be the optimal value of the Benders subproblem (10)–(12) at iteration t when integrality constraints on variables y are imposed. Note that $\bar{\zeta}^u(t) = f^T\bar{x}(t) + d^T\bar{v}(t) + \bar{Z}(t)$ is an upper bound on the Benders decomposition at iteration t, which we call the *approximate upper bound*. Let $\delta(\bar{x}(t)) = \{j \in J | \bar{x}(t) = 0\}$ be a subset of the master problem binary variables x that take the value zero at iteration t.

Algorithm 1 consists of two phases. In the first phase, the algorithm solves the linear programming (LP) relaxation of model (1)–(7) following the method of McDaniel and Devine (1977). This is achieved by alternating between the LP relaxation of the problem (15)–(19) and the generation of classical Benders cuts until no more cuts can be found or the gap between the approximate upper bound $\bar{\zeta}^u(t)$ and the lower bound $\zeta^l(t)$ is small enough (i.e., lower than a tolerance ϵ_1). The generation of classical Benders cuts is improved by adopting the method presented in Algorithm 2 in Papadakos (2008), an alternative to solving the extra auxiliary subproblem introduced in Magnanti and Wong (1981). The algorithm then enters the second phase, where it seeks integer solutions for the master problem by performing B&B. In this phase, the algorithm alternates between solving the Benders master problem with a B&B algorithm and the generation of classical Benders cuts, but when no more of these cuts can be generated (i.e., $\bar{Z}(t) - \bar{\theta}(t)/\bar{Z}(t) \leq \epsilon_2$), the algorithm solves the MIPs of the Benders subproblems. Then, a feasible integer solution $Z(t)$ is computed, and integer Benders cuts (20) are generated if needed, in which case the algorithm restarts with the cycle solving the master problem with B&B/adding classical Benders cuts. Integer Benders cuts are generated to guarantee the optimality of the method because Benders subproblems are MIPs that do not possess the integrality property. Hence, if $Z(t) - \bar{Z}(t)/Z(t) > \epsilon_3$, the following integer Benders optimality cut is added to the master problem:

$$\theta \geq Z(t) - (Z(t) - \bar{\theta}(t))\Big(\sum_{j \in \delta(\bar{x}(t))} x_j\Big) \tag{20}$$

Because the value of θ is to be minimized, the objective of (20) is to eliminate the current solution $\bar{x}(t)$ by changing to one at least one master problem variable x taking the value of zero. However, if the current solution $\bar{x}(t)$ is the optimal solution, the right-hand side is then maximized and is equal to $Z(t)$, the value of the integer Benders subproblem. This optimality cut exploits the fact that exactly r binary variables take value one and are the adaptation to this case of the optimality cut used in the integer L-shaped method for stochastic programming (Laporte & Louveaux, 1993).

Algorithm 1 uses the following Boolean variables: P_0 indicates if the algorithm is in the first (P_0=true) or in the second (P_0=false) phase and Cut indicates if some cuts, classical or integer, have been generated (Cut=true) or not (Cut=false). The algorithm uses three parameters $\epsilon_i \in [0, 1]$, $i = 1, \ldots, 3$, which represent thresholds on different relative gaps: ϵ_1 controls when the first phase (solving the LP relaxation of MP) is stopped using the relative gap between $\bar{\zeta}^u(t)$ and $\zeta^l(t)$, which has to be small enough; ϵ_2 controls the generation of classical Benders cuts in case the relative gap between $\bar{\theta}(t)$ and the upper bound associated with the LP relaxation of the Benders subproblem $\bar{Z}(t)$ is large enough; ϵ_3 controls the generation of integer Benders cuts if the relative gap between the MIP and the LP relaxation bounds of the Benders subproblem is small enough.

4.2 *Algorithm 2: Including Information from the Benders Subproblems in the Master Problem*

In this algorithm, only integer Benders optimality cuts are generated at each iteration. These integer Benders cuts are similar to (20), except that they directly incorporate subproblem variables y instead of the additional variable θ. In addition, the linear relaxation of the subproblem variables is included in the master problem. The mathematical formulations for the Benders subproblem (SP) and the Benders master problem (MP') follow.

$$\text{(SP)} \min c^T y \tag{21}$$

$$\mathbf{W}y = h - \mathbf{T}\bar{x} \tag{22}$$

$$y \in Y \tag{23}$$

$$\text{(MP')} \min f^T x + d^T v + c^T y \tag{24}$$

Data: $t = 0$, $\zeta^l(t) = -\infty$, $\zeta^u(t) = +\infty$, $\bar{\zeta}^u(t) = +\infty$, P_0 = true, Cut = true
Result: Optimal solution x^* and v^* to (1)–(7) and optimal value of the objective function
if P_0=*true* **then**
  while $\overline{\zeta}^u(t) - \zeta^l(t)/\overline{\zeta}^u(t) > \epsilon_1$ *and* Cut=*true* **do**
    $t = t + 1$;
    Solve LP relaxation of the Benders master problem, $\zeta^l(t)$ is the new lower bound and $\bar{x}(t)$, $\bar{v}(t)$, $\bar{\theta}(t)$ the optimal solution;
    Cut = false ;
    Solve the LP relaxation of the Benders subproblem, $\bar{Z}(t)$ is the optimal value ($\bar{Z}(t) = +\infty$ if $\bar{x}(t)$ infeasible);
    if $(\bar{Z}(t) - \bar{\theta}(t))/\bar{Z}(t) > \epsilon_2$ **then**
      Add a classical Benders cut (14) to the Benders master problem, Cut = true;
    end
    $\overline{\zeta}^u(t) = \min\{\overline{\zeta}^u(t), f^T\bar{x}(t) + d^T\overline{v}(t) + \bar{Z}(t)\}$
  end
  Cut=true
end
Declare variables x and $v \in J_2$ as integers in the Benders master problem;
while Cut=*true* **do**
  $t = t + 1$;
  Solve the MILP of the Benders master problem, $\zeta^l(t)$ is the new lower bound and $\bar{x}(t)$, $\bar{v}(t)$, $\bar{\theta}(t)$, the optimal solution;
  Cut = false ;
  Solve the LP relaxation of the Benders subproblem, $\bar{Z}(t)$ is the optimal value ($\bar{Z}(t) = +\infty$ if $\bar{x}$ infeasible);
  if $(\bar{Z}(t) - \bar{\theta}(t))/\bar{Z}(t) > \epsilon_2$ **then**
    Add a classical Benders cut (14) to the Benders master problem, Cut = true;
  end
  if Cut=*false* **then**
    Solve the MILP of the Benders subproblem, return $Z(t)$;
    if $(Z(t) - \bar{Z}(t))/Z(t) > \epsilon_3$ **then**
      Add integer Benders cut (20) to the Benders master problem, Cut=true;
    end
    $\zeta^u(t) = \min\{\zeta^u(t), f^T\overline{x}(t) + d^T\overline{v}(t) + Z(t)\}$
    if $\zeta^u(t) = f^T\bar{x}(t) + d^T\overline{v}(t) + Z(t)$ **then**
      $x^* = \bar{x}(t)$, $v^* = \bar{v}(t)$
    end
  end
end
return x^*, v^*, $\zeta^l(t)$ and $\zeta^u(t)$

Algorithm 1: Decomposition-based algorithm with classical and integer Benders cuts

$$\mathbf{A}x + \mathbf{D}v \leq b \tag{25}$$

$$\mathbf{T}x + \mathbf{W}y = h \tag{26}$$

$$\mathbf{N}y \geq e \tag{27}$$

$$c^T y \geq Z(t) - (Z(t) - c^T \bar{y}(t))\Big(\sum_{j \in \delta(\bar{x}(t))} x_j\Big), t = 0, 1, \ldots \tag{28}$$

$$x \in X, x_i \in \{0, 1\},\ i \in I \tag{29}$$

$$v \in V \tag{30}$$

$$y \geq 0 \tag{31}$$

where $\bar{y}(t)$ denote the optimal values of variables y in the master problem at iteration t, and $\bar{x}(t)$, $\bar{v}(t)$ denote the optimal value of variables x and v at iteration t, respectively. Note that $\zeta^l(t) = f^T\bar{x}(t) + d^T\bar{v}(t) + c^T\bar{y}(t)$ is a lower bound on (1)–(7) at iteration t.

The definition of $\bar{Z}(t)$, $Z(t)$, $\zeta^u(t)$, and $\delta(\bar{x}(t))$ is the same as the one presented in Sect. 4.1. Algorithm 2 consists of only one phase where it seeks integer solutions for the master problem by performing B&B. For each integer solution, the algorithm solves the LP relaxation and the MIP of subproblem (21)–(23) to recover solutions $\bar{Z}(t)$ and $Z(t)$, respectively. If the relative gap between $Z(t)$ and $\bar{Z}(t)$ is larger than a tolerance, cut (28) is added to the master problem. Observe that the integer Benders optimality cut (28) is very similar to cut (20). However, instead of using θ, we propose to include the Benders subproblem variables y multiplied by their cost c to ensure that effective lower bounds are computed.

The algorithm uses the following Boolean variable: *Cut* indicates if some integer cuts have been generated (*Cut*=true) or not (*Cut*=false). This algorithm also uses parameter ϵ_3, which controls the generation of integer Benders cuts in case the relative gap between the MIP and the LP relaxation bounds of the Benders subproblem is small enough.

5 Application to Multi-Commodity Facility Location

Multi-commodity facility location (MCFL) refers to an extension of a facility location problem in which clients can demand different goods (commodities). In this section, we consider a special case of the MCFL, where each commodity can only be served by a limited number of facilities.

Let J, I, and K be the set of facilities, the set of clients, and the set of commodities, respectively. Parameters f_j and c_{ij}^k denote the fixed cost of opening facility $j \in J$ and the non-negative cost associated with serving client $i \in I$ from facility $j \in J$ with commodity $k \in K$, respectively. The demand of each client $i \in I$ for commodity $k \in K$ is represented by b_i^k. The maximum capacity of facility

Data: $t = 0$, $\zeta^l(t) = -\infty$, $\zeta^u(t) = +\infty$, Cut = true
Result: Optimal solution x^* to (1)-(7) and optimal value of the objective function
while Cut=*true* **do**
  $t = t + 1$;
  Solve the MILP of the Benders master problem, $\zeta^l(t)$ is the new lower bound and $\bar{x}(t)$, $\bar{y}(t)$, the optimal solution;
  Cut = false ;
  Solve the LP relaxation of the Benders subproblem, $\bar{Z}(t)$ is the optimal value ($\bar{Z}(t) = +\infty$ if $\bar{x}$ infeasible);
  Solve the MILP of the Benders subproblem, $Z(t)$ is the optimal value;
  if $(Z(t) - \bar{Z}(t))/Z(t) > \epsilon_3$ **then**
    Add integer Benders cut (28) to the Benders master problem, Cut=true;
  end
  $\zeta^u(t) = \min\{\zeta^u(t), f^T\bar{x}(t) + d^T\overline{v}(t) + Z(t)\}$
  if $\zeta^u(t) = f^T\bar{x}(t) + d^T\overline{v}(t) + Z(t)$ **then**
    $x^* = \bar{x}(t)$, $v^* = \bar{v}(t)$
  end
end
return x^*, v^*, $\zeta^l(t)$ and $\zeta^u(t)$

Algorithm 2: Decomposition-based algorithm with integer Benders cuts

$j \in J$ to serve commodity $k \in K$ is denoted by q_j^k. Parameter ρ^k represents the exact number of facilities serving commodity $k \in K$. The model uses three decision variables: x_j is a binary variable that takes value one if facility $j \in J$ is open, and it assumes value zero otherwise; y_{ij}^k is a binary variable that takes value one if client $i \in I$ is served by facility $j \in J$ for commodity $k \in K$, and it assumes value zero otherwise; and v_j^k is a binary variable that takes value one if facility $j \in J$ serves at least one client with commodity $k \in K$. The formulation of the MCFL problem is as follows:

$$\min \sum_{j \in J} f_j x_j + \sum_{i \in I} \sum_{j \in J} \sum_{k \in K} c_{ij}^k y_{ij}^k \tag{32}$$

$$v_j^k \leq x_j, \quad j \in J, \ k \in K \tag{33}$$

$$\sum_{j \in J} v_j^k = \rho^k, \quad k \in K \tag{34}$$

$$\sum_{j \in J} y_{ij}^k = 1, \quad i \in I, \ k \in K \tag{35}$$

$$\sum_{i \in I} b_i^k y_{ij}^k \leq q_j^k v_j^k, \quad j \in J, \ k \in K \tag{36}$$

$$y_{ij}^k \leq v_j^k, \quad i \in I, \ j \in J, \ k \in K \tag{37}$$

$$x_j \in \{0, 1\}, \quad j \in J \tag{38}$$

$$v_j^k \in \{0, 1\}, \quad j \in J, \; k \in K \tag{39}$$

$$y_{ij}^k \in \{0, 1\}, \quad i \in I, \; j \in J, \; k \in K \tag{40}$$

The objective (32) is to minimize the total cost of opening facilities and the total costs of serving the demands of clients for each commodity. Constraints (33) set the link between variables v_j^k and variables x_j (i.e., if facility $j \in J$ is not open, it cannot serve any client with commodity $k \in K$). Constraints (34) guarantee that exactly ρ^k facilities serve clients with commodity $k \in K$. Constraints (35) ensure that each client $i \in I$ is served by exactly one facility $j \in J$ for commodity $k \in K$. Constraints (36) ensure that each facility $j \in J$ does not exceed its capacity for commodity $k \in K$. In a similar way, constraints (37) denote the link between variables y_{ij}^k and variables v_j^k (i.e., if facility $j \in J$ does not have the configuration to serve clients with commodity $k \in K$, then any client $i \in I$ cannot be served for commodity $k \in K$ by facility $j \in J$). Constraints (38)–(40) set the binary nature of the variables x_j, v_j^k and y_{ij}^k.

The set of valid inequalities (41) is introduced in problem (32)–(40) to guarantee that there is always enough capacity to satisfy the demand for each commodity.

$$\sum_{j \in J} q_j^k v_j^k \geq \sum_{i \in I} b_i^k, \quad k \in K \tag{41}$$

5.1 Benders Decomposition Approach

The structure of model (32)–(41) suggests that when variables v_j^k are fixed to particular values $\bar{v}_j^k$, $j \in J, k \in K$, the resulting model decomposes into $|K|$ independent Benders subproblems, one for each commodity. Each subproblem includes the variables associated with the service of clients with facility $j \in J$ for each commodity $k \in K$. The formulation of the Benders subproblem, denoted as $Z^k(\bar{v}^k)$, for commodity k is as follows:

$$Z^k(\bar{v}^k) = \min \sum_{i \in I} \sum_{j \in J} c_{ij}^k y_{ij}^k \tag{42}$$

$$\sum_{j \in J} y_{ij}^k = 1, \quad i \in I \tag{43}$$

$$\sum_{i \in I} b_i^k y_{ij}^k \leq q_j^k \bar{v}_j^k, \quad j \in J \tag{44}$$

$$y_{ij}^k \leq \bar{v}_j^k, \quad i \in I, \; j \in J \tag{45}$$

$$y_{ij}^k \in \{0, 1\}, \quad i \in I, \; j \in J. \tag{46}$$

Since variables y_{ij}^k are required to be integers and Benders subproblems (42)–(46) do not possess the integrality property, the classical Benders decomposition approach needs to be modified to ensure convergence to an optimal solution. As explained in Sect. 4.1, we first generate classical Benders cuts by relaxing the integrality constraints (46) on variables y_{ij}^k giving rise to model $\overline{Z}^k(\bar{v}^k)$. Second, we will generate integer Benders cuts to guarantee the convergence of the method to an optimal solution. Observe that when valid inequalities (41) are added to problem (32)–(40) there is always enough capacity to satisfy the demands and a feasible (fractional) solution always exists. Therefore, when $\overline{Z}^k(\bar{v}^k)$ is solved to obtain classical Benders cuts, no feasibility cuts will be generated.

We introduce the following notation for each commodity $k \in K$ to define optimality cuts. Let π_i^k, α_j^k and ρ_{ij}^k be the dual variables associated with constraints (43), (44) and (45) from $\overline{Z}^k(\bar{v}^k)$, respectively. Let $\mathbf{\Delta}^k$ be the polyhedron obtained from the projection, over the space of variables π_i^k, α_j^k and ρ_{ij}^k, of the set of feasible solutions to the dual of model $\overline{Z}^k(\bar{v}^k)$. Let $E_{\mathbf{\Delta}^k}$ be the set of extreme points of $\mathbf{\Delta}^k$ and let θ^k be a non-negative variable that represents the value of the objective of $\overline{Z}^k(\bar{v}^k)$. The Benders optimality cut for commodity $k \in K$ is then defined as follows

$$\theta^k \geq \sum_{i \in I} \pi_i^k - \sum_{j \in J} \alpha_j^k q_j^k v_j^k - \sum_{i \in I} \sum_{j \in J} \rho_{ij}^k v_j^k \tag{47}$$

Optimality cuts (47) ensure that the value of each variable θ^k is larger than or equal to the LP relaxation value of the Benders subproblem. To derive these cuts, we have assumed that the Benders subproblem is a linear program (i.e., the integrality of variables y_{ij}^k is relaxed). The relaxation of model (32)–(41) obtained by relaxing the integrality of variables y_{ij}^k can thus be reformulated as the following master problem:

$$\min \sum_{j \in J} f_j x_j + \sum_{k \in K} \theta^k \tag{48}$$

$$\theta^k \geq \sum_{i \in I} \pi_i^k - \sum_{j \in J} \alpha_j^k q_j^k v_j^k - \sum_{i \in I} \sum_{j \in J} \rho_{ij}^k v_j^k, \quad k \in K, (\pi^k, \alpha^k, \rho^k) \in E_{\mathbf{\Delta}^k} \tag{49}$$

$$\sum_{j \in J} v_j^k = \rho^k, \quad k \in K \tag{50}$$

$$v_j^k \leq x_j, \; l \; j \in J, \; k \in K \tag{51}$$

$$\sum_{j \in J} q_j^k v_j^k \geq \sum_{i \in I} b_i^k, \quad k \in K \tag{52}$$

$$v_j^k \in \{0, 1\}, \quad j \in J, \, k \in K \tag{53}$$

$$x_j \geq 0, \quad j \in J \tag{54}$$

Observe that replacing $x_j \in \{0, 1\}$ by $x_j \geq 0$ gives an equivalent model. Therefore, branching on variables v_j^k suffices.

Since Benders subproblems are binary problems that do not possess the integrality property, we propose to use the algorithmic strategy presented in Sect. 4.1. Hence, for each k such that $Z^k(t) - \overline{Z}^k(t)/Z^k(t) > \epsilon_3$, we define the following integer Benders cut:

$$\theta^k \geq Z^k(t) - (Z^k(t) - \bar{\theta}^k(t))\Big(\sum_{j \in \delta(\bar{v}^k(t))} v_j^k\Big), \tag{55}$$

where $\bar{\theta}^k(t)$ is the value of variables θ^k at iteration t and $\delta(\bar{v}^k(t)) = \{j \in J | \bar{v}^k(t) = 0\}$. The integer Benders cut (55) aims to eliminate the current solution $\bar{v}^k(t)$ by swapping at least one facility from its client allocation to a different one.

6 Application to Multi-Activity Shift Scheduling

Multi-activity shift scheduling is an extension of shift scheduling where employee schedules can include multiple work activities. In the multi-activity shift scheduling problem, shifts are defined over a planning horizon of one day divided into $|I|$ time intervals of equal length. The multi-activity shift scheduling problem includes a set of work activities J, a set of employees E, and the number of employees b_{ij} required to work in activity $j \in J$ at time interval $i \in I$. Three types of non-negative costs are associated with each work activity: activity allocation costs c_{ij}, activity over-covering costs c_{ij}^+, and activity under-covering costs c_{ij}^-. Multi-activity daily shifts are characterized by their start time, working length, break allocation, and work activity placement at any time interval. The composition of multi-activity shifts is usually constrained by work rules and employee agreements.

When employees have individual preferences and skills, the personalized version of the multi-activity shift scheduling problem appears. The anonymous version of the multi-activity shift scheduling problem corresponds to when employees have identical skills. The application presented in this section considers the anonymous version of the problem, which will be denoted as AMASS in the sequel of this paper. The objective of the AMASS is to determine and to assign one shift to each employee $e \in E$, so that the sum of activity allocation costs, over-covering costs, and under-covering costs is minimized and employee requirements (demands) are guaranteed for each work activity at each time interval in the planning horizon.

The formulation used to model the AMASS is similar to the one presented in Côté et al. (2011b). This formulation uses *context-free grammars* to implicitly include all the work regulations to build multi-activity shifts. A context-free grammar is a set of recursive rewriting rules (or productions) used to generate patterns of strings or in the case of personnel scheduling to generate shifts. A context-free grammar consists of a four-tuple $G = \langle \Sigma, N, S, P \rangle$, where Σ is an alphabet of characters called the terminal symbols, N is a set of non-terminal symbols, $S \in N$ is the starting symbol, and P is a set of productions represented as $A \to \alpha$, where $A \in N$ is a non-terminal symbol and α is a sequence of terminal and non-terminal symbols. The productions of a grammar are used to generate new symbol sequences until terminal symbols have replaced all non-terminal symbols.

A parse tree is a tree where each inner node is labeled with a non-terminal symbol, and each leaf is labeled with a terminal symbol. A grammar recognizes a sequence (shift) if and only if a parse tree exists where the leaves, when listed from left to right, reproduce the sequence. A *DAG* Γ is a directed acyclic graph that embeds all parse trees associated with sequences of a given length $|I|$ recognized by a grammar. The DAG Γ has an and/or structure where the and-nodes represent productions from P, and or-nodes represent non-terminals from N and letters from Σ. An and-node is true if all of its children are true. An or-node is true if one of its children is true. The root node is true if the grammar accepts the sequence encoded by the leaves.

Let O_{il}^{π} be the or-nodes associated with $\pi \in N \cup \Sigma$ (i.e., with non-terminals from N or letters from Σ) that generate a subsequence from time interval i of length l. Note that if $\pi \in \Sigma$, the node is a leaf and l is equal to one. On the contrary, if $\pi \in N$ the node represents a non-terminal symbol and $l \geq 1$. $A_{il}^{\Pi,m}$ is the mth and-node representing production $\Pi \in P$ generating a subsequence from time interval i of length l. There are as many $A_{il}^{\Pi,m}$ nodes as there are ways of using Π to generate a sequence of length l from time interval i. The sets of or-nodes, and-nodes, and leaves of DAG Γ are denoted by O, A, and L, respectively. The root node is described by $O_{1|I|}^{S}$ and its children by $A_{1|I|}^{\Pi,m}$. These children of the root node, also called *shifts shells*, represent all feasible combinations of shift start times and lengths without activity and break allocation. The children of or-node O_{dl}^{π} are represented by $ch(O_{dl}^{\pi})$ and its parents by $par(O_{dl}^{\pi})$. Similarly, the children of and-node $A_{dl}^{\Pi,m}$ are represented by $ch(A_{dl}^{\Pi,m})$ and its parents by $par(A_{dl}^{\Pi,m})$. For more details on using context-free grammar for modeling personnel scheduling, the reader is referred to: Côté et al. (2011b), Boyer et al. (2014), and Restrepo et al. (2018).

In the AMASS, we use grammars to (1) generate sequences of work spanning a minimum and a maximum number of time intervals; (2) generate sequences of breaks spanning a minimum and a maximum number of time intervals; (3) guarantee that each shift has a break between two sequences of work; (4) guarantee that there is a break between activity transitions; (5) define a minimum and a maximum number of consecutive time intervals for each activity (e.g., activity one cannot appear more than four time periods in a row). The formulation for the AMASS includes the following decision variables:

$x_{1|I|}^{\Pi,m}$: variable that denotes the number of employees assigned to the mth children of the root node, representing production Π from DAG Γ producing a sequence from 1 of length $|I|$;
$x_{il}^{\Pi,m}$: variable that denotes the number of employees assigned to the mth and-node, representing production Π from Γ producing a sequence from i of length l;
y_{ij}: variable that denotes the number of employees assigned to leaf O_{i1}^{j}, representing the allocation of activity j at time interval i;
s_{ij}^{+} and s_{ij}^{-}: slack variables that denote over-covering and under-covering of staff requirements of activity j at time interval i, respectively.

The grammar-based formulation of the AMASS is as follows:

$$\min \sum_{i \in I} \sum_{j \in J} c_{ij} y_{ij} + \sum_{i \in I} \sum_{j \in J} (c_{ij}^{+} s_{ij}^{+} + c_{ij}^{-} s_{ij}^{-}) \tag{56}$$

$$\sum_{A_{il}^{\Pi,m} \in ch(O_{il}^{\pi})} x_{il}^{\Pi,m} = \sum_{A_{il}^{\Pi,m} \in par(O_{il}^{\pi})} x_{1|I|}^{\Pi,m}, \quad O_{il}^{\pi} \in ch(A_{1|I|}^{\Pi,m}) \tag{57}$$

$$\sum_{A_{il}^{\Pi,m} \in ch(O_{il}^{\pi})} x_{il}^{\Pi,m} = \sum_{A_{il}^{\Pi,m} \in par(O_{il}^{\pi})} x_{il}^{\Pi,m},$$

$$O_{il}^{\pi} \in O \setminus \{O_{1n}^{S} \cup L \cup ch(A_{1|I|}^{\Pi,m})\} \tag{58}$$

$$y_{ij} - s_{ij}^{+} + s_{ij}^{-} = b_{ij}, \quad i \in I, \ j \in J \tag{59}$$

$$y_{ij} = \sum_{A_{i1}^{\Pi,1} \in par(O_{i1}^{j})} x_{i1}^{\Pi,1}, \quad i \in I, \ j \in J \tag{60}$$

$$\sum_{A_{1|I|}^{\Pi,m} \in ch(O_{1n}^{S})} x_{1|I|}^{\Pi,m} = |E| \tag{61}$$

$$x_{il}^{\Pi,m} \geq 0 \text{ and integer}, \quad A_{il}^{\Pi,m} \in A \tag{62}$$

$$s_{ij}^{+}, s_{ij}^{-} \geq 0, \quad i \in I, \ j \in J \tag{63}$$

$$y_{ij} \geq 0 \text{ and integer}, \quad i \in I, \ j \in J \tag{64}$$

The objective of problem (56)–(64) is to minimize the total staffing cost plus the penalization for over-covering and under-covering of staff requirements. Constraints (57) guarantee, for every child of the shift shells, that the summation of the value of its children is the same as the summation of the value of its parents. Similarly, constraints (58) guarantee for the rest of or-nodes in the grammar, excluding the root node O_{1n}^{S}, the children of shift shells $A_{1|I|}^{\Pi,m}$ and the leaves L, that the summation of the value of its children is the same as the summation of the value of its parents. Observe that constraints (57) and (58) can be seen as flow conservation equations where or-nodes O^{π}_{il} represent "transition nodes". The constraints for those transition nodes guarantee that if n employees are assigned to the productions generating a subsequence associated with node O^{π}_{il}, those n employees have to be distributed along all the possible ways to use π to generate a sequence of length l from position i ($ch(O^{\pi}_{il})$). Constraints (59) ensure that the total number of employees working on time interval i and work activity j, is equal to the demand subject to some adjustments related to under-covering and over-covering. Constraints (60) set the value of variables y_{ij} as the summation of the value of the parents of leaf nodes O^{j}_{i1}. Constraints (61) guarantee that $|E|$ employees are assigned to the shift shells. Constraints (62)–(64) set the non-negativity of variables s^{+}_{ij} and s^{-}_{ij} and the non-negativity and integrality of variables $x^{\Pi,m}_{il}$ and y_{ij}.

Variables x are assumed to be bounded. Hence, as in Sherali and Zhu (2006), we assume that any general bounded integer variable can be transformed using a binary representation.

6.1 Benders Decomposition Approach

After fixing shift shell variables $x^{\Pi,m}_{1|I|}$ to particular values $\overline{x}^{\Pi,m}_{1|I|}$ the resulting model decomposes into one Benders subproblem. This subproblem includes the variables associated with the and-nodes in Γ and with the allocation of work activities and breaks to the shift shells. The formulation of the Benders subproblem is as follows:

$$Z(\bar{x}) = \min \sum_{i \in I} \sum_{j \in J} c_{ij} y_{ij} + \sum_{i \in I} \sum_{j \in J} (c^{+}_{ij} s^{+}_{ij} + c^{-}_{ij} s^{-}_{ij}) \tag{65}$$

$$y_{ij} - s^{+}_{ij} + s^{-}_{ij} = b_{ij}, \ i \in I, \ j \in J \tag{66}$$

$$\sum_{A^{\Pi,m}_{il} \in ch(O^{\pi}_{il})} x^{\Pi,m}_{il} = \sum_{A^{\Pi,m}_{il} \in par(O^{\pi}_{il})} \overline{x}^{\Pi,m}_{1|I|}, \ O^{\pi}_{il} \in ch(A^{\Pi,m}_{1|I|}) \tag{67}$$

$$\sum_{A^{\Pi,m}_{il} \in ch(O^{\pi}_{il})} x^{\Pi,m}_{il} = \sum_{A^{\Pi,m}_{il} \in par(O^{\pi}_{il})} x^{\Pi,m}_{il}, \quad O^{\pi}_{il} \in O \setminus \{O^{S}_{1|I|} \cup L \cup ch(A^{\Pi,m}_{1|I|})\} \tag{68}$$

$$y_{ij} = \sum_{A^{\Pi,1}_{i1} \in par(O^{j}_{i1})} x^{\Pi,1}_{i1}, \ i \in I, \ j \in J \tag{69}$$

$$x^{\Pi,m}_{il} \geq 0, \ A^{\Pi,m}_{il} \in A \tag{70}$$

$$s_{ij}^{+}, s_{ij}^{-} \geq 0, \quad i \in I, \; j \in J \tag{71}$$

$$y_{ij} \geq 0 \text{ and integer}, \quad i \in I, \; j \in J \tag{72}$$

Since variables y_{ij} are required to be integer and Benders subproblem (65)–(72) does not possess the integrality property, the classical Benders decomposition approach needs to be modified to ensure convergence to an optimal solution. As explained in Sect. 4.1, we first generate classical Benders cuts by relaxing the integrality constraints (72) on variables y_{ij} (giving rise to model $\overline{Z}(\bar{x})$). Second, we will generate integer Benders cuts to guarantee the convergence of the method to an optimal solution.

To define optimality cuts, we introduce the following notation. Let ρ_{ij}, γ_{il}^{π} be the dual variables associated with constraints (66) and (67) from $\overline{Z}(\bar{x})$, respectively. Let $\mathbf{\Delta}$ be the polyhedron obtained from the projection, over the space of variables ρ_{ij} and γ_{il}^{π}, of the set of feasible solutions to the dual of model $\overline{Z}(\bar{x})$. Let $E_{\mathbf{\Delta}}$ be the set of extreme points of $\mathbf{\Delta}$. Let θ be a non-negative variable that represents the value of the objective of $\overline{Z}(\bar{x})$. The Benders optimality cuts are then defined as follows:

$$\theta \geq \sum_{i \in I} \sum_{j \in J} \rho_{ij} b_{ij} + \sum_{O_{il}^{\pi} \in ch(A_{1|I|}^{\Pi,m})} \gamma_{il}^{\pi} \sum_{A_{1|I|}^{\Pi,m} \in par(O_{il}^{\pi})} v_{1|I|}^{\Pi,m}, \quad (\boldsymbol{\rho}, \boldsymbol{\gamma}) \in E_{\mathbf{\Delta}} \tag{73}$$

Optimality cuts ensure that the value of each variable θ is larger than or equal to the LP relaxation value of the Benders subproblem. To derive these cuts, we have assumed that the Benders subproblem is a linear program (i.e., the integrality of variables y_{ij} is relaxed). The relaxation of model (56)–(64) obtained by relaxing the integrality of variables y_{ij} can thus be reformulated as the following master problem

$$\min \theta \tag{74}$$

$$\sum_{A_{1|I|}^{\Pi,m} \in ch(O_{1|I|}^{S})} x_{1|I|}^{\Pi,m} = |E| \tag{75}$$

$$\theta \geq \sum_{i \in I} \sum_{j \in J} \rho_{ij} b_{ij} + \sum_{O_{il}^{\pi} \in ch(A_{1|I|}^{\Pi,m})} \gamma_{il}^{\pi} \sum_{A_{1|I|}^{\Pi,m} \in par(O_{il}^{\pi})} x_{1|I|}^{\Pi,m}, \quad (\boldsymbol{\rho}, \boldsymbol{\gamma}) \in E_{\mathbf{\Delta}} \tag{76}$$

$$x_{1|I|}^{\Pi,m} \geq 0 \text{ and integer}, \quad A_{1|I|}^{\Pi,m} \in ch(O_{1|I|}^{S}) \tag{77}$$

In order to derive integer Benders cuts for the problem, shift shell variables $x_{1|I|}^{\Pi,m}$ must be represented as binary variables. Observe that under an iterative Benders decomposition approach these variables will only be generated when needed (i.e., when it will be necessary to generate integer Benders cuts). To simplify the presentation, we define the set M which is composed of the children of the root node $O_{1|I|}^{S}$. Variables $x_{1|I|}^{\Pi,m}$, $A_{1n}^{\Pi,m} \in ch(O_{1n}^{S})$ can then be rewritten as x_m, denoting the number of employees assigned to shift shell $m \in M$. We express each variable x_m

with $|E|$ binary variables x_m^e, which take value 1 if employee $e \in E$ is assigned to the mth shift shell $A_{1|I|}^{\Pi,m}$, and assume value 0 otherwise. We assume that employees are ordered arbitrarily such that $E = \{1, \ldots, |E|\}$. The binary variables are then defined as follows:

$$\sum_{e \in E} x_m^e = x_m, \ m \in M \tag{78}$$

$$x_m^{e+1} \leq x_m^e, \ m \in M, \ e \in E \setminus \{|E|\} \tag{79}$$

$$x_m^e \in \{0, 1\}, \ \ m \in M, \ e \in E \tag{80}$$

Equations (78) guarantee that the sum of the binary variables x_m^e is equal to the number of employees assigned to $m \in M$, while (79) are symmetry breaking constraints. Due to these constraints, each integer variable x_m has a unique representation in terms of the binary variables: if $x_m = n$, then $x_m^e = 1$ for $e \leq n$ and $x_m^e = 0$ for $e > n$. As a consequence, any solution representing an assignment of shifts to $|E|$ employees is represented uniquely with the binary variables x_m^e. Let $\overline{x}_m^e(t)$ be the value of variable x_m^e corresponding to the current solution $\overline{x}_m(t)$ and $\delta(\bar{x}_m^e(t)) = \{(m, e) \in M \times E \mid \overline{x}_m^e(t) = 0\}$. Since Benders subproblem (65) and (64) is an integer problem that does not possess the integrality property, we propose to use the algorithmic strategy presented in Sect. 4.1. Hence, if $Z(t) - \overline{Z}(t)/Z(t) > \epsilon_3$ we add an integer Benders cut to problem (74)–(77) to eliminate the current solution $\overline{x}_m(t)$ by swapping at least one employee from its assigned shift to a different one. The integer Benders cut is as follows:

$$\theta \geq Z(t) - (Z(t) - \bar{\theta}(t)) \Bigg(\sum_{(m,e) \in \delta(\bar{x}_m^e(t))} x_m^e \Bigg) \tag{81}$$

7 Numerical Experiments

In this section, we present the computational experiments we have performed on our implementation of the decomposition-based methods with integer subproblems. The computing environment used for the tests consists of a 1-processor Intel Xeon X5675 with 96 GB of RAM running at 3.07 GHz and operating on a 64-bit GNU/Linux operating system. Algorithms 1 and 2 were implemented in C++. We set a 3-hour time limit to solve the MCFL problem and a 2-hour time limit to solve the AMASS problem. A relative gap tolerance of 0.001 was set as a stopping criterion for solving the MIPs with CPLEX B&B. We used the following values for the tolerance parameters: $\epsilon_2 = \epsilon_3 = 0.00001$ for both problems. We tested the values 0.01, 0.03, 0.05 and 0.1 for the tolerance ϵ_1. The values that showed the best performance of Phase 0 were $\epsilon_1 = 0.03$ for the MCFL and $\epsilon_1 = 0.01$ for the AMASSP.

Table 1 Algorithmic settings for the decomposition-based methods

Setting	P_0	*Classical*	*Iterative*	*Algorithm*
S1	True	True	True	Algorithm 1
S2	False	True	True	Algorithm 1
S3	False	False	True	Algorithm 2
S4	False	False	False	No decomposition

The weights for the computation of a core point to generate stronger Benders optimality cuts (Algorithm 2 in Papadakos, 2008) are set to 0.5 for the value of the masters' problem solution at iteration t ($\bar{x}_t$) and to 0.5 for the value of the core point approximation at iteration $t-1$ (defined as $\hat{x}_{t-1}$). These weights were also set by testing several values: 0.2 for $\bar{x}_t$ and 0.8 for $\hat{x}_{t-1}$; 0.3 for $\bar{x}_t$ and 0.7 for $\hat{x}_{t-1}$; 0.5 for $\bar{x}_t$ and 0.5 for $\hat{x}_{t-1}$; and 0.7 for $\bar{x}_t$ and 0.3 for $\hat{x}_{t-1}$. The chosen values were those that made the method converge faster.

We defined three settings to test the impact of the different algorithmic strategies on the performance of the decomposition-based methods. Table 1 presents the characteristics of these settings. True means that the algorithm strategy was used in the setting. In contrast, false denotes the absence of that algorithmic strategy in the setting. P_0 denotes the use of the method presented in McDaniel and Devine (1977). *Classical* indicates the incorporation of classical Benders cuts. If *Iterative* is true, this means that an iterative method is used to solve the problem. *Algorithm* denotes the type of algorithm used. In addition to these three strategies, we defined a fourth strategy (S4), where the problems are solved without a decomposition method.

7.1 Results on Multi-Commodity Facility Location Instances

We tested our method on instances with 500 clients, 20 commodities, and 200 depots. Parameters ρ^k (number of depots that can serve each commodity) and q_j^k (capacities per depot for each commodity) were randomly generated according to β and λ, respectively. The seed used for the generation of random variables was set to 1. The values for ρ^k were generated with a random variable with integer uniform distribution between $[|J| \times (\beta - 0.1)]$ and $[|J| \times (\beta + 0.1)]$, where β denotes the maximum percentage of depots that can serve each commodity. The values for q_j^k were also randomly generated by adding and removing some capacity perturbation (depending on the value of parameter λ). The instances, and a more detailed description of their generation, are available in https://sites.google.com/site/mariarestreporuiz/instances.

Table 2 Results on the MCFL instances. $\lambda = 1$

Instance	Setting	λ	β	Time (s) P_0	Total	Gap%	T. Cuts	Npb/IBC	Optimal
3 × 2	S1	1	0.3	3389.4	6971.2	50.0	8658.2	0/0	5
	S2	1	0.3	–	4326.8	20.0	1818.5	0/0	8
	S3	1	0.3	–	**1263.3**	0.0	–	0/0	10
	S4	1	0.3	–	1784.2	0.0	–	–	10
4 × 3	S1	1	0.3	6589.0	10, 834.8	100.0	35,566.5	0/0	0
	S2	1	0.3	–	9490.4	80.0	2685.6	0/0	2
	S3	1	0.3	–	**2509.0**	0.0	–	1/3	10
	S4	1	0.3	–	3229.5	0.0	–	–	10
3 × 2	S1	1	0.5	69.6	**301.0**	0.0	539.2	0/0	10
	S2	1	0.5	–	664.0	0.0	1235.7	0/0	10
	S3	1	0.5	–	497.8	0.0	–	0/0	10
	S4	1	0.5	–	742.9	0.0	–	–	10
4 × 3	S1	1	0.5	75.1	6343.8	40.0	2250.1	1/8	6
	S2	1	0.5	–	3747.8	20.0	2524.8	2/11	8
	S3	1	0.5	–	**429.7**	0.0	–	2/11	10
	S4	1	0.5	–	589.8	0.0	–	–	10
3 × 2	S1	1	0.7	41.1	**161.4**	0.0	332.2	0/0	10
	S2	1	0.7	–	177.0	0.0	509.9	0/0	10
	S3	1	0.7	–	188.7	0.0	–	0/0	10
	S4	1	0.7	–	324.2	0.0	–	–	10
4 × 3	S1	1	0.7	24.1	**111.2**	0.0	440.5	3/12	10
	S2	1	0.7	–	536.7	0.0	2469.2	3/12	10
	S3	1	0.7	–	117.5	0.0	–	3/15	10
	S4	1	0.7	–	148.9	0.0	–	–	10
3 × 2	S1	1	1	27.9	106.7	0.0	95.0	0/0	10
	S2	1	1	–	247.0	0.0	145.6	0/0	10
	S3	1	1	–	68.8	0.0	–	0/0	10
	S4	1	1	–	**48.4**	0.0	–	–	10
4 × 3	S1	1	1	13.4	49.7	0.0	104.4	2/10	10
	S2	1	1	–	300.7	0.0	254.0	2/200	10
	S3	1	1	–	38.5	0.0	–	2/10	10
	S4	1	1	–	**25.9**	0.0	–	–	10

7.1.1 Numerical Results

Tables 2, 3, and 4 summarize the results for the numerical experiments on the MCFL problem. Column 1 presents the name of the instance. Column 2 shows the setting used and columns 3 and 4 present the values for parameters λ and β, respectively. Columns 4 and 5 present the average CPU time in seconds to solve the problem with the method presented in McDaniel and Devine (1977) and the total time to solve

Table 3 Results on the MCFL instances. $\lambda = 1.25$

Instance	*Setting*	λ	β	Time (s) P_0	*Total*	*Gap%*	*T. Cuts*	*Npb/IBC*	*Optimal*
3 × 2	S1	1.25	0.3	260.3	4941.8	20.0	1784.0	0/0	8
	S2	1.25	0.3	–	4054.9	10.0	1793.4	0/0	9
	S3	1.25	0.3	–	**1501.5**	0.0	–	0/0	10
	S4	1.25	0.3	–	1801.1	0.0	–	–	10
4 × 3	S1	1.25	0.3	6575.6	10, 848.4	100.0	2933.0	0/0	0
	S2	1.25	0.3	–	9614.3	80.0	3025.7	1/10	2
	S3	1.25	0.3	–	5232.6	0.0	–	7/36	10
	S4	1.25	0.3	–	**4340.7**	7.0	–	–	8
3 × 2	S1	1.25	0.5	75.0	**457.3**	0.0	595.9	0/0	10
	S2	1.25	0.5	–	1704.9	10.0	1703.3	0/0	9
	S3	1.25	0.5	–	578.3	0.0	–	0/0	10
	S4	1.25	0.5	–	684.5	0.0	–	–	10
4 × 3	S1	1.25	0.5	71.8	5862.4	30.0	1929.3	4/37	7
	S2	1.25	0.5	–	2472.1	10.0	1820.4	4/34	9
	S3	1.25	0.5	–	**431.3**	0.0	–	4/30	10
	S4	1.25	0.5	–	1588.8	0.0	–	–	9
3 × 2	S1	1.25	0.7	46.9	**206.1**	0.0	325.7	0/0	10
	S2	1.25	0.7	–	407.3	0.0	643.8	0/0	10
	S3	1.25	0.7	–	261.9	0.0	–	0/0	10
	S4	1.25	0.7	–	317.5	0.0	–	–	10
4 × 3	S1	1.25	0.7	28.5	**136.5**	0.0	392.2	5/46	10
	S2	1.25	0.7	–	561.8	0.0	1886.6	5/50	10
	S3	1.25	0.7	–	176.6	0.0	–	5/35	10
	S4	1.25	0.7	–	171.5	0.0	–	–	10
3 × 2	S1	1.25	1	33.2	112.6	0.0	85.3	0/0	10
	S2	1.25	1	–	232.5	0.0	121.8	0/0	10
	S3	1.25	1	–	80.3	0.0	–	0/0	10
	S4	1.25	1	–	**56.3**	0.0	–	–	10
4 × 3	S1	1.25	1	15.0	62.6	0.0	109.4	3/20	10
	S2	1.25	1	–	676.4	0.0	307.1	4/778	10
	S3	1.25	1	–	56.3	0.0	–	5/26	10
	S4	1.25	1	–	**34.7**	0.0	–	–	10

the problem, respectively. Values in bold represent the smallest computational time within the set of instances. Column *Gap%* presents the gap between the final best upper bound and the final best lower bound for S1, S2, and S3. This gap is computed as: $Gap = (\zeta^u(t) - \zeta^l(t))/\zeta^u(t)$. For S4 this gap represents the integrality gap returned by CPLEX. Columns *T. Cuts* and *Npb/IBC* denote the number of classical Benders cuts added to the problem and the total number of problems that required the incorporation of integer Benders cuts over the total number of integer Benders

Table 4 Results on the MCFL instances. λ = 1.5

				Time (s)					
Instance	*Setting*	λ	β	P_0	*Total*	*Gap%*	*T. Cuts*	*Npb/IBC*	*Optimal*
3 × 2	S1	1.5	0.3	1264.8	3746.0	20.0	1414.7	1/4	8
	S2	1.5	0.3	–	3245.6	10.0	1719.3	2/8	9
	S3	1.5	0.3	–	**1518.6**	0.0	–	2/8	10
	S4	1.5	0.3	–	1630.5	0.0	–	–	10
4 × 3	S1	1.5	0.3	6631.8	10,838.3	100.0	2961.8	0/0	0
	S2	1.5	0.3	–	9839.0	90.0	3115.5	1/20	1
	S3	1.5	0.3	–	4672.4	0.0	–	9/70	10
	S4	1.5	0.3	–	**3128.8**	0.0	–	–	9
3 × 2	S1	1.5	0.5	67.3	538.7	0.0	723.6	0/0	10
	S2	1.5	0.5	–	542.3	0.0	888.3	0/0	10
	S3	1.5	0.5	–	**476.3**	0.0	–	0/0	10
	S4	1.5	0.5	–	666.0	0.0	–	–	10
4 × 3	S1	1.5	0.5	55.7	4225.6	30.0	1700.7	6/55	7
	S2	1.5	0.5	–	2760.9	20.0	1859.4	6/67	8
	S3	1.5	0.5	–	**406.5**	0.0	–	7/60	10
	S4	1.5	0.5	–	1583.4	5.4	–	–	9
3 × 2	S1	1.5	0.7	41.4	**195.2**	0.0	314.0	0/0	10
	S2	1.5	0.7	–	284.0	0.0	484.9	0/0	10
	S3	1.5	0.7	–	230.8	0.0	–	0/0	10
	S4	1.5	0.7	–	294.5	0.0	–	–	10
4 × 3	S1	1.5	0.7	25.5	**135.7**	0.0	416.5	7/88	10
	S2	1.5	0.7	–	700.9	0.0	2375.8	7/92	10
	S3	1.5	0.7	–	156.6	0.0	–	7/78	10
	S4	1.5	0.7	–	155.9	0.0	–	–	9
3 × 2	S1	1.5	1	33.1	116.3	0.0	91.5	0/0	10
	S2	1.5	1	–	303.5	0.0	142.3	0/0	10
	S3	1.5	1	–	72.2	0.0	–	0/0	10
	S4	1.5	1	–	**49.6**	0.0	–	–	10
4 × 3	S1	1.5	1	14.7	69.1	0.0	114.9	5/52	10
	S2	1.5	1	–	1593.2	0.0	494.7	5/2894	10
	S3	1.5	1	–	54.9	0.0	–	6/54	10
	S4	1.5	1	–	**41.4**	0.0	–	–	10

cuts added, respectively. Column *Optimal* presents the number of instances solved to optimality.

Results from Tables 2, 3, and 4 indicate that the performance of the strategies in terms of solution time and quality depends greatly on the values of parameters λ and β. Specifically, when the value of β (the maximum percentage of depots that can serve each commodity) increases from 0.3 to 0.5, to 0.7, and to 1.0, we observe that the strategies are on average 3.8, 18.88, and 31.68 times faster, respectively. We also observe that the increase of λ, given a fixed value of β, usually increases the computational effort.

The four strategies were able to solve all the instances to optimality when $\beta = 0.7$ and $\beta = 1$ for all λs, with the exception of instances in the group 4×3 with $\beta = 0.7$ and $\lambda = 1.5$. However, when $\beta = 0.3$ and $\beta = 0.5$ strategies S1 and S2 had some difficulties to solve the instances to optimality within the computational time limit. This can be due to the fact that most of the problems become more difficult when β decreases, as there are fewer available depots to serve the clients. We also observe that S1 and S2 invest a lot of time in the algorithmic refinement of McDaniel and Devine (1977), especially for the instances in the group 4×3.

The impact of the method from McDaniel and Devine (1977) in the solution time depends on the type of instance solved. For instance, we see that the algorithms perform better when the method of McDaniel and Devine (1977) is not used in instances from the group 3×2 when $\beta = 0.3$ and in instances from the group 4×3 when $\beta = 0.5$. On the contrary, the algorithms are in average 4.8 times faster when the method of McDaniel and Devine (1977) is used when $\beta = 0.7$ and $\beta = 1$.

From Tables 2, 3, and 4, we deduce that strategy S3 works well because the LP relaxation of the SP variables is included in the Benders master problem. In fact, sometimes it was not even necessary to add integer Benders cuts, as the LP relaxation of the subproblem had the same objective value as the objective function value of the MIP subproblem.

The benefits of using Algorithms 1 and 2 (S1 and S3) are evident for instances with $\beta < 1$, as these strategies were able to solve the instances in less computational time (in 16 out of 18 groups of instances) than an approach without using a decomposition method (S4).

7.2 *Results on Multi-Activity Shift Scheduling*

To test our methods on the AMASS problem we generated instances with up to 20 work activities. The planning horizon is divided into 96 time intervals (i.e., 15-minute time intervals). The rules to generate multi-activity shifts are the following:

1. Shifts can start at any time interval during the day, allowing enough time to complete their duration.
2. Three types of shifts are considered: 8-hour shifts with a 1-hour lunch break in the middle and two 15-minute breaks. 6-hour shifts with one 15-minute break and no lunch, and 4-hour shifts with one 15-minute break and no lunch.
3. If performed, the duration of a work activity is at least one hour and at most five hours.
4. A break (or lunch) is necessary between two different work activities.
5. Work activities must be inserted between breaks, lunch, and rest stretches.
6. A fixed number of employees $|E|$ is given, therefore under-covering and over-covering of staff requirements is allowed.

Let a_j be a terminal symbol that defines a time interval of work activity $j \in J$. Let b, l, and r be the terminal symbols that represent break, lunch, and rest periods, respectively. In productions $\Pi \in P$, $\Pi \rightarrow_{[\min, \max]}$ restrict the subsequences generated by a given production to a length between a minimum and a maximum number of periods. The grammar and the productions that define the AMASS are as follows:

$$
\begin{aligned}
G = (&\Sigma = (a_j, j \in J, b, l, r),\\
&N = (S, F, Q, N, W, A_j, j \in J, B, L, R), P, S),\\
&S \rightarrow RFR|FR|RF|RQR|QR|RQ|RNR|NR|RN,\ B \rightarrow b,\ L \rightarrow llll,\\
&F \rightarrow_{[38,38]} NLN,\ Q \rightarrow_{[25,25]} WBW,\\
&N \rightarrow_{[17,17]} WBW,\ R \rightarrow Rr|r,\\
&W \rightarrow_{[4,20]} A_j, j \in J,\ A_j \rightarrow A_j a_j | a_j, j \in J.
\end{aligned}
$$

Activity allocation costs c_{ij} are equal to 15, overcovering costs c^+_{ij} are equal to 20, and undercovering costs are computed as: $c^-_{ij} = 5 \times c_{ij} = 100$. The instances for the AMASS are available in https://sites.google.com/site/mariarestreporuiz/instances.

7.2.1 Numerical Results

Tables 5 and 6 present the results for the numerical experiments on the AMASS problem. Column 1 presents the name of the algorithmic strategy used. Column 2 shows the number of work activities included in the instances. Columns 3–7 present, respectively, the average CPU time in seconds to solve the problem with the method of McDaniel and Devine (1977), to solve the MIP of the master problem, to solve the LP relaxation of the subproblems, to solve the MIP of the subproblems and the total time to solve the problem. Values in bold represent the smallest computational time within the set of instances. Column *Gap%* presents the gap between the best upper bound and the best lower bound for S1 to S3. This gap is computed as: $Gap = (\zeta^u(t) - \zeta^l(t))/\zeta^u(t)$. For S4 this gap represents the integrality gap returned by CPLEX. Columns *T. Cuts* and *Npb/IBC* denote the number of classical Benders cuts added to the problem and the total number of problems that required the incorporation of integer Benders cuts over the total number of integer Benders cuts added, respectively. Column *Optimal* presents the number of instances solved to optimality.

In terms of solution time, results from Table 5 show that strategy four (S4) outperforms the other strategies in most of the instances when the number of activities is small (i.e., two to nine work activities). However, we observe that the total solution time when S4 is used to solve instances with two and three work activities is very similar to the total solution time when strategies S1 or S2 are used

Table 5 Results on the AMASSP instances, one to ten work activities

		Time (s)								
Setting	*Act*	P_0	*MIP MP*	*LP SP*	*MIP SP*	*Total%*	*Gap1%*	*T. Cuts*	*Npb/IBC*	*Optimal*
S1	1	25.13	0.26	31.05	99.97	156.45	0.000	37	1/6	10
S2	1	–	4.65	42.38	99.55	**146.78**	0.000	44	1/6	10
S3	1	–	930.06	0.00	100.37	1030.47	0.000	0	1/6	10
S4	1	–	516.37	–	–	516.40	0.000	–	–	10
S1	2	66.18	0.80	71.58	48.67	187.38	0.000	46	1/28	10
S2	2	–	42.43	91.32	52.40	186.40	0.000	66	1/30	10
S3	2	–	293.02	0.00	49.56	342.62	0.000	–	1/30	10
S4	2	–	141.94	–	–	**141.97**	0.010	–	–	10
S1	3	85.27	1.67	97.46	3.04	187.63	0.000	50	2/4	10
S2	3	–	447.18	166.10	2.78	616.70	0.000	90	1/2	10
S3	3	–	219.30	0.00	1.52	220.87	0.000	–	1/2	10
S4	3	–	177.11	–	–	**177.14**	0.000	–	–	10
S1	4	143.73	140.02	160.11	2.48	446.69	0.000	73	1/4	10
S2	4	–	2691.33	252.95	3.15	2947.86	1.244	117	0/0	8
S3	4	–	430.84	0.00	1.43	432.33	0.000	0	0/0	10
S4	4	–	82.70	–	–	**82.74**	0.009	–	–	10
S1	5	114.66	691.49	202.53	4.19	1013.03	0.498	97	0/0	9
S2	5	–	1526.98	358.73	5.42	1891.77	0.112	153	0/0	9
S3	5	–	194.69	0.00	2.08	196.84	0.000	–	0/0	10
S4	5	–	115.91	–	–	**115.95**	0.000	–	–	10
S1	6	167.17	897.20	211.31	4.38	1280.40	0.155	87	1/2	9
S2	6	–	1538.62	335.37	4.92	1879.50	0.293	123	0/0	9
S3	6	–	306.51	0.00	2.61	309.19	0.000	–	0/0	10
S4	6	–	238.14	–	–	**238.19**	0.009	–	–	10
S1	7	292.23	994.41	462.07	6.40	1756.27	0.118	193	2/4	9
S2	7	–	3568.30	582.36	11.35	4162.97	3.757	197	0/0	5
S3	7	–	164.59	0.00	2.76	167.44	0.000	–	0/0	10
S7	7	–	163.58	–	–	**163.63**	0.000	–	–	10
S1	8	169.08	863.34	378.18	7.20	1418.22	0.000	154	2/4	10
S2	8	–	2521.96	582.16	10.11	3115.20	1.170	190	0/0	7
S3	8	–	267.00	0.00	3.39	270.47	0.000	–	0/0	10
S4	8	–	161.39	–	–	**161.44**	0.000	–	–	10
S1	9	159.44	374.70	346.03	5.89	886.49	0.000	146	0/0	10
S2	9	–	2120.31	686.10	9.68	2817.30	1.199	223	0/0	8
S3	9	–	232.39	0.00	3.58	236.05	0.000	–	0/0	10
S4	9	–	154.16	–	–	**154.22**	0.000	–	–	10
S1	10	235.75	303.24	466.78	9.42	1016.15	0.000	157	0/0	10
S2	10	–	2772.74	626.28	17.98	3417.95	1.572	172	0/0	7
S3	10	–	184.48	0.00	4.33	**188.90**	0.000	–	0/0	10
S4	10	–	189.62	–	–	189.67	0.000	–	–	10

Table 6 Results on the AMASSP instances, eleven to twenty work activities

		Time (s)								
Setting	*Act*	P_0	*MIP MP*	*LP SP*	*MIP SP*	*Total%*	*Gap1%*	*T. Cuts*	*Npb/IBC*	*Optimal*
S1	11	199.63	1.42	459.06	47.46	**707.61**	0.000	67	0/0	10
S2	11	–	3669.51	532.31	86.78	4289.10	1.342	67	0/0	6
S3	11	–	746.90	0.00	336.78	1083.79	0.000	–	0/0	10
S4	11	–	1212.24	–	–	1212.31	0.000	–	–	10
S1	12	218.02	0.77	483.11	63.34	**765.29**	0.000	68	0/0	10
S2	12	–	1678.69	495.07	68.25	2242.47	0.660	60	0/0	8
S3	12	–	1051.35	0.00	304.06	1355.53	0.000	–	0/0	10
S4	12	–	1817.93	–	–	1818.00	0.000	–	–	10
S1	13	269.55	696.05	551.58	42.03	1559.27	0.000	79	0/0	10
S2	13	–	2869.79	606.35	52.54	3529.25	0.695	71	0/0	8
S3	13	–	643.66	0.00	25.28	**669.07**	0.000	–	0/0	10
S4	13	–	941.03	–	–	941.11	0.000	–	–	10
S1	14	238.54	0.68	496.18	33.92	769.38	0.000	73	0/0	10
S2	14	–	4456.98	683.49	28.69	5169.80	0.253	80	0/0	9
S3	14	–	390.43	0.00	30.70	**421.25**	0.000	–	0/0	10
S4	14	–	814.48	–	–	814.55	0.000	–	–	10
S1	15	188.86	2.00	705.31	30.19	926.41	0.000	117	0/0	10
S2	15	–	4343.01	726.79	39.21	5109.82	1.946	92	0/0	5
S3	15	–	406.49	0.00	19.55	**426.18**	0.000	–	0/0	10
S4	15	–	934.08	–	–	934.16	0.000	–	–	10
S1	16	186.50	3.72	836.74	49.13	1076.13	0.000	141	1/2	10
S2	16	–	4433.32	766.36	166.62	5367.15	2.185	95	0/0	6
S3	16	–	349.90	0.00	69.87	**419.91**	0.000	–	0/0	10
S4	16	–	1297.27	–	–	1297.36	0.000	–	–	10
S1	17	215.33	10.11	1260.27	31.84	1517.60	0.000	232	0/0	10
S2	17	–	3567.46	1123.04	93.68	4785.53	3.978	150	0/0	7
S3	17	–	353.75	0.00	21.60	**375.50**	0.000	–	0/0	10
S4	17	–	1036.23	–	–	1036.32	0.000	–	–	10
S1	18	248.23	3.47	761.28	39.74	1052.87	0.000	135	1/2	10
S2	18	–	2169.32	1308.05	39.16	3518.16	0.370	170	0/0	9
S3	18	–	354.96	0.00	25.82	**380.92**	0.000	–	0/0	10
S4	18	–	1225.61	–	–	1225.70	0.000	–	–	10
S1	19	197.38	8.24	1045.99	30.80	1282.46	0.000	189	1/2	10
S2	19	–	2875.45	1901.10	63.56	4842.67	2.561	256	1/2	7
S3	19	–	328.93	0.00	22.33	**351.41**	0.000	–	0/0	10
S4	19	–	1256.95	–	–	1257.06	0.000	–	–	10
S1	20	220.58	13.57	1309.21	32.46	1575.88	0.000	241	2/4	10
S2	20	–	3822.00	1509.09	79.13	5412.28	4.677	203	0/0	6
S3	20	–	403.29	0.00	17.77	**421.24**	0.000	–	0/0	10
S4	20	–	1166.17	–	–	1166.27	0.000	–	–	10

to solve these two instances. The use of the method presented in McDaniel and Devine (1977) (strategy S1) generates a significant decrease in the total running time when compared to using the same algorithm without adding classical Benders cuts to the LP relaxation of the master problem (strategy S2). Specifically, Algorithm 1 is on average 3.34 times faster when the method of McDaniel and Devine (1977) is used. The results in Table 6 indicate that when the number of activities increases, strategy S3 outperforms the rest of the strategies in terms of computational time. More precisely, this strategy is on average 6.4 times faster than the other strategies in instances with 10, and 13–20 work activities. The impact of the use of the method in McDaniel and Devine (1977) is also observed in instances with a larger number of activities (Table 6). With a few exceptions, we can see that the use of this algorithmic refinement helps to significantly reduce the computational time required to solve the problem.

Regarding the distribution of the total time, we see that strategies S2 and S3 are the ones that spend most of the time solving the integer version of the master problem, while strategy S1 spends more time solving the subproblems. The strategy presented in McDaniel and Devine (1977) consumes on average 18% of the total computational time.

Analyzing the number of optimality Benders cuts added to the problem, we see that, with a few exceptions, the algorithms added more of these cuts when strategy S2 was used. Surprisingly only three instances out of 200 needed integer Benders cuts when strategy S3 was used. Just by solving the MIP of the SP we were able to prove optimality.

Similarly to the results from the multi-commodity facility location problem, the benefits of using Benders decomposition Algorithms 1 and 2 (S1 and S3) are evident for instances with a large number of activities (10 up to 20), as these strategies were able to solve the instances in less computational time than an approach without using decomposition (S4).

8 Discussion and Concluding Remarks

In this paper, we presented two decomposition-based algorithms to address a special class of problems having mixed-integer variables in the master problem and subproblems. In these problems, some of the master problem variables are bounded and their use is restricted by cardinality constraints. The first algorithm generates optimality cuts from the LP relaxation of the subproblem and when no more of these relaxed cuts can be added, the algorithm adds a new type of integer Benders optimality cut to ensure the convergence of the method to an optimal solution. The second algorithm adds the relaxation of the Benders subproblem variables in the master problem and includes these variables in the composition of integer Benders optimality cuts.

We tested the two approaches on two different types of problems: the multi-commodity facility location problem and the multi-activity shift scheduling problem. The numerical results show that the second approach has a better performance when solving both problems, since, in most of the instances, it was not necessary to add integer Benders optimality cuts. Hence, including the LP relaxation of the subproblem's decision variables in the master problem seems like a nice idea, as this strategy helps to improve the lower bound of the Benders master problem.

Future research avenues might include testing the performance of the proposed decomposition methods on other type of problems. Also, to implement both algorithms in a single B&B tree instead of using the classical iterative approach.

Acknowledgments The authors would like to thank Serge Bisaillon for generating the MCFLP instances and for providing help in the implementation of the Benders decomposition algorithms.

References

Angulo, G., Ahmed, S., & Dey, S. S. (2016). Improving the integer L-shaped method. *INFORMS Journal on Computing*, *28*(3), 483–499.

Boyer, V., Gendron, B., & Rousseau, L. M. (2014). A branch-and-price algorithm for the multi-activity multi-task shift scheduling problem. *Journal of Scheduling*, *17*(2), 185–197.

Carøe, C. C., & Tind, J. (1998). L-shaped decomposition of two-stage stochastic programs with integer recourse. *Mathematical Programming*, *83*(1), 451–464.

Cordeau, J. F., Stojković, G., Soumis, F., & Desrosiers, J. (2001). Benders decomposition for simultaneous aircraft routing and crew scheduling. *Transportation Science*, *35*(4), 375–388.

Côté, M. C., Gendron, B., & Rousseau, L. M. (2011b). Grammar-based integer programming models for multiactivity shift scheduling. *Management Science*, *57*(1), 151–163.

Elçi, Ö., & Hooker, J. (2022). Stochastic planning and scheduling with logic-based Benders decomposition. *INFORMS Journal on Computing*, *34*(5), 2428–2442.

Gade, D., Küçükyavuz, S., & Sen, S. (2014). Decomposition algorithms with parametric Gomory cuts for two-stage stochastic integer programs. *Mathematical Programming*, *144*(1–2), 39–64.

Gendron, B., Scutellà, M. G., Garroppo, R. G., Nencioni, G., & Tavanti, L. (2016). A branch-and-Benders-cut method for nonlinear power design in green wireless local area networks. *European Journal of Operational Research*, *255*(1), 151–162.

Hooker, J. N., & Ottosson, G. (2003). Logic-based Benders decomposition. *Mathematical Programming*, *96*(1), 33–60.

Kim, K., & Mehrotra, S. (2015). A two-stage stochastic integer programming approach to integrated staffing and scheduling with application to nurse management. *Operations Research*, *63*(6), 1431–1451.

Laporte, G., & Louveaux, F. V. (1993). The integer L-shaped method for stochastic integer programs with complete recourse. *Operations Research Letters*, *13*(3), 133–142.

Li, C., & Grossmann, I. E. (2019). A generalized Benders decomposition-based branch and cut algorithm for two-stage Benders programs with nonconvex constraints and mixed-binary first and second stage variables. *Journal of Global Optimization*, *75*, 247–272.

Magnanti, T. L., & Wong, R. T. (1981). Accelerating Benders decomposition: Algorithmic enhancement and model selection criteria. *Operations Research*, *29*(3), 464–484.

McDaniel, D., & Devine, M. (1977). A modified Benders' partitioning algorithm for mixed integer programming. *Management Science*, *24*(3), 312–319.

Papadakos, N. (2008). Practical enhancements to the Magnanti–Wong method. *Operations Research Letters*, *36*(4), 444–449.

Qi, Y., & Sen, S. (2017). The ancestral Benders' cutting plane algorithm with multi-term disjunctions for mixed-integer recourse decisions in stochastic programming. *Mathematical Programming*, *161*(1–2), 193–235.

Rahmaniani, R., Crainic, T. G., Gendreau, M., & Rei, W. (2017). The benders decomposition algorithm: A literature review. *European Journal of Operational Research*, *259*(3), 801–817.

Restrepo, M. I., Gendron, B., & Rousseau, L. M. (2018). Combining Benders decomposition and column generation for multiactivity tour scheduling. *Computers and Operations Research*, *93*, 151–165.

Sen, S., & Higle, J. L. (2005). The C3 theorem and a D2 algorithm for large scale stochastic mixed-integer programming: Set convexification. *Mathematical Programming*, *104*(1), 1–20.

Sen, S., & Sherali, H. D. (2006). Decomposition with branch-and-cut approaches for two-stage stochastic mixed-integer programming. *Mathematical Programming*, *106*(2), 203–223.

Sherali, H. D., & Fraticelli, B. M. (2002). A modification of Benders' decomposition algorithm for discrete subproblems: An approach for stochastic programs with integer recourse. *Journal of Global Optimization*, *22*(1–4), 319–342.

Sherali, H. D., & Zhu, X. (2006). On solving discrete two-stage stochastic programs having mixed-integer first-and second-stage variables. *Mathematical Programming*, *108*(2), 597–616.

Wolsey, L. A. (2020). *Integer programming*. John Wiley & Sons.

Zhang, M., & Küçükyavuz, S. (2014). Finitely convergent decomposition algorithms for two-stage stochastic pure integer programs. *SIAM Journal on Optimization*, *24*(4), 1933–1951.

Perspectives on Using Benders Decomposition to Solve Two-Stage Stochastic Mixed-Integer Programs

Mike Hewitt and Walter Rei

1 Introduction

Benders decomposition has been established as one of the staple strategies to efficiently solve a wide variety of complex large-scale optimization models. The interested reader is referred to Rahmaniani et al. (2017) for a thorough review of the literature dedicated to this general decomposition strategy. As originally developed by Benders (1962), this strategy enables an optimization model to be partitioned into smaller parts by separating those parts that involve decision variables identified as being *complicating* (i.e., associated with characteristics within the optimization model that render it challenging to solve) from the ones involving decision variables viewed as being *facilitating* (i.e., associated with characteristics in the model that are much easier to address). Through this partition, a master problem, involving the complicating variables, and a subproblem (or subproblems), involving the facilitating variables, are thus created. The Benders algorithm then proceeds by iteratively solving both the master and the subproblem(s), where the solutions of the latter are directly used to improve the master's formulation, through the inclusion of violated cuts. This process is repeated until the optimal solution to the original problem is found. As often remarked, a successful implementation of this algorithm notably rests on the judicious choices that are made regarding which specific decision variables are included in each of the two categories.

M. Hewitt
Quinlan School of Business, Loyola University Chicago, Chicago, IL, USA
e-mail: mhewitt3@luc.edu

W. Rei (✉)
CIRRELT and École des Sciences de la Gestion, Université du Québec à Montréal, Montréal, QC, Canada
e-mail: rei.walter@uqam.ca

T. G. Crainic et al. (eds.), *Combinatorial Optimization and Applications*,
International Series in Operations Research & Management Science 358,
https://doi.org/10.1007/978-3-031-57603-4_12

It should be recalled that the variables in a stochastic program are defined according to the decision stages that characterize the problem under uncertainty that is considered. Specifically, these stages are set according to the information revelation process that dictates how the values associated with the stochastic parameters are observed. This process then defines how decisions are gradually made according to the increasing amount of contextual information divulged. Therefore, two-stage stochastic programs, which are the subject of the present chapter, involve decisions that need to be made a priori while facing all the uncertainty (i.e., the first-stage), which are then followed by what is referred to as the *recourse* decisions that are made once all uncertainty is resolved (i.e., the second-stage).

When applied in the context of solving two-stage stochastic programs, the Benders algorithm, which was coined the L-Shaped method when originally proposed by Van Slyke and Wets (1969), enables the stochastic optimization models to be decomposed according to the decisional stages. Specifically, as initially prescribed and mostly used ever since, the decomposition strategy is implemented in this case by defining the first-stage decisions as the complicating variables and by considering that the recourse decisions are the facilitating ones. By proceeding in this way, the stochastic program not only becomes stage separable (i.e., the master problem is formulated using the first stage decisions, while the recourse decisions define the subproblems), it also becomes divisible according to the realizations of the random variables that formulate the stochastic parameters (i.e., the subproblems are set according to the possible realizations of the stochastic parameters).

Moreover, two-stage stochastic programs are often used to solve problems that appear in the following form: first, a series of investments must be made while facing uncertainty and, second, the investments made are used once the original uncertainty is resolved, see King and Wallace (2012). The Benders decomposition strategy thus enables the investment decisions, which are often modelled as discrete variables, to be separated from the more facilitating continuous variables, which express the decisions of how the investments made are used. An example of such problems, where the usefulness of Benders decomposition as a solution approach has also clearly been established, is the case of stochastic network design problems, see Hewitt et al. (2021).

The traditionally used approach to define the partition of the decision variables between the complicating and facilitating sets has produced good computational results for a wide gamut of stochastic problems, see Rahmaniani et al. (2017). However, in light of the ever-increasing computational power that is available via the use of modern servers and the access to progressively more efficient general purpose mixed-integer programming solvers, one can ask whether or not the original rules of use for Benders decomposition when applied to two-stage stochastic integer programs should remain unchanged.

In the present chapter, we will explore this general question. In doing so, we will highlight the fact that recent studies have shown that the manner in which Benders decomposition is applied to the considered stochastic programs should definitely be revisited. Specifically, two general observations can be made. First, there are clear computational advantages in including information pertaining to the facilitating variables (i.e., the second stage decisions) with the complicating variables (i.e. the first stage decisions), and vice versa. In other words, one can benefit from transferring information either from the subproblems to the master problem, or in the reverse (from the master to the subproblems). Second, to achieve these computational advantages, the original stochastic programs need to be carefully reformulated before applying the decomposition strategy. Thus, a new set of guidelines should be provided for how the Benders approach is applied in the present case.

Therefore, the rest of the chapter is divided as follows. In Sect. 2, we will recall how the Benders decomposition strategy has been traditionally applied to solve the considered two-stage stochastic models. In Sects. 3 and 4, we will present two recently proposed decomposition strategies which, in both cases, seek to apply the general principle of information transfer between the Benders master problem and its subproblems. Thus, Sect. 3 will be dedicated to the partial Benders decomposition strategy, which prescribes the inclusion of information originating from the subproblems into the master problem as a means to strengthen its formulation. As for Sect. 4, it will be devoted to the Benders dual decomposition strategy where the reverse principle is applied. Specifically, information pertaining to the master is added to the subproblems to define a strategy to improve the quality of the cuts that are generated. In Sect. 5, the main challenges involved in implementing these two general decomposition strategies will be highlighted and a general discussion will be provided regarding how these challenges can be efficiently resolved. Finally, in Sect. 6, we will provide some concluding remarks and highlight novel paths of inquiry along the lines of the new general rules of use for the Benders algorithm.

2 Benders Decomposition for Two-Stage Stochastic Models

To ground our discussion, we first present notation. Let $\mathbb{R}^n$ stand for the space of all n-dimensional real vectors and $\mathbb{R}^{m\times n}$ the space of real $m \times n$ matrices. Let $y \in Y \subseteq \mathbb{R}^{n_1}_+$ define a set of first stage decision variables that must take on integer values. These values must also satisfy the constraint set $Ay = b$, with $A \in \mathbb{R}^{m_1\times n_1}$ a known matrix and $b \in \mathbb{R}^{m_1}$ a known vector. Let $f^\top y$, with $f \in \mathbb{R}^{n_1}$, define the first stage costs. Let S be the set of possible scenarios, with probability p_s of scenario $s \in S$ occurring. Associated with each scenario $s \in S$ is a set of second stage decision variables $x^s \in \mathbb{R}^{n_2}_+$ that, together with the first stage decisions y, must satisfy the constraints $B^s y + Dx^s = d^s$ and for which the cost (often referred to as the recourse cost) is defined as $c^\top x^s$, where $c \in \mathbb{R}^{n_2}$. We note that we consider a fixed recourse matrix $D \in \mathbb{R}^{m_2\times n_2}$, with both $B^s \in \mathbb{R}^{m_2\times n_1}$ and $d^s \in \mathbb{R}^{m_2}$ defining the stochastic parameters. Given this notation, we present the following model, *2-SP*, to form the basis of our discussion of Benders decomposition.

$$2\text{-}SP: \min_{y,x^s:s\in S} f^\top y + \sum_{s\in S} p_s c^\top x^s \tag{1}$$

$$\text{s.t. } Ay = b, \tag{2}$$

$$B^s y + Dx^s = d^s, \quad s \in S, \tag{3}$$

$$y \in Y, x^s \geq 0, \quad s \in S. \tag{4}$$

We note that we restrict our discussion to cases wherein *2-SP* is feasible and has a finite optimal objective function value.

Classical applications of Benders view the first stage decision variables y as complicating and the second stage variables as facilitating. Thus, in the first step of the Benders decomposition strategy, (1)–(4) is projected onto the subspace defined by the decision vector y. In doing so, one obtains the following model:

$$\min_y f^\top y + \sum_{s\in S} p_s z^s(y) \tag{5}$$

$$\text{s.t. } Ay = b, \tag{6}$$

$$y \in Y, \tag{7}$$

where the functions $z^s(y), s \in S$ represent the objective function value of the scenario specific subproblem, $SP(y)_s$, formulated as follows:

$$SP(y)_s : z^s(y) = \min_{x^s} c^\top x^s \tag{8}$$

$$\text{s.t. } Dx^s = d^s - B^s y, \tag{9}$$

$$x^s \geq 0. \tag{10}$$

From this first step of the decomposition strategy, one can also see that if y is fixed in model (1)–(4), then it becomes scenario separable (i.e., reducing to $SP(y)_s, s \in S$ as defined by (8)–(10)).

Benders then proceeds with an outer linearization step in which the dual of $SP(y)_s, s \in S$ is taken. Given dual variables $g \in \mathbb{R}^{m_2}$ associated with constraints (9) of $SP(y)_s$, the dual $DSP(y)_s$ is as follows.

$$DSP(y)_s : \quad \max g^\top(d^s - B^s y) \text{ s.t. } g^\top D \leq c.$$

Because the cost vector c does not vary by scenario, the feasible region of $DSP(y)_s$, $Q = \{g : g^\top D \leq c\}$, is the same for all scenarios. Given our assumption that *2-SP* is feasible and has a finite optimal objective function value we can further assume that $Q \neq \emptyset$. Thus, Q has a set of extreme points, indexed by $\mathcal{I}$, we denote extreme points by q^i, $i \in \mathcal{I}$, with cardinality $|\mathcal{I}|$, and extreme rays, indexed by $j \in \mathcal{J}$, we again denote extreme rays by w^j, $j \in \mathcal{J}$, with cardinality $|\mathcal{J}|$. Considering that extreme points (and extreme rays) are unique elements, the mapping between the points (or rays) and their associated indexes are unique. Thus, the sets $\mathcal{I}$ and $\mathcal{J}$ can refer to either the index sets or the sets of extreme points and rays, respectively. A valid reformulation of *2-SP*, which we refer to as the *Benders Master Problem*, *BMP*, is

$$\min \; f^\top y + \sum_{s \in S} p_s z^s \tag{11}$$

$$\text{s.t. } Ay = b, \tag{12}$$

$$q^{i\top}(d^s - B^s y) \leq z^s, \qquad i \in \mathcal{I},\ s \in S, \tag{13}$$

$$w^{j\top}(d^s - B^s y) \leq 0, \qquad j \in \mathcal{J},\ s \in S, \tag{14}$$

$$y \in Y, z^s \geq 0, \qquad s \in S, \tag{15}$$

where constraints (13) and (14) represent the optimality and feasibility cuts, respectively.

Solving the problem (11)–(15) directly requires enumerating all extreme points and rays of Q. Consequently, Benders Decomposition instead solves a relaxation, wherein only a subset of constraints (13) and (14) are considered.

To define this relaxation, we let $I \subseteq \mathcal{I}$ be subsets of extreme points q^i, $i \in I$ and $J \subseteq \mathcal{J}$ be subsets of extreme rays w^j, $j \in J$. Given those sets we let $BMP(I, J)$ denote the relaxed master problem, which is formulated by instantiating BMP, only over the optimality cuts associated with extreme points in I and feasibility cuts associated with extreme rays in J. That $BMP(I, J)$ is a relaxation of BMP follows from the fact that it is formulated with a subset of the constraints defining BMP.

Benders iteratively strengthens the relaxation $BMP(I, J)$ by identifying extreme points not in I and/or extreme rays not in J for which the corresponding constraints should be added to $BMP(I, J)$. To do so, it solves the relaxation to produce a vector of first stage variable values $\overline{y}$ and estimates of scenario costs $\overline{z}^s$, $s \in S$. This solution is used to formulate and solve the dual subproblems $DSP(\overline{y})_s$. Solution information of such subproblems is inspected to identify whether there are extreme points to add to I or extreme rays to add to J.

Specifically, when the optimal objective function value of a dual subproblem $DSP(\overline{y})_s$ is greater than the current estimate $\overline{z}^s$, the corresponding optimal extreme point is added to I to define a new optimality cut to add to $BMP(I, J)$. When a dual subproblem is unbounded the corresponding extreme ray is added to J to define a new feasibility cut to add to $BMP(I, J)$. If no optimality or feasibility cuts are violated, problem *2-SP* has been solved to optimality. Otherwise, the next iteration is executed, starting with the solution of $BMP(I, J)$, albeit strengthened with the newly found optimality and feasibility cuts. The procedure just discussed is often referred to as the *Multi-cut* version of the L-shaped method, see Birge and Louveaux

(1988), as it adds cuts that are specific to each scenario. Pseudo-code is presented in Algorithm 2.

Algorithm 2 Benders decomposition

```
Set I = ∅ and J = ∅
Create BMP(I, J)
while 2 − BP not solved do
   Solve BMP(I, J) to get vector ȳ, estimates z̄
   for s ∈ S do
      Solve dual subproblem DSP(ȳ)_s
      if constraints from sets (13) or (14) are violated then
         Update I, in the case of a violated optimality cut (13)
         Update J, in the case of a violated feasibility cut (14)
      end if
   end for
   if no violated constraints found then
      Stop {Solved 2-SP}
   end if
end while
```

We note that multiple techniques have been developed to reduce the time a Benders-type algorithm like Algorithm 2 needs to converge. One such technique involves strengthening $BMP(I, J)$ with valid inequalities involving only the y variables that are based on structures present in constraints (3) that link the first and second stage decision variables. Another technique involves adjusting the procedure used to identify extreme points or rays to strengthen $BMP(I, J)$ so as to produce stronger cuts. Another is to augment Algorithm 2 with techniques for producing primal solutions and, relatedly, adding a termination criterion involving an optimality gap. A review of these techniques, and others, can be found in Rahmaniani et al. (2017).

3 Partial Benders Decomposition

While the Benders Decomposition procedure presented in Algorithm 2 is guaranteed to converge to an optimal solution, it can be slow to do so computationally-speaking. This can be attributed to multiple factors. First, the problem structure associated with the linking constraints (3) is lost. As a result, many of the valid inequalities (e.g., cover inequalities) that have been developed for structures present in those systems of constraints are inapplicable. Second, the relaxation of constraints (13) and (14) eliminates from $BMP(I, J)$ all links between the first stage decision variables y and the second stage of the problem. As a result, first stage solutions obtained at the beginning of the Benders algorithm can be extremely poor with respect to their recourse cost. They may also be unlikely to induce second stage subproblems that are feasible.

As such, one way to mitigate these factors is to revisit the identification of complicating and facilitating variables that is used to decompose *2-SP* into a master problem and set of subproblems. In particular, to consider including in the Benders master problem information related to variables that would traditionally be labelled as facilitating. Thus, we next present a modified master problem that is strengthened with information related to scenario subproblems.

To do so, we define $\bar{S}$ as a set of scenarios and note we do not presume that $\bar{S} \subseteq S$. In other words, $\bar{S}$ may contain scenarios that are **not** in S. As such, we distinguish the scenarios $s \in S \cap \bar{S}$, which are present in the original problem from the scenarios $s' \in \bar{S}$, which are not (i.e., $s' \notin S$). Such *artificial* scenarios can be created by taking a convex combination of any subset of S, the full set of scenarios. To that effect, for each scenario $s' \in \bar{S}$ and $s \in S \setminus \bar{S}$ we define the non-negative weight $\alpha_s^{s'}$, $s \in S$, wherein $\sum_{s \in S \setminus \bar{S}} \alpha_s^{s'} = 1$. The set $\bar{S}$ and these weights leads to an alternate reformulation of *2-SP*, which we next present.

$$\min \; f^\top y + \sum_{s \in S \cap \bar{S}} p_s c^\top x^s + \sum_{s \in S \setminus \bar{S}} p_s c^\top x^s \tag{16}$$

$$\text{s.t. } Ay = b, \tag{17}$$

$$B^s y + Dx^s = d^s, \qquad s \in \bar{S}, \tag{18}$$

$$c^\top x^{s'} = \sum_{s \in S \setminus \bar{S}} \alpha_s^{s'} c^\top x^s, \qquad s' \in \bar{S}, \tag{19}$$

$$B^s y + Dx^s = d^s, \qquad s \in S \setminus \bar{S}, \tag{20}$$

$$y \in Y, \; x^s \geq 0 \qquad s \in \bar{S}, x^s \geq 0, s \in S \setminus \bar{S}. \tag{21}$$

It should be noted that constraints (19) express the relationships between the variable costs of the artificial scenarios and the scenarios used in their construction. These constraints set the stage for a stronger master problem after the projection step has been applied.

Applying the projection step of Benders, recognizing that information related to the set $\bar{S}$ is to be treated as complicating, yields the following model:

$$\min \; f^\top y + \sum_{s \in S \cap \bar{S}} p_s c^\top x^s + \sum_{s \in S \setminus \bar{S}} p_s z^s(y) \tag{22}$$

$$\text{s.t. } Ay = b, \tag{23}$$

$$B^s y + Dx^s = d^s, \qquad s \in \bar{S}, \tag{24}$$

$$c^\top x^{s'} = \sum_{s \in S \setminus \bar{S}} \alpha_s^{s'} z^s(y), \qquad s' \in \bar{S}, \tag{25}$$

$$y \in Y, \; x^s \geq 0 \qquad s \in \bar{S}.... \tag{26}$$

We see that the variables x^s, $s \in \bar{S}$ are included in the formulation whereas in a traditional application of Benders they would be viewed as facilitating and left to a subproblem.

At this stage, the Benders process can proceed as expected by replacing the explicit solution of subproblems in the formulation with optimality and feasibility cuts, albeit only for scenarios in $S \setminus \bar{S}$. This leads to the following modified relaxed master problem, $BMP(\bar{S}, I, J)$:

$$\min \; f^\top y + \sum_{s \in S \cap \bar{S}} p_s c^\top x^s + \sum_{s \in S \setminus \bar{S}} p_s z^s \tag{27}$$

$$\text{s.t. } Ay = b, \tag{28}$$

$$B^s y + D x^s = d^s, \qquad s \in \bar{S}, \tag{29}$$

$$c^\top x^{s'} = \sum_{s \in S \setminus \bar{S}} \alpha_s^{s'} z^s, \qquad s' \in \bar{S}, \tag{30}$$

$$q^{i\top} d^s \leq q^{i\top} B^s y + z^s, \qquad i \in I, s \in S \setminus \bar{S}, \tag{31}$$

$$w^{j\top} d^s \leq w^{j\top} B^s y, \qquad j \in J, s \in S \setminus \bar{S}, \tag{32}$$

$$y \in Y, \; x^s \geq 0 \qquad s \in \bar{S}, z^s \geq 0, s \in S \setminus \bar{S}. \tag{33}$$

As a subset of scenario subproblem information is retained in the projection step of Benders, the principle underlying the formulation of $BMP(\bar{S}, I, J)$ is referred to as *Partial Benders Decomposition*, see Crainic et al. (2021).

With this new master problem, a Benders algorithm like the one presented in Algorithm 2 can proceed as normal, only recognizing that cuts need not be generated for scenarios in $\bar{S}$. However, retaining the subproblems associated with the scenarios in set $\bar{S}$ in the new master problem has other impacts, computationally-speaking, on the performance of such an algorithm. First, doing so enables a more accurate approximation of the expected recourse cost of a solution $\bar{y}$. Retaining scenario subproblems in the master leads to a smaller relaxation being applied (i.e., a smaller number of subproblems will be outer linearized, reformulated, and finally relaxed). Therefore, the recourse costs associated with retained scenarios are computed exactly, while the rest are approximated via the subset of optimality cuts present in the master's formulation. Second, the retention of scenario subproblems also leads to greater stability with respect to the obtained first stage solutions. The use of the more accurate expected recourse cost approximation enables the algorithm to find better quality solutions in earlier iterations and less oscillations are observed in the search process overall.

Third, by including the linking constraints (29) in $BMP(\bar{S}, I, J)$ for $s \in \bar{S}$, valid inequalities for the polyhedron $P_s = \{(x^s, y) : Ay = b, B^s y + Dx^s = d^s, y \in Y, x^s \geq 0\}$ can be added to $BMP(\bar{S}, I, J)$ to further strengthen the model. Including constraints (29) also eliminates the need to generate feasibility cuts for the scenario subproblems $s \in S \cap \bar{S}$. However, one can also show that employing specific strategies for choosing scenarios from S to include in $\bar{S}$ can obviate the need to generate feasibility cuts for scenarios in $S \setminus \bar{S}$. We next discuss these strategies at a high level.

The first strategy, referred to as **Row Covering**, see Crainic et al. (2021), relies on the concept of one scenario *covering* another, which we next define.

Definition 1 Consider scenarios $s \in \overline{S}$ and $s' \in S \setminus \overline{S}$. We state that s *covers* s' if $y \in Y$ such that there exists a feasible solution to the subproblem $SP(y)_s$ there exists a feasible solution to the subproblem $SP(y)_{s'}$.

The following proposition can be proven based on weak duality. Details of that proof illustrate how one can characterize when one scenario covers another and in turn guide the selection of scenarios to retain.

Proposition 1 *Consider $s \in \bar{S}$ and $s' \in S \setminus \bar{S}$. If s covers s' then the subproblems $SP(y)_{s'}$ will always be feasible.*

The second strategy, referred to as **Convex Hull**, see Crainic et al. (2021), is analogous to the pursuit of convex hulls in integer programming, as having a representation of the convex hull of the feasible region reduces the complexity of solving the resulting integer program. Following this approach, the scenarios added to $\bar{S}$ are the ones that include in their associated convex hull the scenarios in $S \setminus \bar{S}$. To simplify the presentation of the present strategy, it is assumed here that $\bar{S} \subseteq S$. The following proposition, formally proven in Crainic et al. (2021), shows the value of this idea.

Proposition 2 *When there exists a set of weights $\alpha_s^{s'}$ with which one can express the vector $d^{s'}$ and matrix $B^{s'}$ as a convex combination of the vectors d^s and matrices B^s, for $s \in \bar{S}$, then a feasible solution $\bar{y}$ to $BMP(\bar{S}, I, J)$ necessarily induces a feasible scenario subproblem for s' and no feasibility cuts need be generated for $SP(y)_{s'}$.*

We next consider the artificially generated scenarios ($s' \in \bar{S}$ and $s' \notin S$) and the additional constraints (30). As proven in Crainic et al. (2021), a subproblem s', that is a convex combination of the subproblems $s \in S$, can be added to *2-SP* without modifying the objective function value of an optimal solution. Constraints (30) can improve the performance of a Benders algorithm by providing a direct link in the master problem between the variables $x^{s'}$ and the variables z^s, for $s \in S \setminus \bar{S}$, whose values provide a lower bound on the recourse cost in each scenario. Specifically, the inclusion of (30) provides an additional lower bound for the variables z^s, $s \in S$, whose values are now bounded by both the optimality cuts (31) and the value associated with the subproblem corresponding to s'. More precisely, Crainic et al. (2021) prove the following.

Proposition 3 *The lower bound provided on the recourse cost value associated with each subproblem is at least as great when the relaxed master problem is formulated with an artificial scenario as when it is not.*

In the case when only a single artificial scenario, s', is considered, s' can be constructed by computing the mean value scenario. This idea was first proposed in Batun et al. (2011). Crainic et al. (2021) also prove the following.

Proposition 4 *When $\bar{S}$ contains one artificial scenario, s', potentially along with scenarios from S, the bound produced by solving the resulting $BMP(\bar{S}, I, J)$ is greatest when s' is set to the mean value scenario.*

One can strengthen the relaxed master problem with both retained scenarios and an artificial scenario. A Mixed Integer Program for choosing scenarios to retain from S and the constructing an artificial scenario out of the remaining is presented in Crainic et al. (2021). They also show computationally speaking that such a hybrid strategy outperforms both only retaining scenarios from S and only strengthening the relaxed master problem with an artificial scenario.

4 Benders Dual Decomposition

As clearly illustrated in the previous section, there is an undeniable advantage in adjusting the Benders decomposition strategy in such a way as to include information stemming from the scenario subproblems into the master problem. In the present section, we will see that the reverse approach (i.e., including information stemming from the master problem into the scenario subproblems) also defines an overall acceleration strategy for the Benders decomposition method. This approach was originally proposed in Rahmaniani et al. (2020) and was coined the *Benders Dual Decomposition* method. The general idea behind the Benders dual strategy is to explicitly include the first-stage decisions within each scenario subproblem as a means to derive strengthened Benders cuts (both optimality and feasibility) for the relaxed master's formulation. However, to be able to implement this method, a reformulation of the original problem is again needed.

The present reformulation requires the first stage decisions to be duplicated for each scenario $s \in S$. Thus, in addition to $y \in Y$, one defines the following decision vectors: $y^s \geq 0, s \in S$. An equivalent reformulation of the 2-*SP* model (1)–(4) can then be obtained as follows:

$$\min_{y, y^s, x^s : s \in S} f^\top y + \sum_{s \in S} p_s c^\top x^s \tag{34}$$

$$\text{s.t. } Ay = b, \tag{35}$$

$$Ay^s = b, s \in S, \tag{36}$$

$$B^s y^s + dx^s = d^s, s \in S, \tag{37}$$

$$y^s = y, s \in S, \tag{38}$$

$$y \in Y, y^s \geq 0, x^s \geq 0, s \in S. \tag{39}$$

It should be noted that the linking constraints (37), which originally connected the first and second stage decisions, now exclusively involve scenario-specific decision vectors (i.e., y^s and x^s, $s \in S$). The equivalence between the formulations (1)–(4) and (34)–(39) comes from the fact that the non-anticipativity constraints are imposed: i.e., Eqs. (38). The decision vector $y \in Y$ thus serves as the reference solution that, in turn, specifies the values to which the scenario-specific first-stage decisions need to be fixed ($y^s = y, s \in S$). The Benders decomposition approach can now be applied on model (34)–(39).

As prescribed by the first step of the Benders decomposition strategy, (34)–(39) is projected onto the subspace defined by the reference decisions (i.e., decision vector y). In doing so, one obtains the following model:

$$\min_y f^\top y + \inf_y \sum_{s \in S} p_s z^s(y) \tag{40}$$

$$\text{s.t. } Ay = b, \tag{41}$$

$$y \in Y, \tag{42}$$

where the functions $z^s(y), s \in S$, are now formulated as follows:

$$z^s(y) = \min_{y^s, x^s} c^\top x^s \tag{43}$$

$$\text{s. t. } Ay^s = b, \tag{44}$$

$$B^s y^s + Dx^s = d^s, \tag{45}$$

$$y^s = y, \tag{46}$$

$$y^s \geq 0, x^s \geq 0. \tag{47}$$

As indicated previously, by fixing y, model (40)–(42) becomes scenario separable and $z^s(y), s \in S$, define the scenario specific subproblems.

As originally proposed, once the projection is performed, the Benders decomposition then applies the outer linearization step. In the present case, one could again use, for $s \in S$, the standard duals of the models (43)–(47) to perform this step of the decomposition approach. However, when compared to the original Benders strategy, an alternative approach is utilized here. Specifically, Lagrangian relaxation is applied to the non-anticipativity constraints (46). By proceeding in this way, for each $s \in S$, the Lagrangian dual scenario subproblem associated with (43)–(47) is defined as follows:

$$\max_{\lambda \in \mathbb{R}^{n1}} \min_{(y^s, x^s) \in X^s} \{c^\top x^s - \lambda^\top (y^s - y)\}, \tag{48}$$

where, $X^s = \{(y^s, x^s) \in \mathbb{R}_+^{n_1} \times \mathbb{R}_+^{n_2} \mid Ay^s = b, B^s y^s + Dx^s = d^s\}$.

At this point, a few observations can be made. For any given $s \in S$, if the scenario subproblem (43)–(47) is solved while assuming that the first-stage solution is fixed to $\hat{y}$ (e.g., $\hat{y}$ could represent the solution to the master problem at a given iteration), then the following valid optimality cut can be defined:

$$z^s \geq c^\top \hat{x}^s + (y - \hat{y})^\top \hat{\lambda}^s, \tag{49}$$

where the value vectors $\hat{x}^s$ and $\hat{y}^s$ (i.e., $\hat{y}^s = \hat{y}$) represent the optimal solution to the scenario subproblem and $\hat{\lambda}^s$ are the optimal dual multipliers associated with the non-anticipativity constraints (46). The validity of the optimality cut (49) is a direct consequence of the weak duality result associated with the Lagrangian dual scenario subproblem (48).

A similar observation can be made regarding the set of feasibility cuts that can be defined for the Benders master problem. In this case, consider the following scenario subproblem:

$$\min_{y^s, x^s, a_+^s, a_-^s} \quad \mathbf{1}^\top a_+^s + \mathbf{1}^\top a_-^s \tag{50}$$

$$\text{s. t.} \quad Ay^s = b, \tag{51}$$

$$B^s y^s + Dx^s + a_+^s - a_-^s = d^s, \tag{52}$$

$$y^s = y, \tag{53}$$

$$y^s \geq 0, x^s \geq 0, a_+^s \geq 0, a_-^s \geq 0. \tag{54}$$

In this case, the additional vectors $a_+^s \in \mathbb{R}_+^{m_2}$ and $a_-^s \in \mathbb{R}_+^{m_2}$ represent the artificial variables used to assess whether or not Eqs. (52) induce a feasible decision set. It should be noted that $\mathbf{1}$ refers to a value-vector of size m_2 whose components are all set to one. If the first stage solution is again fixed to $\hat{y}$, then the subproblem associated with scenario $s \in S$ can only be feasible if the optimal solution to (50)–(54), i.e., $\hat{x}^s$, $\hat{y}^s$, $\hat{a}_+^s$ and $\hat{a}_-^s$, is such that $\mathbf{1}^\top \hat{a}_+^s + \mathbf{1}^\top \hat{a}_-^s = 0$. Otherwise, a valid feasibility cut can be defined as follows:

$$0 \geq \mathbf{1}^\top \hat{a}_+^s + \mathbf{1}^\top \hat{a}_-^s + (y - \hat{y})^\top \hat{\beta}^s, \tag{55}$$

where $\hat{\beta}^s$ are the optimal dual multipliers associated with the non-anticipativity constraints (53). Again, the validity of the cut (55) resides in the weak duality result of the Lagrangian dual defined for the scenario feasibility subproblem (50)–(54). Lastly, the cuts of types (49) and (55) are often referred to as the *generalized Benders cuts*, see Rahmaniani et al. (2017).

Moving forward, an important point to note is that, in an effort to strengthen the Lagrangian dual's formulation (48), the integrality requirements on the scenario-specific first-stage decisions can be imposed (i.e., defining the decision domain as $(y^s, x^s) \in X^s$ and $y^s \in Y$). This restriction can also be performed partially (i.e., imposing the integrality requirements only on a subset of the decision variables) in an effort to limit the resulting complexity of the dual scenario subproblems.

In turn, this approach, enables the Benders method to be applied with a strengthened set of valid inequalities (both optimality and feasibility cuts). The following proposition, which was originally proposed and proved in Rahmaniani et al. (2020), defines the strengthened optimality cuts that are used in the present context:

Proposition 5 *Given $\hat{y} \in \mathbb{R}^{n_1}$ and $\lambda \in \mathbb{R}^{n_1}$, let $(\bar{x}^s, \bar{y}^s)$ be an optimal solution obtained by solving:*

$$\min_{(y^s, x^s) \in X^s} \{c^\top x^s - \lambda^\top (y^s - \hat{y}) : y^s \in Y\}. \tag{56}$$

Then,

$$z^s \geq c^\top \bar{x}^s + (y - \bar{y}^s)^\top \lambda, \tag{57}$$

defines a valid optimality cut for the recourse cost associated with scenario $s \in S$ for the Benders master problem.

The previous result, when applied on feasible first-stage solutions, i.e., for $\hat{y} \in F \cap Y$ where $F = \bigcap_{s \in S} F^s$, states that the optimality cuts (57) are, at most, as strong as the classical Benders optimality cuts. However, when applied on fractional first-stage solutions, the optimality cuts (57) provide a distinct advantage when compared to the classical version of the cut. This advantage is quantified in the next result:

Theorem 1 *For a given scenario $s \in S$, let $\hat{\lambda} \in \mathbb{R}^{n_1}$ be the dual multiplier values obtained from (43)–(47), then the optimality cut (57) is parallel to cut (49) and at least σ units tighter, where*

$$\sigma = \min_{(y^s, x^s) \in X^s : y^s \in Y} \left\{c^\top x^s - \hat{\lambda}^\top y^s\right\} - \min_{(y^s, x^s) \in X^s} \left\{c^\top x^s - \hat{\lambda}^\top y^s\right\} \geq 0. \tag{58}$$

In Rahmaniani et al. (2020), it was further shown that a similar result also applies in the case of the Benders feasibility cuts.

In all generality, Theorem 1 provides the key insight on how to search for higher quality cuts when applying the Benders algorithm. Specifically, three cut-generation strategies can be defined. In what follows, we present these strategies in the case of searching for violated optimality cuts associated with a solution to the master problem (i.e., identified here as solution $\hat{y}$). The first of these is the *Classical* strategy where, for the given solution $\hat{y}$ and for all $s \in S$, the following two-step approach is applied: *Step 1:* solve model (43)–(47) to obtain $\hat{x}^s$ and $\hat{\lambda}^s$; *Step 2:* using the triplet $(\hat{y}, \hat{x}^s, \hat{\lambda}^s)$ generate an optimality cut of type (49).

Whenever the given solution $\hat{y}$ includes fractional values, the quality of the previous optimality cuts can be improved at the expense of solving an additional model in the cut-generation process, i.e., the *Strengthened* strategy. Specifically, given $\hat{y}$ and for all $s \in S$ apply the following three-step approach: *Step 1:* solve model (43)–(47) to obtain $\hat{\lambda}^s$; *Step 2:* given the pair $(\hat{y}, \hat{\lambda}^s)$ solve model (56) to obtain $\bar{y}^s$ and $\bar{x}^s$; *Step 3:* using the triplet $(\bar{y}^s, \bar{x}^s, \hat{\lambda}^s)$ generate an optimality cut of type (57). As stated in Theorem 1, this approach enables each optimality cut to be lifted by an amount that corresponds to the duality gap.

Lastly, we have the *Lagrangian* strategy, which calls for the Benders cuts to be generated by directly solving (48) for all $s \in S$ (i.e., the Lagrangian dual scenario subproblems). This strategy requires a two-step approach to again be applied. Specifically, given $\hat{y}$ and for all $s \in S$, the following steps are performed: *Step 1:* apply the restriction $y = \hat{y}$ and $y^s \in Y$ and then solve model (48) to obtain $y^{\star s}$, $x^{\star s}$ and $\lambda^{\star s}$ (i.e., the optimal solution to the scenario-specific Lagrangian dual); *Step 2:* using the triplet $(y^{\star s}, x^{\star s}, \lambda^{\star s})$ generate the following optimality cut:

$$z^s \geq c^\top x^{\star s} + (y - y^{\star s})^\top \lambda^{\star s}. \tag{59}$$

It should be noted that all three cut-generation strategies can also be directly applied to find violated feasibility cuts. The only difference resides in the scenario subproblem that is used to define the Lagrangian dual. Specifically, let us recall that for a given $s \in S$, (43)–(47) is used to find optimality cuts, whereas (50)–(54) is applied when searching for feasibility cuts. Furthermore, is important to highlight that applying the *Lagrangian* strategy to obtain cuts of type (59) will require a significant amount of computational effort, i.e., for each $s \in S$, the Lagrange dual subproblem (48) is solved to optimality under the restriction that $y^s \in Y$. To successfully apply the Benders dual decomposition strategy, it is thus necessary to implement computationally efficient methods to generate these cuts, which is a point that will be discussed in the next subsection of the present chapter.

5 Implementation

Implementing the Partial Benders decomposition strategy requires determining whether $\bar{S}$ should include retained scenarios from S, or artificial scenarios, or both. If scenarios are to be retained, one must then determine the number to retain and the strategy used to identify which scenarios to retain, **Row Covering** or **Convex Hull**. The nature of both strategies suggests jointly determining the scenarios to retain is more impactful than making such selections independently. For the **Row Covering** strategy, if both scenarios s and s' cover scenario s'' then retaining both s and s' is likely unnecessary. Similarly, Proposition 2 shows that the impact of the **Convex Hull** strategy depends on the convex combinations of the set of retained scenarios. Thus, the benefit of selecting scenarios to retain jointly is clear.

When artificial scenarios are included in the set $\bar{S}$ one must also determine how many to create and how they should be created. Finally, one must determine whether the choice of scenarios to retain should be made with an explicit recognition that the scenarios not retained will be used to construct one or more artificial scenarios.

Crainic et al. (2021) provide guidance for many of these determinations. First, they prove that if only one artificial scenario is to be created then using the mean value scenario strengthens the resulting master problem the most. The potential of retaining multiple artificial scenarios has not yet been assessed, nor has a mechanism for determining multiple such artificial scenarios been proposed.

Second, for each retention strategy, Crainic et al. (2021) propose a Mixed Integer Program (MIP) for choosing a given number of scenarios to retain from S and (potentially) constructing one artificial scenario that represents the mean value scenario of those remaining. They refer to solving a MIP that retains scenarios with an explicit recognition that an artificial scenario will be created from the remaining as a *hybrid strategy*. They show computationally speaking that a hybrid strategy based on **Row Covering** is the best-performing, and that it outperforms both only retaining scenarios from S via **Row Covering** and only strengthening the relaxed master problem with an artificial scenario. Crainic et al. (2021) study computationally the question of how many scenarios from S to retain and observe that retaining a small fraction of the scenarios in an instance (e.g., retaining two when $|S| = 256$) can have a significant impact on the performance of the Benders algorithm.

Regarding the implementation of the Benders dual decomposition strategy, the main challenge is directly linked to the increased complexity of the scenario subproblems that are solved in order to lift the Benders cuts generated by the algorithm. Here, a few observations are important to highlight before choosing a specific cut-lifting strategy. First, as often remarked when describing the Benders decomposition approach, one of the main drawbacks resides in the fact that once the optimality and feasibility cuts are relaxed, i.e., constraints (13) and (14), the resulting BMP loses all information regarding the subproblems. In turn, this directly leads to instability issues in the beginning of the search process (e.g., the quality of the BMP solutions obtained in the first iterations of the method are oftentimes quite poor).

Second, as highlighted in many studies, see Rahmaniani et al. (2017), the classical Benders cuts can be quite shallow, which can cause long plateaus in the evolution of the overall lower bound (i.e., multiple successive iterations where the observed increase in the lower bound value is limited). In turn, such plateaus can significantly slow down the convergence towards the optimum at the end of the search process.

Therefore, when deciding on how much effort should be dedicated to lifting the Benders cuts, the previous observations tend to justify the fact that one should explicitly consider the evolution of the search process. Specifically, when the search starts, there might not be enough incentive to improve the quality of the cuts that are generated, i.e., the poor quality of the solutions obtained by solving the BMP might not warrant such efforts. However, as the search process progresses and more cuts are added to the BMP's formulation, one might mitigate the issues of the lower bounding plateaus by adding strengthened cuts to accelerate convergence. These general principles were applied in the original implementation of the Benders dual method in Rahmaniani et al. (2020), where a three-phase cut generation approach was proposed to add the Benders cuts to the BMP.

In *Phase 1*, the LP relaxation of the BMP is solved via the use of the classical Benders cuts (i.e., the optimality cuts (49) and feasibility cuts (55) are used in this phase). This approach, which was originally proposed in McDaniel and Devine (1977), provides a warm-start to the solution process, where the BMP's formulation can be quickly improved through the inclusion of easily obtainable cuts. As soon as the LP is solved, *Phase 2* in the cut generation process can begin. In this phase, strengthened Benders cuts, of the form (57), are generated and added to the BMP's formulation. By proceeding in this way, the overall quality of the cuts that are found by the algorithm can be improved thus increasing the efficiency of the solution process. *Phase 3* is only performed when the evolution of the lower bound starts to stagnate towards the end of the search process. At this point, Lagrangian cuts are generated (e.g., optimality cuts of type (59) are generated in this case) to mitigate this phenomena and complete the convergence of the search.

Finally, it should be noted that when the Benders cuts are lifted (e.g., in *Phases 2–3* of the previous approach), the scenario subproblems that are solved are mixed-integer programs, which can require a significant amount of computational effort. Finding efficient strategies to solve these scenario subproblems is thus an important methodological ingredient of any successful implementation of the Benders dual algorithm. In Rahmaniani et al. (2020), a heuristic approach is proposed to solve the scenario-specific subproblems of type (48) that enable Lagrangian cuts to be derived. This heuristic approach first finds the dual multipliers (i.e., λ) by solving an optimization model that seeks to maximize the amount by which the cut is lifted, while considering an inner approximation of the Lagrangian dual subproblem. A relaxed version of the inner minimization problem (56) is then solved to produce the cuts. The effort dedicated to this search was further limited by generating ϵ-optimal cuts. Recently, in Chen and Luedtke (2022), a generalized formulation is proposed to identify the scenario-specific Lagrangian cuts for the Benders algorithm. To further accelerate the cut-generation process, a strategy to efficiently restrict the decisional space of the subproblems is also developed. The extensive computational experiments, involving multiple problem types, conducted by both Rahmaniani et al. (2020) and Chen and Luedtke (2022) have clearly shown the high-potential of this strategy to improve the Benders lower bound and accelerate the overall search process.

6 Conclusion

The Benders algorithm is a well-established strategy for solving two-stage stochastic programs. The classical decomposition strategy used when executing the Benders algorithm involves solving a master problem defined over the first stage decision variables and subproblems defined over the second stage, scenario-specific decision variables. This chapter has reviewed two different decomposition strategies that can significantly improve the performance of the Benders algorithm. One strategy, *Partial Benders Decomposition*, involves including in the master problem the decision variables and constraints of a small set of carefully chosen (or constructed) scenarios. Conversely, the second strategy, *Benders dual decomposition*, involves including first stage decision variables in the subproblems.

Both strategies have positive effects with respect to generating lower bounds on the objective function value of an optimal solution and optimality cuts. Regarding the partial Benders strategy, the inclusion of artificial scenarios and their associated valid inequalities provide an additional lower bound on the expected recourse cost as evaluated in the relaxed master problem. That scenario subproblems are retained in the master limits the extent of the relaxation that is applied to obtain the master problem, leaving a more accurate approximation of the expected recourse cost function. The Benders dual strategy defines a general approach to strengthen the quality of the cuts that are generated by the Benders method. Further, the magnitude of the improvement that is obtained on the recourse cost lower bound when the strengthened optimality cuts are used in lieu of the classical ones can be characterized. The literature has demonstrated the benefits of each strategy, both theoretically and with extensive computational studies. However, there are multiple avenues for future work with respect to each strategy.

Regarding Partial Benders Decomposition, we highlight three areas for enhancements of the strategy we believe have promise. We first recall that the Partial Benders Decomposition strategy involves identifying entire scenarios to retain in the master problem *before* the Benders algorithm begins and then executing the algorithm given that master problem. As including in the master problem additional variables and constraints makes it harder to solve, one could instead consider retaining only portions of scenarios. However, doing so would require a mechanism for identifying which portions of scenarios to retain. Even with the Partial Benders Decomposition strategy the progress of the Benders algorithm can stall, leading it to struggle to converge to a solution that is provably optimal. Hence, one can instead imagine changing the scenario information retained in the master problem during the execution of the Benders algorithm. The final enhancement we propose involves complementing the retained scenarios in the master problem with more than one artificial scenarios. Doing so would necessitate a mechanism for creating multiple artificial scenarios.

As for the Benders Dual Decomposition strategy, a first promising avenue of research would be to partially relax the non-anticipativity constraints as a means to obtain subproblems involving multiple scenarios. Such an approach would entail solving more complex subproblems (each one being defined using a subset of scenarios) but could further strengthen the quality of the cuts generated by the Benders algorithm. Establishing how the multi-scenario subproblems should be defined and assessing the cut-lifting benefits of this approach would certainly be an interesting study to conduct. Another avenue of research worth pursuing would be to develop alternative cutting-plane methods to solve the required scenario-specific integer programs to produce the Lagrangian cuts. Specifically, specialized valid inequalities for the deterministic variants of the considered stochastic models can be leveraged here to accelerate the solution process for the scenario subproblems. Finally, considering the general nature of the Benders dual strategy, one could further investigate how to efficiently apply it to solve even more complex and challenging problems (e.g., non-linear stochastic integer programs).

Acknowledgments Over the years, the authors had the opportunity to have many discussions regarding the field of integer programming, but also life in general, with the late professor Bernard Gendron. The authors are extremely thankful for these discussions and the wisdom imparted by Bernard. In fact, echoes of these discussions reverberate in the present chapter.

References

Batun, S., Denton, B. T., Huschka, T. R., & Schaefer, A. J. (2011). Operating room pooling and parallel surgery processing under uncertainty. *INFORMS Journal on Computing*, *23*(2), 220–237.

Benders, J. F. (1962). Partitioning procedures for solving mixed-variables programming problems. *Numerische Mathematik*, *4*(1), 238–252.

Birge, J. R., & Louveaux, F. V. (1988). A multicut algorithm for two-stage stochastic linear programs. *European Journal of Operational Research*, *34*, 384–392.

Chen, R., & Luedtke, J. (2022). On generating Lagrangian Cuts for Two-Stage Stochastic Integer Programs. *INFORMS Journal on Computing*, *34*(4), 2332–2349.

Crainic, T. G., Hewitt, M., Maggioni, F., & Rei, W. (2021). Partial benders decomposition: general methodology and application to stochastic network design. *Transportation Science*, *55*(2), 414–435.

Hewitt, M., Rei, W., Wallace, S. W. (2021). Stochastic network design. In: TG. Crainic, M. Gendreau, & B. Gendron (Eds.), *Network design with applications to transportation and logistics*. Springer.

King, A. J., & Wallace, S. W. (2012). *Modeling with stochastic programming*. Springer.

McDaniel, D., & Devine, M. (1977). A modified benders' partitioning algorithm for mixed integer programming. *Management Science*, *24*(3), 312–319.

Rahmaniani, R., Ahmed, S., Crainic, TG., Gendreau, M., & Rei, W. (2020). The benders dual decomposition method. *Operations Research*, *68*(3), 878–895.

Rahmaniani, R., Crainic, T., Gendreau, M., & Rei, W. (2017). The Benders decomposition algorithm: A literature review. *European Journal of Operational Research*, *259*, 801–817.

Van Slyke, R. M., & Wets, R. (1969). L-Shaped linear programs with applications to optimal control and stochastic programming. *SIAM Journal on Applied Mathematics*, *17*(4), 638–663.

Decomposition Methods for Choice-Based Optimization Models

Shadi Sharif Azadeh, Meritxell Pacheco Paneque, and Michel Bierlaire

1 Introduction

Incorporating individual's choice in optimization problems has been formalized through choice-based optimization (CBO). These models have received increasing attention in recent years because they permit to explicitly capture the interactions between the decisions to be optimized and the expected demand as explanatory variables of the discrete choice model. In this chapter, we focus on parametric choice models rooted in the random utility maximization (RUM) theory. They assume that each individual associates a utility with every alternative and chooses the choice with the highest utility. Optimization problems that consider these models can be found in different applications such as revenue management, home-delivery services and city logistics, on-demand mobility systems and railway planning to name a few (e.g., Shen and Su, 2007, Ulmer, 2020, Prokhorchuk et al., 2019, Roemer et al., 2023, Azadeh et al., 2022, Binder et al., 2021). Concrete examples include profit maximization in multiple contexts, such as the entertainment industry and retail, and welfare maximization in transportation.

Since the popularization of the logit model McFadden (1974) and thanks to the increasingly availability of abundant and individual-based datasets, RUM models

S. Sharif Azadeh (✉)
Delft University of Technology, Delft, The Netherlands
e-mail: s.sharifazadeh@tudelft.nl

M. Pacheco Paneque
University of Fribourg, Fribourg, Switzerland
e-mail: meritxell.pacheco@unifr.ch

M. Bierlaire
École Polytechnique Fédérale de Lausanne, Lausanne, Switzerland
e-mail: michel.bierlaire@epfl.ch

T. G. Crainic et al. (eds.), *Combinatorial Optimization and Applications*,
International Series in Operations Research & Management Science 358,
https://doi.org/10.1007/978-3-031-57603-4_13

have become the most advanced and operational disaggregate demand models. The main advantage of the logit model is due to its simplicity of its closed-form probability expression. However, the associated assumptions might give rise to unrealistic substitution patterns across alternatives. Advanced models have been suggested to alleviate these assumptions. Nested logit and cross-nested logit models offer a way to consider correlations between options by establishing groups (nests) that bring together alternatives with shared characteristics. Similar to the logit model, these models have a closed-form mathematical expression. Mixtures of logit models, also known as mixed logit models, provide a highly flexible framework that can approximate any RUM model (McFadden & Train, 2000). The appeal of these models stems from their ability to address the primary shortcomings of standard logit models. They achieve this by accommodating random variations in preferences, allowing for unrestricted substitution patterns, and accounting for correlations in unobservable factors over time.

As a result, advanced Random Utility Models (RUM) offer improved predictive capabilities regarding individual behavior. However, their emphasis on behavioral accuracy leads to intricate mathematical expressions that are challenging to incorporate into traditional optimization models. For example, in mixed logit models, choice probabilities lack a closed-form solution because they entail integrals of standard logit probabilities across a parameter density. In particular, the integration of choice probabilities in the mixed-integer linear framework introduces non-convexity and non-linearity in the explanatory variables of the RUM model, and they are not always available in closed form. In the work by Pacheco Paneque et al. (2021) a mixed-integer linear formulation that explicitly incorporates the individual's choices is proposed. In an attempt to deal with the uncertain characteristics of the RUM model, they incorporate scenarios that represent simulated outcomes of the probability distribution associated with its random aspects. These scenarios serve as the basis for approximating individual choice probabilities and, consequently, estimating expected demand using the sample-average approximation (SAA) principle. However, this approach presents the challenge of managing large mixed-integer linear programming (MILP) formulations. While general-purpose MILP solvers can effectively handle small to medium-sized instances, real-world problems of practical significance often involve a larger number of individuals and a substantial volume of scenarios, aiming to improve accuracy.

Each scenario can be perceived as a distinct behavioral context where individuals are contemplating a choice. These scenarios provide a decomposable structure to the problem, an attribute that can be harnessed to manage the computational intensity of the exact method. In this study, we introduce a heuristic method, drawing inspiration from scenario decomposition in the stochastic programming (SP) framework, to address the general CBO problem delineated in Pacheco Paneque et al. (2021). In the context of scenario decomposition, duplicates of the initial-stage decisions of multi-stage SP problems are produced for every scenario. These so-called non-anticipativity constraints, which mandate identical duplicates across scenarios, are dualized. Consequently, the subproblems correlated with each scenario are solved independently, akin to the principles of Lagrangian decomposition or variable splitting in combinatorial optimization. The chief advantage of this technique is the preservation of the interaction between decisions and expected demand within the subproblems. The bound established by scenario decomposition can be enhanced through scenario grouping, though it requires solving larger subproblems (Pacheco Paneque et al., 2022). To facilitate this, we examine various strategies founded on similarity metrics such as Euclidean distance.

We enhance the suggested approach by introducing an algorithm that addresses a constrained Mixed-Integer Linear Programming (MILP) formulation of the original problem. This formulation incorporates information from the solutions of subproblems obtained during each iteration of the subgradient method, which is employed to solve the Lagrangian dual problem. The algorithm takes advantage of the characteristics of the CBO framework to generate feasible solutions efficiently. Our computational experiments demonstrate that the proposed methodology surpasses the performance of the evaluated commercial solver regarding the highest-quality feasible solution discovered within a given time budget. Simultaneously, the generation of feasible solutions at each subgradient method iteration facilitates the calculation of the duality gap. This gap enables the assessment of solution quality and thereby, the method's performance, since optimal solutions for larger instances are unattainable.

The remainder of this chapter is structured as follows. A review of the current literature is provided in Sect. 2. Section 3 provides the general framework of the CBO mathematical model. Section 4 presents the scenario decomposition approach followed by computational results in Sect. 5. Section 6 discusses the reasons behind our choice of decomposition method and possibilities for future work. Finally, the chapter is concluded in Sect. 7.

2 Literature Review

Because of the probabilistic nature of RUM models, the formulations resulting from their integration in optimization problems are typically non-convex and non-linear. Researchers have therefore focused on the development of reformulations and efficient algorithms to solve practical problems. In Sect. 2.1, we provide an overview of the solution methodologies that have been recently proposed to address transportation problems. We investigate the decomposable structure provided by the simulation scenarios in order to develop a solution methodology for CBO problems. As a result, we review the literature on scenario decomposition and scenario grouping in Sect. 2.2. To conclude, we outline the main contributions of this chapter in Sect. 2.3.

2.1 Solution Methods for Transport Problems

In this section, we shortly present resolution methods for CBO problems in the context of transportation. Models of demand that focus on customer behavior are frequently utilized in the field of revenue management. These models integrating are known as choice-based revenue management models, and have gained popularity over the last years. They generally aim at maximizing both revenue and customer satisfaction by deciding about pricing while controlling for product availability. A general and relevant application is introduced in Gallego et al. (2004). They propose a dynamic programming (DP) model for a choice-based, generalized network revenue management problem. They illustrate that the optimal value of this model can be closely mirrored by the optimal solution of a suitably devised linear programming (LP) model. Since then, scholarly attention has been centered on various approximations of the foundational DP formulation, providing bounds on the optimal expected revenue rather than assurances of performance on the optimality gap (Strauss et al., 2018).

In several transportation networks, travelers may modify their travel arrangements (e.g., departure time, route) depending on the level of service of the network and/or the price associated with the available alternatives. One example can be found in the toll setting problem, which was introduced by Labbé et al. (1998), and consists of a bilevel model where the authority (upper level) wants to maximize its revenues from a taxation scheme on a transportation network at the same time that the users (lower level) minimize their generalized travel costs (i.e., the tolls and the travel costs) while allowing for those tax levels. In Gilbert et al. (2014a), the users are assumed to follow paths determined by the logit model. The derived optimization model is non-convex and non-linear, with the potential for multiple local optima. A comprehensive algorithmic analysis of this problem is conducted in a subsequent study by Gilbert et al. (2014b). They apply a mixed-integer approximation approach to solve the problem, establishing starting points from which local search can reach near-optimal outcomes. Further, they assign paths based on a mixed logit model, the probabilities of which do not possess a closed form. A comparable solution method is implemented to address the problem. The large duality gaps obtained for some of the tested instances (for a given time budget) confirm the fact that the combinatorial approximation of the problem is hard to solve.

Yan et al. (2008) address the problem on fleet routing and flight scheduling with both stochastic market demands (via scenarios) and market shares (via a logit model). To solve the two-stage SP model, they propose two heuristic algorithms that fix a market demand scenario and approximate the resulting mixed-integer nonlinear programming (MINLP) model by fixing the decisions that are explanatory variables of the logit model (see Wei et al., 2020; Justin et al., 2021).

Another pertinent application in the context of transportation is assortment optimization. We witness a tendency in the literature to address this type of models via Markov chain choice models. These models formulate the substitution behavior of customers by state transitions in a Markov chain. In assortment optimization,

Blanchet et al. (2016) address the problem of identifying the *right* model capturing substitution patterns with Markov chain choice models, since they provide a simultaneous approximation for all RUM models. The authors conclude that these provide a good approximation to the true choice probabilities under mild assumptions, and they feature a polynomial-time algorithm that compute the optimal assortment for a mixed logit model. Likewise, Feldman and Topaloglu (2017) provide an LP formulation for the assortment optimization, single resource revenue management, and network revenue management problems under the Markov chain choice model. The computed experiments show that the logit model might be preferable because it prevents the fact that the Markov chain choice model may suffer from overfitting and it is able to compute the mean utility of a product without having to reestimate the parameters of the choice model.

2.2 Scenario Decomposition for Choice-Based Optimization

Scenario decomposition is employed on the deterministic equivalent formulation that emerges from taking into account a finite set of scenarios from a multi-stage SP problem. An algorithm for multi-stage SP problems, encompassing mixed-integer variables across all stages, was introduced by Carøe and Schultz (1999). Their approach involves creating duplicates of the first-stage variables, which are independent of the random data, for each scenario. The non-anticipativity constraints are then relaxed in a Lagrangian fashion. This technique is integrated into a branch-and-bound algorithm to aid in achieving convergence. Numerical tests incorporating up to 10 scenarios illustrate that the Lagrangian dual yields substantially superior lower bounds than the LP relaxation, and that feasible solutions are identified within 0.2% of the optimal outcome.

Scenarios can be clustered, with a replica of the first-stage variables introduced for each cluster. The pioneering work by Escudero et al. (2013) laid the groundwork for such a concept. They presented a Lagrangian decomposition approach based on clusters for the two-stage stochastic programming mixed 0–1 problem. Each cluster, which is a randomly assembled group of scenarios, inherently considers the non-anticipativity constraints. Computational experiments conducted on instances with up to 500 scenarios show that this approach outperforms the traditional Lagrangian decomposition for individual scenarios in terms of both the quality of the bounds and computational effort. This concept has also been extended to address multi-stage stochastic programming problems, as evidenced in works by Escudero et al. (2016) for binary cases and Gade et al. (2016) for integer cases (see Jiang et al., 2021; Bakir et al., 2020; Jiang et al., 2022). The concept of scenario grouping has also been evaluated in conjunction with scenario reduction methods by van Ackooij et al. (2018) to dynamically modify the groups throughout the iterative resolution process.

Crainic et al. (2014) conduct an assessment of various grouping strategies for the two-stage network design issue. These strategies are defined based on their splitting type (cover or partition), grouping method (random or influenced by a similarity-dissimilarity measure rooted in the κ-means clustering algorithm), and the characteristics of the scenario on which similarity-dissimilarity is gauged. Numerical analyses, carried out for up to 32 scenarios on instances with 10 nodes, as many as 83 arcs, and up to 50 commodities, indicate that the covering strategy yields the highest-quality solutions. More recently, an innovative technique was presented by Ryan et al. (2020), which formulates an optimization problem to group scenarios for maximizing bound enhancement. This method offers superior initial relaxation bounds compared to random and κ-means clustering, albeit at the expense of heightened computational time. The primary benefit of this method is its capability to be integrated into any standard scenario decomposition algorithm as an initial processing step.

In addition to scenario grouping, variance reduction methods, such as the classical approaches of control variates or importance sampling, can be carried out in order to reduce the number of scenarios while preserving a certain degree of precision of the CBO model. The novel approach proposed in Bertsimas and Mundru (2023) for scenario reduction for two-stage SP problems shows a significantly better performance for 1–2% of the total number of scenarios in relation to other state-of-the-art methods (e.g., importance sampling, Monte Carlo sampling, and Wasserstein scenario reduction with squared Euclidean norm). Another interesting alternative can be found in Zhang et al. (2023). The idea revolves around the formulation of a scenario subset selection model that optimizes the recourse approximation over a pool of solutions. Then, a scenario assortment optimization approach is conceived to compute a lower bound by relaxing the non-anticipativity constraints across scenario *bundles*. To solve it, a new column-evaluation-and-generation algorithm is developed, which provides a generalizable method for optimization problems featuring many decision variables and hard-to-estimate objective parameters. The results show that the proposed method outperforms scenario reduction benchmarks and provide tighter lower bounds.

2.3 Chapter Contributions

To tackle the above-mentioned complexity of solving CBO problems, especially for large instances, in this chapter, we rely on the mixed-integer linear formulation of RUM models introduced in Pacheco Paneque et al. (2021). The MILP model that formulates the CBO problem is sensitive from a computational point of view. Indeed, the exact method fails at solving instances with a large number of scenarios, alternatives and/or individuals, which usually arise in real-life problems. To tackle such instances, we introduce a heuristic solution method that takes advantage of the simulation-based linearization of the MILP model that formulates a general CBO problem. This problem aims at deciding on (some of) the features (e.g., price) of a given set of alternatives such that an objective function determined by the planner (e.g., revenue) is optimized. Notice that these decisions are the same across scenarios. They play the role of first-stage variables in multi-stage SP problems. The proposed solution method exploits the principles of scenario decomposition and scenario grouping.

The proposed method is also general with respect to the RUM model. This is an important feature because we observe that, in most cases, the proposed methods are specific to the assumed model, which is typically the logit model because of the simplicity of its closed-form probability formula. This is not the case of advanced models like the mixed logit model, whose individual choice probabilities do not present a closed form. As a consequence, obtaining a convex formulation with advanced RUM models is unattainable, which significantly complicates their use in exact methods of optimization. In the CBO framework, we avoid the use of the probability expressions of RUM models by specifying the associated preference structure directly in terms of the utility functions (structural equations of RUM models). We numerically test our method on a revenue maximization problem under a mixed logit model. The results show that it outperforms by a considerable margin a general-purpose MILP solver for a given time budget and that the obtained duality gaps are as well restricted.

3 Choice-Based Optimization Framework

The optimization problem considered in this paper embeds a RUM model whose parameters have been exogenously estimated to represent the expected demand. Let N be the number of individuals in the sample (population) and J the number of alternatives that can be chosen by the individuals according to the RUM model. For each alternative i, we assume a capacity denoted by $c_i \geq 1$, which indicates the maximum number of individuals who can choose it.

In a RUM model, U_{in} represents the utility that individual n obtains from choosing alternative i. It consists of a systematic component and a random component that captures everything that has not been explicitly included. The behavioral assumption is that each individual wants to maximize their own utility. We denote by D_i the expected demand of alternative i. This quantity is obtained by aggregating the individual choice probabilities $P_n(i)$ associated with alternative i, i.e.,

$$D_i = \sum_{n=1}^{N} P_n(i). \tag{1}$$

We denote by R the number of scenarios. They correspond to draws from the probability distribution of the random component of the RUM model (e.g., Gumbel, normal). They are generated outside of the optimization problem (known as exterior approach in the SP framework). For each scenario r, we introduce a deterministic utility U_{inr} that has the same systematic component as U_{in} but replaces the random component by a realization of its probability distribution (scenario). The expected demand D_i is approximated with a mixed-integer linear formulation based on the SAA principle in the space of the utilities. To characterize this approximation, we define the following auxiliary variables. The binary variables y_{inr} represent the availability of alternatives, i.e.,

$$y_{inr} = \begin{cases} 1 & \text{if the capacity of alternative } i \text{ has not been reached} \\ & \text{for individual } n \text{ in scenario } r\text{, and} \\ 0 & \text{otherwise.} \end{cases} \tag{2}$$

The auxiliary variables z_{inr} prevent an unavailable alternative from obtaining the highest utility, i.e.,

$$z_{inr} = \begin{cases} U_{inr} & \text{if alternative } i \text{ is available to individual } n \\ & \text{in scenario } r \text{ in scenario } r\text{, and} \\ \ell_{nr} & \text{otherwise,} \end{cases} \tag{3}$$

where ℓ_{nr} is a value that is low enough (see formal definition after Model 7 is introduced). The behavioral assumption of the RUM model is expressed in terms of z_{inr}, i.e., individual n chooses alternative i in scenario r if $i = \text{argmax}_{1\leq j\leq J} z_{jnr}$. The variables

$$U_{nr} = \max_{1\leq j\leq J} z_{jnr} \tag{4}$$

capture the highest utility for individual n in scenario r, and the binary variables w_{inr} indicate the choice, i.e.,

$$w_{inr} = \begin{cases} 1 & \text{if } U_{nr} \text{ is achieved at alternative } i\text{, and} \\ 0 & \text{otherwise.} \end{cases} \tag{5}$$

The expected demand of alternative i is then approximated by

$$D_i^R = \frac{1}{R}\sum_{n=1}^{N}\sum_{r=1}^{R} w_{inr}. \tag{6}$$

For a better understanding of the introduced variables, consider the following illustrative example. Consider an individual n that can choose from $I = 3$ alternatives, each of them with a capacity of $c_i = 10$ units, for $i = 1, 2, 3$. For a given scenario r, the capacity of alternative $i = 2$ has been reached, whereas there are some units left (at least one) of alternatives $i = 1$ and $i = 3$. Hence, $y_{2nr} = 0$ and $y_{1nr} = y_{3nr} = 1$. As alternative $i = 2$ is not available, $z_{2nr} = \ell_{nr}$, whereas $z_{1nr} = U_{1nr}$ and $z_{3nr} = U_{3nr}$. We now *forget* about the U_{inr} variables and formulate the behavioral assumption directly in terms of z_{inr}, i.e., we look for the highest value among the z_{inr} variables: $U_{nr} = \max\{z_{1nr}, z_{2nr}, z_{3nr}\}$. Imagine this value corresponds to z_{3nr}. Then, w_{3nr} is equal to 1 while the other two choice variables, w_{1nr} and w_{2nr} are equal to 0.

To formulate the optimization problem, we denote the K decisions associated with alternative i and individual n (to be made by the planner) by $x_{in1}, \ldots, x_{inK}$. These variables are assumed to be bounded, i.e., $x_{ink} \in [a_{ink}, b_{ink}], \forall i, n, k$, and might be binary. The objective function relates these decisions to the expected demand. We denote it by $f(x, D^R)$, where x is the $J \times N \times K$-dimensional vector containing the variables x_{ink} and D^R is the J-dimensional vector that contains the demands D_i^R. We assume it is linear in x and the choice variables w_{inr}. Without loss of generality, we assume $f(x, D^R)$ is to be maximized. We define Model 7 for the general CBO problem considered in this chapter.

$$z^{\text{MILP}} = \max f(x, D^R) \tag{7a}$$

$$\text{s.t. } U_{inr} = \sum_{k=1}^{K} \beta_{ink} x_{ink} + d_{inr} \qquad i, n, r \tag{7b}$$

$$\ell_{nr} \leq z_{inr} \qquad i, n, r \tag{7c}$$

$$z_{inr} \leq \ell_{nr} + M_{inr} y_{inr} \qquad i, n, r \tag{7d}$$

$$U_{inr} - M_{inr}(1 - y_{inr}) \leq z_{inr} \qquad i, n, r \tag{7e}$$

$$z_{inr} \leq U_{inr} \qquad i, n, r \tag{7f}$$

$$z_{inr} \leq U_{nr} \qquad i, n, r \tag{7g}$$

$$U_{nr} \leq z_{inr} + M_{nr}(1 - w_{inr}) \qquad i, n, r \tag{7h}$$

$$\sum_{i} w_{inr} = 1 \qquad n, r \tag{7i}$$

$$w_{inr} \leq y_{inr} \qquad i, n, r \tag{7j}$$

$$\sum_{m=1}^{n} w_{imr} \leq (c_i - 1)y_{inr} + (n-1)(1 - y_{inr}) \quad i\ ,\ n > c_i\ ,\ r \tag{7k}$$

$$c_i(1 - y_{inr}) \leq \sum_{m=1}^{n} w_{imr} \quad i\ ,\ n > 1\ ,\ r \tag{7l}$$

$$Ax \leq b \tag{7m}$$

$$a_{ink} \leq x_{ink} \leq b_{ink} \quad i\ ,\ n\ ,\ k \tag{7n}$$

$$y_{inr}, w_{inr} \in \{0, 1\} \quad i\ ,\ n\ ,\ r \tag{7o}$$

Equations (7b) define U_{inr} as the sum of the systematic component that depends on the planner's decisions and a constant term denoted by d_{inr}. The coefficients β_{ink} in the term that depends on the planner's decisions are the parameters of the RUM model, which are estimated beforehand (i.e., outside of the optimization scheme). The constant term d_{inr} includes the r-th scenario associated with alternative i and individual n and other elements of the systematic component that are constant to the optimization problem. As $x_{ink}, \forall i, n, k$, is bounded and $d_{inr}, \forall i, n, r$, is constant, we can derive bounds on U_{inr}: $\ell_{inr} \leq U_{inr} \leq m_{inr}$.

Constraints (7c)–(7f) provide the linear formulation of the variables z_{inr}, where $\ell_{nr} = \min_{1 \leq j \leq J} \ell_{jnr}$ is the smallest of the lower bounds across all alternatives and $M_{inr} = m_{inr} - \ell_{nr}$ is a suitable big-M value. Constraints (7g) and (7h) provide the linear formulation of U_{nr}, where $m_{nr} = \max_{1 \leq j \leq J} m_{jnr}$ largest of the upper bounds across alternatives and $M_{nr} = m_{nr} - \ell_{nr}$ is a suitable big-M value. Constraints (7i) ensure that one and only one alternative is chosen per individual and scenario. Constraints (7j) prevent an unavailable alternative from being chosen. Constraints (7k) and (7l) handle capacity constraints on the alternatives. We assume that the access of individuals to the alternatives for each scenario is modeled by an exogenously given priority list. This list determines the access order of individuals, i.e., if individual n does not have access to the alternative in that scenario, neither does individual $n + 1$, and consequently the upcoming individuals. Constraints (7k) forbid the access to an alternative for a scenario when its capacity has been reached, whereas constraints (7l) ensure its availability otherwise.

Lastly, constraint (7m) represents the set of linear constraints that identify the feasible configurations of the decision variables x (where A and b are, respectively, a matrix and a vector of appropriate dimensions). Consider for instance a company running movie theaters and needs to decide what movies to schedule at what time, in what theater and at what price. The show time, the choice of the movie and the price are all decisions and explanatory variables of the RUM model. Decisions about staff and equipment are *only* decision variables of the optimization problem.

The objective function computes the expected profit, calculated as the difference between the expected revenue generated by the ticket sales and the operational costs. Operational constraints may include the fact that two movies cannot be displayed in the same theater at the same time, that a given theater cannot fit more spectators than the number of seats, etc. In retail, we can model the shelf space allocation problem at the product level. The objective can be defined as the allocation of a certain number of facings of each product in order to maximize profits under some constraints, such as the availability of the shelf space and the space effects on consumer demand. In transportation, we can consider the design of a train timetable by a train operating company, whose objective can be defined as a measure of welfare that needs to be maximized subject to operational (e.g., availability of trains and breaks and working hours of the drivers) and budgetary (i.e., the required investment must not exceed the available budget) requirements. Decision variables such as the frequency and level of comfort have an impact on the passengers' choice.

Model 7 aims at optimizing an aggregate performance of the system on the decisions to be made by a planner while incorporating the associated demand responsiveness. The representation of the demand via a RUM model brings uncertainty to the optimization problem. As we handle this uncertainty with the generation of possible outcomes (scenarios) of the random component of the RUM model, the general philosophy of the CBO framework is inspired by SP. Indeed, the idea of relying on scenarios to represent uncertainty is exploited in SP, and in particular in the two-stage SP model (Shapiro et al., 2021). For each scenario, and because of utility maximization as the decision rule, we assume that each individual solves an optimization problem in order to come up with a choice. The introduced framework is therefore also related to bilevel optimization since multiple problems at the lower level (utility maximization) are nested within another problem at the upper level (planner's decisions). In particular, the CBO framework can be seen as a Stackelberg game in which the (single) planner plays the leader role and the individuals collectively play the follower role. Indeed, the leader knows how the follower will respond to any potential decision being made thanks to the behavioral assumption as provided by the RUM model.

As shown in the numerical experiments performed in Pacheco Paneque et al. (2021), Model 7 lacks efficiency for solving large instances with a general-purpose MILP solver. Indeed, the individual representation of the expected demand and the simulation-based linearization of the choice model, which involves the presence of the so-called big-M constraints, i.e., (7c)–(7f) and (7g) and (7h), contribute to the complexity of the formulation. In Sect. 4, we characterize a solution methodology that exploits the decomposability provided by the underlying mathematical structure of Model 7 given by the scenarios.

4 Scenario Decomposition Method

We introduce a heuristic solution method that exploits the properties of the choice-based optimization problem described in Sect. 3. It involves a subgradient procedure that generates an upper bound and a feasible solution (lower bound) to Model 7 at each iteration. Section 4.1 describes the Lagrangian decomposition approach based on scenario groups and outlines the scenario grouping strategies. Section 4.2 presents the algorithm to generate feasible solutions from the Lagrangian solution, and Sect. 4.3 describes the subgradient method that is applied to solve the Lagrangian dual to find the best possible upper bound.

4.1 Decomposition by Scenario Groups

Let S be the number of scenario groups (indexed by s) and R_s the set that contains the scenarios belonging to group s. Notice that the scenarios are divided into groups, i.e., each scenario belongs to one and only one group. Both S and R_s, $\forall s$, are defined according to one of the grouping strategies described in the following. The copy of x_{ink} associated with group s is denoted by x^s_{ink}. Constraints (8) represent the non-anticipativity constraints imposing that the copied variables x^s_{ink} must be equal across groups. Constraints (8b) are redundant, but help achieving tighter upper bounds, as presented by preliminary experiments performed on the instances explained in Sect. 5. Note that other characterizations of the non-anticipativity constraints are possible, such as relating all copies with each other (e.g., Escudero et al., 2013).

$$x^s_{ink} - x^{s+1}_{ink} = 0 \qquad i\ ,\ n\ ,\ k\ ,\ s < S \tag{8a}$$

$$x^s_{ink} - x^{1}_{ink} = 0 \qquad i\ ,\ n\ ,\ k\ ,\ s = S \tag{8b}$$

The Lagrangian relaxation of constraints (8) with associated multipliers $\alpha^s_{ink} \in \mathbb{R}, \forall i, n, k, s$, yields independent subproblems for each group s. As the objective function $f(x, D^R)$ is linear, it can be split into the functions denoted by $f^s(x^s, D^{R,s})$ such that $f(x, D^R) = \sum_{s=1}^{S} f^s(x^s, D^{R,s})$, where x^s is the $J \times N \times K$-vector containing the variables x^s_{ink}, $D^R = \sum_{s=1}^{S} D^{R,s}$ and $D^{R,s}$ is the J-vector containing the demands $D^{R,s}_i$:

$$D^{R,s}_i = \frac{1}{R} \sum_{n=1}^{N} \sum_{r \in R_s} w_{inr}. \tag{9}$$

The subproblem associated with group s is formulated in Model 10. It is essentially Model 7 for a reduced number of scenarios whose objective function is modified.

As shown in Pacheco Paneque et al. (2021), the computational complexity of Model 7 increases exponentially with respect to R. Therefore, for a given number of individuals, it is usually more efficient to solve multiple problems with a small number of scenarios each rather than a single problem including multiple scenarios. This trade-off is analyzed in Sect. 5. We denote its optimal value by $z_s^{\text{UB}}(\alpha^s, \alpha^{s-1})$.

$$\max f^s(x^s, D^{R,s}) + \sum_{i \in C} \sum_{n=1}^{N} \sum_{k=1}^{K} (\alpha^s_{ink} - \alpha^{s-1}_{ink}) x^s_{ink} \tag{10a}$$

$$\text{s.t. } U_{inr} = \sum_{k=1}^{K} \beta_{ink} x^s_{ink} + d_{inr} \qquad i\ ,\ n\ ,\ r \in R_s \tag{10b}$$

$$\ell_{nr} \le z_{inr} \qquad i\ ,\ n\ ,\ r \in R_s \tag{10c}$$

$$z_{inr} \le \ell_{nr} + M_{inr} y_{inr} \qquad i\ ,\ n\ ,\ r \in R_s \tag{10d}$$

$$U_{inr} - M_{inr}(1 - y_{inr}) \le z_{inr} \qquad i\ ,\ n\ ,\ r \in R_s \tag{10e}$$

$$z_{inr} \le U_{inr} \qquad i\ ,\ n\ ,\ r \in R_s \tag{10f}$$

$$z_{inr} \le U_{nr} \qquad i\ ,\ n\ ,\ r \in R_s \tag{10g}$$

$$U_{nr} \le z_{inr} + M_{nr}(1 - w_{inr}) \qquad i\ ,\ n\ ,\ r \in R_s \tag{10h}$$

$$\sum_{i \in C} w_{inr} = 1 \qquad n\ ,\ r \in R_s \tag{10i}$$

$$w_{inr} \le y_{inr} \qquad i\ ,\ n\ ,\ r \in R_s \tag{10j}$$

$$\sum_{m=1}^{n} w_{imr} \le (c_i - 1) y_{inr} + (n-1)(1 - y_{inr}) \qquad i\ ,\ n > c_i\ ,\ r \in R_s \tag{10k}$$

$$c_i(1 - y_{inr}) \le \sum_{m=1}^{n} w_{imr} \qquad i\ ,\ n > 1\ ,\ r \in R_s \tag{10l}$$

$$Ax^s \leq b \tag{10m}$$

$$a_{ink} \leq x^s_{ink} \leq b_{ink} \qquad i\ ,\ n\ ,\ k \tag{10n}$$

$$y_{inr}, w_{inr} \in \{0, 1\} \qquad i\ ,\ n\ ,\ r \in R_s \tag{10o}$$

The Lagrangian problem associated with Model 7 is defined as follows:

$$z^{\text{UB}}(\alpha) = \sum_{s=1}^{S} z_s^{\text{UB}}(\alpha^s, \alpha^{s-1}). \tag{11}$$

It provides an upper bound on z^{MILP} for any values of α. Notice that α^0_{ink} (obtained when $s = 1$) refers to $\alpha^S_{ink}, \forall i, n, k$. We explore different strategies for grouping scenarios, drawing inspiration from the grouping methods presented in Crainic et al. (2014) for the two-stage network design problem. These strategies are determined by criteria of similarity or dissimilarity, which are evaluated based on descriptive statistics derived from the scenarios. In our case, the descriptive statistics for each scenario r are given by the $J \times N$-dimensional vector that contains the realizations associated with r.

Random (RAN) This strategy serves as a benchmark to evaluate the more refined strategies described below. To ensure an equitable distribution of scenarios among groups, we define σ groups with $\lceil R/S \rceil$ scenarios and $S - \sigma$ groups with $\lfloor R/S \rfloor$ scenarios, where σ is the remainder in the Euclidean division of R by S.

Similar (SIM) This method takes inspiration from the k-means clustering algorithm, where scenario distances are determined by calculating the Euclidean distance between their respective descriptive vectors. In addition to creating the scenario groups, this strategy determines the number of groups by assuming a lower ($\underline{S}$) and an upper bound ($\overline{S}$) on S and selecting $S \in \{\underline{S}, \ldots, \overline{S}\}$ such that the difference between the error associated with S and $S - 1$ is maximized. Since this strategy might generate very unbalanced groups with respect to the number of scenarios, we set $\lceil R/S \rceil$ as the maximum number of scenarios in a group. Once this threshold is attained, the group ceases to receive updates. Scenarios are allocated to groups based on the disparity between their proximity to the nearest and farthest groups. Algorithm 1 outlines the pseudocode of this strategy.

Similar Without Dissimilar Scenarios (SIM-D) This approach initiates by forming *similar* groups through the use of SIM and subsequently generates extra *dissimilar* groups, each including a single scenario from every similar group. The chosen scenario within each dissimilar group is the one that exhibits the closest proximity to its respective mean.To balance the number of scenarios, we also set a maximum of $\lceil R/S \rceil$ scenarios per group. As soon as this value is reached, a new dissimilar group is created.

Algorithm 1: Similar grouping

Initialize error_diff $= -\infty$, max_error_diff $=$ error_diff, $S^* = S$;
for $S \in \{\underline{S}, \ldots, \overline{S}\}$ **do**
  Initialize one group with a randomly chosen scenario r_1: $g_1 = \{r_1\}$, mean$(g_1) = r_1$;
  for $s = 2 \ldots S$ **do**
    Initialize $g_s = \{r_s\}$, r_s selected at random with a higher probability for distant scenarios with respect to r_{s-1}, mean$(g_s) = r_s$;
  for $r \notin \cup_{s=1}^{S} g_s$ **do**
    Assign r to the group with the closest mean;
  while *groups change* **do**
    Update mean$(g_s) = r_s, \forall s$;
    for $r = 1 \ldots R$ *according to distance difference* **do**
      Assign r to the group with the closest mean;
      if $|g_s| \geq \lceil R/S \rceil$ **then**
        g_s becomes unavailable;
  Calculate error$(S) = \sum_{r=1}^{R}$ dist$(r$, mean$(g_s) | r \in g_s)$;
  if $S > \underline{S}$ **then**
    error_diff $=$ error(S) $-$ error$(S-1)$;
  if *error_diff* $>$ *max_error_diff* **then**
    max_error_diff $=$ error_diff;
    Update $S^* = S$ and store the associated scenario groups $g_s, \forall s = 1 \ldots S^*$;
Return S^* and the associated scenario groups $g_s, \forall s = 1 \ldots S^*$;

Dissimilar (DIS) In contrast to SIM, scenarios are allocated to the groups that have the greatest distance from the mean. They are ordered in ascending fashion based on the disparity between their proximity to the nearest and farthest group. The selected number of groups $S \in \{\underline{S}, \ldots, \overline{S}\}$ is the one with the largest average of the distances between each group mean and the scenario closest to that mean.

4.2 Generation of Feasible Solutions

The decomposition technique described in Sect. 4.1 does not necessarily yield a feasible solution to Model 7. A feasible solution provides a lower bound to the optimal solution of Model 7. It allows to compute the *duality* gap at each iteration of the subgradient method. This gap is defined according to the relative difference between the decomposition's upper bound and the generated lower bound.

To efficiently generate a feasible solution, we solve the restricted MILP formulation of Model 7 that fixes the decisions x to the values these variables obtain in the Lagrangian solution. When the decisions are fixed, the resulting problem can be iteratively solved with Algorithm 2. It iterates over the scenarios and the individuals in the order provided by the priority list. For each individual and scenario, it generates the associated choice and updates the occupancy level (i.e., number of individuals that have chosen it) of the chosen alternative.

Notice that S values from the Lagrangian solution are available to each variable x_{ink} (as we have solved S Lagrangian subproblems). Let $\bar{x}^s_{ink}, \forall i, n, k, s$, be the value associated with the copied variable x^s_{ink} in the optimal solution of Model 10. We evaluate all combinations that can be characterized from $\bar{x}^s_{ink}, \forall i, n, k, s$, as shown in Algorithm 3. This results into $S^{J \times N \times K}$ combinations, where J represents the number of alternatives, N the number of individuals and K the number of decisions. For each combination, Algorithm 2 is performed, and the feasible solution with the largest objective function value is selected. Despite Algorithm 3 being efficient (see Sect. 5), evaluating all combinations might prompt a considerable computational burden. In this case, Algorithm 3 can be simplified by only evaluating one combination per group, which generates S feasible solutions.

Algorithm 2: Solution method for Model 7 with $x_{ink} = \bar{x}_{ink}, \forall i, n, k$

---

for $r = 1 \dots R$ **do**
    Initialize occupancy level $o_{ir} = 0, \forall i$, and availability variables $y_{inr} = 1, \forall i, n$;
    for $n = 1 \dots N$ **do**
        for $i = 1 \dots J$ **do**
            if $o_{ir} < c_i$ **then**
                Calculate $U_{inr} = \sum_{k=1}^{K} \beta_{ink} \bar{x}_{ink} + d_{inr}$ and set $z_{inr} = U_{inr}$;
            else
                Set the alternative unavailable: $y_{inr} = 0$;
                Set the discounted utility to the corresponding lower bound: $z_{inr} = \ell_{nr}$;
        Determine $U_{nr} = \max_{\{1 \le i \le J \mid y_{inr}=1\}} z_{inr}$ and $j = \operatorname{argmax} U_{nr}$;
        Set $w_{jnr} = 1$ and $w_{inr} = 0, \forall i \neq \{j\}$;
        Update the occupancy level $o_{jr} = o_{jr} + 1$;
Obtain D^R by calculating $D_i^R = \frac{1}{R} \sum_{n=1}^{N} \sum_{r=1}^{R} w_{inr}, \forall i$;
Obtain the objective function $f(\bar{x}, D^R)$;

Algorithm 3: Method to generate feasible solutions to Model 7 with $\bar{x}^s_{ink}, \forall i, n, k, s$

Initialize the best lower bound $z^{\text{LB}} = -\infty$;
for *all* $S^{J \times K \times N}$ *combinations from* $\bar{x}^s_{ink}, \forall i, n, k, s$ **do**
    Solve Model 7 with $x_{ink}, \forall i, n, k$ fixed to the corresponding combination $(\bar{x})$ with Algorithm 2 and obtain $z^{\text{LB}}_{\text{current}} = f(\bar{x}, D^R)$;
    if $z^{LB}_{current} > z^{LB}$ **then**
        Update $z^{\text{LB}} = z^{\text{LB}}_{\text{current}}$ and keep the associated feasible solution;
Return the best lower bound z^{LB} and the associated feasible solution;

4.3 Subgradient Method

To obtain the best possible upper bound to Model 7, we need to solve the Lagrangian dual

$$z^{\text{LD}} = \min \left\{ z^{\text{UB}}(\alpha) \ : \ \alpha \in \mathbb{R}^{J \times N \times K \times S} \right\}. \tag{12}$$

The Lagrangian function $z^{\text{UB}}(\alpha)$ is non-differentiable. However, subgradient directions can be easily generated. The subgradient method constructs a sequence $\{\alpha^\nu\}_\nu$ using

$$\alpha^{\nu+1} = \alpha^\nu + \gamma^\nu d^\nu, \tag{13}$$

where ν denotes the iteration, γ^ν is a positive scalar called step size and d^ν is a vector representing the direction of motion called step direction.

The step direction d^ν can be directly defined as a subgradient direction. The $J \times N \times K \times S$-dimensional vector v defined by $v^s_{ink} = x^s_{ink} - x^{s+1}_{ink}$, $\forall i, n, k, s$, and evaluated at the Lagrangian solution, is a subgradient of $z^{\text{UB}}(\alpha)$ at any value of α. Nevertheless, preliminary experiments showed that the angle between v^ν and $v^{\nu-1}$ is obtuse in multiple occasions. This leads to a next Lagrangian multiplier $\alpha^{\nu+1}$ that is close to the previous one, which slows down the convergence of the procedure. This effect is known as zigzagging of kind I, and can be overcome by deflecting the step direction (Camerini et al., 1975):

$$d^\nu = -(v^\nu + \zeta^\nu d^{\nu-1}), \tag{14}$$

where $\zeta^\nu \in \mathbb{R}_{\geq 0}$ is a suitable scalar called deflection parameter (notice the negative sign in the definition of the step direction d^ν because the Lagrangian dual is to be minimized). By defining this parameter as

$$\zeta^\nu = \begin{cases} -\tau \frac{v^\nu d^{\nu-1}}{\|d^{\nu-1}\|^2} & \text{if } v^\nu d^{\nu-1} < 0, \\ 0 & \text{otherwise,} \end{cases} \tag{15}$$

with $1 \leq \tau < 2$, the step direction is forced to always form an acute angle with the preceding direction, which eliminates the zigzagging of kind I. Notice that in the absence of zigzagging of kind I (i.e., $v^\nu d^{\nu-1} \geq 0$), the step direction d^ν is equal to $-v^\nu$ (the negative subgradient). We consider the step size most commonly used in practice (Held et al., 1974):

$$\gamma^\nu = \lambda^\nu \frac{z^{\text{UB}}(\alpha^\nu) - z^{\text{LB,best}}}{\|d^\nu\|^2}, \tag{16}$$

where $\|d^{\nu}\|^2$ is the norm of the step direction, i.e., $\|d^{\nu}\|^2 = \sum_{i=1}^{J} \sum_{n=1}^{N} \sum_{k=1}^{K} \sum_{s=1}^{S} (d_{ink}^{s,\nu})^2$, and λ^{ν} is a step size decreasing parameter satisfying $0 < \lambda^{\nu} \leq 2$. Fisher (1973) suggests to halve this value whenever $z^{\mathrm{UB}}(\alpha^{\nu})$ has failed to decrease in a given number θ of consecutive iterations.

Algorithm 4 presents the pseudocode of the deflected subgradient method. Notice that unlike the ordinary gradient method, the subgradient method is not a descent method. This is why we keep track of the best upper and lower bounds found throughout the process. We set a time budget as stopping criterion.

Concerning the convergence of the approach, when R tends to infinity, the sequence of optimal solutions is obtained for the general CBO problem (Model 7) to an optimal solution of the same problem (Pacheco Paneque, 2020). The computational performance and theoretical convergence of the subgradient method are assessed in Held et al. (1974), who show that the method converges to the optimal value of the Lagrangian dual if the sequence of step multipliers tends to 0 and the associated series tend to infinity. Practically, unless we obtain a set of Lagrangian multipliers such that the optimal objective function of the Lagrangian dual is equal to that of a known feasible solution, optimality cannot be proven. The method should terminate upon meeting a certain stopping criterion (e.g., number of iterations).

Algorithm 4: Deflected subgradient method

Initialize $\nu = 0$, $z^{\mathrm{UB,best}} = +\infty$, $z^{\mathrm{LB,best}} = -\infty$, α^0, λ^0 and $d^0 = 0$;
while *stopping criterion* **do**
    for $s = 1 \ldots S$ **do**
        Solve Model 10;
    Obtain $\bar{x}_{ink}^{s}$, $\forall i, n, k, s$, and $z^{\mathrm{UB}}(\alpha^{\nu})$;
    if $z^{UB}(\alpha^{\nu}) < z^{UB,best}$ **then**
        Update $z^{\mathrm{UB,best}} = z^{\mathrm{UB}}(\alpha^{\nu})$;
    Run Algorithm 3 and obtain z^{LB};
    if $z^{LB} > z^{LB,best}$ **then**
        Update $z^{\mathrm{LB,best}} = z^{\mathrm{LB}}$;
    if $z_{UB}(\alpha^{\nu})$ *has not improved in the last* θ *consecutive iterations* **then**
        Set $\lambda^{\nu} \leftarrow \lambda^{\nu}/2$;
    Calculate γ^{ν} according to (16);
    Update $\alpha_{ink}^{s,\nu+1} = \alpha_{ink}^{s,\nu} + \gamma^{\nu} d_{ink}^{s,\nu}$, $\forall i, n, k, s$, where $d_{ink}^{s,\nu} = -(v_{ink}^{s,\nu} + \zeta^{\nu} d_{ink}^{s,\nu-1})$ and ζ^{ν} is calculated according to (15);
    Set $\nu \leftarrow \nu + 1$;
    Check stopping criterion;

5 Computational Experiments

The goal of the experiments carried out in this section is threefold. First, we evaluate the performance of the decomposition method against a general-purpose MILP solver for a given time budget and compare the different scenario grouping strategies against each other (Sect. 5.2). Second, we assess the trade-off between the solution quality and duality gap and the number of scenarios per group (Sect. 5.3). Third, we analyze the impact in the algorithm's performance of scaling up the problem size by increasing the dimensions not involved in the decomposition (Sect. 5.4 for an increase in the number of alternatives and Sect. 5.5 for an increase in the number of individuals). Before that, we characterize the revenue maximization problem. This CBO problem yields a concrete MILP formulation that we use to run the experiments. Section 5.1 describes the case study and the calibration of parameters for the subgradient method.

Consider a planner that aims at finding the best pricing strategy in order to maximize its revenue by offering services to a market. Each service in the set of services C has a given associated capacity (denoted by c_i). The market is composed of N individuals, which are assumed to be heterogeneous and price elastic. In a revenue maximization context, we need to model competition so that individuals are not captive. To this end, we incorporate an opt-out option into the model to capture individuals leaving the market, either because they choose a competitor's service or because they do not choose anything at all. The opt-out option is denoted by $i = 0$ and is always available to all individuals.

We assume that the price is the only planner's decision. We define $p_i \in [a_i, b_i]$ as the price to be paid to access service $i \in C$. Thus, the planner needs to perform one decision for each alternative, and this decision (the price) is assumed to be the same across individuals (i.e., $x_{ink} = x_{in} = p_i$). The expected revenue obtained from the services in C is calculated as

$$f(p, D^R) = \sum_{i \in C} p_i D_i^R = \frac{1}{R} \sum_{i \in C} \sum_{n=1}^{N} \sum_{r=1}^{R} p_i w_{inr}. \tag{17}$$

Model 18 comprises the MILP formulation of the revenue maximization problem. Notice that the product of the price (continuous variable) and the choice (binary variable) can be linearized because bounds on the former are assumed. In particular, we introduce the variables $\eta_{inr} = p_i w_{inr}$ to capture the product of the two, with linearizing constraints (18e)–(18h). We denote by $\bar{C} = C \cup \{0\}$ the set containing all services and the opt-out option. The utilities associated with the opt-out option for each individual and scenario are included in (18b), and since the opt-out option is always available, we add constraints (18d) to enforce the availability variables y_{inr} to be equal to 1 when $i = 0$.

$$z^{\text{MILP}} = \max \frac{1}{R} \sum_{i \in C} \sum_{n=1}^{N} \sum_{r=1}^{R} \eta_{inr} \tag{18a}$$

$$\text{s.t.}\quad U_{0nr} = d_{0nr} \qquad n\,,\ r \tag{18b}$$

$$U_{inr} = \beta_{in} p_i + d_{inr} \qquad i \in C\,,\ n\,,\ r \tag{18c}$$

(7c)–(7l)

$$y_{0nr} = 1 \qquad n\,,\ r \tag{18d}$$

$$a_i w_{inr} \leq \eta_{inr} \qquad i \in C\,,\ n\,,\ r \tag{18e}$$

$$\eta_{inr} \leq b_i w_{inr} \qquad i \in C\,,\ n\,,\ r \tag{18f}$$

$$p_i - (1 - w_{inr}) b_i \leq \eta_{inr} \qquad i \in C\,,\ n\,,\ r \tag{18g}$$

$$\eta_{inr} \leq p_i - (1 - w_{inr}) a_i \qquad i \in C\,,\ n\,,\ r \tag{18h}$$

$$y_{inr}\,,\ w_{inr} \in \{0, 1\} \qquad i \in \bar{C}\,,\ n\,,\ r \tag{18i}$$

5.1 Experimental Setting

We consider the case study on parking services in a park-and-ride context used in Pacheco Paneque et al. (2021). The RUM model for parking choices (mixed logit model) is specified and estimated in Ibeas et al. (2014). They interview 197 individuals to model their preferences with respect to three parking services: paid on-street parking (PSP), paid parking in an underground car park (PUP) and free on-street parking (FSP). We assume that FSP represents the opt-out option because it does not provide any revenue to the parking manager. As pointed out in Sect. 1, the individual probabilities $P_n(i)$ of mixed logit models do not have a closed form.

For the integration of the RUM model in Model 18, we replace $\beta_{in}, \forall i \in \bar{C}, n,$ with their estimates, and we compute $d_{inr}, \forall i \in \bar{C}, n, r,$ with the values of the other explanatory variables of the RUM model in the data and the scenario ξ_{inr}. We determine the capacity associated with PSP and PUP (the opt-out option, FSP, is assumed to have unlimited capacity) according to the number of individuals in the instance, so that it is appropriate for the sample under consideration but restrictive enough so that some users are forced to choose FSP because PSP and/or PUP become unavailable. For the sake of simplicity, we assume $c_{\text{PSP}} = c_{\text{PUP}}$, and denote it simply by c. We assume lower price bounds for PSP than PUP, i.e., $p_{\text{PSP}} \in [0.5, 0.65]$ and $p_{\text{PUP}} \in [0.7, 0.85]$.

The instances used in the upcoming sections are characterized from the available data by setting specific values for N, R and $J = |\bar{C}|$. More precisely, for each configuration of these parameters, we generate various instances by randomly selecting N individuals from the whole dataset, together with the associated R scenarios for the J parking services. The configurations are labeled as NX_RY_JZ_cT, where X indicate the number of individuals, Y the number of scenarios, Z the number of services and T the capacity. The calibration of the parameters of the deflected subgradient method (Algorithm 4) has been performed on different configurations of N, R and S. We consider $\lambda^0 = 0.5$ as larger values result in a higher increase of the upper bound in the first iterations and $\theta = 5$. The reason is that for some of the performed experiments a low number of iterations is expected. For the deflection parameter, we follow the literature recommendations and we set $\tau = 1.5$. The code is implemented in C++ using ILOG Concert Technology to access CPLEX 12.8, and all the instances were run using 12 threads in a 3.33 GHz Intel Xeon X5680 server running a 64-bit Ubuntu 16.04.2.

5.2 *Performance and Scenario Grouping Strategies*

In this experiment, we compare the results obtained with the decomposition method for a given time budget against the exact approach while evaluating the different grouping strategies. We consider two configurations: N50_R100_J3_c20 and N50_R200_J3_c20. As discussed in Pacheco Paneque et al. (2021), the number of scenarios to be considered in order to obtain accurate estimates is closely related to the variance of the random components of the RUM model. That is, if the RUM model is highly deterministic (i.e., low variance of the random components), the choices will mostly be driven by the systematic component of the utility function, and a low number of draws will be required. Despite the considerable level of stochasticity of the RUM model for parking choices (which is measured by the ratio of the choices estimated by the model that do not match the deterministic choice, i.e., the choice obtained in the absence of the random component, to the total number of estimated choices), a relative low number of scenarios ensures a low variability of the approximated quantities (e.g., revenue, prices). Indeed, such quantities are stable enough (i.e., experience a low variability) for $R \geq 25$. Furthermore, we need to restrict the number of individuals and scenarios to be able to solve to optimality Model 18 with CPLEX. Nevertheless, the decomposition method can be applied for larger values of R. Note that in this case a larger time budget might be considered to ensure that the algorithm runs for a few iterations.

To increase statistical significance, we allow for 10 instances of each configuration (see Table 1 for computational details). Furthermore, we determine 10% and 25% of the average computational time (reported by CPLEX) as two time budget values to evaluate the performance of the deflected subgradient method (Algorithm 4) at different stages of the procedure. Notice that the computational time of an iteration of the subgradient method is unknown before being executed. At every iteration, we compute an average computational time per iteration to avoid exceeding the time budget, which is calculated by keeping track of the computational times per iteration and by averaging them over the number of iterations carried out so far. The subgradient method is then terminated as soon as the expected computational time after completing the current iteration, i.e., the sum of the computational time so far and the average time per iteration, is larger than the time budget.

Table 1 Computational details on the tested configurations (10 instances per configuration)

					CPLEX time (min)			Time budget (min)	
Configuration	N	R	J	c	Average	Lowest	Highest	10%	25%
N50_R100_J3_c20	50	100	3	20	191.4	112.8	318.2	19.1	47.9
N50_R200_J3_c20	50	200	3	20	1038.8	725.2	1508.0	103.9	259.7

Given the size of the instances, we consider scenario groups that have approximately five scenarios each. We could consider a larger number of scenarios per group while still being able to solve the subproblems within the given time budget at the expense of performing a smaller number of iterations. In the case of random grouping, we define S such that the number of scenarios per group does not exceed five, i.e., $S = \lceil R/5 \rceil$. For the other strategies, we define the lower and upper bound on S as $\underline{S} = \lceil R/5 \rceil - 2$ and $\overline{S} = \lceil R/5 \rceil + 2$, respectively (see Sect. 4.1). For example, when $R = 100$, $S \in \{18, 19, 20, 21, 22\}$. Table 2 summarizes the computational results obtained for the four grouping strategies and the two time budget values. We report the average, lowest and highest values of the duality and optimality gaps over the 10 instances, which are calculated as follows:

$$\text{gap}_{\text{opt}} = \frac{z^{\text{MILP}} - z^{\text{LB,best}}}{z^{\text{RM}}}, \tag{19a}$$

$$\text{gap}_{\text{dual}} = \frac{z^{\text{UB,best}} - z^{\text{LB,best}}}{z^{\text{LB,best}}}. \tag{19b}$$

Furthermore, we include the results on the best feasible solution found by CPLEX for the given time budget. Note that in this case the duality gap (gap_{dual}) is replaced by the gap reported by CPLEX, which is computed as the relative difference between the best integer solution and the solution associated with the best node of the branch-and-bound tree.

We observe that the average computational time for any grouping strategy is lower than the time budget because of the above-mentioned stopping criterion. The average number of scenario groups S is larger in SIM-D because more scenario groups are created in addition to the ones generated by SIM. Even though more subproblems are solved, each of them has on average less scenarios, which results in a lower average UB time per iteration. This has also an impact on the average LB time, as more price combinations are evaluated. In any case, the algorithm to generate feasible solutions is much faster compared to solving the Lagrangian subproblem. Notice that more iterations are performed for $R = 200$ than $R = 100$ within each time budget. This is due to the fact that the time budget values are obtained from the average computational time of the exact method over 10 instances considered per configuration, that does not increase linearly with respect to R (see Table 1). In the case of $R = 200$, we observe a decrease both in the average UB time and LB time per iteration from 10% to 25% time budget. These results indicate that the subproblems and the restricted MILP formulations are more rapidly solved

Table 2 Performance by scenario grouping strategies and comparison with the best feasible solution provided by CPLEX for a given time budget (10 instances per configuration)

	10% time budget					25% time budget				
	RAN	SIM	SIM-D	DIS	CPLEX	RAN	SIM	SIM-D	DIS	CPLEX
N50_R100_J3_c20										
Avg. time (min)	**15.26**	17.81	17.84	16.02	19.16	45.77	**42.84**	44.99	44.73	47.90
Avg. UB time per iter. (min)	7.61	7.60	5.05	6.79	–	7.75	7.90	4.75	6.55	–
Avg. LB time per iter. (s)	0.09	0.10	0.14	0.11	–	0.09	0.10	0.14	0.11	–
Avg. num. iter.	2.1	2.4	3.6	2.4	–	6.0	5.6	9.8	7.1	–
Avg. iter. best UB is reached	2.1	2.4	2.9	2.3	–	5.1	5	6.6	5.7	–
Avg. iter. best LB is reached	1.6	2	1.9	1.7	–	3.4	3.1	4.7	3.1	–
Avg. S	20	20.3	24.6	21.9	–	20	20.3	24.6	21.9	–
gap_{opt} (%)										
Average	0.08	0.07	**0.05**	0.10	5.77	0.05	0.05	**0.04**	0.08	0.10
Lowest	0.00	0.00	0.00	0.05	2.35	0.00	0.00	0.00	0.02	0.25
Highest	0.16	0.12	0.10	0.16	8.78	0.12	0.12	0.08	0.13	3.91
gap_{dual} (%)										
Average	1.88	**1.85**	1.93	2.03	25.69[a]	**1.62**	1.65	1.74	1.73	16.48[a]
Lowest	1.52	1.55	1.47	1.63	20.25[a]	1.29	1.25	1.42	1.42	11.46[a]
Highest	2.25	2.10	2.30	2.31	30.26[a]	1.97	1.98	2.03	2.03	20.16[a]
N50_R200_J3_c20										
Avg. time (min)	98.47	99.35	100.4	**94.79**	103.9	251.0	253.7	256.3	**249.1**	259.6
Avg. UB time (min)	16.35	16.79	11.67	18.90	–	15.02	16.12	8.69	16.32	–
Avg. LB time (s)	0.79	0.80	1.18	0.78	–	0.77	0.77	1.10	0.73	–
Avg. num. iter.	6.1	6.0	8.7	5.3	–	16.9	15.9	29.9	15.7	–
Avg. iter. best UB is reached	5.6	5.6	7.8	4.3	–	15.3	14	28.7	13.8	–
Avg. iter. best LB is reached	2.6	4.6	4.5	3.2	–	6.5	6.2	8.7	7.8	–
Avg. S	40	40.2	48.4	39.5	–	40	40.2	48.4	39.5	–
gap_{opt} (%)										
Average	0.03	0.03	**0.03**	0.03	1.46	**0.02**	0.02	0.02	0.03	0.54
Lowest	0.01	0.00	0.01	0.02	0.22	0.00	0.00	0.01	0.02	0.08
Highest	0.05	0.06	0.04	0.05	4.16	0.04	0.03	0.04	0.04	1.83
gap_{dual} (%)										
Average	1.82	**1.82**	2.03	1.84	19.05[a]	**1.62**	1.64	1.73	1.64	16.71[a]
Lowest	1.53	1.57	1.54	1.62	17.43[a]	1.36	1.38	1.38	1.46	13.05[a]
Highest	2.07	2.13	2.39	2.13	22.57[a]	1.85	1.90	2.00	1.83	19.02[a]

[a] In this case, gap_{dual} corresponds to the gap reported by CPLEX (i.e., relative difference between the best integer solution and the solution associated with the best node of the branch-and-bound tree)

in the long run. We also notice that the best LB is reached earlier than the best UB. This means that the method is able to find high-quality feasible solutions at a relatively early stage whereas the best UB gets refined throughout the iterations. Thus, as long as the duality gap is low, the subgradient method can be terminated after a small number of iterations without improvement of the best LB.

The optimality gaps are very low for any strategy (below 0.16% in all cases), and they decrease as the time budget increases. We notice that they are lower on average for $R = 200$ than $R = 100$ because of the larger time budget values. On the contrary, the optimality gaps with respect to the best feasible solutions provided by CPLEX are much higher and more dispersed. This is especially the case for $R = 100$ and a 10% time budget, where the optimality gap fluctuates between 2.35% and 8.78% with an average value equal to 5.77%. The duality gaps are larger and of the same order of magnitude across grouping strategies (they are below 2.31% in all cases). As expected, the average duality gaps decrease as the time budget increases by a relatively low margin. For instance, for N50_R200_J3_c20 and RAN, it decreases from 1.82% for a 10% time budget to 1.62% for a 25% time budget. This reveals the slow convergence behavior of the subgradient method.

Regarding the grouping strategies, we observe an impact on the obtained results but we do not identify a clear winner. This is statistically supported by the ANOVA test, that evaluates whether optimality and duality gap means are equal across strategies. This is indeed the case, and post hoc tests to evaluate these means for each pair of strategies show that there is no significant difference between them. In general, DIS (dissimilar) reports the largest optimality and duality gaps. As opposed to Crainic et al. (2014), the obtained results are not superior to RAN (random). They also show that SIM-D (similar without dissimilar scenarios) often leads to higher quality solutions at the expense of longer computational times. In our case, this strategy provides the smallest optimality gaps for $R = 100$ but not for $R = 200$. With respect to the duality gap, other strategies yield lower values. As analyzed in Sect. 5.3, this has to do with the fact that the subproblems have on average less scenarios, and despite the method is able to perform more iterations, it does not contribute to reducing the duality gap. RAN and SIM report reasonable duality gaps for both values of R. SIM reports a lower average optimality gap for the 10% time budget, whereas RAN reports lower values for the 25% time budget. Therefore, we cannot conclude that any of the refined grouping strategies outperforms random grouping. RAN is easier to implement and shows slightly faster solution times with relatively lower duality gaps.

5.3 Size of the Scenario Groups

To analyze the trade-off between the size of the scenario groups and the performance of the method, we consider various values of S and run the configuration N50_R200_J3_c20 (5 instances) for a 2-hour time budget. Table 3 includes the obtained results for groups with 1–5, 10, 15, 20 and 25 scenarios each. As expected,

Table 3 Impact of the size of scenario groups for the configuration N50_R200_J3_c20 with a 2-hour time budget (5 instances)

	Number of scenario groups (S)								
	200	100	66	50	40	20	13	10	8
Number of scenarios per group	1	2	3[a]	4	5	10	15[b]	20	25
Avg. time (min)	119.2	118.6	116.9	116.1	110.7	109.5	96.0	85.8	72.1
Avg. UB time per iter. (min)	0.86	2.78	4.60	7.83	12.41	32.62	43.97	56.05	72.14
Avg. LB time per iter. (s)	19.13	4.75	2.05	1.15	0.75	0.18	0.08	0.05	0.03
Avg. num. iter.	101.2	41.6	25.8	15	9	3.4	2.2	1.6	1
Avg. iter. best UB is reached	100	41.2	24.2	14.2	8.2	3	3	1.4	1
Avg. iter. best LB is reached	7.8	7.4	8.8	8.4	3.6	1	1	1	1
gap_{opt} (%)									
Average	0.02	0.01	0.02	0.03	0.03	0.04	0.04	0.06	0.06
Lowest	0.00	0.00	0.00	0.02	0.01	0.03	0.00	0.03	0.03
Highest	0.04	0.03	0.04	0.05	0.05	0.06	0.08	0.08	0.09
gap_{dual} (%)									
Average	3.67	2.80	2.30	2.02	1.83	1.31	0.99	0.90	0.78
Lowest	3.56	2.68	2.12	1.85	1.73	1.22	0.87	0.83	0.61
Highest	3.84	2.97	2.38	2.18	1.96	1.38	1.09	1.02	0.98

[a] 2 groups have 4 scenarios
[b] 5 groups have 16 scenarios

the larger the number of scenarios per group, the larger the average UB time per iteration, resulting in a lower average number of iterations within the time budget.

Concerning the optimality gap, we notice an increasing trend in the obtained values as S decreases. Indeed, as the number of scenarios per group increases, the number of subproblems to solve decreases, and thus less price combinations are evaluated. Nevertheless, the average optimality gaps remain below 0.06% in all cases. The decrease in S has a more noticeable impact on the duality gaps, which decrease as S decreases. We test for each instance the evolution of the best UB obtained with respect to the number of scenarios per group. Indeed, we observe a decrease in their values as the number of scenarios per group increases.

In the case of $S = 200$ (1 scenario per group), even though the method performs on average 101.2 iterations, it can only reach an average duality gap of 3.67%. For larger groups (i.e., 15, 20 and 25 scenarios per group), the algorithm runs for a very limited number of iterations, but the resulting duality gaps are below 1.02% on average. We observe that despite the lower number of iterations and combinations for the generation of feasible solutions, it is beneficial to define subproblems with a large number of scenarios per group as long as they can be efficiently solved within the considered time budget. Notice that we might need to consider groups of lower size in other settings that involve more individuals, more scenarios, a lower time budget, etc.

5.4 Impact of the Number of Alternatives

We now evaluate the performance of the decomposition method when the problem is scaled up by increasing the number of services J. To do so, we segment the prices associated with PSP and PUP by residency, which is one of the explanatory variables of the RUM model. This means that different prices are proposed to the residents in the study area and to non-residents. It can be seen as extending $\bar{C}$ from $J = 3$ to $J = 5$ services where residents have only access to PSP and PUP associated with the resident price (and the opt-out option FSP) and analogously for non-residents. Since non-residents are willing to pay larger prices than residents, we modify the price bounds as follows: $p_{\mathrm{PSP}}^{\mathrm{res}}, p_{\mathrm{PSP}}^{\mathrm{non\text{-}res}} \in [0.5, 0.75]$ and $p_{\mathrm{PUP}}^{\mathrm{res}}, p_{\mathrm{PUP}}^{\mathrm{non\text{-}res}} \in [0.75, 1.0]$.

Table 4 shows the results for three configurations (5 instances) for a 2-hour time budget. Due to the complexity induced by the increase in J, solely the MILP formulation for N50_R25_J5_c20 could be solved to optimality in a reasonable time (18.72 h on average). We therefore consider 3 scenarios per group in the decomposition method. The average UB time per iteration is much larger for $J = 5$. For instance, it increases from 7.8 min for N50_R100_J3_c20 to 34.4 min for N50_R100_J5_c20. The average LB time per iteration is also higher but remains below 0.3 s for all configurations. The optimality gaps are larger than the ones obtained for $J = 3$ in Sect. 5.2 but they remain below 0.5%. As expected, the duality gaps are also larger and they increase as R increases for the given time budget.

Table 4 Impact of the size of scenario groups for the configuration N50_R200_J3_c20 with a 2-hour time budget (5 instances)

	N50_R25_J5_c20	N50_R50_J5_c20	N50_R100_J5_c20
Avg. time (min)	116.8	110.9	114.9
Avg. UB time per iter. (min)	7.61	20.03	34.38
Avg. LB time per iter. (s)	0.00	0.03	0.28
Avg. num. iter.	16.2	5.8	3.4
Avg. iter. best UB is reached	13.8	4.8	1
Avg. iter. best LB is reached	4.2	3.4	1.4
S	8	16	33
$\mathrm{gap}_{\mathrm{opt}}$ (%)			
Average	0.38	N/A[a]	N/A[a]
Lowest	0.21	N/A[a]	N/A[a]
Highest	0.49	N/A[a]	N/A[a]
$\mathrm{gap}_{\mathrm{dual}}$ (%)			
Average	3.34	4.57	5.19
Lowest	2.79	3.72	4.69
Highest	3.91	5.09	5.59

[a] Instances could not be solved to optimality with CPLEX in a reasonable time

Table 5 Impact of an increase in the number of individuals for a 2-hour time budget (5 instances per configuration)

	N50_R100 _J3_c20	N100_R100 _J3_c40	N150_R100 _J3_c60	N197_R100 _J3_c80
Avg. time (min)	119.5	110.2	98.2	193.2[a]
Avg. UB time per iter. (min)	1.21	15.11	61.58	193.18
Avg. LB time per iter. (s)	0.59	1.21	1.80	2.89
Avg. num. iter.	98.2	7.4	1.6	1
Avg. iter. best UB is reached	96.6	6	1.6	1
Avg. iter. best LB is reached	18.2	3.6	1.2	1
S	50	50	50	50
gap_{opt} (%)				
Average	0.03	N/A[b]	N/A[b]	N/A[b]
Lowest	0.00	N/A[b]	N/A[b]	N/A[b]
Highest	0.07	N/A[b]	N/A[b]	N/A[b]
gap_{dual} (%)				
Average	2.38	1.98	1.91	1.56
Lowest	2.05	1.91	1.75	1.49
Highest	2.63	2.05	2.25	1.63

[a] The first iteration of the solution method is always performed (even if it exceeds the given time budget)

[b] Instances could not be solved to optimality with CPLEX in a reasonable time

5.5 Impact of the Number of Individuals

In this experiment we test the method on configurations with larger values of N, i.e., $N \in \{50, 100, 150, 197\}$, with capacities $c \in \{20, 40, 60, 80\}$, respectively. Table 5 shows the results for these four configurations (5 instances) with $R = 100$ and $J = 3$ for a 2-hour time budget. We consider scenario groups of 2 scenarios each. The average UB time per iteration grows exponentially. In the case of $N = 197$, it even exceeds the 2-hour time budget (the first iteration is always performed). The average number of iterations follows the opposite trend, as it decreases on average from 98.2 iterations for $N = 50$ to a single iteration for $N = 197$. Interestingly, we observe that the duality gaps decrease as N increases.

6 Decomposition for CBO

The CBO framework, and particularly the revenue maximization problem introduced in Sect. 5, has two sets of binary variables: the choice variables (w_{inr}), which are the result of the linearization of the variable capturing the largest utility (as given by the variables z_{inr}) for a given individual n and scenario r, and the availability variables (y_{inr}), which are involved in the linearization of the variables z_{inr} and the capacity constraints. Furthermore, both variables are related in the constraints that ensure that an alternative that is not available cannot be chosen.

Furthermore, various complicating constraints can be identified in this framework. The equations that characterize the utility for each scenario (U_{inr}) work as the link between the RUM model (individuals) and the MILP model (planner). The capacity constraints complicate the solution approach and prevent a potential decomposition by individual. The model also relies on big-M constraints to linearize the definition of z_{inr} and the maximization of its highest value.

Since the big-M constraints are the tightest possible to our formulation, and the constraints mentioned above do not allow to exploit the two decomposition dimensions (i.e., the individuals and the scenarios), we resolve to rely on Lagrangian relaxation to take advantage of the decomposable structure of the CBO framework. Notice that recent works, such as the one proposed in Rahmaniani et al. (2020), combine the complimentary advantages of the two decomposition methods considered typically considered for MILP (i.e., Lagrangian relaxation and Benders decomposition), which allow us not to have to opt for one of the two.

To provide some context on the development of the proposed method, we originally identified two subproblems: the subproblem on the planner (supply side) and the subproblem on the individuals (demand side), and we relaxed the constraints linking both subproblems (constraints defining the utility functions for each scenario). These constraints were seen as complicating constraints and relaxed in a Lagrangian fashion to be able to separate the planner and the individual subproblem. These subproblems are much easier to solve. Nevertheless, this relaxation leads to very loose bounds. This is why we decided to rely on Lagrangian decomposition, or variable splitting. This technique artificially induces decomposition by introducing copies of the original variables for a subset of constraints of the original problem. It then dualizes the constraints that impose that these variables should be identical to the original ones. This approach is of special interest for problems whose constraint set is the intersection of several specially structured constraint sets, which is the case of the CBO framework.

More precisely, in the CBO framework, many variables and constraints are defined per individual (n) and scenario (r). The only variables that are defined to be the same across scenarios are the planner's decisions x_{ink}. In the revenue maximization problem introduced in Sect. 5, the decisions (price) even need to be the same across individuals. Initially, we introduced R copies of the price variables p_i, namely $p_{ir}, \forall r = 1, \ldots, R$, and relax the constraints that relate them, i.e., $p_{i1} = \ldots = p_{iR}$. This decomposition leads to a large number of subproblems whose resulting price values reach the bounds in several cases (i.e., irrelevant values). This is the reason why we group some of the price duplicates and introduce copies for groups of scenarios and not for independent scenarios (p_{is}, where s is a scenario group). We relax the constraint that relates these copies and apply the subgradient method to approximate the Lagrangian dual to find the best possible upper bound on the original problem. A lower bound is obtained by accommodating a method that rapidly generates feasible solutions from the solution to the Lagrangian subproblem.

7 Conclusions

In this work, we introduce a tailored heuristic solution approach for choice-based optimization (CBO) based on scenario decomposition and scenario grouping. It is built upon the CBO model introduced in Pacheco Paneque et al. (2021) and Pacheco Paneque et al. (2022). This decomposition approach enables the retention of the intricate supply-demand relationships explicitly modeled in the original problem since all the constraints from the original optimization problem are incorporated into the Lagrangian subproblem. Additionally, we have devised an algorithm for the efficient generation of feasible solutions to the original problem using the solutions obtained from the Lagrangian subproblem. To gather the scenarios into groups, we adapt several strategies from Crainic et al. (2014). The grouping strategies are implemented as a preprocessing step of the subgradient method, which iteratively generates upper bounds (by solving the Lagrangian subproblem) and lower bounds (by generating feasible solutions) to the optimization problem.

To empirically assess the effectiveness of the Lagrangian decomposition approach, we formulate a revenue maximization problem in a case study concerning parking choices. Our findings demonstrate that near-optimal solutions can be achieved across all scenario grouping strategies and for different numbers of scenarios, all within significantly reduced computational time compared to the exact method. These solutions are typically produced in the initial phases of the subgradient method, enabling the algorithm to be promptly terminated if the best feasible solution has not shown improvement over a certain number of consecutive iterations. This termination condition is applied as long as the duality gap remains below an acceptable threshold. We cannot identify a scenario grouping strategy that outperforms the others, so we resolve to rely on the random partition for the performed experiments. As long as the Lagrangian subproblems are computationally manageable, a large number of scenarios per group is recommended, as it leads to smaller duality gaps for a given computational time. We also observe low duality gaps for configurations with a larger number of alternatives and individuals.

In summary, Lagrangian decomposition offers a valuable approach to tackle the computational challenges posed by CBO problems. Furthermore, the method we've introduced can be integrated with other decomposition techniques as needed to meet the specific demands of the model. For instance, if the capacity of the alternatives is included as a decision variable (that is formulated with a reduced set of binary variables), Benders decomposition could be as well explored. Furthermore, parallelization routines that allow to solve multiple subproblems simultaneously could be implemented within the subgradient method in order to reduce the total run time (i.e., the time between the start and the end of the method). These approaches allow to further enhance and expand the applicability of the method to the application areas discussed in introduction, such as revenue management and transportation-related problems.

Acknowledgments This research was conducted in collaboration with Professor Bernard Gendron who visited EPFL in 2015 for his sabbatical. Given his vast and deep knowledge on large scale optimization, together with him, we investigated different decomposition methods for the CBO framework. We eventually decided to choose the Lagrangian Decomposition. He will be remembered not only for his scientific work but for his professionalism, friendship, sense of humor and his mentorship.

References

Azadeh, S. S., van der Zee, J., & Wagenvoort, M. (2022). Choice-driven service network design for an integrated fixed line and demand responsive mobility system. *Transportation Research Part A: Policy and Practice*, *166*, 557–574.

Bakir, I., Boland, N., Dandurand, B., & Erera, A. (2020). Sampling scenario set partition dual bounds for multistage stochastic programs. *INFORMS Journal on Computing*, *32*(1), 145–163.

Bertsimas, D., & Mundru, N. (2023). Optimization-based scenario reduction for data-driven two-stage stochastic optimization. *Operations Research*, *71*(4), 1343–1361.

Binder, S., Maknoon, M., Azadeh, S. S., & Bierlaire, M. (2021). Passenger-centric timetable rescheduling: A user equilibrium approach. *Transportation Research Part C: Emerging Technologies*, *132*, 103368.

Blanchet, J., Gallego, G., & Goyal, V. (2016). A Markov chain approximation to choice modeling. *Operations Research*, *64*(4), 886–905. https://doi.org/10.1287/opre.2016.1505

Camerini, P. M., Fratta, L., & Maffioli, F. (1975). On improving relaxation methods by modified gradient techniques. *Nondifferentiable Optimization*, *3*, 26–34.

Carøe, C. C., & Schultz, R. (1999). Dual decomposition in stochastic integer programming. *Operations Research Letters*, *24*(1–2), 37–45.

Crainic, T. G., Hewitt, M., & Rei, W. (2014). Scenario grouping in a progressive hedging-based meta-heuristic for stochastic network design. *Computers & Operations Research*, *43*, 90–99.

Escudero, L. F., Garín, M. A., Pérez, G., & Unzueta, A. (2013). Scenario cluster decomposition of the lagrangian dual in two-stage stochastic mixed 0–1 optimization. *Computers & Operations Research*, *40*(1), 362–377.

Escudero, L. F., Garín, M. A., & Unzueta, A. (2016). Cluster Lagrangean decomposition in multistage stochastic optimization. *Computers & Operations Research*, *67*, 48–62.

Feldman, J. B., & Topaloglu, H. (2017). Revenue management under the Markov chain choice model. *Operations Research*, *65*(5), 1322–1342. https://doi.org/10.1287/opre.2017.1628

Fisher, M. L. (1973). Optimal solution of scheduling problems using lagrange multipliers: Part I. *Operations Research*, *21*(5), 1114–1127.

Gade, D., Hackebeil, G., Ryan, S. M., Watson, J. P., Wets, R. J. B., & Woodruff, D. L. (2016). Obtaining lower bounds from the progressive hedging algorithm for stochastic mixed-integer programs. *Mathematical Programming*, *157*, 47–67.

Gallego, G., Iyengar, G., Phillips, R., & Dubey, A. (2004). Managing flexible products on a network. *Technical report*, Columbia University, New York.

Gilbert, F., Marcotte, P., & Savard, G. (2014a). Logit network pricing. *Computers & Operations Research*, *41*, 291–298.

Gilbert, F., Marcotte, P., & Savard, G. (2014b). Mixed-logit network pricing. *Computational Optimization and Applications*, *57*, 105–127.

Held, M., Wolfe, P., & Crowder, H. P. (1974). Validation of subgradient optimization. *Mathematical Programming*, *6*, 62–88.

Ibeas, A., Dell'Olio, L., Bordagaray, M., & Ortúzar. J. d. D. (2014). Modelling parking choices considering user heterogeneity. *Transportation Research Part A: Policy and Practice*, *70*, 41–49.

Jiang, X., Bai, R., Ren, J., Li, J., & Kendall, G. (2022). Lagrange dual bound computation for stochastic service network design. *European Journal of Operational Research, 302*(3), 1097–1112.

Jiang, X., Bai, R., Wallace, S. W., Kendall, G., & Landa-Silva, D. (2021). Soft clustering-based scenario bundling for a progressive hedging heuristic in stochastic service network design. *Computers & Operations Research, 128*, 105182.

Justin, C. Y., Payan, A. P., & Mavris, D. (2021). Demand modeling and operations optimization for advanced regional air mobility. In *AIAA aviation 2021 forum* (p. 3179).

Labbé, M., Marcotte, P., & Savard, G. (1998). A bilevel model of taxation and its application to optimal highway pricing. *Management Science, 44*(12-part-1), 1608–1622. https://doi.org/10.1287/mnsc.44.12.1608, https://pubsonline.informs.org/doi/pdf/10.1287/mnsc.44.12.1608

McFadden, D. (1974). Conditional logit analysis of qualitative choice behavior. In: *Frontiers in econometrics* (pp. 105–142). New York: Academic Press.

McFadden, D., & Train, K. (2000). Mixed MNL models for discrete response. *Journal of Applied Econometrics, 15*(5), 447–470.

Pacheco Paneque, M. (2020). *A general framework for the integration of complex choice models into mixed integer optimization*. Tech. rep., EPFL.

Pacheco Paneque, M., Bierlaire, M., Gendron, B., & Sharif Azadeh, S. (2021). Integrating advanced discrete choice models in mixed integer linear optimization. *Transportation Research Part B: Methodological, 146*, 26–49.

Pacheco Paneque, M., Gendron, B., Azadeh, S. S., & Bierlaire, M. (2022). A Lagrangian decomposition scheme for choice-based optimization. *Computers & Operations Research, 148*, 105985.

Prokhorchuk, A., Dauwels, J., & Jaillet, P. (2019). Stochastic dynamic pricing for same-day delivery routing. Preprint. arXiv:191202946.

Rahmaniani, R., Ahmed, S., Crainic, T. G., Gendreau, M., & Rei, W. (2020). The benders dual decomposition method. *Operations Research, 68*(3), 878–895.

Roemer, N., Müller, S., & Voigt, G. (2023). A choice-based optimization approach for contracting in supply chains. *European Journal of Operational Research, 305*(1), 271–286.

Ryan, K., Ahmed, S., Dey, S. S., Rajan, D., Musselman, A., & Watson, J. P. (2020). Optimization-driven scenario grouping. *INFORMS Journal on Computing, 32*(3), 805–821.

Shapiro, A., Dentcheva, D., & Ruszczynski, A. (2021). *Lectures on stochastic programming: modeling and theory*. SIAM.

Shen, Z. J. M., & Su, X. (2007). Customer behavior modeling in revenue management and auctions: A review and new research opportunities. *Production and Operations Management, 16*(6), 713–728.

Strauss, A. K., Klein, R., & Steinhardt, C. (2018). A review of choice-based revenue management: Theory and methods. *European Journal of Operational Research, 271*(2), 375–387.

Ulmer, M. W. (2020). Dynamic pricing and routing for same-day delivery. *Transportation Science, 54*(4), 1016–1033.

van Ackooij, W., de Oliveira, W., & Song, Y. (2018). Adaptive partition-based level decomposition methods for solving two-stage stochastic programs with fixed recourse. *Informs Journal on Computing, 30*(1), 57–70.

Wei, K., Vaze, V., & Jacquillat, A. (2020). Airline timetable development and fleet assignment incorporating passenger choice. *Transportation Science, 54*(1), 139–163.

Yan, S., Tang, C. H., & Fu, T. C. (2008). An airline scheduling model and solution algorithms under stochastic demands. *European Journal of Operational Research, 190*(1), 22–39.

Zhang, W., Wang, K., Jacquillat, A., & Wang, S. (2023). Optimized scenario reduction: Solving large-scale stochastic programs with quality guarantees. *INFORMS Journal on Computing, 35*, 886–908.

Part II
Application-Oriented Developments

The Static Elevator Dispatching Problem with Destination Control

Camille Richer, Michel Bierlaire, and Fabian Torres

1 Introduction

Elevators are essential modes of transportation used for vertical mobility within buildings. The earliest records of elevators date back to the third century BCE, when they were powered by humans or animals. The invention of hydraulic power and the construction of larger buildings during the nineteenth century paved the way for modern elevators. At first, an operator was required to drive the elevator and assist passengers. However, with the proliferation of elevators and increased urbanization, controls were subsequently introduced within the elevator, enabling passengers to select their destination floor independently. In most modern elevators, passengers call the elevator from the lobby by pushing a button indicating the intended direction of travel and once inside, they choose their destination floor.

In some buildings, there are multiple elevators, forming an elevator group. It usually consists of several single-deck elevators with the same capacity, which are close to each other and share the same call buttons: when a passenger presses a button, the call is not meant for a specific elevator and the system must decide how to dispatch elevators to requests. In some large buildings with an elevator group, a destination control system is employed. In this system, instead of having two buttons to indicate the direction of the request, there is a keypad on which passengers select their destination floor in the lobby before entering the elevator (and possibly the number of passengers for the request). The interface then directs the passenger to the appropriate elevator to take (see Fig. 1) and they wait for their assigned elevator to arrive. The elevators' movements are constrained to avoid

C. Richer · M. Bierlaire · F. Torres (✉)
Transport and Mobility Laboratory, Ecole Polytechnique Fédérale de Lausanne, Lausanne, Switzerland
e-mail: camille.richer@epfl.ch; michel.bierlaire@epfl.ch; fabian.torres@epfl.ch

T. G. Crainic et al. (eds.), *Combinatorial Optimization and Applications*,
International Series in Operations Research & Management Science 358,
https://doi.org/10.1007/978-3-031-57603-4_14

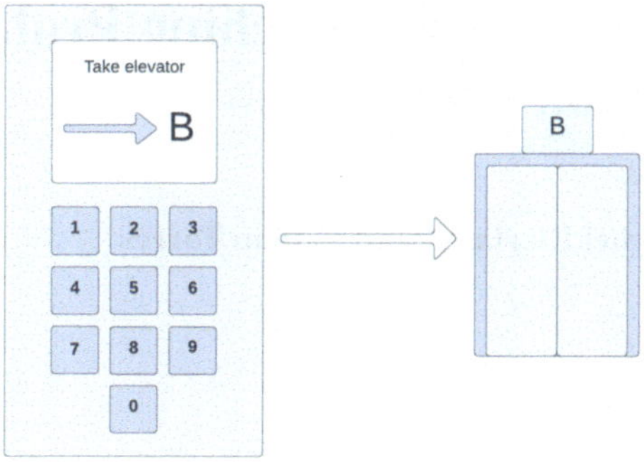

Fig. 1 With destination control, passengers choose the destination floor at the lobby and the user interface tells the passenger what elevator to take. The interface instructs the passenger to take elevator B when it arrives

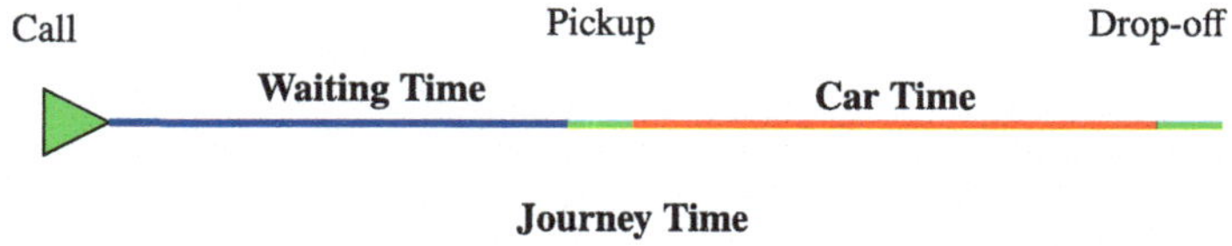

Fig. 2 The waiting time end right before pickup and the journey time ends right before drop-off

inconveniencing passengers, i.e. passengers can get frustrated if the elevator travels away from their intended destination floor, thus elevator movements that results in any passenger moving in the opposite direction to their destination floor is strictly forbidden. If a passenger in the elevator should be going up to reach their destination floor, the elevator cannot go down, and conversely if a passenger in the elevator needs to go down, the elevator cannot go up.

This problem is by nature dynamic, as passenger arrival times and destinations are not known in advance. However, we consider the static case, where perfect information is assumed about the future arrival of all passengers: arrival time, origin floor and destination floor. Most studies focuses on the dynamic case and heuristic methods to solve it, because they are applicable in practical cases (Beamurgia et al., 2016; Tyni & Ylinen, 2006; Vodopija et al., 2022). In this chapter, we use an exact method based on mathematical programming to solve the problem. Elevators are also a significant source of energy consumption in buildings, but only a few studies consider energy. Waiting time or journey time are the main concern. The *waiting time* of a passenger is the time from the instant the passenger registers the request until the instant the passenger can start boarding the elevator at the origin floor. The *journey time* of a passenger is the time from the instant the passenger registers the request, until the instant the passenger can start exiting the elevator at the destination floor (see Fig. 2). They are the metric for quality of service, so taking into account energy leads to multi-objective optimization and increases the complexity of the problem. As a result, studies that consider both objectives often turn the problem into a single-objective one instead of doing true multi-objective optimization (Beamurgia et al., 2022; Hu et al., 2010).

This chapter focuses on the static Elevator Dispatching Problem with Destination Control (EDPDC). There are two objectives to minimize: (1) the average journey time of passengers, and (2) the energy consumption of the elevators. While energy consumption might be a natural objective to minimize in order to reduce costs, passengers would be unsatisfied with the performance of the elevators if the average journey time was too high. The model chosen for energy is that of a counterweight elevator, which is the most common type of elevator found in high buildings.

The main contributions of this chapter are:

- we introduce the model of the EDPCD with a counterweight elevator energy model and a mathematical description of the set of feasible arcs in the graph representation;
- we introduce the first explicit mathematical formulation for the static EDPDC with two objectives, i.e. average journey time and total energy consumption;
- we provide valuable insight about the runtime of the algorithm with respect to the parameters by doing a sensitivity analysis and provide the Pareto frontier for small instances.

The remainder of the chapter is organized as follows: In Sect. 2, we review appropriate literature on the EDP. In Sect. 3, we provide a description of the problem. In Sect. 4, we present a graph model of the problem and a mathematical formulation that includes two objectives. In Sect. 5, we present the results of the sensitivity analysis on the parameters used. In Sect. 6, we make final remarks and show different research directions.

2 Literature Review

The EDPDC belongs to a family of combinatorial optimization problems called Vehicle Routing Problems (VRP), where a set of vehicles are used to satisfy delivery requests. Many variants of this problem exist: some VRPs consider time window constraints where requests must be satisfied after an earliest arrival time, but not before a latest arrival time (Dayarian et al., 2015; Torres et al., 2022a,b; Boyer et al., 2014); other VRP variants consider pick-up and delivery of parcels (or passengers) that must be picked up at a specific location and dropped off at their specific destination (Lahrichi et al., 2015; Fontaine et al., 2021). In the case of the EDPDC, time windows start at the moment a passenger makes the call for the elevator and a pick-up and delivery is done with each passenger similarly to the VRP with Pickup and Delivery or to the Dial-a-Ride Problem (DARP). In Cordeau (2006), the authors propose a three-index formulation for the DARP that is similar to the one presented in this chapter. However, the objective of a DARP is to minimize the total routing cost while respecting a maximum ride time. The cost does not depend on the load of the vehicle and the problem is less constrained than the EDPDC, as there is no rule preventing a vehicle from picking up some combination of passengers.

The static EDPDC has received limited attention in the literature, with most studies focusing on the dynamic problem and using heuristics as solution methods due to the difficulty of the problem. In practice, passengers arrive at unknown random times and the elevator group control system must decide in real time how to dispatch elevators, updating its decision regularly. On top of that, it must decide quickly so as not to make passengers wait. The usual computational time limit allowed is in the order of 0.5 s (Sorsa et al., 2009; Beamurgia et al., 2016; Ruokokoski et al., 2015; Tyni & Ylinen, 2006), which rules out most exact methods and shifts the focus towards metaheuristics.

For instance, in Beamurgia et al. (2016), the authors tackle the EDP with traditional control, which implies that the problem cannot be modeled as a VRP (as destinations are not known before the passenger has been assigned and is inside an elevator). The system receives two types of calls: landing calls, made from a lobby and indicating the origin floor and the direction of travel (up or down), and car calls, made inside an elevator car and indicating a destination floor. They build on the work of Sorsa et al. (2009) and consider the dispatching problem as a two-level problem, the first level dealing with the assignment of landing calls to elevators, and the second level generating the route of each elevator for a given set of car calls and assigned landing calls. The assignment problem is solved with a genetic algorithm, where an individual of the population is an assignment of landing calls to elevators. The fitness function is the inverse of the squared estimated waiting time (so that the fitness function is maximized). The estimated waiting time is the sum of the passengers waiting time in this assignment where each elevator route is computed by a greedy algorithm that takes into account the current state of the system (elevator location, movement, on-board passengers). The time limit for each run must be less than 0.5 s, meaning the number of iterations and the population size depend on the hardware. The algorithm is run continuously, and an assignment can be reconsidered until an elevator stops at the origin floor to serve it.

While heuristic methods are practical and implemented in practice, exact methods can provide valuable insights about the mathematical properties of the problem. Ruokokoski et al. (2015) introduce the first mixed-integer linear programming formulation for the EDPDC, giving the constraints in explicit mathematical form. They model the problem as a VRP with a single objective, minimizing the average waiting time of passengers, and propose a two-index formulation that has an exponential number of constraints. The three-index formulation we propose allows for a polynomial number of constraints. They also make some additional assumptions about the boarding or alighting order of passengers that have the same origin or destination floor. Moreover, they do not consider all passengers from the beginning, and instead, the static EDPDC is solved repeatedly with up-to-date information after the registration of one or more new requests. It means that the formulation takes into account the initial state of the system, where some elevators may be moving, with passengers inside, and some requests to pick up passengers may have already been assigned to an elevator and will not be reconsidered. They propose several improvements to reduce the number of variables (at the expense of the number of constraints), and they identify valid inequalities to strengthen the formulation. They analyze the polyhedral structure of the formulation in the case of down-peak traffic and identify some facet inducing inequalities among the proposed valid inequalities. However, they do not present computational results.

In an extension, Ruokokoski et al. (2016) point out that in many methods used to solve the EDP, decisions specify only request-to-elevator assignment, and given a set of requests assigned to an elevator, its route is determined by a heuristic (Beamurgia et al., 2016; Tyni & Ylinen, 2006). They propose to compare this two-level approach with the routing formulation they previously developed in Ruokokoski et al. (2015) in the case with destination control. They introduce the assignment formulation, that has the assignment of requests to elevator as decision variables, and for a given set of requests, there is a unique route defined by a heuristic called collective control together with the assumptions on boarding or alighting order they did in their previous work. Since not all routes are considered, the assignment formulation is a restricted version that only considers a smaller set of routes and thus is not guaranteed to be optimal. In both cases, the problem is solved by repeatedly solving the updated static problem with one or more new unassigned requests. They compare the two approaches on small instances and for two objectives: the average waiting time and the average journey time. The objective in Ruokokoski et al. (2015) is easily adaptable to the journey time. They prove that the optimal solution for the assignment formulation is also optimal for the routing formulation for elevators of unlimited capacity and up-peak traffic, i.e. all passengers start on the entrance floor and go up, but not necessarily for limited capacity or other types of traffic. Naturally, they observe that the assignment formulation is faster than the routing formulation but gives worse solutions for instances with traffic that is not up-peak, and they notice that minimizing the average journey time makes instances more difficult for both formulations. For both methods, the computation time quickly exceed the acceptable 0.5 s even on reasonably small instances.

In the works on the EDP previously mentioned, the objective function is only time related, to ensure quality of service. However, energy consumption in high buildings is an increasing concern and it is desirable to try to optimize it in elevator routing. The EDP then becomes a multi-objective problem. The two objectives are usually conflicting and there are various methods to take them both into account.

Tyni and Ylinen (2006) introduce a bi-objective EDP for the case with traditional control. The two objectives considered are the average waiting time of passengers and the energy consumption. They use a genetic algorithm in the same way as Beamurgia et al. (2016). The energy model they use for the counterweight elevator is very similar to the one used in our paper. Due to computation time concerns, they use a weighted aggregation method to transform the problem into a single objective problem, and they adjust the relative weights of the two objectives dynamically when solving. On a very small instance with three elevators and three passengers, the conflicting objectives are obvious: when optimizing waiting time only, all three elevators are used, but when optimizing energy only, one elevator is used for all requests.

In Beamurgia et al. (2022), the authors extend their previous work in Beamurgia et al. (2016). They keep the same framework, i.e. traditional control, assignment of requests to elevators by a genetic algorithm and routes generated by a heuristic and add the energy consumption in the objective. It uses a counterweight model but the explicit formula with respect to the load and trip distance is not given. As in Tyni and Ylinen (2006), the objective function is a weighted aggregation of the passenger waiting time and the energy consumption, but the weights are computed beforehand. The same conflict between the objectives is observed.

In both cases, the multi-objective problem is reduced to a single objective problem to keep the computation time low. On the basis of this observation, in a recent study, Vodopija et al. (2022) present a multi-objective EDP with traditional controls that performs true multi-objective optimization using simulation-based optimization. Extending their previous work in Vodopija et al. (2018), they propose a constrained multi-objective problem, with minimizing the number of waiting passengers and the number of elevator stops as objectives. The constraint limits the number of elevator skips (an elevator skips a floor with a passenger waiting to go in the same direction as the elevator). The solutions are evaluated with a simulation on several different configurations of buildings and passenger arrival rates, and the solution space is explored by metaheuristics. They test five different multi-objective algorithms that implement the constraint to explore the solution space. Comparing their method to weighted-sum approaches like the one used in Tyni and Ylinen (2006), they conclude that true multi-objective approximation gives better results.

Another way of handling the two objectives is given in Hu et al. (2010). They develop a multi-objective genetic algorithm for the EDP with destination control, considering simultaneously energy consumption and acceptable levels of waiting time. The problem is solved on a rolling time horizon and the energy consumption is the main objective: they force a maximum average waiting time by adding it as a constraint. They use a genetic algorithm to assign calls to elevators but do not explain how the route of an elevator is determined. They compare their solution with the nearest elevator heuristic (at every new call, it is assigned to the nearest elevator with adequate direction) on a 10-floor building with 4 elevators and different passenger arrival rates. They find that compared to the heuristic, they can reduce the energy consumption by 23.6% on average but at the expense of increasing the average waiting time by 64.9% and the average journey time by 39.5%.

Table 1 summarizes the articles about the EDP cited in this review with some of their characteristics. We can observe from the literature that there is no exact method proposed for the multi-objective problem with destination control that considers both the average journey time and the energy consumption.

3 Problem Description

In this Section, we describe the EDPDC as a VRP with additional constraints and we explain the energy model.

Table 1 Literature on EDPDC

Study	Control	Type of data	Objective(s)	Method
Beamurgia et al. (2016)	Traditional	Dynamic	Waiting time	Heuristics
Ruokokoski et al. (2015)	Destination	Static with rolling time horizon	Waiting time	Exact
Ruokokoski et al. (2016)	Destination	Static with rolling time horizon	Waiting time/journey time	Exact and heuristic
Tyni and Ylinen (2006)	Traditional	Dynamic	Aggregated waiting time and energy consumption	Heuristics
Beamurgia et al. (2022)	Traditional	Dynamic	Aggregated waiting time and energy consumption	Heuristics
Vodopija et al. (2022)	Traditional	Dynamic	Waiting passengers and elevator stops	Heuristics
Hu et al. (2010)	Destination	Dynamic	Energy consumption with constrained waiting time	Heuristics
This chapter	Destination	Static	Journey time and energy consumption	Exact

3.1 Additional Rules and Service Order

Unlike the usual VRP, where vehicles can move freely in a two-dimensional horizontal space, elevator movements are restricted to up and down movements in a vertical space. In addition, passengers are sensitive and can feel stress or frustration if the elevator moves in a direction that does not reduce the distance to their destination. With the sensitivity of the passengers in mind (Closs, 1970) defines a set of rules that prevent elevators from taking a passenger in the opposite direction to their intended travel. Hence, the elevator system operates according to the following rules (see Closs, 1970):

- If no passenger enters or exits a floor, the elevator will skip it.
- The elevator will not skip a floor if a passenger wants to get off there.
- Passengers cannot enter an elevator that is going in the opposite direction to their desired travel and that already has passengers.
- If there is at least one passenger inside, the elevator cannot change direction.

We also use some assumptions on service order at a floor similar to those described in Tanaka et al. (2005) for the case of a single elevator. They determine an order when several passengers alight or board the same elevator on the same floor.

- Passengers cannot board the elevator before all on-board passengers who are going to leave at that stop finish alighting.

- If there are two or more passengers who are going to board the elevator at a floor, they should board it in the ascending order of their arrival times.
- If there are two or more passengers who are going to leave the elevator at a floor, they should alight in the descending order of how they boarded the elevator.

3.2 Problem Data

In the static EDPDC, we consider a building with F floors, numbered $0, 1, \ldots, F-1$, 0 being the ground floor where the entrance of the building is located. The building is equipped with an elevator group K of identical elevators denoted $0, \ldots, m-1$. The capacity of an elevator is expressed in number of people, under the assumption that a person weighs on average 75 kg. All elevators have the same capacity Q, door opening time o, door closing time c and travel times between floors. For any pair of floors (f, g), the *travel time* from f to g is the time an elevator takes to go from floor f to floor g without any stop in between, once the door is closed on f and before the door opens on g. It is known for all pairs of floors and equal to 0 if $f = g$. All elevators can serve every floor in the building and their shafts are right next to each other. The elevator call device is located right next to the shafts on every floor, so a passenger is immediately ready to board any elevator after making a call.

We assume that there is exactly one passenger per call. Before the first passenger arrives, all elevators are standing at the ground floor, door closed. All passengers need the same *service time* s to board or to leave the elevator on a floor. It might happen that two passengers are waiting at the same time on the same floor and are served by the same elevator but not picked up at the same time, the elevator could pick up a passenger on that floor and not the waiting passenger. The elevator then comes back to pick up the passenger later. There are n passengers denoted $p_1, \ldots, p_n$. Each passenger p_i is a tuple (a_i, α_i, β_i) where a_i is the arrival time (the instant the call is registered), α_i the origin floor and β_i the destination floor. The direction of the trip (up or down) can be easily deduced. We assume that $\alpha_i \neq \beta_i$, so that the direction is always well defined.

3.3 Objectives

The two performance criteria that we use are the average journey time of passengers and the total energy consumption of the elevators.

The *energy consumption* of an elevator is the net amount of energy that the elevator uses from the instant it moves to serve a first request, to the instant it alights its last passenger. We only consider the energy used by the elevator to move up and down at a constant speed: our model does not take into account the energy required to open or close the door, nor the acceleration and braking energy.

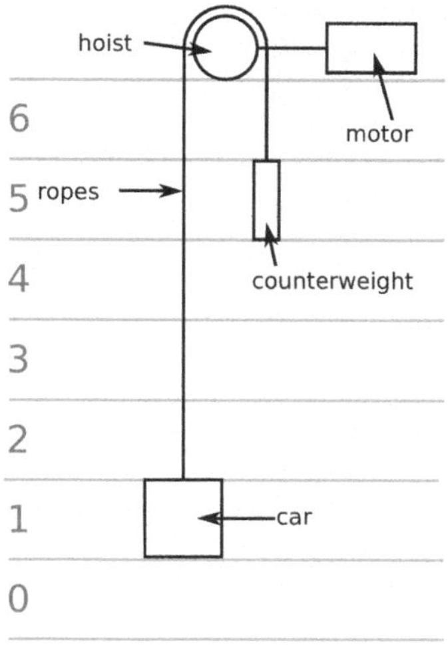

Fig. 3 Simple illustration of an elevator with a counterweight

The assumptions on service order define a unique order when an elevator serves several passengers at the same stop, avoiding considering equivalent solutions. They do not affect these two objectives: what happens when an elevator stops does not affect energy consumption, and the average journey time does not depend on the order in which passengers alight or board.

3.4 Energy Model

Most elevators in high buildings work with a counterweight (see Fig. 3). The mass of the counterweight is equal to the mass of the elevator car plus half the maximum payload of the car, so that the system is balanced when the half-loaded car and the counterweight are standing at half the height of the shaft. When the car goes up, the counterweight goes down and conversely. This system is meant to save energy. If the mass of the loaded car is lighter than the counterweight, and the car is going up, then the system produces energy instead of consuming. Conversely, if the loaded car is going down and lighter than the counterweight, the system uses energy to lift the counterweight. Evidently, if the car is more than half-loaded, the opposite is true, i.e., the system consumes energy when the car is going up and produces energy when it is going down (see Fig. 4). The energy formula and the values for efficiency parameters were provided by Schindler.

Let E_ℓ denote the energy (in kJ) needed by the elevator to go from a floor to another with a certain load if they were no energy loss. E_ℓ is a function of the actual load of the elevator in kilograms (i.e., q_{kg}) and of the trip distance in meters (i.e., d_m) between the floors in the building. The following equation (provided by Schindler)

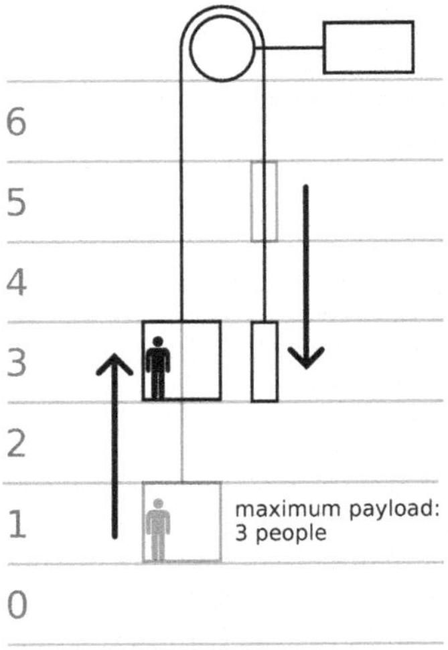

Fig. 4 Example of a situation where the system produces energy

gives the total energy consumed without inefficiency:

$$E_\ell(q_{\text{kg}}, d_m) = \left(q_{\text{kg}} - \frac{Q_{\text{kg}}}{2}\right) dg$$

where the constant Q_{kg} is the maximum payload in *kg*, the trip distance d counted positive if the elevator is going up and negative if it is going down, and g is the gravity acceleration. In most commercial buildings, the height of all floors is identical, thus, to know the total distance traveled, it is sufficient to consider the total floors traveled by the elevator and multiply it by the constant height of each. If we denote f_c the number of the current floor (recall floors are numbered from 0 to $F - 1$), from which the elevator leaves, and f_ℓ the number of the landing floor, then the distance travelled is proportional to $f_\ell - f_c$. In practise, passengers' weights can vary significantly, however elevator companies standardize the weight of passengers to 75 kg to calculate the maximum capacity. Hence, assuming also that a person weighs on average 75 kg, we can consider that the load q_{kg} and the maximum payload Q_{kg} are proportional to the load q and the capacity Q in number of people. $E_\ell(q, d)$ is proportional to the floors difference and the load in number of people, and we can omit the constant factors (storey height, average weight and g). The formula can be simplified as follows:

$$E_\ell(q, f_c, f_\ell) = \left(q - \frac{Q}{2}\right)\left(f_\ell - f_c\right).$$

The energy produced by the system can be stored to use it when it is needed. The efficiency of such a system also needs to be considered since storing 100% of the energy produced by the elevator is unlikely. In addition, energy losses occur due to friction and transmission losses. We consider an inefficiency factor of 0.75, thus if

$E_\ell > 0$, the elevator consumes energy and the energy actually consumed is $\frac{1}{0.75}E_\ell$. If $E_\ell < 0$, the elevator produces excess energy, which is fed back into the grid with efficiency 0.75, i.e., the energy fed back is $0.75E_\ell$. The energy consumption for a direct trip between two floors is then:

$$E(q, f_c, f_\ell) = \begin{cases} \frac{1}{0.75}E_\ell & \text{if } E_\ell > 0 \\ 0.75E_\ell & \text{if } E_\ell \leq 0. \end{cases}$$

The total consumption of an elevator route is the sum of the energy consumption over every direct trip of the elevator, but the coefficient varies according to the load of the trip, which leads to a nonlinear objective function.

4 The Static Elevator Dispatching Problem with Destination Control

In this Section, we introduce the network representation of the problem as a VRP, and then we introduce the resulting routing formulation.

4.1 *The Network Representation*

The problem can be represented by a weighted directed graph $G = (V, A)$. The vertices of the graph (i.e., V) represent the requests to be fulfilled by the elevators rather than the floors of the building. A request is either a passenger to pick up or a passenger to drop off. The set of arcs (i.e., A) is a set that connects the vertices. An arc exists between two vertices if an elevator can fulfill these two requests consecutively. The graph is described in the following subsections.

4.1.1 Vertices of *G*

Each passenger p_i is associated with two vertices $i \in V$ and $n + i \in V$ corresponding to the pick-up and drop-off requests respectively. A vertex associated with a pick-up request is called a pick-up vertex and the set of pick-up vertices is denoted $P = \{1, \ldots, n\}$. Likewise, we denote $D = \{n + 1, \ldots, 2n\}$ the set of drop-off vertices. There are two more vertices denoted 0 and $2n+1$ that are the source and the sink of the graph. The vertex 0 corresponds to the starting point of all elevators and $2n + 1$ is the end point of all elevators. The vertex set is the union of all these, i.e., $V = \{0\} \cup P \cup D \cup \{2n + 1\}$ (see Fig. 5).

Each vertex i of the graph has a demand d_i that represents the number of passengers boarding the elevator when it visits the vertex. The demand is then

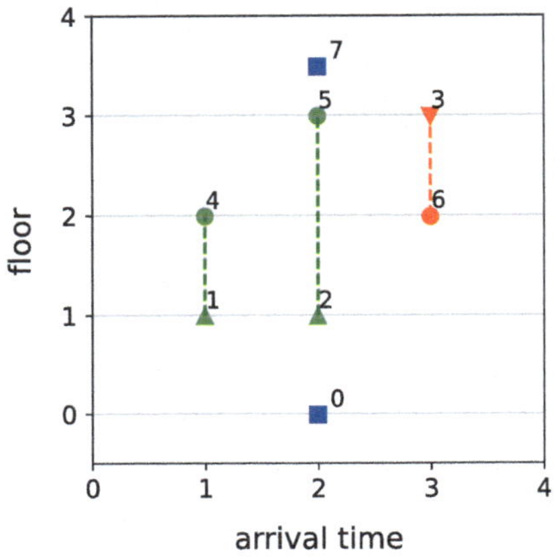

Fig. 5 The vertices of the graph model. The two requests of a passenger linked by a dashed line and positioned at their arrival time on the x-axis. Pick-up requests are represented by a triangle pointing up or down and drop-off requests are represented by a circle. Up requests are green and down requests are red. The blue squares represent the source and the sink of the graph. The vertices that have an associated floor are represented on the line of the floor

$d_i = 1$ for pick-up vertices, $d_i = -1$ for drop-off vertices and $d_0 = d_{2n+1} = 0$. Each vertex $i \in P \cup D$ also has a floor f_i and a direction u_i that describes whether the passenger is going up or down. For each passenger p_i, the associated pick-up vertex i has floor $f_i = \alpha_i$, i.e., the origin floor of p_i, and the associated drop-off vertex $i + n$ has floor $f_{i+n} = \beta_i$, i.e., the destination floor of p_i. Both i and $i + n$ have the direction of p_i, i.e., $u_i = u_{i+n} =$ up if $\alpha_i < \beta_i$ and $u_i = u_{i+n} =$ down if $\alpha_i > \beta_i$.

Different vertices can have the same associated floor if two passengers share a common floor. Vertex 0 has no direction but has floor $f_0 = 0$ (ground floor). Naturally, this assumption can be relaxed and the elevators can start on different floors by replacing the start vertex with a vertex per elevator and placing the vertices at the desired floors. Vertex $2n + 1$ has no direction and no floor since we assume that an elevator ends its route at the destination floor of its last passenger. This assumption can also be modified with simple changes to the model, e.g., if all elevators have to return to the ground floor.

An example of a vertex set for a small instance is given in Fig. 5.

4.1.2 Arcs of *G*

Arc set A consists of an arc (i, j) if the request j can be served by an elevator immediately after it has served the request i. It means that a single elevator serves the request i on f_i (i.e., either picking up or dropping off a passenger), goes from f_i to f_j without any stop in between and serves the request j on f_j (i.e., either picking up or dropping off a passenger). According to the directions u_i and u_j and the floors

f_i and f_j, serving i then j may require that the elevator changes direction in i or j (i.e., up to down or down to up). For instance, if the elevator goes up to floor f_i, alights a passenger, then goes up to floor f_j to pick up a passenger who wants to go down, we consider that the elevator changes direction in j. If the elevator goes up to floor f_i, alights a passenger, then goes down to floor f_j to pick up a passenger, the elevator changes direction in i. We say that an arc (i, j) requires a change of direction if the passengers associated with i and j are distinct and if the elevator must change direction in i or in j. However, elevators cannot change direction with passengers inside (Closs rules). Hence, we have to choose the set of arcs carefully so that there is no infeasible arc.

We denote A_1 the set of feasible arcs that do not require a change of direction, and A_2 the set of feasible arcs that require a change of direction. Thus $A = A_1 \cup A_2$ and these sets are disjoint. For simplicity, we consider arcs that leave the start vertex (i.e., 0) or arcs that end at the sink (i.e., $2n+1$) as elements of A_2. Given two vertices i and j, we will denote $[f_i, f_j]$ the floors located between f_i and f_j included, regardless of which one is higher. Thus, the set $[f_i, f_j]$ is equal to the following:

$$[f_i, f_j] = \begin{cases} \{f_i, f_i + 1, \ldots, f_j\} & \text{if } f_i < f_j \\ \{f_j, f_j + 1, \ldots, f_i\} & \text{if } f_i > f_j \\ \{f_i\} & \text{if } f_i = f_j. \end{cases}$$

We use in the same way the half-open intervals $[f_i, f_j[$ or $]f_i, f_j]$ where respectively f_j or f_i is excluded from the interval, and the open interval $]f_i, f_j[$ where both are excluded.

Arcs of A_1

An arc $(i, j) \in A_1$ exists if an elevator can serve consecutively requests i and j without changing direction. By convention, there is no arc in A_1 incident to the source or sink (i.e., 0 or $2n + 1$). A necessary condition for an arc (i, j) in A_1 is:

(i) $i \neq j$ and $u_i = u_j$.

Condition (i) ensures that there is no loop, and that the two vertices have the same direction. Otherwise, there would be a change of direction in i or j. Assuming (i) is verified, there are different cases.

If $i, j \in P$, then $(i, j) \in A_1$ if one of these propositions is true:

(ii) $f_i = f_j$ and $a_i < a_j$.
(iii) $f_j \in]f_i, f_{i+n}[$

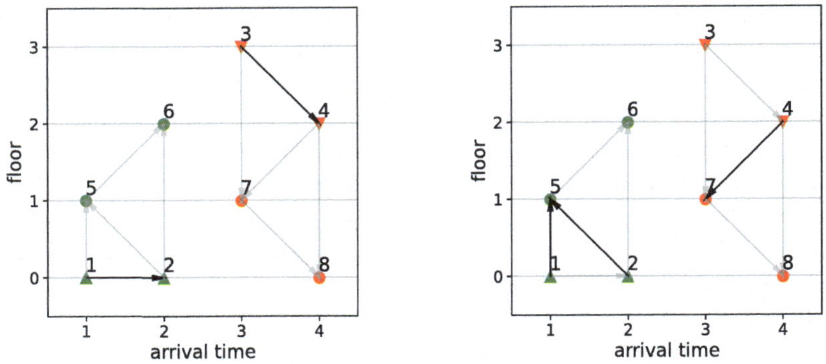

Fig. 6 On the left, (1, 2) verifies (ii) and (3, 4) verifies (iii). On the right, (1, 5) verifies (iv), (2, 5) verifies (v) and (4, 7) verifies (vi). All arcs verify (i)

If the two passengers are picked up at the same elevator stop, (ii) enforces service order. If they are picked up consecutively on different floors by the same elevator, it means that p_j is waiting on a floor strictly between the origin and destination floors of p_i, as expressed by (iii). Figure 6 illustrates this case.

If $i \in P, j \in D$*, then* $(i, j) \in A_1$ *if one of these propositions is true:*

(iv) $j = i + n$.
(v) $f_i = f_{j-n}$ and $a_{j-n} < a_i$ and $f_j \in]f_i, f_{i+n}[$.
(vi) $f_i \in]f_{j-n}, f_j[$ and $f_j \in]f_i, f_{i+n}[$.

Condition (iv) states that if i and j are respectively the pick-up and drop-off requests of a same passenger, the arc exists. Otherwise they are requests of two different passengers and p_{j-n} is waiting on a floor between the origin and destination floors of p_i. If they are picked up on the same elevator stop, (v) enforce service order with the additional condition that p_{j-n} must alight strictly before p_i. Condition (vi) covers the case where they are not waiting on the same floor. Figure 6 illustrates this case.

If $i \in D, j \in P$*, then* $(i, j) \in A_1$ *if one of these propositions is true:*

(vii) $u_i =$ up and $f_i \leq f_j$.
(viii) $u_i =$ down and $f_i \geq f_j$.

Even if the trips of two passengers do not overlap, they can be served by the same elevator without changing direction. Conditions (vii) and (viii) cover this cases, respectively when the passengers are going up and down. Figure 7 illustrates this case.

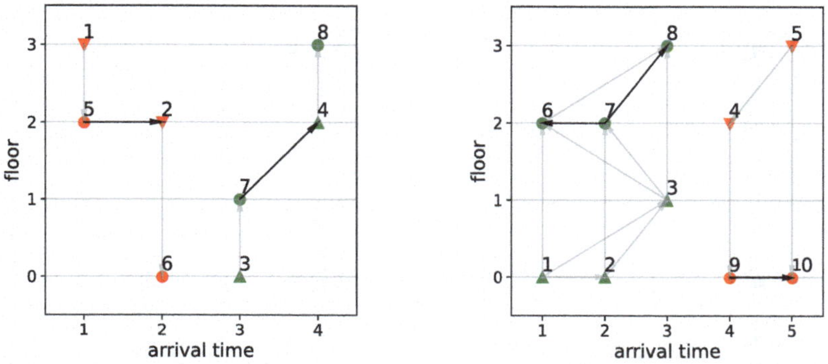

Fig. 7 On the left, (5, 2) verifies (viii) and (7, 4) verifies (vii). On the right, (7, 6) verifies (ix), (9, 10) verifies (x) and (7, 8) verifies (xi). All arcs verify (i)

If $i, j \in D$, then $(i, j) \in A_1$ if one of these propositions is true:

(ix) $f_i = f_j$ and $f_{i-n} = f_{j-n}$ and $a_{j-n} < a_{i-n}$.
(x) $f_i = f_j$ and $f_{i-n} \in]f_{j-n}, f_j[$.
(xi) $f_i \in]f_{j-n}, f_j[$.

If the two passengers are dropped off at the same elevator stop, (ix) and (x) enforce service order. If they are dropped off consecutively by the same elevator on different floors, it means that p_{i-n} is stopping on a floor strictly between the origin and destination floors of p_{j-n}, as expressed by (xi). Figure 7 illustrates this case.

Arcs of A_2

An arc $(i, j) \in A_2$ exists if an elevator can serve consecutively requests i and j with a change of direction, i.e., arcs in A_2 can only be active if the elevator is empty.

Among these arcs are those incident to the source and sink. They correspond to one of these sets:

(xii) $\{(0, j) : j \in P\}$.
(xiii) $\{(i, 2n + 1) : i \in D\}$.
(xiv) $\{(0, 2n + 1)\}$.

All elevators start empty, thus all arcs from the source to a pick-up vertex exists in A_2, as expressed by (xii). Any passenger of an elevator could be its last, thus after any drop-off vertex, the elevator can go to the sink, as expressed by (xiii). It is also possible that an elevator serves no request, hence (xiv). Figure 8 illustrates these sets.

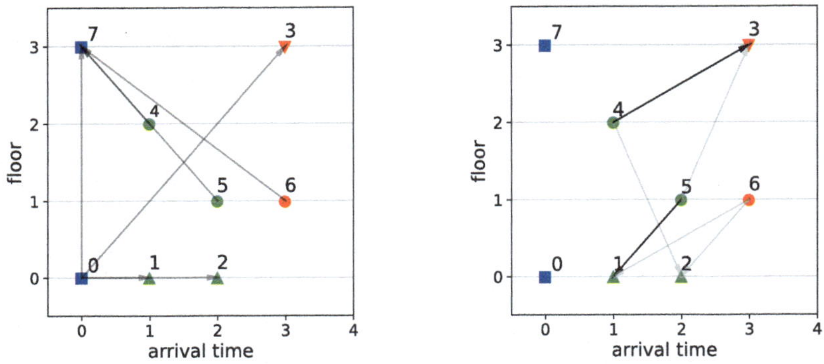

Fig. 8 On the left are arcs incident to the source or sink, as in (xii), (xiii) and (xiv). On the right, (4, 3) verifies (xv) and (5, 1) verifies (xvi)

Otherwise, since an elevator cannot change direction unless it is empty, an arc $(i, j) \in A_2$ must verify $i \in D, j \in P$. If $i \in D, j \in P$, $(i, j) \in A_2$ if one of the following conditions is true:

(xv) $u_i \neq u_j$
(xvi) $u_i = u_j =$ up and $i \neq n + j$ and $f_j < f_i$
(xvii) $u_i = u_j =$ down and $i \neq n + j$ and $f_j > f_i$.

If passengers p_{i-n} and p_j have opposite directions, the elevator will necessarily change direction in i or j to serve the two passengers consecutively, leading to (xv). For each drop-off vertex, all arcs going to a pick-up vertex of the opposite direction exist in A_2. Conditions (xvi) and (xvii) cover the last possibility: if p_{i-n} and p_j have the same direction, the arc (i, j) is in A_2 only if the elevator must change direction to travel from the destination floor of p_{i-n} to the origin floor of p_j. In that case, it will change direction both in i and j. Figure 8 illustrates these rules.

4.1.3 Time Costs on A

Each arc $(i, j) \in A$ is associated with time τ_{ij} which is the time spent between these two requests. The service time s is considered spent in vertices and not arcs. For any arc $(i, j) \in A$, the associated time spent between the two requests i and j is the following:

$$\tau_{ij} = \begin{cases} 0 & \text{if } j = 2n + 1 \\ t_{0j} + o & \text{if } i = 0 \text{ and } j \neq 2n + 1 \\ 0 & \text{if } i, j \in P \cup D \text{ and } f_i = f_j \\ c + t_{ij} + o & \text{if } i, j \in P \cup D \text{ and } f_i \neq f_j \end{cases}$$

where c is the door closing time, o the door opening time, and t_{ij} is the travel time from floor f_i to floor f_j as previously defined. After alighting its last passenger, the empty elevator stays on the same floor, thus, going to vertex $2n+1$ is instantaneous. Initially, the elevators have closed doors so when they leave vertex 0, there is no door closing time. If an elevator serves consecutively several passengers on the same floor (e.g., a passenger leaves the elevator and another enters), the door does not close and reopen between passengers and the elevator does not change floors. Otherwise, τ_{ij} includes the door closing time, the travel time and the door opening time.

4.1.4 Time Windows

Every vertex $i \in P$ is associated with a time window $[a_i, \infty[$ because it can only be served after the call has been registered, when the passenger is waiting on the origin floor. Every vertex $i \in D$ is associated with a time window $[a_{i-n} + \tau_{i-n,i}, \infty[$ because $\tau_{i-n,i}$ is the amount of time needed to go from the pick-up request $i - n$ to the drop-off request i. For $i \in D$, we will denote $a_i = a_{i-n} + \tau_{i-n,i}$, so that the time window of every request is $[a_i, \infty[$.

4.2 The Optimization Model

We define the following decision variables:

- $x_{ij}^k \in \{0, 1\}\ \forall k \in K, (i, j) \in A$: x_{ij}^k is equal to 1 if and only if elevator k travels along (i, j).
- $Q_i \in [\max(0, d_i), \min(Q, Q+d_i)]\ \forall i \in P \cup D$: Q_i holds the load of an elevator when it leaves vertex i (it takes integer values but can be considered continuous).
- $T_i \geq a_i\ \forall i \in P \cup D$: T_i holds the time an elevator arrives in vertex i before serving that request.
- $C_i \in \mathbb{R}\ \forall i \in P \cup D$: C_i holds the energy consumption of an elevator from vertex 0 to vertex i.
- $C_{2n+1}^k \in \mathbb{R}\ \forall k \in K$: C_{2n+1}^k holds the total energy consumption of route k.

We denote $\Gamma^-(i)$ as the predecessors of i in the graph, $\Gamma^+(i)$ as the successors of i, $\Gamma_2^+(i) = \{j \in \Gamma^+(i) : (i, j) \in A_2\}$ as the successors of i restricted to arcs in A_2 and $A' = \{(i, j) \in A : i, j \in P \cup D\}$ as the arcs of A incident to two requests. For clarity, we denote $E_{ij} = (Q_i - Q/2)(f_j - f_i)$ and $E_{0j} = (-Q/2)f_j$

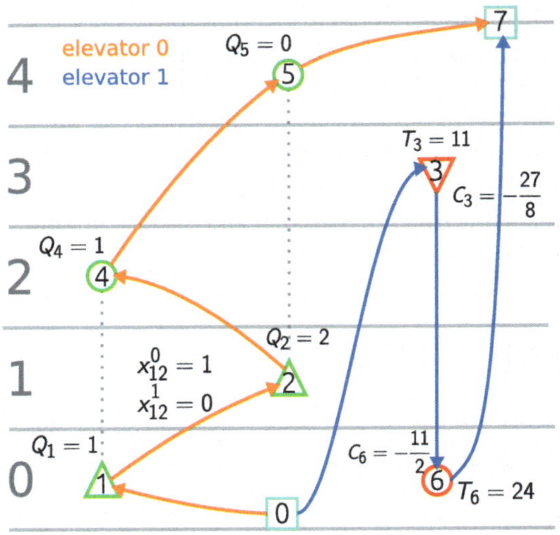

Fig. 9 In this instance, there are 3 passengers and 2 elevators of capacity $Q = 3$. All passengers arrive at time 0. The travel time between two floors is simply 3 s times the absolute difference of floor numbers. The door opening and closing times are 2 s and the service time is 1 s. The load variables are illustrated on the route of elevator 0, and the time and energy consumption variables are illustrated on the route of elevator 1

in the formulation below, but they are not decision variables in the model. Figure 9 illustrates the decision variables on a small instance. The formulation of the problem is:

$$\min \quad \left(\frac{1}{n}\sum_{i\in P}\left(T_{i+n} - a_i\right), \quad \sum_{k\in K} C^k_{2n+1}\right) \tag{1}$$

$s.t.$

$$\sum_{k\in K}\sum_{j\in\Gamma^-(i)} x^k_{ji} = 1 \qquad \forall i \in P \cup D \tag{2}$$

$$\sum_{j\in\Gamma^-(i)} x^k_{ji} = \sum_{j\in\Gamma^+(i)} x^k_{ij} \qquad \forall k \in K, i \in P \cup D \tag{3}$$

$$\sum_{j\in\Gamma^+(i)} x^k_{ij} = \sum_{j\in\Gamma^+(i+n)} x^k_{i+n,j} \qquad \forall k \in K, i \in P \tag{4}$$

$$\sum_{j\in\Gamma^+(0)} x^k_{0j} = 1 \qquad \forall k \in K \tag{5}$$

$$\min(Q, Q + d_i)(1 - \sum_{j\in\Gamma^+_2(i)} x^k_{ij}) \geq Q_i \qquad \forall k \in K, i \in P \cup D \tag{6}$$

$$T_i + \tau_{i,i+n} \leq T_{i+n} \quad \forall i \in P \tag{7}$$

$$(\min(Q, Q + d_i) + d_j)(1 - x_{ij}^k) + Q_j \geq Q_i + d_j \quad \forall k \in K, (i, j) \in A' \tag{8}$$

$$(T + \tau_{ij} + s)(1 - x_{ij}^k) + T_j \geq T_i + \tau_{ij} + s \quad \forall k \in K, (i, j) \in A' \tag{9}$$

$$\tau_{0j}(1 - x_{0j}^k) + T_j \geq \tau_{0j} \quad \forall k \in K, j \in P \tag{10}$$

$$(2C + e_{ij}^{\max})(1 - x_{ij}^k) + C_j \geq C_i + \frac{4}{3}E_{ij} \quad \forall k \in K, (i, j) \in A' \tag{11}$$

$$(2C + e_{ij}^{\max})(1 - x_{ij}^k) + C_j \geq C_i + \frac{3}{4}E_{ij} \quad \forall k \in K, (i, j) \in A' \tag{12}$$

$$(C + e_{0j}^{\max})(1 - x_{0j}^k) + C_j \geq \frac{3}{4}E_{0j} \quad \forall k \in K, j \in P \tag{13}$$

$$2C(1 - x_{i,2n+1}^k) + C_{2n+1}^k \geq C_i \quad \forall k \in K, i \in D \tag{14}$$

$$C(1 - x_{0,2n+1}^k) + C_{2n+1}^k \geq 0 \quad \forall k \in K \tag{15}$$

$$x_{ij}^k \in \{0, 1\} \quad \forall k \in K, (i, j) \in A$$

$$Q_i \in [\max(0, d_i), \min(Q, Q + d_i)] \quad \forall i \in P \cup V$$

$$T_i \in [a_i, T] \quad \forall i \in P \cup V$$

$$C_i, C_{2n+1}^k \in [-C, C] \quad \forall i \in P \cup D, \ \forall k \in K$$

There are two objective functions to minimize: the first one is the average journey time and the second one is the total energy consumption. The journey time of passenger p_i is given by $(T_{i+n} - a_i)$ and the energy consumption of elevator k is given by C_{2n+1}^k. The average journey time represents quality of service is the primary objective, and is optimized first. Then the second objective is optimized, allowing some deterioration of the average journey time or not.

Constraints (2)–(7) are the set of constraints defining feasible routes. Constraints (2) ensure that all requests are served by exactly one elevator. (3) are flow conservation constraints for each elevator, and constraints (4) ensure that the pick-up and drop-off requests of a passenger are served by the same elevator. Constraints (5) set the number of elevators to $|K|$ indirectly. Constraints (6) ensure that an elevator cannot change direction unless empty. Constraints 7 ensure precedence: a passenger is picked up before it is dropped off. Due to (2) and (3), all routes automatically end in $2n + 1$.

Constraints (8) monitor the load of an elevator along its route: if k uses (i, j), it means that k visits i and j, and the constraint becomes:

$$Q_j \geq Q_i + d_j$$

where Q_i and Q_j are the load of k when it leaves i and j respectively. It will be active because the load has to be zero in the last drop-off vertex of the route to have a feasible route. Otherwise, the constraint is deactivated. In particular, if k uses (i, j), then for the other elevators, $x_{ij}^{k'} = 0$ and the constraint is deactivated.

Constraints (9)–(10) are the set of constraints monitoring time variables along the route of an elevator. Constraints (10) address the special case of arcs incident to 0. T is suitable constant so that the constraint is deactivated when $x_{ij}^{k} = 0$. It should an upper bound for T_i. Since T_i is the arrival time of an elevator in i, it is not greater than the total time of an elevator route that has visited at most $2n - 1$ vertex. Finding the longest route in G is a difficult problem so we replace it with:

$$T = (2n - 1)s + \min\left[\sum_{i \in V} \max_{j \in \Gamma^+(i)} \tau_{ij}, \sum_{i \in V} \max_{j \in \Gamma^-(i)} \tau_{ji}\right].$$

The first term is the maximum time spent in vertices and the second term is the sum over V of every longest leaving arc or the sum of every longest entering arcs, according to which one is smaller. If there is no arc leaving (resp. entering) a vertex, we consider that the max is zero. When elevator k travels along (i, j) (resp. $(0, j)$), (9) and (10) become:

$$T_j \geq T_i + \tau_{ij} + s$$
$$T_j \geq \tau_{0j}.$$

The objective function tend to minimize time variables, so T_j will take the smallest value compatible with (9) or (10) and with the bounds of T_j.

Constraints (11)–(15) are the set of constraints monitoring energy consumption in a similar way as time variables. The constant $e_{ij}^{\max}$ is an upper bound of $\frac{4}{3}E_{ij}$ and C is an upper bound of C_i and C_{2n+1}^k, so that the constraints are deactivated when $x_{ij}^k = 0$. If k uses (i, j), (11) and (12) become:

$$C_j \geq C_i + \frac{4}{3}(Q_i - \frac{Q}{2})(f_j - f_i)$$
$$C_j \geq C_i + \frac{3}{4}(Q_i - \frac{Q}{2})(f_j - f_i).$$

If the elevator consumes energy to go from i to j, i.e. $E_{ij} > 0$, then $\frac{3}{4}E_{ij} < \frac{4}{3}E_{ij}$, so constraint (12) is dominated by (11) and (11) is active because of minimization of (1). Conversely, if $E_{ij} \leq 0$, then $\frac{3}{4}E_{ij} \geq \frac{4}{3}E_{ij}$ so constraint (11) is dominated by (12) and (12) is active. That way, C_j always takes the expected value when k uses (i, j). Constraints (13) are specific versions of the same constraints for arcs incident to the source and behave in the same way. There is no energy consumption on an arc incident to $2n + 1$ since we consider that the elevators stay where they are. Constraints (14)–(15) ensure that C^k_{2n+1} takes the value of C_i, with i the last visited vertex of k, and zero if k does not serve any request.

A suitable upper bound $e^{\max}_{ij}$ of $\frac{4}{3}E_{ij}$ for arc $(i, j) \in A'$ is:

$$e^{\max}_{ij} = \frac{4}{3}\left|\frac{Q}{2}(f_j - f_i)\right| = \frac{2Q}{3}|f_j - f_i|.$$

If $i = 0$, it is:

$$e^{\max}_{0j} = -\frac{3}{4}\frac{Q}{2}f_j = -\frac{3Q}{8}f_j.$$

And if $j = 2n + 1$, $e^{\max}_{i,2n+1} = 0\ \forall i \in D$. C is an upper bound on C^k_i that we can derive in a similar way as what was done for T. We set:

$$C = \min\left[\sum_{i \in V} \max_{j \in \Gamma^+(i)} e^{\max}_{ij}, \sum_{i \in V} \max_{j \in \Gamma^-(i)} e^{\max}_{ij}\right].$$

4.3 Symmetry Breaking and Cuts

To break symmetries, we impose that passenger 1 must be served by elevator 0, passenger 2 by elevator 0 or elevator 1, and so on. In practice, it is simply implemented by setting some arc variables to zero: the variables associated with arcs incident to a request are set to zero for the elevators that cannot serve this request. To speed up computation, we also add the following cuts:

$$\sum_{k \in K} x^k_{ij} \leq 1 \qquad \forall (i, j) \in A.$$

5 Experiments and Results

In this Section, we perform computations to study various aspects of the problem. The algorithm used was implemented in Python and uses Gurobi Optimizer version 9.5.2. For all computations, the numerical accuracy of the optimizer was slightly increased as some issues could arise otherwise. The variable branching strategy was set on pseudo shadow prices because it seemed the most efficient on small initial tests. The other parameters have retained their default values. The experiments were performed on a computer equipped with a 2.40 GHz processor and 7 GB of memory. The instances used were either randomly generated or extracted from real instances provided by Schindler. Each experiment is conducted for three types of traffic: up-peak (all passengers are coming from the ground floor), down-peak (all passengers are going to the ground floor) and mixed (any request is possible).

5.1 Efficiency of Symmetry Breaking and Cuts

The first computational experiment is meant to assess the usefulness of the symmetry breaking cuts and the other simple cuts. The building has 8 floors and 3 elevators of capacity 8. We consider three instances (one for each traffic type) of 11 passengers with similar empirical arrival rates. Their arrival rate are close to $\lambda = 0.08$.

The multi-objective optimization is hierarchical: first the average journey time is optimized and then the total energy consumption is optimized without deteriorating the value of the first objective. For each step, a limit of 600 s for CPU time was set. The symmetry detection of Gurobi was deactivated for this experiment only. Four cases are considered: no symmetry breaking and no cuts (*none*), symmetry breaking only (*sym*), cuts only (*cuts*), and both (*all*).

Tables 2 and 3 report the initial gap, the final gap, the time to find the best solution, the total time and the objective values for each step, each instance and each case. The main difficulty of the problem seems to be the optimality proof: in all cases, optimal solutions are found quickly, closing the gap takes most of the time. It is not surprising given the numerous big-M constraints used in the formulation. The gap value is meaningful for the time objective, but for the energy objective it takes huge values until, because the variables can be negative and the bounds are loose.

Note that here we do not allow sub-optimal average journey time, so the second step does not require an optimizer. It would be enough to retrieve the optimal or best solution found at step 1 and compute the energy consumption of this solution.

The different traffic types seem to have different levels of difficulty: the mixed traffic instance is solved faster than the other two in all cases, and up- or down-peak instances are similar. Symmetry breaking significantly improves the runtime, as it avoid exploring entire subtrees of the exploration tree. It does not improve the root gap, since it only deletes a few variables. Cuts do not have much effect on the root gap or the runtime.

Table 2 Computational results: optimizing journey time

Case	Traffic type	Root gap (%)	Final gap (%)	CPU time to best (s)	Total CPU time (s)	Objective
None	Mixed	4.75	0.0	1.1	3.22	22.7555
	Up	11.4	0.0	9.3	20.56	27.404
	Down	11.2	0.0	4.6	21.43	32.8674
Sym	Mixed	5.63	0.0	0.1	2.06	22.7555
	Up	10.6	0.0	2.1	8.15	27.404
	Down	13.6	0.0	3.0	9.46	32.8674
Cuts	Mixed	7.30	0.0	0.1	2.87	22.7555
	Up	10.3	0.0	6.3	20.93	27.404
	Down	19.7	0.0	4.1	24.67	32.8674
All	Mixed	3.61	0.0	0.7	2.56	22.7555
	Up	10.7	0.0	3.9	8.35	27.404
	Down	17.0	0.0	2.8	8.81	32.8674

Table 3 Computational results: optimizing energy consumption

Case	Traffic type	Root gap (%)	Final gap (%)	CPU time to best (s)	Total CPU time (s)	Objective
None	Mixed	5502	0.0	0.2	13.82	21.4167
	Up	5807	0.0	0.5	120.11	26.9167
	Down	57,233	0.0	0.4	23.10	−3.33333
Sym	Mixed	4972	0.0	0.5	3.70	21.4167
	Up	5807	0.0	0.7	76.50	26.9167
	Down	57,233	0.0	0.6	18.33	−3.33333
Cuts	Mixed	5554	0.0	1.2	7.04	21.4167
	Up	5807	0.0	1.1	114.96	26.9167
	Down	57,233	0.0	0.9	40.43	−3.33333
All	Mixed	5502	0.0	1.1	2.00	21.4167
	Up	5807	0.0	2.5	58.16	26.9167
	Down	57,233	0.0	1.1	18.56	−3.33333

5.2 *Size of the Model*

In this experiment and subsequent ones, the symmetry detection of Gurobi is set to default, and symmetry breaking and cuts are always part of the model.

To assess the effects of the different parameters on the number of variables and constraints, we randomly generate instances of various sizes and traffic types. For the generation of mixed traffic, passengers have a probability of 0.4 to be going to or coming from the ground floor. Aside from the ground floor, all floors have equal probability. The arrival times were generated with a Poisson process with parameter $\lambda = 0.1$, which represent the arrival rate of passengers per second (0.1 is 30 passengers every 5 min), and the elevator capacity is $Q = 13$ but these two parameters have no influence on the graph model, and thus no influence on the size of the mathematical model.

The parameters whose influence on the size of the model is analyzed are: the number of passengers n, the number of elevators m, the number of floors F, and the type of traffic. In each case only one parameter among (F, m, n) varies. For each type of traffic, the number of variables that are not forced to zero is plotted against the varying parameter and the number of constraints is plotted against that same parameter. We also represent the theoretical size of the model where G is a complete graph. In that case, the number of variables would be:

$$\begin{aligned} &|K||A| + |P \cup D| + |P \cup D| + |P \cup D| + |K| \\ &= m(2n+2)^2 + 6n + m \end{aligned}$$

and the number of constraints would be:

$$\begin{aligned} &4|K||A'| + 2|K||P \cup D| + 3|K||P| + |K||D| + |P \cup D| + 2|K| + |A| + |P| \\ &= 4m(2n)^2 + 2m(2n) + 3mn + mn + (2n+2)^2 + 3n + 2m \\ &= 16mn^2 + 8mn + (2n+2)^2 + 2m + 3n. \end{aligned}$$

Figures 10, 11 and 12 show the results for each parameter. Overall, the rules for constructing the set of feasible arcs allows for a significant reduction in the number of variables and constraints. Indeed, the greatest factor for the number of variables and constraints is $|A|$.

The number of variables and constraints is quadratic with the number of passengers n and linear with the number of elevators m, consistently with the theoretical formula. The number of floors do not directly affect the number of variables or constraints because of the graph representation and service order assumptions: whether two vertices share the same floor or not, there is at most one arc between them. If they do share the same floor, there is exactly one, which explains why there is more arcs in up-peak and down-peak instances.

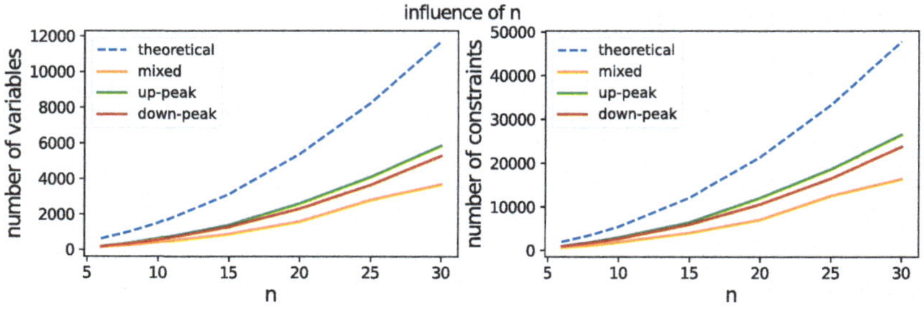

Fig. 10 Number of variables and constraints in the models generated according to different values of n with $F = 15, m = 3$

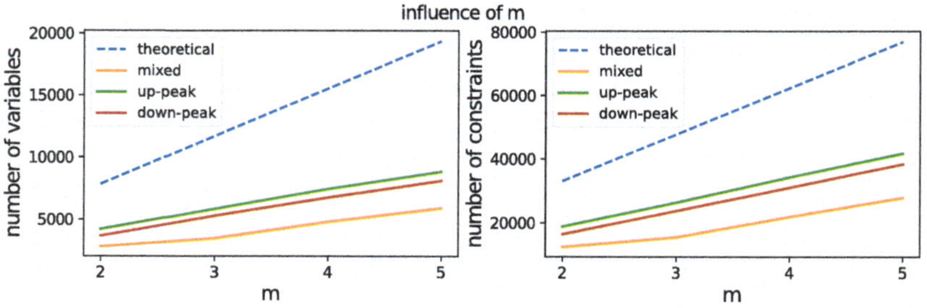

Fig. 11 Number of variables and constraints in the models generated according to different values of m with $n = 30, F = 20$

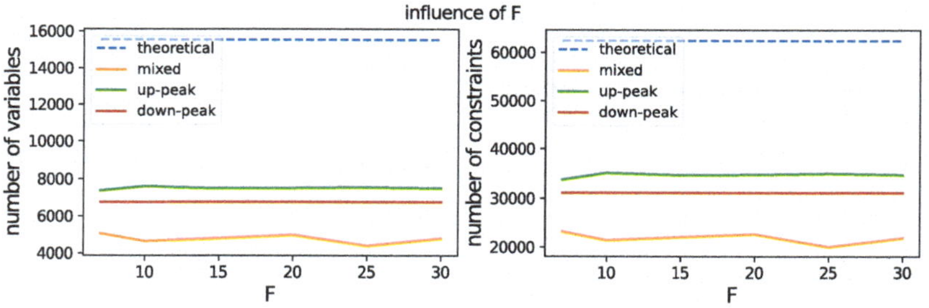

Fig. 12 Number of variables and constraints in the models generated according to different values of F with $n = 30, m = 4$

The up-peak and down-peak instances models are significantly larger than mixed traffic instances. It can explain why they are harder to solve. In these two traffic types, all passengers share the same origin floor (resp. destination floor), which leads to many more arcs than in the case of mixed traffic. Up-peak instances have slightly more variables and constraints than down-peak instance. These results are consistent with the observations of the first experiment.

5.3 Pareto Frontiers

To assess how much energy can be saved when allowing for a sub-optimal average journey time, we compute the Pareto frontier of instances for the three traffic types. To test if some characteristics or situations allow for a greater flexibility in energy saving, we also try it for two values of elevator capacity and two arrival rates. For a given instance, solutions are computed by optimizing the journey time first and then the energy consumption in a second step. The first step is the same for all rounds, but when optimizing the energy consumption, some relative deviation from the optimum of the first step is allowed. Solutions were computed for a relative deviation allowed of 0%, 5%, 10%, 15%, 20%, 25%, 30%, 35%, 40% and 50%. Allowing a deviation too high would be undesirable for quality of service. If the deviation allowed is 15%, the algorithm will return the solution that has the minimum energy consumption among the solutions it found with an average journey time that is within 15% of the value found in the first step. All instances have $n = 15$ passengers. A time limit of 200 s is set for the first step and 1000 s for the second step. Since closing the gap is very difficult, the focus was shifted on finding good feasible solutions in Gurobi parameters.

For each traffic type, one instance is solved for two values of Q. Figure 13 reports the Pareto frontiers obtained. The three instances were chosen to have arrival rates close to $\lambda = 0.09$.

The increase in elevator capacity has a similar effect on the Pareto frontier in the cases of mixed and down-peak traffic: frontier is shifted up as capacity increases. Maybe there were not enough requests that could be pooled together to save energy and account for a heavier elevator car. In the case of up-peak traffic, the two frontiers are close to each other, although higher capacity allow to achieve slightly better journey time for a similar consumption. In all cases, the slope is steeper for lower values of deviation, indicating that significant energy gains can be made with little deterioration of the average journey time.

Figure 14 reports the results when solving two instances of each type with the same elevator capacity $Q = 13$ and different arrival rates λ. Instances were chosen so that for a given type of traffic, one arrival rate is low and the other is high. A low arrival rate is around $\lambda = 0.05$ passenger per second (15 passengers every 5 min)

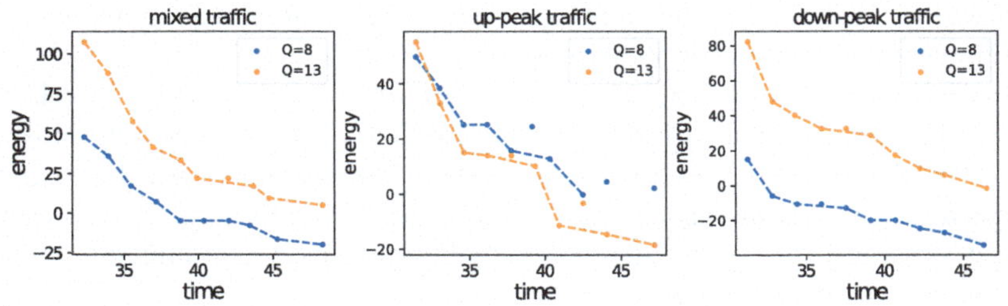

Fig. 13 The estimated Pareto frontier for three instances and two elevator capacities

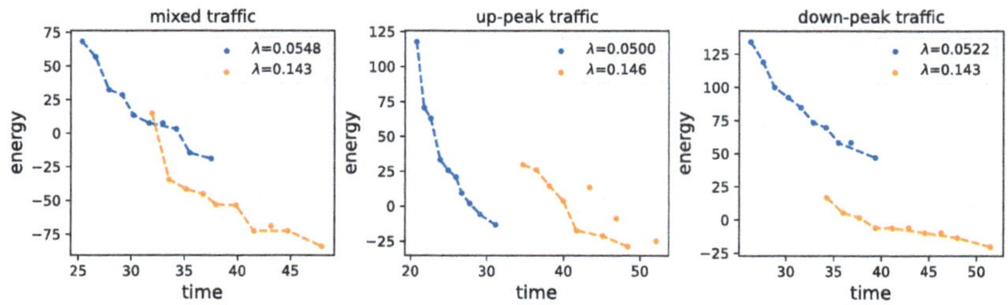

Fig. 14 The estimated Pareto frontier of six instances with low and high arrival rates

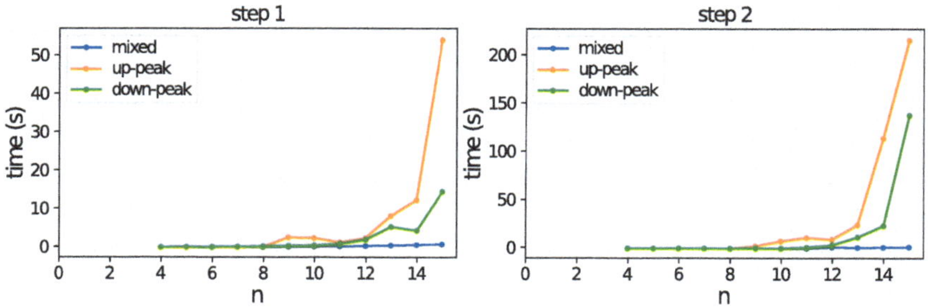

Fig. 15 The runtime to optimality for different traffic types w.r.t n

and a high arrival rate is around $\lambda = 0.14$ passenger per second (42 passengers every 5 min).

The frontier is shifted right and down when the arrival rate increases, meaning that the average journey time is higher and the energy consumption is lower. More requests can be pooled together, but it makes the completion of individual requests longer. We can also notice a steep slope for lower values of deviation for low arrival rates.

5.4 *Running Time According to Parameters*

The purpose of the following experiments is to assess the time complexity of the problem w.r.t. traffic type and some parameters: the number of passengers n, the arrival rate λ and the number of elevators m.

To study the time complexity with n, instances for each type of traffic are selected so that they have similar arrival rates over the first 15 passengers, very close to $\lambda = 0.06$. They are solved to optimality for the n first passengers, $n \in \{4, \ldots, 15\}$. There are 3 elevators of capacity 13. The first step is still optimizing the average journey time, and the second step is optimizing the energy consumption without deteriorating the first objective.

Figure 15 shows the runtime for each instance and for the two steps according to n. A time limit of 1 h was set for each step. The runtime increases exponentially

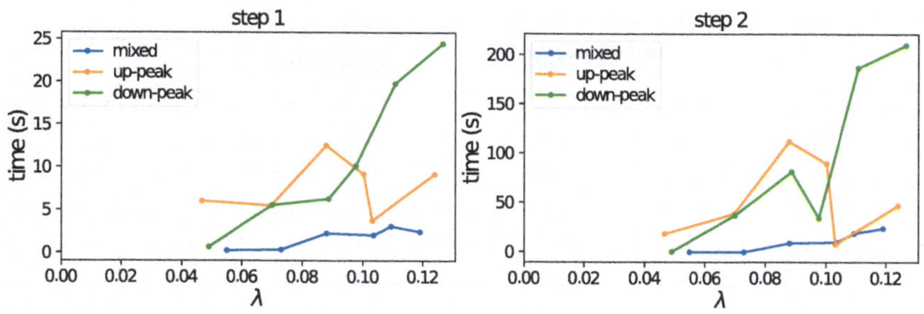

Fig. 16 The runtime to optimality for different traffic types w.r.t λ

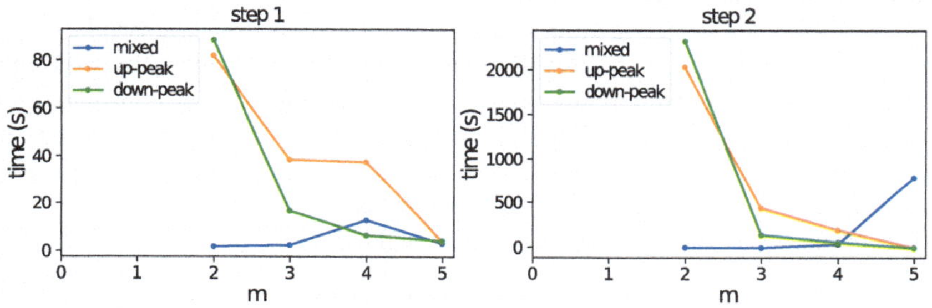

Fig. 17 The runtime to optimality for different traffic types w.r.t m

with n, and it increases faster for down-peak and up-peak traffics than mixed traffic. The second step is much longer than the first to close the gap. The up-peak instance is more difficult than the down-peak one, and mixed traffic is much easier.

The second parameter studied is the arrival rate: instances of the three traffic types with $n = 10$ passengers are selected for various values of λ and their running time is plotted against λ. There is also 3 elevators of capacity 8.

Figure 16 shows the runtime for each instances and arrival rate. A time limit of 1 h was set for each step. The curves are not smooth but we can see that the time increases faster for up-peak and down-peak instances than for mixed instances.

The last parameter studied is the number of elevators m. Instances chosen have an arrival rate λ close to 0.1, $n = 10$ and the elevator capacity is 8 regardless of m.

Although having more elevators increases the size of the model, it makes it easier to solve. The case $m = 2$ is the most difficult one, except for mixed traffic (see Fig. 17).

Overall, there are clear differences between traffic types: the case of mixed-traffic is easier than the two others which are similar. These observations are consistent with the theoretical findings of Ruokokoski et al. (2015). The runtime is extremely sensitive to parameters n and λ.

5.5 Effect of Saturated Elevator Capacity

Elevator capacity is usually slightly too large for a normal traffic, so that elevators can handle special cases like a wheelchair or ambulance stretcher, meaning that elevators are often not saturated unless there is a lot of traffic. The purpose of this experiment is to observe what happens when elevator capacity is saturated. For each type of traffic, an instance with $n = 11$ passengers and arrival rate λ close to 0.11 is chosen. A higher arrival rate ensure than several passengers will travel in the same elevator at the same time. There are 8 floors and 3 elevators. We consider three cases: elevator capacity $Q = 8$ (*default*), decreased elevator capacity $Q = 2$ (*small* Q) and increased passengers' demands $d_i \in \{3, -3\}$ with $Q = 8$ (*large* d).

Table 4 gives the objective values and runtime for each step. The two last cases give the same solutions, as the elevator can carry two passengers at the same time in both cases. The average journey time is the same, but the energy consumption is different because it depends on the elevator capacity and load. Between the default case and the other two, the solution changes. The energy consumption is heavily influenced the average journey time slightly increases.

For instance, the solution for mixed traffic in the default case is:

$$0 \xrightarrow{0} 1 \xrightarrow{1} 12 \xrightarrow{0} 5 \xrightarrow{1} 16 \xrightarrow{0} 10 \xrightarrow{1} 21 \xrightarrow{0} 23$$

$$0 \xrightarrow{0} 2 \xrightarrow{1} 3 \xrightarrow{2} 13 \xrightarrow{1} 14 \xrightarrow{0} 9 \xrightarrow{1} 11 \xrightarrow{2} 22 \xrightarrow{1} 20 \xrightarrow{0} 23$$

$$0 \xrightarrow{0} 4 \xrightarrow{1} 15 \xrightarrow{0} 6 \xrightarrow{1} 7 \xrightarrow{2} 8 \xrightarrow{3} 19 \xrightarrow{2} 18 \xrightarrow{1} 17 \xrightarrow{0} 23$$

where the number of passengers in the elevator is indicated above the arc. The route of elevator 2 is no longer possible if $Q = 2$ or $d_i \in \{-3, 3\}$. The solution becomes:

$$0 \xrightarrow{0} 1 \xrightarrow{1} 12 \xrightarrow{0} 6 \xrightarrow{1} 7 \xrightarrow{2} 18 \xrightarrow{1} 17 \xrightarrow{0} 23$$

Table 4 Computational results: elevator capacity

		Step 1		Step 2	
Case	Traffic type	Time (s)	Objective	Time (s)	Objective
Default	Mixed	2.53	33.8363	10.55	18.8333
	Up	140.07	31.0405	437.06	10
	Down	31.83	34.3043	96.62	18.3333
Small Q	Mixed	4.08	35.1918	21.32	3.08333
	Up	61.92	31.1068	212.76	32
	Down	16.68	35.4801	57.20	−33
Large d	Mixed	6.02	35.1918	24.90	8.66137
	Up	89.95	31.1068	398.19	85.1667
	Down	24.14	35.4801	146.43	−86.1667

$$0 \xrightarrow{0} 2 \xrightarrow{1} 3 \xrightarrow{2} 13 \xrightarrow{1} 14 \xrightarrow{0} 9 \xrightarrow{1} 10 \xrightarrow{2} 21 \xrightarrow{1} 20 \xrightarrow{0} 23$$

$$0 \xrightarrow{0} 4 \xrightarrow{1} 15 \xrightarrow{0} 8 \xrightarrow{1} 19 \xrightarrow{0} 5 \xrightarrow{1} 16 \xrightarrow{0} 11 \xrightarrow{1} 22 \xrightarrow{0} 23.$$

All elevator routes have been modified compared to the default case, both in assignment and order, but it keeps the average journey time very close to the previous one for the three traffic types.

Regarding runtime, decreasing capacity or increasing demands does not affect mixed traffic much, but it improves the runtime for up- and down-peak instances, as it reduces the number of possible routes. However reducing Q is more efficient than increasing d_i, since it tightens the bounds in big-M constraints.

6 Conclusions and Future Research

We extended the classical flow formulation for the EDPDC to include the energy consumption of the elevators while keeping a linear formulation, and we applied it to small instances. The model for the energy consumption is that of a counterweight elevator, which is the most common type of elevator, and it takes into account energy losses during transmission. We perform true multi-objective optimization and study the Pareto frontier of several instances to try to identify factors that influence energy savings, concluding that significant energy gains can be made with an acceptable increase of journey time.

We also studied the difficulty of the problem according to some parameters, identifying key differences between traffic types and arrival rates. These finding support the need for different strategies that adapt to the situation when designing elevator group control systems that optimize both time and energy. Analyzing the detailed changes in solutions when allowing for a small deterioration of the journey time could help to devise new rules for multi-objective heuristics.

While the presented model provides a useful benchmark for the EDPDC, the assumption of complete information is not realistic and an exact method is not applicable in practice. The instances solved to optimality had at most 14 passengers. Developing and implementing advanced solution methods to solve this problem could lead to solving larger and more instances, and have a more precise understanding of the characteristics of this problem. For example, valid inequalities can be developed to use in a branch-and-cut algorithm, as their usefulness has been demonstrated on a closely related problem (Cordeau, 2006). The EDPDC is a highly constrained problem, thus Constraint Programming could be used to solve larger instances.

In a final note, Bernard Gendron and Michel Bierlaire have been close friends since 1993 when they were working on their PhD. Their paths have crossed many times, mainly during conferences around the world. It is only recently that they had the opportunity to work together on a research project, during Bernard's academic visit to EPFL in 2016. Some of this research is published in Pacheco et al. (2021, 2022), and is also the subject of another chapter in this book. Although Bernard was not directly involved in it, we chose the elevator project because we think it exemplifies the type of problems Bernard liked to solve. We would have loved to have his opinion.

Acknowledgments This research was funded by Schindler, which provided the baseline data.

References

Beamurgia, M., Basagoiti, R., Rodríguez, I., & Rodriguez, V. (2016). A modified genetic algorithm applied to the elevator dispatching problem. *Soft Computing, 20*, 3595–3609.

Beamurgia, M., Basagoiti, R., Rodríguez, I., & Rodríguez, V. (2022). Improving waiting time and energy consumption performance of a bi-objective genetic algorithm embedded in an elevator group control system through passenger flow estimation. *Soft Computing, 26*(24), 13673–13692.

Boyer, V., Gendron, B., & Rousseau, L. M. (2014). A branch-and-price algorithm for the multi-activity multi-task shift scheduling problem. *Journal of Scheduling, 17*, 185–197.

Closs, G. D. (1970) The computer control of passenger traffic in large lift systems. PhD thesis, University of Manchester Institute of Science and Technology.

Cordeau, J. F. (2006). A branch-and-cut algorithm for the dial-a-ride problem. *Operations Research, 54*(3), 573–586.

Dayarian, I., Crainic, T. G., Gendreau, M., & Rei, W. (2015). A column generation approach for a multi-attribute vehicle routing problem. *European Journal of Operational Research, 241*(3), 888–906.

Fontaine, P., Crainic, T. G., Jabali, O., & Rei, W. (2021). Scheduled service network design with resource management for two-tier multimodal city logistics. *European Journal of Operational Research, 294*(2), 558–570.

Hu, Z., Liu, Y., Su, Q., & Huo, J. (2010). A multi-objective genetic algorithm designed for energy saving of the elevator system with complete information. In *2010 IEEE International Energy Conference* (pp. 126–130). IEEE.

Lahrichi, N., Gabriel Crainic, T., Gendreau, M., Rei, W., & Rousseau, L. M. (2015). Strategic analysis of the dairy transportation problem. *Journal of the Operational Research Society, 66*(1), 44–56.

Pacheco, M., Bierlaire, M., Gendron, B., & Sharif Azadeh, S. (2021) Integrating advanced discrete choice models in mixed integer linear optimization. *Transportation Research Part B: Methodological, 146*, 26–49. Accepted Feb 09, 2021. https://doi.org/10.1016/j.trb.2021.02.003

Pacheco, M., Gendron, B., Sharif Azadeh, S., & Bierlaire, M. (2022). A lagrangian decomposition scheme for choice-based optimization. *Computers & Operations Research, 148*, 105985. Accepted Jul 29, 2022. https://doi.org/10.1016/j.cor.2022.105985

Ruokokoski, M., Ehtamo, H., & Pardalos, P. M. (2015). Elevator dispatching problem: A mixed integer linear programming formulation and polyhedral results. *Journal of Combinatorial Optimization, 29*, 750–780.

Ruokokoski, M., Sorsa, J., Siikonen, M. L., & Ehtamo, H. (2016). Assignment formulation for the elevator dispatching problem with destination control and its performance analysis. *European Journal of Operational Research 252*(2), 397–406.

Sorsa, J. S., Ehtamo, H., Siikonen, M., Tyni, T., & Ylinen, J. (2009). The elevator dispatching problem. Retrieved from https://www.academia.edu/28505639/The_Elevator_Dispatching_Problem

Tanaka, S., Uraguchi, Y., & Araki, M. (2005). Dynamic optimization of the operation of single-car elevator systems with destination hall call registration: Part i. Formulation and simulations. *European Journal of Operational Research, 167*(2), 550–573.

Torres, F., Gendreau, M., & Rei, W. (2022a). Crowdshipping: An open VRP variant with stochastic destinations. *Transportation Research Part C: Emerging Technologies, 140*, 103677.

Torres, F., Gendreau, M., & Rei, W. (2022b). Vehicle routing with stochastic supply of crowd vehicles and time windows. *Transportation Science, 56*(3), 631–653.

Tyni, T., & Ylinen, J. (2006). Evolutionary bi-objective optimisation in the elevator car routing problem. *European Journal of Operational Research, 169*(3), 960–977.

Vodopija, A., Stork, J., Bartz-Beielstein, T., & Filipic, B. (2018). Model-based multiobjective optimization of elevator group control. In *International Conference on High-Performance Optimization in Industry, HPOI* (pp. 43–46).

Vodopija, A., Stork, J., Bartz-Beielstein, T., & Filipič, B. (2022). Elevator group control as a constrained multiobjective optimization problem. *Applied Soft Computing, 115*, 108277.

Flow-Based Robustness in Consistent Home Care Service Delivery

Paola Cappanera and Maria Grazia Scutellà

1 Introduction

The design of Home Care Services is receiving strong interest and investment, both in the academic literature and also in national and international projects dedicated to the subject. This is due to the ever-increasing age of the population and the consequent need to reduce hospitalization costs. Cost containment, however, is not the only objective pursued by private and public companies. In fact, also the design of patient-centered services appears to be very relevant from several perspectives. The term *Home Care Services* refers to medical, paramedical, and social services that may be delivered to patients' homes, and it addresses many complex decisions such as caregiver-to-patient assignment, scheduling of patient requests, and caregiver routing. Typically, such decisions have been considered separately in the related literature, usually using heuristic approaches (e.g., Hertz & Lahrichi, 2009; Bertels & Fahle, 2006; Bredström & Rönnqvistr, 2008; Elbenani et al., 2008; Eveborn et al., 2006; Fikar & Hirsch, 2015; Liu et al., 2014; Mankowska et al., 2014; Nickel et al., 2012; Rest et al., 2012). On the other hand, a few exact approaches have been proposed that deal simultaneously with the different planning aspects in Home Care. They include Gamst and Jensen (2012); Rasmussen et al. (2012) (on a daily perspective), Trautsamwieser and Hirsch (2014), and Cappanera and Scutellà (2015) where the concept of *pattern*, i.e., a template for scheduling multiple visits, has been introduced to coordinate diverse decisions in a multiple-day

P. Cappanera
Dipartimento di Ingegneria dell'Informazione, Università degli Studi di Firenze, Firenze, Italy
e-mail: paola.cappanera@unifi.it

M. G. Scutellà (✉)
Dipartimento di Informatica, Università di Pisa, Pisa, Italy
e-mail: maria.grazia.scutella@unipi.it

T. G. Crainic et al. (eds.), *Combinatorial Optimization and Applications*,
International Series in Operations Research & Management Science 358,
https://doi.org/10.1007/978-3-031-57603-4_15

planning horizon. Patterns proved to be useful also in the design of decomposition algorithms (Yalçındağ et al., 2016). We refer to Fikar and Hirsch (2017), Grieco et al. (2021), and Di Mascolo et al. (2021) for recent reviews on both heuristic and exact approaches to planning decisions in Home Care.

Data uncertainty in Home Care has been taken into account quite recently, but the literature on this subject is now rich. In particular, managing demand uncertainty (Cappanera & Scutellà, 2017; Cappanera et al., 2018) appears to be very relevant in Home Care since both the set of patients under treatment and their care plan can change dynamically over time, causing patient or visit cancellations as well as additional patient visits. On the other hand, a few works have addressed *consistency* in Home Care. Consistency means service regularity and it is crucial also in some related problems. According to the classification proposed in Kovacs et al. (2014), consistency involves at least three features: (i) arrival time consistency, (ii) person-oriented consistency, and (iii) delivery consistency. Among these, arrival time consistency and person-oriented consistency have a prominent role in Home Care. In fact, schedules which are consistent with respect to the caregiver arrival time, i.e., such that almost all the visits are performed in the same time slot, are very much appreciated by patients who, in this setting, are often fragile and thus particularly sensitive to changes in their daily routines. An interesting kind of arrival time consistency is addressed in Feillet et al. (2014), where the authors minimize the maximum number of different time slots a customer is visited in. Moreover, person-oriented consistency positively impacts on service quality both from the patient and the caregiver side. Indeed, person-oriented consistency, often referred to as care continuity in the Home Care literature (Maya Duque et al., 2015; Wirnitzer et al., 2016), guarantees that almost the same set of caregivers take care of a patient in the whole planning horizon. Care continuity simplifies the communication between patients and caregivers, thus making it possible to better organize working activities and to provide a high-quality service. Consequently, both the patient and the caregiver satisfaction appear to be improved (Frey et al., 2013).

In Cappanera and Scutellà (2022), demand uncertainty, arrival-time consistency and person-oriented consistency in Home Care have been managed simultaneously, while jointly optimizing assignment, scheduling and routing decisions over a multiple-day time horizon. The resulting problem, named the Consistent Robust Home Care Problem, is very complex indeed because strongly interrelated decisions are made in the presence of context-specific requirements, side constraints, and in an uncertain environment. For this reason, an approximate formulation of the addressed form of robustness was proposed in Cappanera and Scutellà (2022), rather than an exact one.

The aim of this paper is to present an exact formulation of the Consistent Robust Home Care Problem. Specifically, a non-standard cardinality-constrained model, which is based on a constrained flow formulation of the uncertainty aspects, is proposed to model exactly the problem. Precisely, the uncertainty paradigm is modeled in terms of a constrained longest path on a suitably defined auxiliary graph. Although more structured than the traditional models used so far in cardinality-constrained robust formulations, the constraint matrix of the proposed flow-based inner model is proved to be still totally unimodular, as in more classical formulations, thus allowing to derive an exact Mixed Integer Linear Programming (MILP) formulation to the Consistent Robust Home Care Problem. This testifies the power and the relevance of network design (Crainic et al., 2021) in addressing applications related to the fields of logistics, transportation, and, more specifically, health care (Beaulieu et al., 2000).

The plan of the paper is the following. After a review of the literature focusing on uncertainty and consistency issues in Home Care, provided in Sect. 2, the Consistent Robust Home Care Problem is described in Sect. 3. Section 4 presents the exact robust mathematical model we propose to formulate the problem. Some cuts and valid inequalities, which proved to be effective in enhancing the flow-based formulation, are also presented. Section 5 reports the results of a preliminary computational experience, where a first comparison with the approximate robust approach in Cappanera and Scutellà (2022) is performed. Finally, Sect. 6 concludes the paper by outlining some promising avenues of research.

2 Literature: Uncertainty and Consistency in Home Care

The relevance of addressing data uncertainty in Home Care is testified by the significant number of papers that have been published on the topic in the last years. Recently, multi-period staffing, assignment, routing, and scheduling of caregivers for home care services have been addressed in Naderi et al. (2023) by considering uncertainty in service and travel times via a hybridized form of interval and polyhedral robustness. A novel logic-based Benders branching-decomposition algorithm is proposed with the aim of minimizing the caregiver fixed and overtime costs. Home Care scheduling and routing with stochastic service times have been addressed also in Yuan et al. (2014). A stochastic programming model with recourse is proposed, which is then transformed into a set partitioning problem and an elementary shortest path problem with resource constraints, by approximating the expected penalty function for late arrival at a patient. A branch and price approach is designed and validated by means of numerical experiments. Stochastic service times are considered also in Rodriguez et al. (2015), where the authors propose a two-stage approach, based on integer linear stochastic programming, to deal with staff dimensioning in Home Care, and in Liu et al. (2019), where a branch and price algorithm is presented by considering stochastic travel times in addition to stochastic service times. See also Du and Zhang (2022), for a cross-regional scheduling and routing problem in Home Care with stochastic service times in a single-day planning horizon, and Shiri et al. (2021), for a three-phase methodology for Home Care routing and scheduling dealing with uncertain service times.

Following the work in Cappanera et al. (2018), demand uncertainty in Home Care is addressed in Zheng et al. (2021) in the context of a joint service authorization and capacity planning problem. The problem is formulated as a two-stage stochastic programming with recourse that maximizes the expected total revenue, where service authorization and capacity planning are modeled by the first-stage decisions, while the service hour allocation is modeled in terms of second-stage decisions taking into account the demand realizations. A supergradient-based nested decomposition algorithm that exploits the problem structure is then proposed. Demand uncertainty is also addressed in Restrepo et al. (2020), where a two-stage stochastic model for employee staffing and scheduling in Home Care is presented. The first-stage decisions model the staffing and the scheduling of caregivers in geographic districts, while the second-stage decisions are related to the temporary reallocation of caregivers to neighboring districts, in order to allow caregivers to work also on days off thus covering under- and over-demand. Uncertainty on the availability of the caregivers is considered also in Nguyen et al. (2015), where a matheuristic is proposed to handle the case where caregivers are unexpectedly unable to operate. Moreover, uncertainty related to social aspects is considered in Shiri et al. (2023), where the opening of efficient health centers, and the routing and scheduling of home health care services by taking into account corporate social responsibility and efficiency, are addressed by means of a robust-fuzzy approach.

Just considering Home Care assignment and scheduling decisions under uncertainty, the problem of assigning a set of patients to a set of caregivers over a time horizon is addressed in Carello and Lanzarone (2014) by imposing that exactly one caregiver must be assigned to each patient and assuming that the amount of working time required by each patient is uncertain. A cardinality-constrained robust assignment model is proposed. Also, a robust assignment model which includes the time-dependency of the patient working times, and which is based on the implementor-adversarial framework, is proposed in Carello et al. (2017). An implementor/adversary algorithm is used also in Holte and Mannino (2013) for a cyclic scheduling problem arising in health care, while a dynamic scheduling framework to assist in the assignment of health practitioners to patients who arrive stochastically over time is presented in Cire and Diamant (2022). There, the assignment decisions are modeled in terms of a discrete-time, rolling horizon, infinite-stage Markov decision process, which is then approximated via a dynamic programming approach. We also mention Du et al. (2019), where unexpected events are addressed via an effective real-time scheduling model aiming at minimizing the total required scheduling time.

By considering time consistency issues, firstly introduced in Groër et al. (2009) in the more general context of VRP, the goal is to visit a patient at about the same time as often as they need a service, and preferably by the same group of caregivers. To the best of our knowledge, only a few papers have addressed a consistency different from the care continuity one in Home Care, which on the contrary has been widely taken into account. Regarding care continuity, see for instance Moosavi et al. (2022), where the problem of minimizing the number of distinct staff members serving each room in residential care facilities is addressed by considering the Covid19 pandemic context, and Savaşer and Kara (2022), where care continuity is addressed in mobile healthcare services to rural areas. Considering other forms of time consistency, in Bennett and Erera (2011) the authors studied a dynamic periodic nurse routing and scheduling problem with visit time consistency constraints. A set of patients, revealed dynamically, must be visited according to a prescribed weekly frequency over a prescribed number of weeks, and each visit must be assigned a precise appointment time chosen from a fixed menu of allowable appointments. The consistency requirements impose that weekly visits, which should occur according to an allowable visit-day combination for the patient, must repeat on the same days and times throughout the patient's service duration. A rolling horizon myopic planning approach for the single nurse variant is proposed and experimented with. The approach relies on a capacity-based insertion heuristic, which explicitly considers the remaining available time in the nurse's schedule when inserting the current patient requests. On the other hand, in Chun-Cheng et al. (2018) the authors address the case where each patient is associated with a list of tasks to be accomplished each week, and they have to indicate a fixed time slot of a certain day of each week within which all their tasks must be executed. Specifically, two time slots are considered: morning (8:00 a.m.-2:00 p.m.) and afternoon (2:00 p.m.-8:00 p.m.). This form of consistency can be viewed as a special case of the one addressed in Cappanera and Scutellà (2022), since for each patient all the visits occur in a unique time slot, which is fixed according to the patient's indications. On the other hand, in Cappanera and Scutellà (2022) the patient time slot is optimally selected via a robust matheuristic approach dealing with the scheduling of the patient visits, some of which can be uncertain, in a given time horizon, with the assigning of adequate caregivers to the patients, and with the determination of the routing of the caregivers for each day of the planning horizon. Finally, a scheduling problem in Home Care where patients arrive dynamically over time and, for the sake of service continuity, they have to be visited on the same day and time each week during their episode of care, is heuristically approached in Demirbilek et al. (2019).

3 The Consistent Robust Home Care Problem

The Consistent Robust Home Care Problem, introduced in Cappanera and Scutellà (2022), extends the Robust Home Care Problem (Cappanera et al., 2018) by managing consistency in addition to robustness. The problem considers a set of n patients, asking for service in a multiple-day planning horizon, say W. Each patient requires a set of services (hereinafter called requests), each characterized by a skill. To perform them, a set of skilled caregivers, based at a "depot", is available. The set of skills, say K, is assumed to be hierarchically organized, so that a caregiver with skill k can work requests with skill up to k.

Demand uncertainty is taken into account. Specifically, some patient requests are certain, i.e., they will certainly occur and therefore must be satisfied, whereas others are uncertain. Each patient j is thus associated with both a *certain care plan* $\bar{r}_j$, specifying, for each skill, the number of the certain requests of j in the planning horizon, and with an *uncertain care plan* $\tilde{r}_j$, indicating the number of their uncertain requests in the planning horizon, again for each skill. In particular, $\tilde{r}_j$ or $\bar{r}_j$ could be missing for a patient j: the first case verifies if j is a deterministic patient, whereas the latter can be used to model j as a new patient. To reflect this double nature of the patient requests, i.e., certain or uncertain, in Cappanera and Scutellà (2022) each patient j is modeled by a pair of nodes in the logistic network: by a node $j \in \bar{N} = \{1, \ldots, n\}$, modeling the certain nature of the patient, and therefore having associated the related certain care plan $\bar{r}_j$, and by a node $\tilde{j} \in \tilde{N} = \{\tilde{1}, \ldots, \tilde{n}\}$, modeling the uncertain nature of the patient, and therefore having associated the related uncertain care plan $\tilde{r}_j$. Hereafter, $N = \bar{N} \cup \tilde{N}$ will denote the set of nodes of the resulting logistic network corresponding to patients. Notice that this network is composed of up to $2n$ nodes plus the depot.

The requests of each patient, expressed by the associated certain and uncertain care plans, have to be scheduled in the planning horizon. Furthermore, caregivers of appropriate skill have to be assigned to patients to serve their requests, and the caregiver routing has to be decided as well. As in Cappanera et al. (2018), scheduling, assignment, and routing decisions are coordinated by means of the concept of *pattern*, introduced in Cappanera and Scutellà (2015), which denotes a template for scheduling the patient requests in W. More precisely, it is assumed that, for each $j \in N$, the requests expressed by the associated care plan can be operated according to a set P_j of *a priori* given patterns. Assuming that each patient requires at most one visit per day, each pattern $p \in P_j$ is such that $p(d) = 0$ if no service is offered on day d, while $p(d) = k$ indicates that a request of skill k will be scheduled on day d according to pattern p. Pattern p has to be compatible with the care plan associated with j, guaranteeing the required number of visits of appropriate skill.

In the Consistent Robust Home Care problem, a pattern $p \in P_j$ is selected for each $j \in N$, thus scheduling the requests of j in the planning horizon. Indeed, certain or uncertain requests are scheduled depending on whether $j \in \bar{N}$ or $j \in \tilde{N}$. Hereafter, $P = \cup_{j \in N} P_j$ will denote the set of all the available patterns. These patterns are determined via an auxiliary pattern generation optimization model and are provided to the Consistent Robust Home Care problem as input data.

Example To make the notation clearer, we introduce the following toy example. The same example will be used consistently hereinafter, to exemplify the main concepts. Consider a set of four patients: the first three of them have both certain and uncertain requests, which are expressed in terms of a certain and an uncertain care plan respectively, while the fourth patient has only a certain care plan. Thus, the logistic network consists of 8 nodes: the depot, i.e., node 0, 4 nodes representing the certain nature of patients, i.e., set $\bar{N} = \{1, 2, 3, 4\}$, and 3 nodes representing the uncertain nature of patients, i.e., set $\tilde{N} = \{\tilde{1}, \tilde{2}, \tilde{3}\}$. Suppose that there are two types of requests: requests involving skill 1 and requests involving skill 2. Thus, $K = \{1, 2\}$. Consider a 5-day planning horizon, in which day $d = 0$ corresponds to Monday and $d = 4$ corresponds to Friday. Thus, $W = \{0, 1, 2, 3, 4\}$. Figure 1 provides the care plan, both certain and uncertain, associated with the nodes of the logistic network and a set $P = \{p_1, \ldots, p_7\}$ of possible patterns to schedule the patient requests in W. For each node, the patterns in P which are compatible with the care plan associated with the node are also reported.

	k = 1	k = 2	Compatible Patterns
$\bar{r}_1$	1	0	$p_3\,p_4\,p_5\,p_6\,p_7$
$\bar{r}_2$	1	2	$p_1\,p_2$
$\bar{r}_3$	1	0	$p_3\,p_4\,p_5\,p_6\,p_7$
$\bar{r}_4$	1	0	$p_3\,p_4\,p_5\,p_6\,p_7$

	k = 1	k = 2	Compatible Patterns
$\tilde{r}_1$	1	2	$p_1\,p_2$
$\tilde{r}_2$	1	0	$p_3\,p_4\,p_5\,p_6\,p_7$
$\tilde{r}_3$	1	0	$p_3\,p_4\,p_5\,p_6\,p_7$

	d = 0	d = 1	d = 2	d = 3	d = 4
p_1	1	0	0	2	2
p_2	2	0	1	0	2
p_3	1	0	0	0	0
p_4	0	1	0	0	0
p_5	0	0	1	0	0
p_6	0	0	0	1	0
p_7	0	0	0	0	1

Fig. 1 Care plans and patterns

The care plans express the number of requests for each skill. For example, the first patient has a certain care plan $\bar{r}_1$ characterized by one request of skill 1, which is associated with node 1 in the logistic network, and an uncertain care plan $\tilde{r}_1$ characterized by one request of skill 1 and two requests of skill 2, which is associated with node $\tilde{1}$. The request in the certain care plan will be performed with certainty, whereas the three requests in the uncertain care plan, being uncertain, may be realized partly or totally, i.e., 0, 1, 2, or 3 uncertain visits will be performed according to the optimal solution of the problem. Patterns p_1 and p_2 can be used to schedule three visits, one involving skill 1 and 2 involving skill 2. Thus, they are compatible with care plans $\tilde{r}_1$ and $\bar{r}_2$, and they can be assigned to node $\tilde{1}$ and to node 2. According to pattern p_1, the request of skill 1 is scheduled on Monday, whereas the two requests of skill 2 are scheduled on Thursday and Friday respectively. The assignment of patterns to nodes must also guarantee that no more than one visit be assigned to a patient on each day in the planning period, and to make the assignment feasible the certain and uncertain nature of the patient have to be considered jointly. For example, considering patient 2, if p_1 is assigned to node 2, only patterns p_4 or p_5 can be assigned to $\tilde{2}$, whereas pattern p_2 is compatible only with p_4 or p_6. The same holds true for patient 1. Patterns p_1 and p_3 cannot be assigned to the two nodes corresponding to the same patient, because in that case, two visits would occur on the same day (Monday). Patterns p_3, p_4, p_5, p_6, and p_7 can be assigned to nodes 1, $\tilde{2}$, 3, $\tilde{3}$, and 4 which require one visit of skill 1. The same pattern cannot be assigned to nodes 3 and $\tilde{3}$, since otherwise the unique certain request and the unique uncertain request of the third patient would be scheduled on the same day.

Concerning the kind of robustness addressed to manage demand uncertainty, the Consistent Robust Home Care Problem considers, at the same time, (i) Γ-robustness (Bertsimas & Sim, 2004), (ii) *sequence-preserving robustness* (Cappanera et al.,

2018), and (iii) *consistency-aware robustness* (Cappanera & Scutellà, 2022). Specifically, all the certain and all the uncertain requests of the patients are scheduled in the planning horizon according to the selected patterns, they are assigned to the caregivers, and the tours of the caregivers are determined. However, in order to achieve a compromise between the degree of robustness and solution cost, it is assumed that, whereas all the certain requests must be satisfied, at most Γ uncertain requests (or, equivalently, at most Γ nodes from set $\tilde{N}$) in each tour may realize and therefore have to be satisfied, where Γ is a given parameter. Γ-robustness is thus addressed. Given a tour, the subtours obtained from that tour by considering all its certain requests (i.e., all its nodes in $\bar{N}$), and at most Γ nodes of $\tilde{N}$, have been denoted as Γ-tours, and they are said to be generated from that tour. Indeed, special Γ-tours are considered in the Consistent Robust Home Care problem, namely the sequence-preserving Γ-tours, thus addressing also sequence-preserving robustness. Specifically, a sequence-preserving Γ-tour is a Γ-tour whose nodes are visited in the same order given by the tour that generated it. Besides being computationally tractable, the sequence-preserving robustness is also greatly appreciated in practice by caregivers who experiment with limited deterioration of their planned tours in case a disruption occurs.

Referring to the previous example, sequence-preserving Γ-tours are illustrated in Fig. 2. Suppose that nodes $\tilde{1}, \tilde{2}, \tilde{3}$ and 4 are visited by the same caregiver on the same day, in that order, as depicted in Fig. 2 on the left. On the right, four possible sequence-preserving Γ-tours are represented. They visit nodes in the same order as the tour from which they are generated and contain at most Γ nodes in $\tilde{N} = \{\tilde{1}, \tilde{2}, \tilde{3}\}$.

By combining Γ-robustness and sequence-preserving robustness, a set of tours C, including all the certain and all the uncertain requests of the patients, is a *robust set of tours* with respect to Γ if, for each tour $\tau \in C$, the length of each sequence-

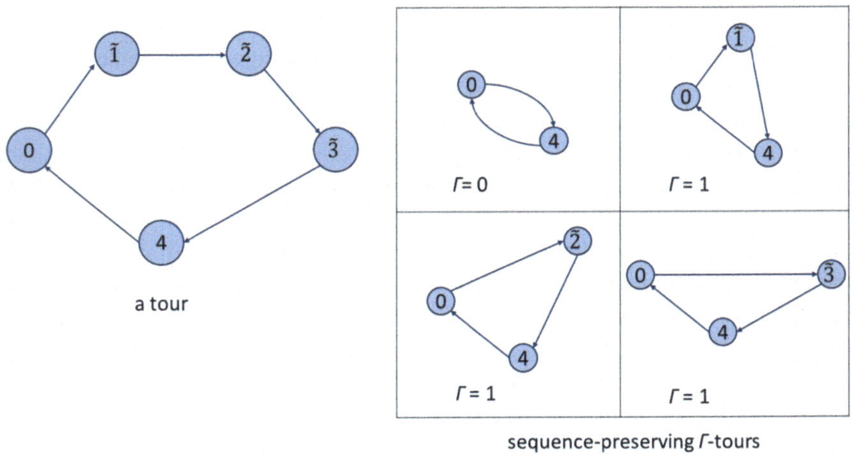

Fig. 2 A tour (on the left) and four possible sequence-preserving Γ-tours for $\Gamma = 0$ and $\Gamma = 1$ (on the right)

preserving Γ-tour generated from τ, taking into account the travel time along the arcs of the tour and the service time at the patients, does not exceed the workday length of the caregiver assigned to τ. That is, C is a robust solution with respect to Γ if, for each tour $\tau \in C$, the caregiver assigned to τ is able to perform the tour whatever the (at most Γ) uncertain requests of τ will possibly be realized, without changing the planned order of the visits. Equivalently, the maximum length of the sequence-preserving Γ-tours generated from τ must be less than or equal to the workday length of the caregiver.

Robustness is further empowered with consistency, giving rise to a consistency-aware robustness. Two forms of *consistency* are addressed in Cappanera and Scutellà (2022) along the lines outlined in Kovacs et al. (2014): (i) care continuity (or person-oriented consistency) and (ii) arrival time consistency. Care continuity constraints bound the number of different caregivers who can visit each patient during the time horizon. Precisely, each patient can be visited by at most T different caregivers over W, where T is given. Arrival time consistency guarantees that all the visits required by a patient, at most one per day, must be performed always at the same daily time slot throughout the planning horizon. As an example, either in the morning or in the afternoon.

In order to handle arrival time consistency, each caregiver ω is modeled in terms of a certain number of copies of it. Specifically, the number of copies depends on the caregiver's workday duration and on the length of the time slot given by patients when they express their time consistency requirements. As an example, if the daily workday length of a caregiver is 8 hours and the time slot given by the patients is 2 hours (for example all the visits must occur from 8 a.m. to 10 a.m., or from 10 a.m. to 12 p.m., or from 2 p.m. to 4 p.m., or from 4 p.m. to 6 p.m.), then each caregiver is replaced by 8/2=4 copies of it, each representing the availability of the caregiver in each time slot a workday is organized in. As in Cappanera and Scutellà (2022), we address the case in which a day is organized in two time slots, i.e., a morning slot and an afternoon slot. This setting, which is often used in practice, see, for example, Chun-Cheng et al. (2018), is also the one our experimentation is based on. Thus, each original caregiver ω is modeled in terms of two copies of it, one relative to the morning shift, say ω_m, and the other relative to the afternoon shift, say ω_a. Hereafter, O will denote the duplicated set of caregivers with ω denoting its generic element. When necessary, however, ω_m and ω_a will be used to specify which copy we are referring to. Furthermore, for each $\omega \in O$, D_ω will indicate the workday length of the caregiver. Specifically, D_ω will indicate the morning workday length in case ω refers to the morning shift, while it will indicate the afternoon workday length in case ω refers to the afternoon shift. Arrival time consistency is guaranteed by assigning caregivers of the same type to each patient, i.e., all morning or all afternoon caregivers.

Summarizing, the *Consistent Robust Home Care Problem* jointly addresses: (i) *care plan scheduling*, assigning a pattern both to the node modeling the certain nature of each patient j, thus scheduling the certain requests expressed by $\bar{r}_j$ in the planning horizon, and also to the node modeling the uncertain nature of each patient j, thus scheduling the uncertain requests expressed by $\tilde{r}_j$ in the planning horizon;

(ii) *caregiver assignment*, assigning caregivers to patients while guaranteeing arrival time and person-oriented consistency and skill compatibility between patient requests and caregivers, and (iii) *daily caregiver routing*, determining the tour of each (morning and afternoon) caregiver on each day of the planning horizon, so that the resulting set of tours constitutes a *robust set of tours* with respect to Γ. The objective function guiding the three groups of decisions is the caregiver workload balancing. This social equity criterion, also investigated in Cappanera and Scutellà (2015), Yalçındağ et al. (2016), and Cappanera et al. (2018), minimizes the maximum caregiver utilization factor over the planning horizon. In the robust framework under study, the *caregiver utilization factor* is expressed as the total workload of the caregiver during the planning horizon in the worst scenario, i.e., by summing up the lengths of the longest sequence-preserving Γ-tours generated from the (morning and afternoon) tours of the caregiver in the planning horizon, over the maximum possible workload of the caregiver in the planning horizon.

For example, by referring to the previous 5-day planning horizon, in the case of a caregiver with a workday of 180 minutes, with a workload on both Monday and Tuesday of 140 minutes at worst, and a workload equal to 150 minutes at worst on each of the next 3 days, the utilization factor is $(2 \cdot 140 + 3 \cdot 150)/(5 \cdot 180)$, that is, approximately 0.81.

We conclude by listing the notation used in the next section to state the proposed mathematical model to the problem:

W	planning horizon
A	set of arcs in the (complete) logistic network
O	set of skilled caregivers
$O_d \subseteq O$	set of caregivers available on day d, for each $d \in W$
ω_m	morning copy of caregiver ω
ω_a	afternoon copy of caregiver ω
D_ω	workday length of caregiver ω, $\omega \in O$
K	set of skills
$s_\omega \in K$	skill of caregiver ω, $\omega \in O$
$\bar{N} = \{1, \ldots, n\}$	set of nodes modeling the certain nature of the patients
$\tilde{N} = \{\tilde{1}, \ldots, \tilde{n}\}$	set of nodes modeling the uncertain nature of the patients
r_{jk}	number of requests for j in W involving skill k, $j \in \bar{N} \cup \tilde{N}$, $k \in K$
t'_j	service (or assistance) time at patient j, $j \in \bar{N} \cup \tilde{N}$
t_{ij}	travel time from node i to node j, $(i, j) \in A$
T	maximum number of caregivers who can visit each patient in W
P_j	set of patterns for j, $j \in \bar{N} \cup \tilde{N}$
P	overall set of patterns.

4 A Flow-Based Mathematical Model

Addressing the uncertainty of the patient requests jointly with assignment, scheduling, and routing decisions is indeed very complex. Moreover, the consistency requirements make the Robust Home Care Problem addressed in Cappanera et al. (2018) still harder. This is why in Cappanera and Scutellà (2022) we started our study of the Consistent Robust Home Care Problem by considering an approximate formulation of the addressed form of robustness rather than an exact one.

Here, we present the flow-based mathematical formulation we propose to state the addressed form of robustness in an exact way. Also, the objective function is formally presented, while the main families of constraints characterizing the model are informally described. The complete mathematical formulation is then reported in the Appendix. For completeness, we also report the approximate formulation of the robustness paradigm which has been considered and experimented with in Cappanera and Scutellà (2022). The overall mathematical model in Cappanera and Scutellà (2022) and the one here proposed differ in such constraints, i.e., in the way robustness is expressed: via approximation in Cappanera and Scutellà (2022), exactly in this paper.

In both cases, the robustness constraints are based on the following family of variables, which model the routing decisions in the Consistent Robust Home Care:

$$x_{ij}^{\omega d} = \begin{cases} 1 & \text{if caregiver } \omega \text{ travels along } (i, j) \\ & \text{on day } d \\ 0 & \text{otherwise.} \end{cases} \qquad (i, j) \in A, d \in W, \omega \in O_d$$

The robustness constraints guarantee that the set of tours described by the design variables $x_{ij}^{\omega d}$ is a *robust set of tours* with respect to Γ. In an abstract and intuitive form, they can be stated as follows:

$$R^{\omega d} \leq D_{\omega} \qquad d \in W, \omega \in O_d \tag{1}$$

where, for each day d and each caregiver available on day d, i.e., $\omega \in O_d$, $R^{\omega d}$ represents the maximum length amongst all the sequence-preserving Γ-tours generated from the tour of ω on day d, say τ. By imposing that such a maximum length cannot exceed the workday duration of ω, i.e., D_{ω}, we ensure that all the sequence-preserving Γ-tours generated from τ satisfy the daily length of ω. Thus, the solution represents a robust set of tours, as required.

Constraints (1) are used also to state the objective function. As indicated before, the objective function represents the maximum caregiver utilization factor, i.e., the maximum total workload of the caregivers during the planning horizon, in the worst scenario, over their maximum possible workload, which we want to minimize.

The overall mathematical model, which includes constraints that assign exactly one pattern to each node modeling a patient, constraints which guarantee that the design variables $x_{ij}^{\omega d}$ define a set of tours which include all the certain and all the uncertain patient requests, flow conservation constraints to avoid subtours, care continuity constraints and constraints modeling the consistency requirements, can then be outlined as follows:

$$
\begin{aligned}
&\min m \\
&\quad (1) \\
&\quad (22)\text{–}(46) \\
&\quad \frac{\sum_{d \in W} (R^{\omega_m d} + R^{\omega_a d})}{|W| \cdot (D_{\omega m} + D_{\omega a})} \leq m, \text{ for all original caregivers } \omega.
\end{aligned}
$$

In the model, (22)–(46) refer to the Appendix, where the previously outlined families of constraints are formally presented, while the auxiliary variable m defines an upper bound to each caregiver utilization factor, and therefore to the maximum one, which the model aims to minimize.

4.1 Approximate Robust Formulation

In Cappanera and Scutellà (2022), an approximation of the robustness constraints (1) has been achieved by replacing each $R^{\omega d}$ expression with the upper approximation suggested in Cappanera et al. (2018).

Specifically, given a tour τ, let $A(\tau)$ denote the set of the arcs of τ. Furthermore, let $t_j^{max} = \max_{(i,j) \in A} \{t_{ij}\}$ be the maximum travel time amongst all the arcs entering j, for each node j. Then the following property holds true (Cappanera et al., 2018):

Proposition 1 *Given a tour τ and a sequence-preserving Γ-tour π generated from τ, the following expression is an upper bound on the workload due to π:*

$$
U(\pi) = \sum_{(i,j) \in A(\pi) \cap A(\tau)} (t_{ij} + t'_j) + \sum_{(i,j) \in A(\pi):(i,j) \notin A(\tau)} (t_j^{max} + t'_j). \tag{2}
$$

The motivation for distinguishing between the arcs in $A(\pi) \cap A(\tau)$ and the rest, is that the former are fully characterized within the mathematical model by the design variables $x_{ij}^{\omega d}$, and their contribution to the workload due to the sequence-preserving Γ-tour π can then be expressed using such a characterization. Indeed, to further simplify the upper bound computation, in Cappanera et al. (2018) only the arcs in $A(\pi) \cap A(\tau)$ that are incident to at least one node in $\bar{N}$ have been evaluated exactly, i.e., via the values t_{ij}, whereas the travel times of the others, i.e., the ones for which end nodes both belong to $\tilde{N}$, have been over-estimated too, via the values t_j^{max}.

The problem of estimating the maximum workload due to the sequence-preserving Γ-tours according to the just stated upper approximation has been formalized in Cappanera et al. (2018) by introducing auxiliary decision variables $v_j^{\omega d}$, $j \in \tilde{N}$, $d \in W$, $\omega \in O_d$. These are binary variables taking on value 1 if node $j \in \tilde{N}$ is visited by caregiver ω on day d (i.e., if the uncertain request, scheduled on day d, of the patient modeled by node j is realized, and thus j belongs to a sequence-preserving Γ-tour of ω on day d), and 0 otherwise:

$$v_j^{\omega d} = \begin{cases} 1 \text{ if node } j \text{ is visited by caregiver } \omega \text{ on day } d \\ 0 \text{ otherwise.} \end{cases} \qquad j \in \tilde{N}, d \in W, \omega \in O_d$$

Using the variables above, the following upper approximation $\bar{R}^{\omega d}$ of $R^{\omega d}$ has been considered in Cappanera et al. (2018) and Cappanera and Scutellà (2022) for each day d and each caregiver $\omega \in O_d$:

$$\begin{aligned}\bar{R}^{\omega d} = & \sum_{\substack{(i,j)\in A:\\ i,j\in \bar{N}\cup\{0\}}} (t_{ij} + t'_j) x_{ij}^{\omega d} + \sum_{\substack{(i,j)\in A:\\ i\in\tilde{N}, j\in\bar{N}\cup\{0\}}} (t_j^{max} + t'_j) x_{ij}^{\omega d} \\ & + \max \left\{ \sum_{\substack{(i,j)\in A:\\ i\in\bar{N}\cup\{0\}, j\in\tilde{N}}} (t_{ij} + t'_j) x_{ij}^{\omega d} v_j^{\omega d} + \sum_{\substack{(j,i)\in A:\\ i\in\bar{N}\cup\{0\}, j\in\tilde{N}}} (t_{ji} - t_i^{max}) x_{ji}^{\omega d} v_j^{\omega d} \right. \\ & \left. + \sum_{\substack{(i,j)\in A:\\ i\in\tilde{N}, j\in\tilde{N}}} (t_j^{max} + t'_j) x_{ij}^{\omega d} v_j^{\omega d} \right\} \end{aligned} \tag{3}$$

with

$$\sum_{j\in\tilde{N}} v_j^{\omega d} \leq \Gamma \tag{4}$$

$$v_j^{\omega d} \in \{0, 1\} \qquad j \in \tilde{N}. \tag{5}$$

Given a tour τ, modeled by the design variables $x_{ij}^{\omega d}$, and a sequence-preserving Γ-tour π generated from it, whose selected nodes in $\tilde{N}$ are defined by (4)–(5), the objective function in (3) estimates the workload associated with π by considering exactly the contribution due to the arcs linking nodes in $\bar{N}$ as well as the depot (first summation in (3)), and also the contribution of the arcs linking nodes in $\bar{N}$ and depot to selected nodes in $\tilde{N}$, i.e., to nodes $j \in \tilde{N}$ such that $v_j^{\omega d} = 1$ (third and fourth summation in (3)). Note that all such arcs belong to $A(\tau)$. On the other hand, for the arcs $(i, j) \notin A(\tau)$, and for the arcs in $A(\tau)$ linking selected nodes in $\tilde{N}$, the travel time of (i, j) is estimated from above via t_j^{max}. Therefore, problem (3)–(5) looks for the longest sequence-preserving Γ-tour based on the just-stated over-estimation of the workload. From Proposition 1, we get $\bar{R}^{\omega d} \geq R^{\omega d}$.

The model obtained by replacing constraints (1) with:

$$\bar{R}^{\omega d} \leq D_\omega \quad d \in W, \omega \in O_d, \tag{6}$$

thus represents an approximate formulation of the Consistent Robust Home Care Problem. Since the constraint matrix of model (3)–(5) is totally unimodular (TU) (Wolsey, 1998) for each day d and each caregiver $\omega \in O_d$ then, by treating the design variables $x_{ij}^{\omega d}$ as constants, a standard technique in Robust Optimization, each inner ILP model (3)–(5) can be replaced by its Linear Programming relaxation and so, by strong duality, by its Linear Programming dual, leading to a MILP approximate formulation to the problem.

4.2 Exact Robust Formulation

In order to formulate constraints (1) in an exact way, the idea is to model exactly the sequence-preserving Γ-tours generated from the tour of a caregiver ω on a day d, by introducing auxiliary decision variables, denoted as $\alpha_{ij}^{\omega d}$, which specify what pairs of nodes are visited by ω on day d according to the order given by their tour, therefore indicating what arcs can belong to its sequence-preserving Γ-tours. Precisely, by considering pairs of nodes i and j modeling patients:

$$\alpha_{ij}^{\omega d} = \begin{cases} 1 & \text{if } i \text{ is visited before } j \\ & \text{by caregiver } \omega \text{ on day } d \\ 0 & \text{otherwise.} \end{cases} \qquad i, j \in N, d \in W, \omega \in O_d$$

In other words, $\alpha_{ij}^{\omega d} = 1$ if and only if i precedes j along the tour of ω on day d, and 0 otherwise. Equivalently, $\alpha_{ij}^{\omega d} = 1$ if and only if i can be the immediate predecessor of j along a sequence-preserving Γ-tour generated from the tour of ω on day d, and therefore the arc (i, j) may be part of such a sequence-preserving Γ-tour. Notice that, whereas the design variables $x_{ij}^{\omega d}$ define the direct precedence of pairs of nodes i and j along the tour of ω on day d, variables $\alpha_{ij}^{\omega d}$ define a precedence relationship, along that tour, which is not necessarily direct.

Regarding the relationship between the depot, i.e., node 0, and the patients along the tour, for modeling motivations we assume that node 0 precedes any patient visited in the tour and that at the same time, node 0 follows any visited patient. Consequently, we introduce:

$$\alpha_{0j}^{\omega d} = \alpha_{j0}^{\omega d} = \begin{cases} 1 & \text{if } j \text{ belongs to the tour of} \\ & \text{caregiver } \omega \text{ on day } d \\ 0 & \text{otherwise.} \end{cases} \qquad j \in N, d \in W, \omega \in O_d$$

Figure 3 exemplifies the relationship between variables $x_{ij}^{\omega d}$ and $\alpha_{ij}^{\omega d}$. Always referring to the example previously introduced, the tour depicted on the left shows that caregiver ω on day d visits nodes 4, $\tilde{2}$ and $\tilde{1}$ in that order: variables $x_{ij}^{\omega d}$ are 1 for all the arcs in this tour. The graph on the right identifies the corresponding $\alpha_{ij}^{\omega d}$

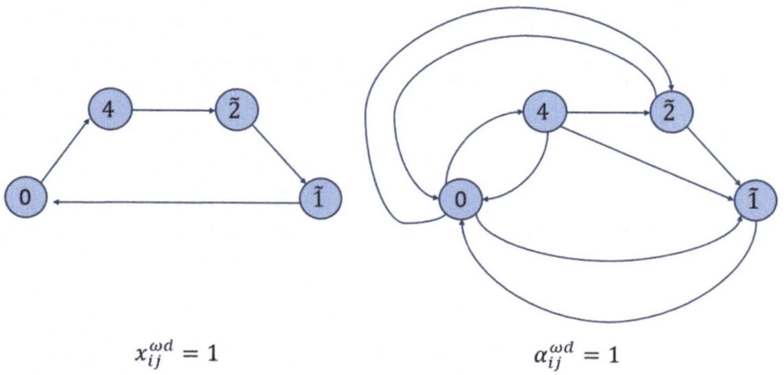

Fig. 3 A tour identified by design variables $x_{ij}^{\omega d}$ (on the left), and corresponding $\alpha_{ij}^{\omega d}$ variables set to one according to the definition (on the right)

variables set to one: each node is connected to the depot in both directions to allow each node to be either the first or the last node in a sequence-preserving Γ-tour. The rest of the arcs identify precedence relationships between nodes: as an example, 4 precedes $\tilde{2}$ (direct precedence), and $\tilde{1}$ (undirect precedence).

If i and j denote patients, then variables $\alpha_{ij}^{\omega d}$ can be defined as the linear statement of the following logical conditions, stating that i is visited before j along the tour if either j is the immediate successor of i along the tour, i.e., the caregiver visits i and then j (in this case the routing variable $x_{ij}^{\omega d}$ is 1), or there exists an intermediate node k such that i is visited before k and k is visited before j:

$$\alpha_{ij}^{\omega d} = x_{ij}^{\omega d} \bigvee_{k} (\alpha_{ik}^{\omega d} \wedge \alpha_{kj}^{\omega d}).$$

We shall express such logical conditions by exploiting some auxiliary flow variables that have been introduced in the Consistent Robust Home Care formulation in order to avoid subtours, also reported in the Appendix, that is:

$y_{ij}^{d} \geq 0$: auxiliary flow variable which represents the number of patients visited after patient i by the caregiver moving along (i, j) on day d, $(i, j) \in A, i, j \in N$, $d \in W$ ($y_{ij}^{d} = 0$ if no caregiver moves along (i, j) on day d).

Using the y^d flow variables, $\alpha_{ij}^{\omega d}$ must be set to 1 if both i and j belong to the tour of caregiver ω on day d, and the total y^d flow entering i is greater than the one entering j. That is to say, the following constraints must be satisfied:

$$\sum_{k \in N \cup \{0\}} \left((n-1)x_{kj}^{\omega d} - y_{kj}^{d} + (n-1)x_{ki}^{\omega d} + y_{ki}^{d} \right) - (n-1)\alpha_{ij}^{\omega d} \leq 2(n-1)$$

$$i, j \in N, d \in W, \omega \in O_d. \qquad (7)$$

In order to fully specify the $\alpha_{ij}^{\omega d}$ variables, we introduce also constraints forcing the $\alpha_{ij}^{\omega d}$ variables to zero if any one of the conditions above is not satisfied. Precisely:

$$\alpha_{ij}^{\omega d} \leq \sum_{(k,i)\in A} x_{ki}^{\omega d} \qquad i, j \in N, d \in W, \omega \in O_d \tag{8}$$

$$\alpha_{ij}^{\omega d} \leq \sum_{(k,j)\in A} x_{kj}^{\omega d} \qquad i, j \in N, d \in W, \omega \in O_d \tag{9}$$

$$(1 - \alpha_{ij}^{\omega d}) \geq (\sum_{(k,j)\in A} y_{kj}^{d} - \sum_{(k,i)\in A} y_{ki}^{d})/2n \quad i, j \in N, d \in W, \omega \in O_d. \tag{10}$$

Regarding constraints (10), notice that, since the total number of nodes other than the depot in the logistic network is at most $2n$, then $-1 \leq (\sum_{(k,j)\in A} y_{kj}^{d} - \sum_{(k,i)\in A} y_{ki}^{d})/2n \leq 1$ for any d. Hence, if patient j precedes patient i along the tour of caregiver ω on day d, then the right-hand-side of (10) is positive, and therefore the variable $\alpha_{ij}^{\omega d}$ is forced to 0.

Regarding the relationship between the depot and the patients along the tour, we introduce:

$$\alpha_{0j}^{\omega d} \leq \sum_{(k,j)\in A} x_{kj}^{\omega d} \qquad j \in N, d \in W, \omega \in O_d \tag{11}$$

$$\alpha_{j0}^{\omega d} \leq \sum_{(k,j)\in A} x_{kj}^{\omega d} \qquad j \in N, d \in W, \omega \in O_d. \tag{12}$$

Specifically, constraints (11)–(12) force the $\alpha_{0j}^{\omega d}$ and $\alpha_{j0}^{\omega d}$ variables to 0 if j does not belong to the tour of caregiver ω on day d.

For each caregiver ω and each day d, the $\alpha_{ij}^{\omega d}$ variables previously introduced are used to define, in an implicit way, a directed graph containing all the possible arcs that ω can use on day d, according to the considered robustness paradigm, to move along a sequence-preserving Γ-tour, as exemplified in Fig. 3. Precisely, the idea is to model the longest (simple) path from/to the depot in such an implicitly defined graph, i.e., a sequence-preserving tour of maximum length, which is constrained in the sense that it can include at most Γ nodes in $\tilde{N}$.

In order to model sequence-preserving Γ-tours in the graph implicitly defined by the $\alpha_{ij}^{\omega d}$ variables, we introduce the following binary flow variables:

$$f_{ij}^{\omega d} = \begin{cases} 1 & \text{if } (i,j) \text{ is in the constrained} \\ & \text{path of caregiver } \omega \text{ on day } d \\ 0 & \text{otherwise.} \end{cases} \qquad i, j \in (N \cup \{0\}), d \in W, \omega \in O_d$$

For each caregiver ω and each day d, we then consider the following flow-based Integer Linear Programming (ILP) model:

$$\tilde{R}^{\omega d} = \max \sum_{(i,j)\in A} (t_{ij} + t'_j) \cdot \alpha_{ij}^{\omega d} \cdot f_{ij}^{\omega d} \tag{13}$$

$$- \sum_{(j,0)\in A} f_{j0}^{\omega d} = -1 \tag{14}$$

$$\sum_{(0,j)\in A} f_{0j}^{\omega d} = 1 \tag{15}$$

$$\sum_{(i,j)\in A} f_{ij}^{\omega d} - \sum_{(j,i)\in A} f_{ji}^{\omega d} = 0 \qquad i \in N \tag{16}$$

$$\sum_{(i,j)\in A: j\in \tilde{N}} f_{ij}^{\omega d} \leq \Gamma \tag{17}$$

$$f_{ij}^{\omega d} \in \{0, 1\} \qquad (i, j) \in A, \tag{18}$$

where (13)–(18) model the longest path from/to the depot, i.e., node 0, subject to the complicating constraint (17), which imposes that at most Γ nodes in $\tilde{N}$ belong to the path. Notice that, by construction, each sequence-preserving Γ-tour is feasible to (13)–(18), and its objective function value is exactly the total time spent along this tour, i.e., the corresponding workload, since the $\alpha_{ij}^{\omega d}$ variables corresponding to the arcs of sequence-preserving Γ-tours are all set to 1 by constraints (7). However, also paths, from/to node 0, including arcs (i, j) which can not belong to a sequence-preserving Γ-tour, could be feasible to (13)–(18), although the travel time and the service time along such undesired arcs are not counted in the objective function since the corresponding $\alpha_{ij}^{\omega d}$ variables are forced to be 0. Moreover, the model could admit, as a feasible solution, also a path from/to node 0 plus some subtours. Therefore, in the presented form, also $\tilde{R}^{\omega d}$ can be just an upper approximation of $R^{\omega d}$.

Figure 4 exemplifies the case above described. Suppose that the service time of each request is 45 minutes and that the travel times are the ones associated with the arcs in the figure. The length of the sequence-preserving Γ-tour depicted on the left

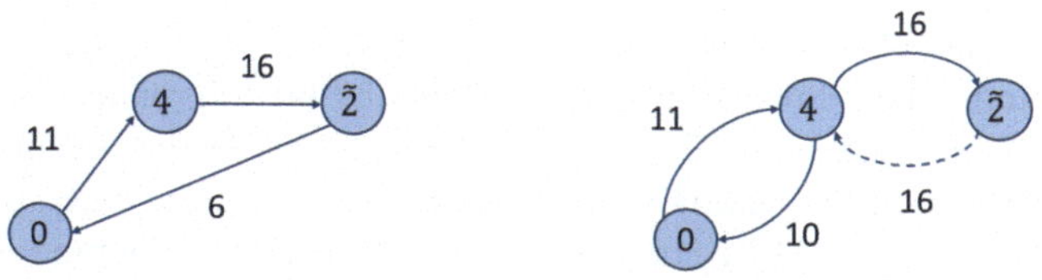

Fig. 4 A sequence-preserving Γ-tour (on the left), and a feasible longer path which uses an arc corresponding to an $\alpha_{ij}^{\omega d}$ variable set to zero (on the right, dotted arc)

is thus 123 (45 + 45 + 11 + 16 + 6). However, the non-simple path from node 0 to node 0, depicted on the right, is a feasible solution to model (13)–(18), too, and it is longer than the sequence-preserving Γ-tour. In fact, using an arc for which the corresponding $\alpha_{ij}^{\omega d}$ variable is set to zero (dotted arc in the figure), an additional arc can be part of the path (arc (4,0) in the figure), and thanks to this arc the path length is 127 (45+45+11+16+10).

However, as formally proved below, under the reasonable assumption that, for each pair of nodes i and j, the travel time t_{ij} associated with (i, j) represents the minimum time to move from i to j in the underlying physical logistic network, it is possible to convert (13)–(18) into an exact formulation of $R^{\omega d}$, by associating a very large negative value, in place of value 0, with the arcs which can not be part of a sequence-preserving Γ-tour. That is to say, using the following variables:

$$\tilde{\alpha}_{ij}^{\omega d} = \begin{cases} 1 & \text{if } i \text{ is visited before } j \\ & \text{by caregiver } \omega \text{ on day } d \qquad i, j \in N, d \in W, \omega \in O_d \\ -M & \text{otherwise,} \end{cases}$$

in place of the binary variables $\alpha_{ij}^{\omega d}$, where M denotes a very large value. Analogously:

$$\tilde{\alpha}_{0j}^{\omega d} = \tilde{\alpha}_{j0}^{\omega d} = \begin{cases} 1 & \text{if } j \text{ belongs to the tour} \\ & \text{of caregiver } \omega \text{ on day } d \qquad j \in N, d \in W, \omega \in O_d \\ -M & \text{otherwise.} \end{cases}$$

Indeed, the $\tilde{\alpha}_{ij}^{\omega d}$ variables can be expressed in terms of the binary variables $\alpha_{ij}^{\omega d}$, previously introduced, as follows:

$\tilde{\alpha}_{ij}^{\omega d} = -M + (1 + M)\alpha_{ij}^{\omega d}$.

In fact, $\tilde{\alpha}_{ij}^{\omega d} = 1$ if and only if $\alpha_{ij}^{\omega d} = 1$, and $\tilde{\alpha}_{ij}^{\omega d} = -M$ if and only if $\alpha_{ij}^{\omega d} = 0$. Therefore, by replacing the expression $\tilde{\alpha}_{ij}^{\omega d} = -M + (1 + M)\alpha_{ij}^{\omega d}$ in place of $\alpha_{ij}^{\omega d}$ within (13), we get:

$$R^{\omega d} = \max \sum_{(i,j)\in A} (t_{ij} + t'_j) \cdot (-M + (1 + M)\alpha_{ij}^{\omega d}) \cdot f_{ij}^{\omega d}. \qquad (19)$$

Theorem 1 *If t_{ij} is the minimum time to move from i to j in the underlying physical logistics network, for all $(i, j) \in A$, then model (14)–(19) is an exact formulation of $R^{\omega d}$, for each caregiver ω and each day d.*

Proof Constraints (14)–(18) model a directed path from/to the depot node, plus possible subtours, i.e., directed cycles not including node 0, under the additional constraint that no more than Γ nodes in $\tilde{N}$ are present in the solution. For the nonlinear presence of the auxiliary binary variables $\alpha_{ij}^{\omega d}$ in the objective function definition, that is regulated by constraints (7)–(12), each arc linking nodes which are not sequentially visited by caregiver ω on day d has a very negative cost in the objective function, due to the factor $-M$. Therefore, the presence of this kind of arc at the optimal solution level is excluded during the optimization process.

In particular, no subtour can be present in an optimal solution, since it should necessarily include at least one arc connecting pairs of nodes that do not satisfy the precedence relationship given by the visit order along the tour of ω on day d. For the same motivation, in any optimal solution, this kind of arc is also excluded from the directed path from/to the depot node, which is therefore simple. It follows that any optimal solution is a simple directed path from/to the depot node, which is composed only of arcs (i, j) such that i precedes j in the tour of caregiver ω on day d. Moreover, due to the assumption on the arc travel times, there exists an optimal solution that includes all the nodes in $\bar{N}$ visited by ω on day d, in the same order of visit. That is, all the certain requests scheduled on day d and assigned to caregiver ω are satisfied. The thesis follows. □

Referring to the example provided in Fig. 4, the use of $\tilde{\alpha}_{ij}^{\omega d}$ variables instead of $\alpha_{ij}^{\omega d}$ guarantees that the path depicted on the right can not be an optimal solution to the model since the coefficient of the dotted arc in the objective function is a big negative value, i.e., $-M$.

Proposition 2 *The matrix associated with (14)–(18) is TU.*

Proof It follows from the results in Wolsey (1998). □

From Proposition 2, in constraints (1) it is possible to replace each inner ILP model (14)–(19), defining $R^{\omega d}$ for each caregiver ω and each day d, by its Linear Programming relaxation and so, by the corresponding Linear Programming dual problem, leading to an exact MILP formulation to the Consistent Robust Home Care problem. The complete model is reported in the Appendix.

4.3 Approximate vs Exact Formulation: An Example

Consider again the example introduced in Sect. 3, but limiting the attention only to requests of skill 1. Moreover, consider a 2-day planning horizon, and suppose that the requests have been scheduled as follows: nodes $\tilde{1}, \tilde{2}, \tilde{3}$, and 4 are assigned to pattern (1,0), i.e., their requests are scheduled on day $d = 0$, whereas nodes 1,2, and 3 are assigned to pattern (0,1), i.e., their requests are scheduled on day $d = 1$. Suppose also that there are two caregivers skilled to perform visits of skill 1, and that each of them is available only on 1 day, with a workday length equal to 180 minutes. Let the service time of each request be equal to 45 minutes, and the travel times associated with the arcs of the logistic network be the ones given in Table 1.

Suppose that the visits have been sequenced as described in Fig. 5 and consider $\Gamma = 2$. On day $d = 0$, there are three possible sequence-preserving Γ-tours with exactly 2 nodes in $\tilde{N}$: (0-$\tilde{2}$-$\tilde{3}$-4-0), that excludes node $\tilde{1}$, (0-$\tilde{1}$-$\tilde{3}$-4-0), that excludes node $\tilde{2}$, and (0-$\tilde{1}$-$\tilde{2}$-4-0), that excludes node $\tilde{3}$. According to the travel times given in Table 1, all of them have a length of 171 minutes. Thus, when the exact model is used, all of these tours are feasible. Consequently, the caregiver working on day

Table 1 Travel times

	0	1	2	3	4
0	0	7	7	7	11
1	6	0	3	3	16
2	6	3	0	3	16
3	6	3	3	0	16
4	10	16	16	16	0

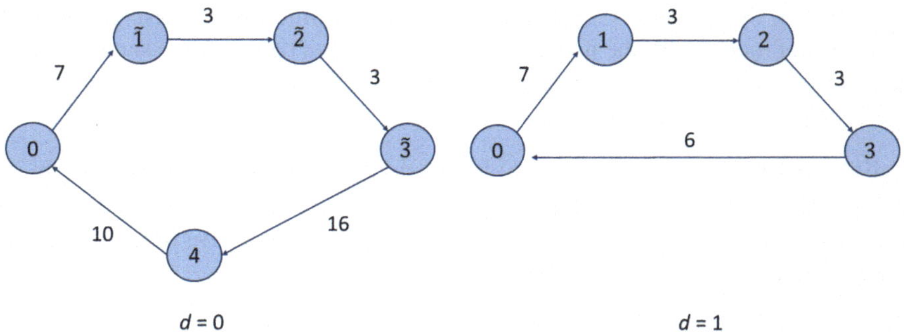

Fig. 5 A solution for a 2-day planning horizon: arcs with $x_{ij}^{\omega d}$ set to 1

$d = 0$ has a workload of 171 minutes, while the workload of the caregiver available on day $d = 1$ is 154, and the maximum workload is correctly computed as equal to 171 in the optimal solution. On the other hand, when the approximate formulation is used, the travel time between two nodes i and j in $\tilde{N}$ is approximated to the maximum travel time of the arcs entering j, which is 16, and therefore all of the three sequence-preserving Γ-tours have a total length of 184, which exceeds the daily availability of the caregivers thus resulting in no feasible solution.

Notice that, due to the characteristics of the considered example, the requests of all the patients must be performed in the same time slot, i.e., either in the morning slot or in the afternoon slot, in order to satisfy the consistency requirements. Consequently, the tours depicted in Fig. 5 both refer to the same slot.

4.4 Cuts and Valid Inequalities

Here, we present three families of constraints that proved to enhance the lower bound provided by the linear programming relaxation of the proposed robust flow-based formulation reported in the Appendix.

Specifically, the first family comprises the following valid inequalities:

$$\alpha_{ij}^{\omega d} \geq x_{ij}^{\omega d} \qquad (i, j) \in A, d \in W, \omega \in O_d. \tag{20}$$

Constraints (20) state that, if (i, j) belongs to the tour of caregiver ω on day d, i.e., $x_{ij}^{\omega d} = 1$, then the corresponding auxiliary variable $\alpha_{ij}^{\omega d}$ must be set to 1, since (i, j) may be part of a sequence-preserving Γ-tour of ω on day d.

The other two families of constraints stem from simple observations on the value of $R^{\omega d}$, i.e., the length of the longest sequence-preserving Γ-tour for caregiver ω on day d. Specifically, when no patients are assigned to a given caregiver on a given day, the value of the longest sequence-preserving Γ-tour should be zero. To model this situation, we add a zero-cost arc from the depot to the depot itself in the inner problem (14)–(19). The corresponding flow variable, namely $f_{00}^{\omega d}$, is added to the flow conservation constraints regarding the depot node ((14)–(15)) and gives rise to an additional constraint for each operator ω and for each day d when the inner problem is dualized. Note that, alternatively, we could force $R^{\omega d}$ to zero adding the following constraints:

$$R^{\omega d} \leq D_{\omega} \sum_{j \in N} x_{0j}^{\omega d} \qquad d \in W, \omega \in O_d$$

$$R^{\omega d} \geq 0 \qquad d \in W, \omega \in O_d$$

which, however, when variables $x_{0j}^{\omega d}$ assume fractional values in the linear programming relaxation, are weaker than inserting an extra arc in the inner problem.

The third family of constraints is based on the following observation. For each caregiver ω and each day d, according to its definition, any sequence-preserving Γ-tour must include all the nodes, modeling the certain nature of the patients, which are visited by ω on d, in the same order of visit. Consequently, in case some visits are assigned to ω on day d, then the optimal objective function value of model (14)–(19) must satisfy:

$$R^{\omega d} \geq \sum_{j \in \bar{N} \cup \{0\}} \left(\sum_{\substack{(i,j) \in A: \\ i \in \bar{N}}} (t_j' + t_{ij}) x_{ij}^{\omega d} + \sum_{\substack{(i,j) \in A: \\ i \in \tilde{N}}} (t_j' + t_j^{min}) x_{ij}^{\omega d} \right) \quad d \in W, \omega \in O_d \tag{21}$$

where, as previously introduced, $\bar{N}$ and $\tilde{N}$ denote respectively the subsets of nodes modeling the certain and the uncertain nature of the patients, and t_j^{min} represents the minimum travel time required to reach node j. Observe that, in addition to the service time coming from certain visits, the travel time required to reach each node in $\bar{N}$ or the depot is considered in (21) to increase the lower bound on $R^{\omega d}$. Specifically, if the immediate predecessor of a node j in $\bar{N} \cup \{0\}$ along the tour is a node i in $\bar{N}$, then the travel time t_{ij} is accounted for. Otherwise, the minimum time t_j^{min} to reach j is considered. We want to emphasize that (21) are not valid inequalities since they cut away all the feasible solutions of model (14)–(19) with a negative cost. That is, all the solutions involving arcs that can not be part of a sequence-preserving Γ-tour are disregarded. In other words, the proposed cuts force optimization on the subset of the sequence-preserving Γ-tours, as desired. These types of constraints are sometimes referred to in the literature as super-valid inequalities.

5 Computational Results

5.1 Instances, Experimensts and Computing Environment Description

We generated 18 consistent robust instances from the data set publicly available at `https://commalab.di.unipi.it/datasets/HC/` and used for the first time in Cappanera and Scutellà (2015). The original data set comes from a real public provider operating in the North of Italy. The data set used in these experiments comprises 2 groups of 9 instances each, generated by original instances with respectively 40 and 60 patients over a 5-day planning horizon (a week). Each instance is characterized by 2 types of hierarchical skills, 4 caregivers, travel times obtained via Google Maps, a service time of 45 minutes for each visit, and $\Gamma = 1$. On each day of the week, in each session (morning or afternoon), one caregiver for each skill is available. Caregivers working during the morning session are not on duty in the afternoon session and vice-versa. The number of uncertain requests in each instance is about 30% of all the visits requests and is roughly equal to the maximum number of uncertain requests that can be covered, which is given by $\Gamma \cdot |W| \cdot |O| = 20$ in this dataset. Instance names range from 1 to 9 in each group (either 40 or 60 patients). Consistent robust instances are generated from the nominal ones by randomly selecting a subset, of the desired size, of requests, and making them uncertain. Specifically, for a given nominal instance, patients are ordered according to the total number of requests they need regardless of the skill. Then, starting from the patients with the highest number of requests, at each iteration a patient is randomly selected and part of their requests are randomly classified as uncertain provided that for the selected patient there exists at least a certain request and an uncertain one.

In the following sections, we present some preliminary computational results. In particular, the experiments consist of solving both the approximate and the exact robust models where scheduling decisions are fixed according to a selected pattern generation policy. The motivation is to simplify the comparison among the two models, by fixing a priori the patterns, i.e., the scheduling decisions, according to the same kind of policy. Specifically, two of the consistency-aware policies defined in Cappanera and Scutellà (2022), namely, *C-Bal* and *C-BalU* have been considered. Both of them are driven by a balancing criterion and minimize the difference between the maximum and the minimum number of requests occurring each day over W. The difference between the two policies concerns the type of requests considered: in *C-Bal* both certain and uncertain requests occurring in a day are considered, whereas in *C-BalU* only uncertain requests are taken into account.

The numerical analysis has been performed on a PC equipped with an AMD® Ryzen 9 3950D× 16-core processor ×32 and 32 Gb of RAM. The optimization models have been coded in C++ and solved using the IBM ILOG Cplex 20.1 solver imposing a time limit equal to 7200 seconds and a memory limit for the branch and bound tree equal to 8Gb.

5.2 Approximate vs Exact Robust Formulation

In this section, we provide a preliminary comparison of the two robust models, i.e., the approximate and the exact one (equipped with valid and super-valid inequalities), both in terms of solution quality and computational efficiency.

With regard to the quality of the solutions, we report the value m of the objective function, namely the workload of the busiest caregiver, and the number v of realizations of uncertain requests. Specifically, Table 2 provides for each KPI (either m or v) the average, minimum, and maximum values obtained over the 9 instances, separately for each group of instances (either 40 or 60 patients), for each robust model (either approximate or exact) and for each pattern generation policy (either C-Bal or C-BalU). Regardless of the problem size and the pattern generation policy adopted, we observe that the average value of m (avg m) is greater in the exact model than in the approximate model, which is an unexpected result, while the average value of v (avg v) is greater in the exact model than in the approximate model. The unexpected result can be explained in the view of what we will see below on the computational efficiency of the two models. We observe that, for the 40-patient instances, the policies C-Bal and C-BalU are almost equivalent in terms

Table 2 Solution quality: for each robust model (either approximate or exact), for each pattern generation policy (either C-Bal or C-BalU), and for each group of instances (either 40 or 60 patients), average, minimum, and maximum values across the 9 instances of objective function and realization of uncertain requests

		Approx		Exact	
Size	KPI	C-Bal	C-BalU	C-Bal	C-BalU
40	avg m	0,68	0,66	0,71	0,68
	min m	0,63	0,63	0,65	0,64
	max m	0,73	0,69	0,75	0,74
	avg v	10,22	9,44	11,89	11,56
	min v	9	8	10	9
	max v	12	12	15	14
60	avg m	0,71	0,73	0,75	0,76
	min m	0,68	0,69	0,72	0,71
	max m	0,76	0,79	0,81	0,82
	avg v	9,67	11,00	12,89	13,44
	min v	7	8	12	12
	max v	13	14	14	15

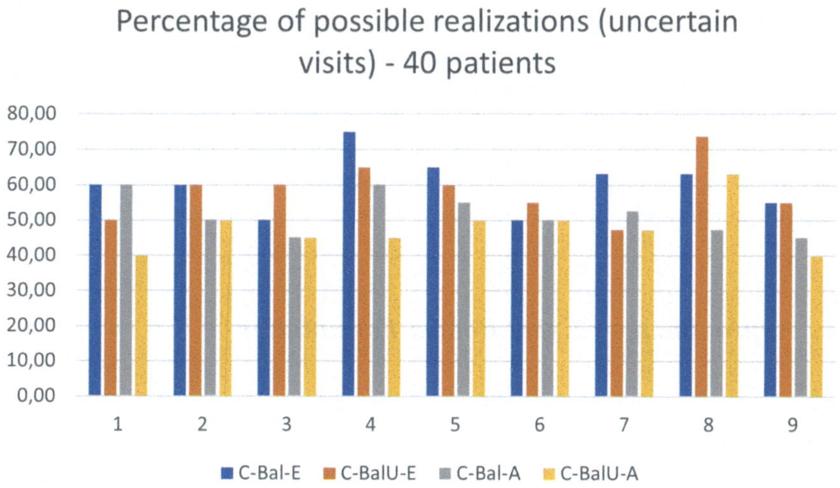

Fig. 6 Realization of uncertain requests: exact (E) vs approximate (A)—C-Bal vs C-BalU—40 patients

of avg v when we use the exact model, whereas C-Bal is preferable to C-BalU when the approximate model is used. Contrarily, for the 60-patient instances the C-Bal-U policy performs better than the C-Bal one regardless of the model used, and the increase of avg v when comparing the two models with the same pattern generation policy is slightly greater for C-Bal than for C-BalU. A more in-depth analysis on the value of v is shown in Figs. 6 and 7 which report, respectively for 40 and 60 patients, the percentage of realizations of uncertain requests obtained from the two models for each policy. From the figures, the exact model is better than the approximate one in terms of the number of realizations, and the increase can be significant on some instances. An explanation could be related to the upper approximation of the length of the sequence-preserving Γ-tours considered by the approximate model, for each day and each operator available on the day. This length depends both on the number of uncertain requests that can be satisfied and on the travel time along the arcs of the tours. Since the travel time along some arcs is over-estimated via the values called t_j^{max} in Sect. 4.1, the solver may be encouraged to concentrate many uncertain requests in a few tours since at most Γ uncertain requests in each tour will be satisfied. The exact model, on the contrary, due to the exact estimation of all the travel times, probably is less sensible to this form of concentration, thus spreading the uncertain requests among more tours, with a consequent greater rate of uncertain request realizations. This is just a guess, however, which will deserve a deeper investigation.

With regards to the efficiency, Table 3 shows separately for each model, for each pattern policy and for each problem size, average time (expressed in seconds) and average optimality gap obtained on the 9 instances in each group, as well as the number of instances (over 9) for which the solution is certified as optimal (n. opt) and the number of instances for which the time limit has been reached (n. TL). We

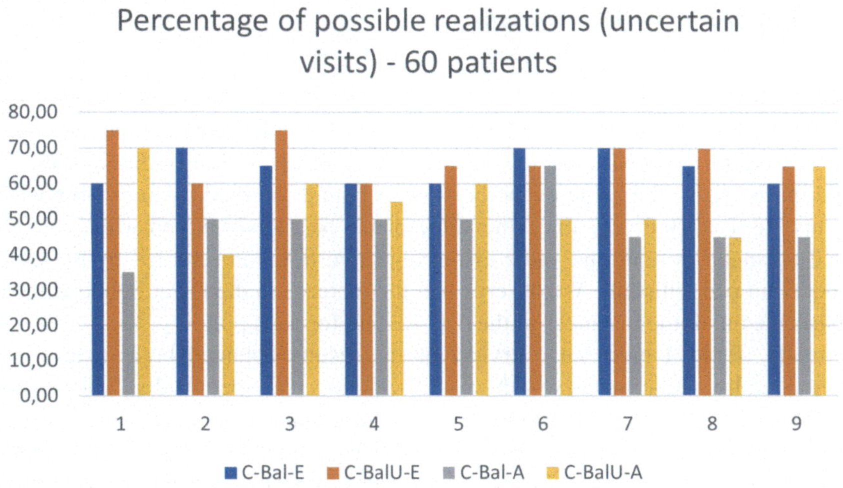

Fig. 7 Realization of uncertain requests: exact (E) vs approximate (A)—C-Bal vs C-BalU—60 patients

Table 3 Efficiency: for each robust model (either approximate or exact), for each pattern generation policy (either C-Bal or C-BalU), and for each group of instances (either 40 or 60 patients), average values of computational time and optimality gap, number of instances solved to optimality (n.opt), number of instances for which the time limit is exceeded (n. TL)

		Approx		Exact	
Size	KPI	C-Bal	C-BalU	C-Bal	C-BalU
40	avg time	2223,71	1920,63	7200,00	7109,76
	avg gap	0,00	0,00	0,22	0,22
	n. opt	7	7	0	0
	n. TL	2	2	9	8
60	avg time	5078,74	6573,07	7200,00	7200,00
	avg gap	0,00	0,01	0,21	0,21
	n. opt	4	1	0	0
	n. TL	5	8	9	9

observe that the exact model always reaches the time limit but in one case with 40 patients, where it exceeds the memory limit, and never certifies the optimality of the solution obtained. Also, the optimality gaps obtained with the exact model are quite high and stable regardless of the number of patients and of the policy adopted. Such high values of the gaps thus seem to justify the bigger value of the objective function obtained with the exact model with respect to the one obtained with the approximate one. The exact model is certainly much more computationally complex than the approximate one. The reported preliminary results therefore suggest the need to search for efficient techniques for solving the exact model that may better highlight the advantages of using the exact value of the caregiver' tour length rather than an upper estimate.

5.3 Impact of Valid Inequalities

In this section, we report the effects of introducing valid and super-valid inequalities on the instances characterized by 60 patients, which are the most challenging in the data set. Figure 8 shows separately for the two policies, C-Bal on the left and C-BalU on the right, the value of the best lower bound returned by the solver when no cuts are used (the gray line), the value of the best lower bound when cuts are added (orange line) and the value of the objective function returned by the exact model equipped with cuts. We observe that the increase in the lower bound value is remarkable on all the instances in the data set and for each policy.

Finally, Table 4 gives an overview of the results obtained both in terms of solution quality and efficiency without and with the cuts. Specifically, with regards to solution quality, the table reports average, minimum, and maximum values of objective function value (m) and lower bound (LB), while with regards to efficiency, the table shows the average gap and the number of instances for which no feasible

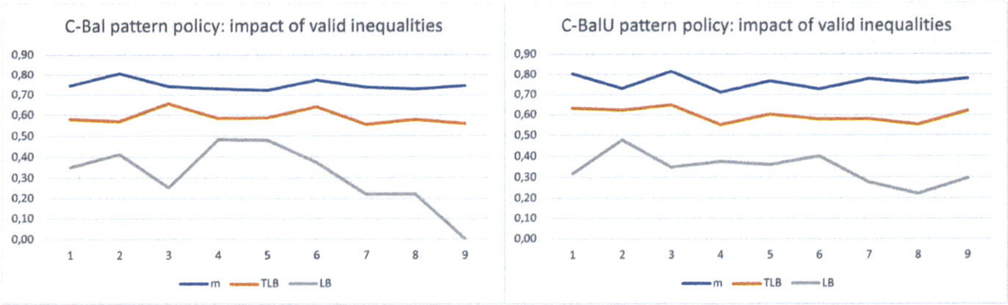

Fig. 8 Values of m, tightened LB and LB for each of the 9 instances with 60 patients: on the left for C-Bal policy, on the right for C-BalU policy

Table 4 Impact of valid inequalities on solution quality and efficiency - 60 patients: without and with cuts and for each pattern generation policy (either C-Bal or C-BalU), average, minimum and maximum values of objective function and LB, average optimality gap, number of not solved instances. The average value of the objective function is computed across the solved instances while the average LB is computed across all of the instances since the LB is available for each instance

	Without cuts		With cuts	
	C-Bal	C-BalU	C-Bal	C-BalU
avg m	0,78	0,80	0,75	0,76
min m	0,73	0,79	0,72	0,71
max m	0,81	0,81	0,81	0,82
avg LB	0,31	0,34	0,59	0,60
min LB	0,00	0,22	0,56	0,55
max LB	0,48	0,48	0,66	0,65
avg gap	0,49	0,53	0,21	0,21
not solved	5	7	0	0

solution is obtained within the memory and time limits imposed. The average gap decreases from 49 to 21% when C-Bal is used and from 53 to 21% when C-BalU is used. Significant is the number of instances for which no feasible solution is found when cuts are not used. The proposed cuts have therefore proved to be relevant both in breaking down optimality gaps and in finding feasible solutions.

6 Conclusions

In this paper, we have presented an exact model to the Consistent Robust Home Care Problem, which is based on a new flow-based formulation of the form of robustness addressed. The results of a preliminary comparison with an approximate formulation from the literature are also presented. We plan to enhance the resolution of the proposed exact model, via additional cuts and/or ad hoc methodological approaches, and also apply the presented flow-based robust modeling technique to other relevant application contexts.

Appendix

The complete model makes use of the following families of variables:

$$z_{jp} = \begin{cases} 1 \text{ if pattern } p \text{ is assigned to } j \\ 0 \text{ otherwise} \end{cases} \qquad j \in \bar{N} \cup \tilde{N}, p \in P_j$$

$$u_{\omega j} = \begin{cases} 1 \text{ if caregiver } \omega \text{ is assigned to } j \\ 0 \text{ otherwise} \end{cases} \qquad \omega \in O, j \in \bar{N} \cup \tilde{N}$$

$$x_{ij}^{\omega d} = \begin{cases} 1 \text{ if caregiver } \omega \text{ travels along } (i, j) \text{ on day } d \\ 0 \text{ otherwise} \end{cases} \qquad (i, j) \in A, d \in W, \omega \in O_d$$

$y_{ij}^{d} \geq 0$ number of nodes visited after i by a caregiver moving along (i, j) on day d, $d \in W$

$$Cons_j = \begin{cases} 1 \text{ if all requests of patient } j \text{ are scheduled in the morning} \\ 0 \text{ otherwise (i.e., they are scheduled in the afternoon).} \end{cases} \qquad j \in \bar{N}$$

Moreover, it uses the families of variables $\gamma_0^{\omega d}$, $\eta^{\omega d}$, $\gamma_i^{\omega d}$ and $\xi^{\omega d}$, i.e., the dual variables associated with constraints (14)–(17), respectively, and the family $\pi_{ij}^{\omega d}$, i.e., the dual variables associated with $f_{ij}^{\omega d} \leq 1$ in the linear relaxation of (18). Precisely, such dual variables are used to formulate the dual of the linear relaxations of inner problems (14)–(18), thus linearizing the constraints (1) previously presented.

Using the variables and the notation above, the Consistent Robust Home Care problem can be modeled exactly in terms of the following MILP formulation, where constraints (22)–(34) have already been presented in Cappanera and Scutellà (2022), while constraints (35)–(47) are the ones characterizing the proposed exact formulation. For a matter of convenience, in the following, we will use $\bar{t}_{ij}$ to denote the sum of the traveling time from i to j plus the service time of node j, i.e., $\bar{t}_{ij} = t_{ij} + t'_j$.

$$\min \quad m$$

$$\sum_{(i,j)\in A} \sum_{\omega\in O_d} x_{ij}^{\omega d} \leq \sum_{\substack{p\in P_j:\\ p(d)\geq 1}} z_{jp} \qquad j \in \bar{N} \cup \tilde{N}, d \in W \tag{22}$$

$$\sum_{(i,j)\in A} \sum_{\substack{\omega\in O_d:\\ s_\omega\geq k}} x_{ij}^{\omega d} \geq \sum_{\substack{p\in P_j:\\ p(d)=k}} z_{jp} \qquad j \in \bar{N} \cup \tilde{N}, d \in W, k \in K \tag{23}$$

$$\sum_{p\in P_j} z_{jp} = 1 \qquad j \in N \tag{24}$$

$$\sum_{\omega\in O_d} \left(\sum_{(i,j)\in A} x_{ij}^{\omega d} + \sum_{(i,\tilde{j})\in A} x_{i\tilde{j}}^{\omega d} \right) \leq 1 \qquad j \in \bar{N}, d \in W \tag{25}$$

$$\sum_{\omega\in O} (u_{\omega j} + u_{\omega\tilde{j}}) \leq T \qquad j \in \bar{N} \tag{26}$$

$$x_{ij}^{\omega d} \leq u_{\omega j} \qquad (i, j) \in A, j \in \bar{N} \cup \tilde{N}, \; d \in W, \omega \in O_d \tag{27}$$

$$u_{\omega j} \leq \sum_{(i,j)\in A} \sum_{d\in W} x_{ij}^{\omega d} \qquad j \in \bar{N} \cup \tilde{N}, \omega \in O \tag{28}$$

$$\sum_{(i,j)\in A} x_{ij}^{\omega d} = \sum_{(j,i)\in A} x_{ji}^{\omega d} \qquad j \in \bar{N} \cup \tilde{N}, d \in W, \omega \in O_d \tag{29}$$

$$\sum_{(0,j)\in A} y_{0j}^{d} = \sum_{j\in N} \sum_{\substack{p\in P_j:\\ p(d)\geq 1}} z_{jp} \qquad d \in W \tag{30}$$

$$\sum_{(i,j)\in A} y_{ij}^{d} - \sum_{(j,i)\in A} y_{ji}^{d} = \sum_{\substack{p\in P_j:\\ p(d)\geq 1}} z_{jp} \qquad j \in \bar{N} \cup \tilde{N}, d \in W \tag{31}$$

$$y_{ij}^{d} \leq 2n \sum_{\omega\in O_d} x_{ij}^{\omega d} \qquad (i, j) \in A, d \in W \tag{32}$$

$$Cons_j \leq 1 - \left(\sum_{\omega_a\in O} (u_{\omega_a j} + u_{\omega_a \tilde{j}})/T \right) \qquad j \in \bar{N} \tag{33}$$

$$Cons_j \geq \sum_{\omega_m \in O} (u_{\omega_m j} + u_{\omega_m \bar{j}})/T \qquad j \in \bar{N} \tag{34}$$

$$\sum_{k \in N \cup \{0\}} ((n-1)x_{ki}^{\omega d} + y_{ki}^{d}) - (n-1)\alpha_{ij}^{\omega d}) + \tag{35}$$

$$+ \sum_{k \in N \cup \{0\}} ((n-1)x_{kj}^{\omega d} - y_{kj}^{d}) \leq 2(n-1)$$

$$\alpha_{ij}^{\omega d} \leq \sum_{(k,i)} x_{ki}^{\omega d} \qquad i, j \in N, d \in W, \omega \in O_d$$

$$d \in W, \omega \in O_d, \tag{36}$$

$$\alpha_{ij}^{\omega d} \leq \sum_{(k,j)} x_{kj}^{\omega d} \qquad (i,j) \in A, i, j \in N$$

$$d \in W, \omega \in O_d, \tag{37}$$

$$1 - \alpha_{ij}^{\omega d} \geq \left(\sum_{(k,j)} y_{kj}^{d} - \sum_{(k,i)} y_{ki}^{d} \right) /2n \qquad (i,j) \in A, i, j \in N$$

$$d \in W, \omega \in O_d, \tag{38}$$

$$\alpha_{0j}^{\omega d} \leq \sum_{(k,j) \in A} x_{kj}^{\omega d} \qquad (i,j) \in A, i, j \in N$$

$$d \in W, \omega \in O_d, \tag{39}$$

$$\alpha_{j0}^{\omega d} \leq \sum_{(k,j) \in A} x_{kj}^{\omega d} \qquad (0,j) \in A$$

$$d \in W, \omega \in O_d, \tag{40}$$

$$\gamma_i^{\omega d} - \gamma_j^{\omega d} + \pi_{ij}^{\omega d} \geq \bar{t}_{ij}(-M + (1+M)\alpha_{ij}^{\omega d}) \qquad (j,0) \in A$$

$$d \in W, \omega \in O_d, \tag{41}$$

$$(i,j) \in A :$$

$$\gamma_i^{\omega d} - \gamma_j^{\omega d} + \pi_{ij}^{\omega d} + \xi^{\omega d} \geq \bar{t}_{ij}(-M + (1+M)\alpha_{ij}^{\omega d}) \qquad i \in N, j \in \bar{N} \cup \{0\}$$

$$d \in W, \omega \in O_d, \tag{42}$$

$$(i,j) \in A :$$

$$\eta^{\omega d} - \gamma_j^{\omega d} + \pi_{0j}^{\omega d} \geq \bar{t}_{0j}(-M + (1+M)\alpha_{0j}^{\omega d}) \qquad i \in N, j \in \tilde{N}$$

$$d \in W, \omega \in O_d, \tag{43}$$

$$\eta^{\omega d} - \gamma_j^{\omega d} + \pi_{0j}^{\omega d} + \xi^{\omega d} \geq \bar{t}_{0j}(-M + (1+M)\alpha_{0j}^{\omega d}) \quad (0,j) \in A : j \in \bar{N}$$

$$d \in W, \omega \in O_d, \quad (44)$$

$$(0,j) \in A : j \in \tilde{N}$$

$$R^{\omega d} = -\gamma_0^{\omega d} + \eta^{\omega d} + \Gamma\xi^{wd} + \sum_{(i,j)\in A} \pi_{ij}^{\omega d} \leq D_\omega \quad d \in W, \omega \in O_d \quad (45)$$

$$\frac{\sum_{d\in W}(R^{\omega_m d} + R^{\omega_a d})}{|W|(D_{\omega_m} + D_{\omega_a})} \leq m \quad \omega \in O \quad (46)$$

$$\alpha_{ij}^{\omega d} \in \{0,1\} \quad (i,j) \in A, \quad (47)$$

$$\pi_{ij}^{\omega d} \geq 0 \quad d \in W, \omega \in O_d$$

$$(i,j) \in A, \quad (48)$$

$$\xi^{\omega d} \geq 0 \quad d \in W, \omega \in O_d$$

$$d \in W, \omega \in O_d. \quad (49)$$

Constraints (22) state that at most one caregiver per day can visit a patient j, if a visit has been scheduled on that day for node j. Constraints (23) impose that, on day d and for each skill k, exactly one caregiver of adequate skill must visit j, if a service of that skill has been scheduled for j on day d. Constraints (24) ensure that exactly one pattern is assigned to each node modeling a patient. Specifically, one pattern is assigned to each node modeling the certain nature of the patient and also to the node modeling their uncertain nature, thus scheduling all the certain and all the uncertain requests of each patient. Consequently, the design variables $x_{ij}^{\omega d}$ model a set of tours that include all the certain and all the uncertain requests. Constraints (25) impose that, on each day and for each patient j, at most one of the two nodes modeling the patient, i.e., j and $\tilde{j}$, can be visited. Constraints (26) express the care continuity requirement, i.e., they bound the number of caregivers who can visit each patient during the time horizon. In particular, the two nodes modeling each patient are considered together. Constraints (27) guarantee that a caregiver can visit a patient only if assigned to that patient, while constraints (28) force $u_{\omega j}$ to zero if caregiver ω never visits patient j. Constraints (29) are the classical flow conservation constraints on the routing variables. Constraints (30) and (31) are the flow conservation constraints on the auxiliary y variables, which are introduced to avoid subtours. They also guarantee the correct linking between scheduling and flow variables. Constraints (32) link together routing and flow variables. Constraints (33)–(34) model the consistency requirements. In fact, they guarantee that all the caregivers assigned to a patient are of the same type, either morning or afternoon. As for the care continuity constraints, in constraints (33)–(34) the two nodes modeling each patient are considered together.

Constraints (35)–(40), which involve the auxiliary variables $\alpha_{ij}^{\omega d}$, have been introduced and commented in Sect. 4.2, while (41)–(44) are the dual constraints related to the inner problems (14)–(19). Moreover, constraints (42) guarantee that the set of tours described by the design variables $x_{ij}^{\omega d}$ is a robust set of tours with respect to Γ. Notice that such constraints are constraints (1) once the maximum length of the sequence-preserving Γ-tours, i.e., $R^{\omega d}$, is replaced by the objective function of the dual of the linear relaxation of (14)–(19), which gives its exact statement, as previously proved. Then, constraints (43), thanks to the auxiliary variable m, allow one to state the objective function of the overall model, i.e., the maximum caregiver utilization factor, which we want to minimize. Finally, constraints (44)–(46) define the variable domain.

References

Beaulieu, H., Ferland, J. A., Gendron, B., & Michelon, P. (2000). A mathematical programming approach for scheduling physicians in the emergency room. *Health Care Management Science, 3*, 193–200.

Bennett, A. R., & Erera, A. L. (2011) Dynamic periodic fixed appointment scheduling for home health. *IIE Transactions on Healthcare Systems Engineering, 1*(1), 6–19.

Bertels, S., & Fahle, T. (2006). A hybrid setup for a hybrid scenario: combining heuristics for the home health care problem. *Computers and Operations Research, 33*(10), 2866–2890.

Bertsimas, D., & Sim, M. (2004). The price of robustness. *Operations Research, 52*(1), 35–53.

Bredström, D., & Rönnqvistr, M. (2008). Combined vehicle routing and scheduling with temporal precedence and synchronization constraints. *European Journal of Operational Research, 191*(1), 19–31.

Cappanera, P., & Scutellà, M. G. (2015). Joint assignment, scheduling, and routing models to home care optimization: A pattern-based approach. *Transportation Science, 49*(4), 830–852.

Cappanera, P., & Scutellà, M. G. (2017). Pattern generation policies to cope with robustness in home care. In: P. Cappanera, J. Li, A. Matta, E. Sahin, N. Vandaele, & F. Visintin (Eds.), *Health care systems engineering. ICHCSE 2017*. Springer proceedings in mathematics & statistics (Vol. 210, pp. 257–268). Springer.

Cappanera, P., & Scutellà, M. G. (2022). Addressing consistency and demand uncertainty in the home care planning problem. *Flexible Services and Manufacturing Journal, 34*, 1–39.

Cappanera, P., Scutellà, M. G., Nervi, F., & Galli, L. (2018). Demand uncertainty in robust home care optimization. *Omega, 80*, 95–110.

Carello, G., & Lanzarone, E. (2014). A cardinality-constrained robust model for the assignment problem in home care services. *European Journal of Operational Research, 236*(2), 748–762.

Carello, G., Lanzarone, E., Laricini, D., & Servilio, M. (2017). Handling time-related demands in the home care nurse-to-patient assignment problem with the implementor-adversarial approach. In: P. Cappanera, J. Li, A. Matta, E. Sahin, N. Vandaele, & F. Visintin (Eds.), *Health care systems engineering. ICHCSE 2017*. Springer proceedings in mathematics & statistics (Vol. 210, pp. 87–97). Springer.

Chun-Cheng, L., Lun-Ping, H., Wan-Yu, L., & Ming-Chun, T. (2018). Jointly rostering, routing, and rerostering for home health care services: A harmony search approach with genetic, saturation, inheritance, and immigrant schemes. *Computers & Industrial Engineering, 115*, 151–166.

Cire, A. A., & Diamant, A. (2022). Dynamic scheduling of home care patients to medical providers. *Production and Operations Management, 31*(11), 4038–4056.

Crainic, TG., Gendreau, M., & Gendron, B. (Eds) (2021). *Network design with applications in transportation and logistics*. Springer.

Demirbilek, M., Branke, J., & Strauss, A. (2019). Dynamically accepting and scheduling patients for home healthcare. *Health Care Management Science, 22*, 140–155.

Di Mascolo, M., Martinez, C., & Espinouse, M. L. (2021). Routing and scheduling in home health care: a literature survey and bibliometric analysis. *Computers & Industrial Engineering, 158*, 1–48, 107255.

Du, G., & Zhang, J. (2022). Cross-regional manpower scheduling and routing problem with stochastic service times in home health care. *Computers & Industrial Engineering, 173*, 108668.

Du, G., Zheng, L., & Ouyang, X. (2019). Real-time scheduling optimization considering the unexpected events in home health care. *Journal of Combinatorial Optimization, 37*(1), 196–220.

Elbenani, B., Ferland, J. A., & Gascon, V. (2008). Mathematical programming approach for routing home care nurses. In: *IEEE international conference on industrial engineering and engineering management* (pp. 107–111)

Eveborn, P., Flisberg, P., & Rönnqvist, M. (2006). Laps care-and operational system for staff planning of home care. *European Journal of Operational Research, 171*(3), 962–976.

Feillet, D., Garaix, T., Lehuédé, F., Péton, O., & Quadri, D. (2014). A new consistent vehicle routing problem for the transportation of people with disabilities. *Networks, 63*, 211–224.

Fikar, C., & Hirsch, P. (2015). A matheuristic for routing real-world home service transport systems facilitating walking. *Journal of Cleaner Production, 105*, 300–310.

Fikar, C., & Hirsch, P. (2017). Home health care routing and scheduling: A review. *Computers & Operations Research, 77*, 86–95.

Frey, R. V., Bayón, T., & Totzek, D. (2013). How customer satisfaction affects employee satisfaction and retention in a professional services context. *Journal of Service Research, 16*(4), 503–517.

Gamst, M., & Jensen, T. S. (2012). A branch-and-price algorithm for the long-term home care scheduling problem. In: D. Klatte, H. J. Lthi, K. Schmedders (Eds.), *Operations Research Proceedings 2011* (pp. 483–488), Springer Berlin Heidelberg.

Grieco, L., Utley, M., & Crowe, S. (2021). Operational research applied to decisions in home health care: A systematic literature review. *Journal of the Operational Research Society, 72*(9), 1960–1991.

Groër, C., Golden, B. L., & Wasil, E. (2009). The consistent vehicle routing problem. *Manufacturing & Service Operations Management, 11*(4), 630–643.

Hertz, A., & Lahrichi, N. (2009). A patient assignment algorithm for home care services. *Journal of the Operational Research Society, 60*(4), 481–495.

Holte, M., & Mannino, C. (2013). The implementor/adversary algorithm for the cyclic and robust scheduling problem in health-care. *European Journal of Operational Research, 226*, 551–559.

Kovacs, A. A., Golden, B. L., Hartl, R. F., & Parragh, S. N. (2014). Vehicle routing problems in which consistency considerations are important: A survey. *Networks, 64*(3), 192–213.

Liu, R., Xie, X., & Garaix, T. (2014). Hybridization of tabu search with feasible and infeasible local searches for periodic home health care logistics. *Omega, 47*, 17–32.

Liu, R., Yuan, B., & Jiang, Z. (2019). A branch-and-price algorithm for the home-caregiver scheduling and routing problem with stochastic travel and service times. *Flexible Services and Manufacturing, 31*, 989–1011.

Mankowska, D., Meisel, F., & Bierwirth, C. (2014). The home health care routing and scheduling problem with interdependent services. *Health Care Management Science, 17*(1), 15–30.

Maya Duque, P. A., Castro, M., Sörensen, K., & Goos, P. (2015). Home care service planning. The case of Landelijke Thuiszorg. *European Journal of Operational Research, 243*(1), 292–301.

Moosavi, A., Ozturk, O., & Patrick, J. (2022). Staff scheduling for residential care under pandemic conditions: The case of COVID-19. *Omega, 112*, 102671.

Naderi, B., Begen, M., Zaric, G., & Roshanaei, V. (2023). A novel and efficient exact technique for integrated staffing, assignment, routing, and scheduling of home care services under uncertainty. *Omega, 116*, 102805.

Nguyen, T. V. L., Toklu, N. E., & Montemanni, R. (2015). Matheuristic optimization for robust home health care services. *Lecture Notes in Management Science, 7*, 1–7.

Nickel, S., Schröder, M., & Steeg, J. (2012). Mid-term and short-term planning support for home health care services. *European Journal of Operational Research, 219*(3), 574–587.

Rasmussen, M. S., Justesen, T., Dohn, A., & Larsen, J. (2012). The home care crew scheduling problem: Preference-based visit clustering and temporal dependencies. *European Journal of Operational Research, 219*(3), 598–610.

Rest, K. D., Trautsamwieser, A., & Hirsch, P. (2012). Trends and risk in home health care. *Journal of Humanitarian Logistics and Supply Chain Management, 2*(1), 34–53.

Restrepo, M., Rousseau, L., & Vallée, J. (2020). Home healthcare integrated staffing and scheduling. *Omega, 95*, 102057.

Rodriguez, C., Garaix, T., Xie, X., & Augusto, V. (2015). Staff dimensioning in homecare services with uncertain demands. *International Journal of Production Research, 53*(24), 7396–7410.

Savaşer, S., & Kara, B. (2022). Mobile healthcare services in rural areas: an application with periodic location routing problem. *OR Spectrum, 44*(3), 875–910.

Shiri, M., Ahmadizar, F., & Mahmoudzadeh, H. (2021). A three-phase methodology for home healthcare routing and scheduling under uncertainty. *Computers & Industrial Engineering, 158*, 107416.

Shiri, M., Ahmadizar, F., Thiruvady, D., & Farvaresh, H. (2023). A sustainable and efficient home health care network design model under uncertainty. *Expert Systems with Applications, 211*, 118185.

Trautsamwieser, A., & Hirsch, P. (2014). A branch-price-and-cut approach for solving the medium-term home health care planning problem. *Networks, 64*(3), 143–159.

Wirnitzer, J., Heckmann, I., Meyer, A., & Nickel, S. (2016). Patient-based nurse rostering in home care. *Operations Research for Health Care, 8*, 91–102.

Wolsey, L. A. (1998). *Integer programming*. Wiley-Interscience series in discrete mathematics and optimization. John Wiley & Sons.

Yalçındağ, S., Cappanera, P., Scutellà, M. G., Matta, A., & Şahin, E. (2016). Pattern-based decompositions for human resource planning in home health care services. *Computers & Operations Research, 73*, 12–26.

Yuan, B., Liu, R., & Jiang, Z. (2014). Home health care crew scheduling and routing problem with stochastic service times. In: *Proceedings of 2014 IEEE international conference on automation science and engineering* (pp. 564–569)

Zheng, C., Wang, S., Li, N., & Wu, Y. (2021). Stochastic joint homecare service and capacity planning with nested decomposition approaches. *European Journal of Operational Research, 295*(1), 203–222.

Production Inventory Technician Routing Problem: A Bi-objective Post-sales Application

Alline Zanette, Michel Gendreau, and Walter Rei

1 Introduction

Supply chain management has become essential for companies to obtain a competitive advantage. Common practices applied by managers in the past, such as trade-offs against competing dimensions—namely cost, flexibility, quality and delivery—cannot be used anymore (Erengüç et al., 1999). The increasing competition of the market forces companies to simultaneously optimize multiple processes of the supply chain in order to improve their performance (Liu et al., 2021).

In this sense, the study of integrated problems has received increased attention, since they allow for improvements in more than one function of the supply chain at the same time. Amongst these integrated problems, we can highlight the production routing problem (PRP), which, in addition to the decisions of quantity to deliver to each customer, time of delivery and routing plans, also includes decisions about set ups and amounts to be produced. The PRP has broad application and has proved very successful in theory and practice, being able to reduce the logistic costs of several companies (e.g., Çetinkaya et al., 2009; Qiu et al., 2018a,b; Li et al., 2019b; Neves-Moreira et al., 2019; Qiu et al., 2019).

A. Zanette (✉)
MAGI, Polytechnique Montréal, Montréal, QC, Canada
e-mail: alline-2.zanette@polymtl.ca

M. Gendreau
CIRRELT and Département de mathématiques et de génie industriel, Polytechnique Montréal, Montreal, QC, Canada
e-mail: michel.gendreau@cirrelt.net

W. Rei
CIRRELT and School of Management, Université du Québec à Montréal, Montréal, QC, Canada
e-mail: rei.walter@uqam.ca

T. G. Crainic et al. (eds.), *Combinatorial Optimization and Applications*,
International Series in Operations Research & Management Science 358,
https://doi.org/10.1007/978-3-031-57603-4_16

The problem herein studied is inspired by a real case of post-sales maintenance services and it can be considered as an integration of the PRP with the technician routing and scheduling problem (TRSP), here called the production inventory technician routing and scheduling problem (PITRSP). In addition to the usual PRP decisions related to production and inventory, we assume the existence of a heterogeneous set of technicians who are assigned to serve customer inquiries and need to perform routes to do so.

To the best of our knowledge, after-sales aspects have not been studied in the context of the PRP, but they are highly relevant, since they can significantly impact profitability (de Souza et al., 2011; Choudhary, 2015; Murthy et al., 2004). When providing post-sales services, the inventory management of spare parts is considered fundamental to avoid delays, and to enable the companies to fulfill more orders in a shorter period of time (Syahrial et al., 2019; de Souza et al., 2011). Moreover, responsiveness is considered as a source of competitive advantage (Choudhary, 2015; Hallikas et al., 2014), and, along with competence, is one of the determinants of perceived service quality (Parasuraman et al., 1985). Thus, it is essential to provide prompt service and with skilled staff in order to guarantee higher levels of customer satisfaction.

In this chapter we study a PITRSP with a single capacitated production facility that produces multiple types of products. The supply chain is also made up of a set of geographically dispersed distribution centers (DCs), service centers (SCs) and tasks that need to be served. A set of technicians is available in each SC to perform the incoming tasks, which can be emergency or preventive. In order to achieve high levels of responsiveness, the field technicians need to provide fast and accurate service and need to have the minimum required skills to do so. A bi-objective formulation is proposed, aiming at minimizing costs and maximizing customer satisfaction. The final model is solved using Benders decomposition (BD) and trade-off and sensitivity analyses are performed.

The remainder of the chapter is organized as follows. In Sect. 2, we introduce the company and the case that inspired this work, as well as the problem description. The literature review is presented in Sect. 3, followed by the problem formulation in Sect. 4. Section 5 is dedicated to explaining the solution method, while Sect. 6 describes the instances, computational experiments performed and results. Conclusions are drawn in Sect. 7.

2 Problem Description

The company that inspired this general problem produces multiple types of tracked vehicles, and provides spare parts and maintenance service to its customers. Due to the type of vehicles produced, proper maintenance and fast repair are essential to reduce the costs incurred by the clients, thus, the company aims at providing an excellent, flexible and quick service. All the maintenance inquiries can be solved by the technical service team, and an emergency line is available to ensure that services are provided during holidays.

The main focus of this work is on the supply chain of spare parts, herein called products, as well as the maintenance service provided to the customers. In this context, the enterprise has a continental covering network, which contains a single production plant, two DCs and five SCs. The suppliers provide resources to the production plant and the DCs, and products are directly shipped between the facilities - from production plant to DCs, from DCs to SCs, and from SCs to clients - by third parties. In the cases of facilities located in the same site, the transportation is internal, and made by the company itself. Maintenance inquiries are made by the clients, and are served by the company's technicians, who execute routes using the company's own fleet of vehicles. Each inquiry requires the execution of one or two maintenance tasks, which demand products.

Currently, the enterprise uses one system to manage the sourcing/procurement phase and another one to control the distribution of products from DCs to SCs. The company wants to integrate the management of the production, inventory and distribution, incorporating the routing of technicians when providing maintenance services. Their goal is to minimize the operational costs while guaranteeing customer satisfaction.

Based on the information provided by the company, we propose a general application problem setting. The context to be addressed considers only the phases of production, inventory and distribution, including the routing of vehicles/technicians from SCs to serve the maintenance inquiries of each client. The sourcing/procurement stage is not covered here.

We assume that each type of spare part is a different product, and, instead of clients, we need to serve maintenance tasks, which can be geographically in the same location, demanded by the same customer. For facilities located in the same site, we consider that all the inventory is allocated in only one of them. The consequence is twofolded: the DCs do not hold any inventory and can be disconsidered from the network; and the shipment of products is now allowed between SCs. The flow of products and technicians then occurs as shown in Fig. 1.

The problem can be defined on a complete graph $\mathcal{G} = (\mathcal{N}, \mathcal{A})$, where $\mathcal{A} = \{(i, j) : i, j \in \mathcal{N}, i \neq j\}$ is the set of arcs, and $\mathcal{N}$ is the set of nodes, which consist of a production plant $\mathcal{P} = \{0\}$, a set of SCs $\mathcal{S} = \{1, \dots, S\}$ and a set of tasks $\mathcal{C} = \{S + 1, \dots, S + C\}$. Multiple products $\mathcal{M} = \{1, \dots, M\}$ are produced by the plant, with production capacity $C_m, m \in \mathcal{M}$, along a finite discrete time horizon $\mathcal{T} = \{1, \dots, l\}$. When production takes place in a period t, we assume that there are no delays, and products are immediately fabricated and distributed in the same period; moreover, a fixed setup cost f_m and a variable production cost u_m per unit are incurred for each product $m \in \mathcal{M}$. The products are shipped from the production plant to the SCs, and can be further transported between SCs. All the SCs supply the necessary products to perform a task in the period it is served, and we assume that the items arrive at the task's site in time for its execution. All the shipment of products is made by a third party, incurring a cost $c_{ijm}, i \in \mathcal{N} \backslash \mathcal{C}, j \in \mathcal{N} \backslash \mathcal{P}, m \in \mathcal{M}$. The products can be stored in the production facility and SCs up to their inventory

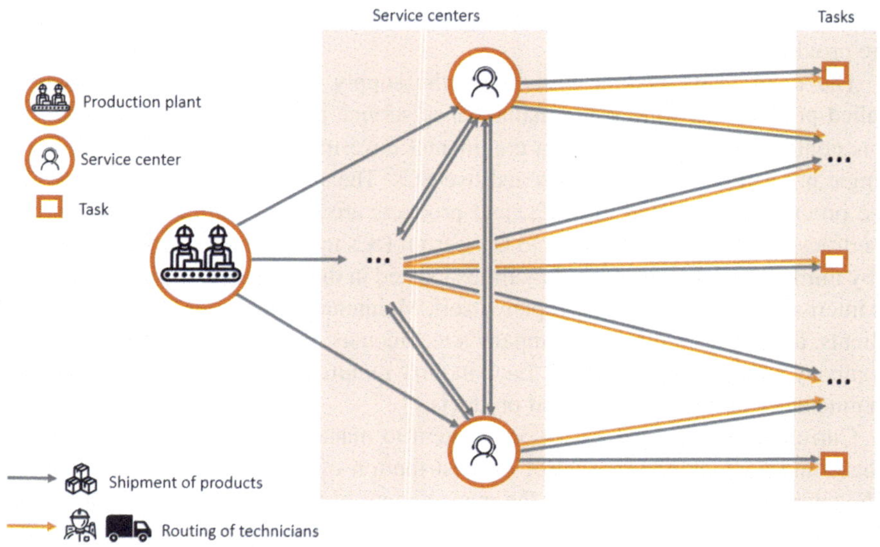

Fig. 1 Representation of the considered supply chain network

limit L_i, incurring a unit inventory holding cost of h_{im}, $i \in \mathcal{N} \setminus C, m \in \mathcal{M}$. No inventory is kept at the tasks' sites.

A limited number of heterogeneous technicians $\mathcal{K}_i = \{1, \ldots, K_i\}, i \in \mathcal{S}$, is available in each SC to provide maintenance services, and they do that by executing routes. Each route is a circuit in $\mathcal{G}$ and is served by a single technician. Each technician can perform at most one route per period, which must be completed within their working hours W. To perform the routes, each SC has a homogeneous fleet of vehicles, and their capacity is disconsidered, as all the products are directly shipped from SCs using a third party service. During the routing phase, when a vehicle travels directly from i to j, a cost r_{ij} is incurred, $i, j \in \mathcal{S} \cup C$, and the transportation time of arc (i, j) is $w_{ij}, i, j \in \mathcal{S} \cup C$.

Two types of maintenance inquiries are considered, namely scheduled and emergency. In the scheduled inquiries, appointments need to be requested for the services to be provided, which is usually the case for preventive maintenance. In this situation, only one task needs to be performed by the technical staff, and we can say that the task i belongs to a set C^{Δ} of tasks without successors or predecessors. This type of task has a demand d_{im} of products for its execution, $i \in C^{\Delta}, m \in \mathcal{M}$. For emergency inquiries, immediate technical service is provided, and two visits/tasks are needed. Although a single SC must be responsible for both visits, they can be performed by different employees. In the first task, the technician only makes a diagnosis of the work that needs to be done, thus there is no demand of products. These tasks belong to the set C^{Γ} of tasks with successors. In the second task, the maintenance service is actually executed, and it can start only after the first one is completed. We can then define a set Π_i containing the tasks that cannot be started before task $i \in C^{\Gamma}$ is finished, and a set C^{Π} of tasks with predecessors. Each task $i \in C^{\Pi}$ has a demand of products d_{im}, $i \in C^{\Pi}$, $m \in \mathcal{M}$.

Each task has an execution time given by e_i, $i \in C$, and it requires one or more skills from the technicians. Each employee can have multiple skills with different proficiency levels, thus, only those with the minimum level of skills to perform a task can be assigned to it. This is addressed by the parameter a_i^k, $i \in C$, $k \in \mathcal{K}_j$, $j \in \mathcal{S}$, which has value 1 if technician k has the required skills to perform task i and zero otherwise.

In order to guarantee the customers satisfaction, we assume that each task has a target period α_i, $i \in C$, to be performed. Any deviation from this target increases the customer dissatisfaction. In the specific case of emergency inquiries, the sooner they are executed, the better; thus, if the first and second tasks are served on the first and second periods, respectively, the customer dissatisfaction is minimum.

The goal of the problem is two folded, i.e., we need to execute the tasks within the given time horizon, such that the costs are minimized and the customers dissatisfaction is also minimized. The latter is addressed by minimizing how far from the target periods the tasks are actually served.

3 Literature Review

The amount of literature that deals with the optimization of integrated problems has significantly increased in the past decades. This is highly motivated by the perceived benefits from the use of this approach, as well as by the advances in the optimization methods, information systems, and computer capabilities (Díaz-Madroñero et al., 2015; Mostafa & Eltawil, 2015; Absi et al., 2018; Avci & Yildiz, 2019; Ji et al., 2019). However, to the best of our knowledge, there are no studies that comprise decisions on production, inventory, distribution, and assignment and routing of technicians simultaneously. These decisions can be made by combining the PRP and the TRSP. In this section, we explore the relevant literature for the studied problem, which includes the PRP, the TRSP and BD applications on integrated problems.

3.1 Production Routing Problem

The PRP is a generalization of the inventory routing problem (IRP). Considering its main characteristics in the available literature, it is possible for us to define its general structure. It usually involves a production-distribution network with a single production facility and a set of geographically dispersed customers/retailers, in which all the nodes have a dedicated storage area for finished goods. In each period of the planning horizon, the demands at the clients have to be completely fulfilled, without backordering, and the plant has to decide whether production occurs or not. If so, the lot size has to be determined, according to the available production capacity. A single product is then distributed to the customers by a limited number of homogeneous capacitated vehicles. Generally the goal is to minimize the operational costs, which include set up and unit production costs, holding costs in every node, and transportation costs.

Most studies focus on the deterministic version of the PRP. Some of its variants include, but are not limited to, the following aspects: multiple production facilities (Lei et al., 2006); uncapacitated plants (Shiguemoto & Armentano, 2010); multi-stage production (Vahdani et al., 2017); multiple products (Shiguemoto & Armentano, 2010); perishable products (Li et al., 2016); unlimited number of vehicles (Adulyasak et al., 2014b); heterogeneous fleet (Çetinkaya et al., 2009); uncapacitated vehicles (Absi et al., 2018); transshipment (Fattahi et al., 2017); physical internet (Ji et al., 2019); time windows (Kumar et al., 2016); replenishment lead time (Liu et al., 2019); visit spacing (Avci & Yildiz, 2019); backordering (Brahimi & Aouam, 2016); outsourcing (Li et al., 2019a); startup costs (Qiu et al., 2018b); machine scheduling (Du et al., 2019); multiple time grids (Zhang et al., 2017); uncertain data (Adulyasak et al., 2015a); closed-loop supply chain (Zhang et al., 2018); multi-objective optimization (Fattahi et al., 2017); and reduction of emissions (Kumar et al., 2016). More details concerning the available studies can be found in the survey of Adulyasak et al. (2015b).

As the PRP is an NP-hard problem, most studies have focused on the use of heuristic methods to solve it. However, exact algorithms have been successfully applied to deterministic and stochastic versions of the problem. The majority of the authors implement branch-and-cut (B&C) algorithms (e.g., Ruokokoski et al., 2010; Adulyasak et al., 2014a; Qiu et al., 2018a), but decomposition methods, such as BD are also applied in some studies. Due to the characteristics of our problem, this is the chosen approach to solve the proposed formulation, and it will be further discussed in Sect. 3.3.

3.2 Technician Routing and Scheduling Problem

In our problem, the vehicle routing part of the PRP is substituted by the TRSP. The latter can be seen as an extension of the vehicle routing problem (VRP), in which technicians play the role of vehicles, clients are associated with inquiries, and no vehicle capacity is considered. In this case, a set of tasks needs to be served, generally by a skilled professional, exactly once in the time horizon. As the number of tasks is usually greater than the number of technicians, they need to perform routes to visit all the locations. Differently from the pick-up and delivery problems, the tasks in the TRSP have a significant time of execution, which should be considered when performing the routes (Castillo-Salazar et al., 2016; Chen et al., 2016; Pillac et al., 2013). Many variants of the TRSP exist and it is a problem that has had applications in different fields, e.g., health, maintenance, telecommunications, public utilities, and security (Castillo-Salazar et al., 2016).

Similarly to our study, some of the literature on the TRSP is based on real case problems. The TRSP in the context of emergency electrical breakdown maintenance services is studied by Weintraub et al. (1999), in which random requests are considered. The quality of the service is measured considering the response time, and the authors assume a priority level for each request, which depends on how long it has been since the breakdown occurred. This study shows improvements of 16% in average weighted time in comparison to the manual routing and assignment of technicians. Maintenance services for car breakdowns are studied by Grötschel et al. (2003). The authors use column generation (CG) to solve a TRSP with soft windows, aiming at minimizing costs. In order to reduce the service time, it is assumed that contractors are available to serve the requests that cannot be performed by the company. A TRSP with soft time windows is also addressed by Cortés et al. (2014), and is solved using a branch-and-price (B&P) algorithm. This work analyzes the post-sales maintenance network of a real company. Target service times are assumed, depending on the importance of each client, and customer satisfaction is maximized if travel times and the difference between the actual service time and the target service time are minimal. A TRSP with stochastic travel times is addressed by Hadjiconstantinou and Roberts (2002) in the context of maintenance services for a utility company. A two-stage stochastic programming formulation aiming at minimizing operational costs is proposed, and emergency, planned and unplanned tasks are considered. A paired search tree algorithm is proposed, and the authors obtain reductions in the travel time, travel distance and number of overtime work hours.

The skill set of technicians is also an important aspect of some studies on TRSP. In the work of Kovacs et al. (2012), the authors address a TRSP in which a set of tasks needs to be served by skilled technicians within a given time window. Each technician has a proficiency level in each existing skill. Similarly, each task has a minimum skill level requirement to be executed (Cordeau et al., 2010; Kovacs et al., 2012). The technicians can be teamed up or not, and if there are no professionals (or teams) with the minimum skills to serve a task, the service is outsourced. The goal is to minimize costs and an adaptive large neighborhood search (ALNS) is used to solve the proposed formulations. Chen et al. (2016) assume that each task requires a given skill level, and the service time depends on the experience of each technician on that specific skill, i.e., the lower the technician's skill level, the longer it takes for the task to be executed. After finishing a task, the technicians gain experience on its associated skill. The authors use a rolling horizon approach and a local search algorithm to solve the problem, which aims at minimizing the completion time of the last task.

Customer satisfaction is also relevant in post-sales services, as customers have high expectations for both products and services provided by manufacturers (Syahrial et al., 2019). We can understand customer behavior under the light of prospect theory (Tversky & Kahneman, 2019). According to it, an individual selects a reference point in the beginning of the decision process, and evaluates outcomes relative to that reference point, classifying them as gains or losses. In the context of technicians scheduling, customer satisfaction depends directly on the time the service is provided (Wu et al., 2021). Customers have arrival time expectations, and if the actual service time differs from that, a loss is incurred. Some studies related to technician scheduling that aim at reducing the service response time assume the tasks as the focal point. Cordeau et al. (2010) and Xu and Chiu (2001) set a priority for each task and try to serve first the most important ones. In the study of Tang et al. (2007), the priority of each task is characterized by time-dependent rewards, which need to be maximized; the more urgent a task is, the higher the rewards associated with it. Some authors, on the other hand, consider the perspective of the customer to improve the service provided. Jia et al. (2008) define a piecewise linear satisfaction function, based on the actual service time and on a tolerable window, to maximize the customer satisfaction. Similarly, Wu et al. (2021) propose a model that minimizes a dissatisfaction function inspired by prospect theory, which also depends on the actual time the customer is served and on a tolerable time interval.

3.3 Benders Decomposition

The BD algorithm has been successfully used to solve complex problems in several fields of application (Rahmaniani et al., 2017). Amongst them, some integrated problems, like the one herein studied, can be highlighted. The aircraft routing and crew scheduling problem is solved by Cordeau et al. (2001) by means of an heuristic procedure. The authors use BD and CG to solve a relaxed version of the model; the integrality requirements are then sequentially reintroduced in the master problem (MP) and in the subproblem (SP), which are solved again. Results show that the proposed solution method performs better than traditional CG, and that the integrated approach allows for cost savings of up to 9.4 %, when compared with solving the problems sequentially. Restrepo et al. (2018) also use BD and CG to solve a multi-activity tour scheduling problem, aiming at minimizing costs while respecting the demand of employees and work rules. The problem is decomposed using BD, and the MP is solved using CG, due to the large number of tour related variables. Results demonstrate that the solution approach is able to produce solutions with small gaps that outperform those of the B&P method. Alkaabneh et al. (2020) solve an IRP considering perishable products and greenhouse gas emissions. Besides the traditional BD, the authors also implement some acceleration strategies, namely, valid inequalities, Pareto optimal cuts, knapsack inequalities, trust region, and a warm start based on the greedy randomized adaptative search procedure (GRASP). Results show that, in general, the BD with acceleration strategies outperforms both the traditional BD and the commercial solver GAMS. An IRP is also addressed by Wang and Chu (2022), who propose a decomposition-based heuristic to solve it. A relaxed version of the problem is initially solved using a logic-based BD algorithm (Hooker, 2019), which includes inequalities for diversification of the solutions and for greedy search, in order to accelerate the solution process. A restricted version of the model is then solved, in order to explore the most promising part of the solution space previously obtained. The solution method used has better performance than some state of the art heuristics, both in terms of number of optimal solutions obtained and gaps.

To the best of our knowledge, the seminal work that uses BD in the context of the PRP is the paper by Adulyasak et al. (2015a). The authors study a general PRP with stochastic demand, and propose two formulations, one with two stages and another with multiple stages. To solve the problem, the authors implement a Benders-based branch-and-cut (BBC) algorithm and incorporate computational enhancements for both formulations, which allow for the attainment of better results. In Zhang et al. (2021), a multi-vehicle PRP is solved using a logic-based BD algorithm. In their study, the dual linear problem of the traditional BD is generalized into an inference dual and its solution takes the shape of a logical deduction that yields valid Benders cuts. When the decomposition is performed, in addition to the MP, the authors end up with a capacitated VRP for each period, which are solved by an exact algorithm based on the set partitioning formulation. Finally, the BD method is also used by Farghadani-Chaharsooghi et al. (2022) to solve a PRP with workforce planning, perishable products, time windows and uncertain demand and travel times. The authors use progressive hedging (PH) to decompose the two-stage stochastic formulation into multiple independent scenario-based problems, which are then solved using a BD algorithm. The latter is accelerated by using a genetic algorithm to generate initial solutions for the routing variables.

Considering the problem that is tackled here, we propose a bi-objective formulation for the PITRSP, aiming at minimizing the operational costs and minimizing customer dissatisfaction. As a special feature, we consider that each task is associated with a matrix of skill requirements, and each technician has a given set of skill levels. This is done by adopting a similar approach to the one used in the studies of Cordeau et al. (2010) and Kovacs et al. (2012). Moreover, the customer satisfaction level depends on a given target period for a task to be served and the actual period it is performed.

4 Problem Formulation

In this section, we describe the parameters and variables used to formulate the problem herein studied, as well as the model itself.

Taking into consideration the assumptions and description of the problem mentioned in Sect. 2, we can formulate it as follows:

Sets/Subsets

$C^{\Delta} \subseteq C$	set of tasks without successors or predecessors
$C^{\Gamma} \subseteq C$	set of tasks with successors
$C^{\Pi} \subseteq C$	set of tasks with predecessors
$\Pi_i \subseteq C$	set of tasks that cannot start before task $i \in C^{\Gamma}$ is finished
$\mathcal{K}$	set of technicians in the system

Parameters

S	number of SCs
C	number of tasks
M	number of products
l	number of periods
C_m	production capacity of product m, $m \in \mathcal{M}$
f_m	fixed setup cost for production of product m, $m \in \mathcal{M}$
u_m	unit production cost of product m, $m \in \mathcal{M}$
d_{im}	demand of task i for product m, $i \in C$, $m \in \mathcal{M}$
c_{ijm}	shipping cost of product m from i to j, $i \in \mathcal{N} \setminus C$, $j \in \mathcal{N} \setminus \mathcal{P}$, $m \in \mathcal{M}$
L_i	inventory limit at node i, $i \in \mathcal{N} \setminus C$
h_{im}	unit inventory holding cost of product m at node i, $i \in \mathcal{N} \setminus C$, $m \in \mathcal{M}$
I_{im0}	initial inventory of product m at node i, $i \in \mathcal{N} \setminus C$, $m \in \mathcal{M}$
K_i	number of technicians available at SC i, $i \in \mathcal{S}$
a_i^k	binary parameter which is equal to 1 if technician k has the required skills to perform task i, and 0 otherwise, $i \in C$, $k \in \mathcal{K}$
W	maximum number of hours that each technician can work in a period
r_{ij}	travel cost over arc (i, j), $i, j \in \mathcal{S} \cup C$
w_{ij}	travel time over arc (i, j), $i, j \in \mathcal{S} \cup C$
e_i	time needed to execute task i, $i \in C$
β_i	target period for a task i to be served, $i \in C$

Decision Variables

x_{ijt}^k	binary variable which is equal to 1 if technician k serves node i immediately before node j in period t, and 0 otherwise, $i, j, \in \mathcal{S} \cup C$, $k \in \mathcal{K}$, $t \in \mathcal{T}$
z_{st}	number of vehicles that leave SC s in period t, $s \in \mathcal{S}$, $t \in \mathcal{T}$
z_{it}	binary variable which is equal to 1 if task i is visited in period t, and 0 otherwise, $i \in C$, $t \in \mathcal{T}$
y_{mt}	binary variable which is equal to 1 if a set up to produce item m occurs in period t, and 0 otherwise, $m \in \mathcal{M}$, $t \in \mathcal{T}$
p_{mt}	quantity of product m produced in period t, $m \in \mathcal{M}$, $t \in \mathcal{T}$
q_{ijmt}	quantity of product m sent from node i to node j in period t, $i \in \mathcal{N} \setminus C$, $j \in \mathcal{N} \setminus \mathcal{P}$, $m \in \mathcal{M}$, $t \in \mathcal{T}$
I_{imt}	quantity of product m in inventory at node i at the end of period t, $i \in \mathcal{N} \setminus C$, $m \in \mathcal{M}$, $t \in \mathcal{T}$

$[PITRSP]$:

$$f_1 = \min_{\substack{x,y,z \\ p,q,I}} \sum_{t \in \mathcal{T}} \left(\sum_{m \in \mathcal{M}} \left(f_m y_{mt} + u_m p_{mt} + \sum_{i \in \mathcal{N} \setminus C} h_{im} I_{imt} + \sum_{i \in \mathcal{N} \setminus C} \sum_{j \in \mathcal{N} \setminus \mathcal{P}} c_{ijm} q_{ijmt} \right) + \sum_{i,j \in \mathcal{S} \cup C} \sum_{k \in \mathcal{K}} r_{ij} x_{ijt}^k \right) \tag{1}$$

$$f_2 = \min_{\substack{x,y,z\\p,q,I}} \quad \frac{100}{A}\sum_{t\in\mathcal{T}}\sum_{i\in C}|tz_{it} - \beta_i| \tag{2}$$

$$\text{s.t. } I_{0,m,t-1} + p_{mt} = I_{0mt} + \sum_{i\in\mathcal{S}} q_{0imt}, \qquad m\in\mathcal{M}, t\in\mathcal{T} \tag{3}$$

$$I_{i,m,t-1} + q_{0imt} + \sum_{\substack{j\in\mathcal{S}\\i\neq j}} q_{jimt}$$

$$= I_{imt} + \sum_{\substack{j\in\mathcal{S}\\i\neq j}} q_{ijmt} + \sum_{j\in C} q_{ijmt}, \qquad i\in\mathcal{S}, m\in\mathcal{M}, t\in\mathcal{T} \tag{4}$$

$$\sum_{m\in\mathcal{M}} I_{0mt} \leq L_0, \qquad t\in\mathcal{T} \tag{5}$$

$$\sum_{m\in\mathcal{M}} (I_{i,m,t-1} + q_{0imt} + \sum_{\substack{j\in\mathcal{S}\\i\neq j}} q_{jimt}) \leq L_i, \qquad i\in\mathcal{S}, t\in\mathcal{T} \tag{6}$$

$$p_{mt} \leq C_m y_{mt}, \qquad m\in\mathcal{M}, t\in\mathcal{T} \tag{7}$$

$$\sum_{t\in\mathcal{T}} z_{it} = 1, \qquad i\in C \tag{8}$$

$$\sum_{t\in\mathcal{T}}\sum_{i\in\mathcal{S}} q_{ijmt} = d_{jm}, \qquad j\in C, m\in\mathcal{M} \tag{9}$$

$$z_{it}d_{im} \leq \sum_{j\in\mathcal{S}} q_{jimt}, \qquad i\in C, m\in\mathcal{M}, t\in\mathcal{T} \tag{10}$$

$$\sum_{j\in\mathcal{S}\cup C}\sum_{k\in\mathcal{K}_s} x_{ijt}^k$$

$$\leq \sum_{t'=t+1}^{l}\sum_{j\in\mathcal{S}\cup C}\sum_{k\in\mathcal{K}_s} x_{ijt'}^k, \qquad i\in C^\Gamma, i'\in\Pi_i, s\in\mathcal{S}, t\in\mathcal{T}\setminus\{l\} \tag{11}$$

$$z_{il} = 0, \qquad i\in C^\Gamma \tag{12}$$

$$z_{i0} = 0, \qquad i\in C^\Pi \tag{13}$$

$$\sum_{j\in\mathcal{S}\cup C} x_{ijt}^k \leq a_i^k, \qquad i\in C, k\in\mathcal{K}, t\in\mathcal{T} \tag{14}$$

$$\sum_{t\in\mathcal{T}}\sum_{j\in\mathcal{S}\cup C}\sum_{k\in\mathcal{K}} x_{ijt}^k = 1, \qquad i\in C \tag{15}$$

$$\sum_{j\in C} x_{ijt}^k \leq 1, \qquad i\in\mathcal{S}, k\in\mathcal{K}_i, t\in\mathcal{T} \tag{16}$$

$$\sum_{j\in\mathcal{S}\cup C} x_{ijt}^k \leq \sum_{i'\in C}\sum_{j\in\mathcal{S}\cup C} x_{i'jt}^k, \qquad i\in\mathcal{S}, k\in\mathcal{K}_i, t\in\mathcal{T} \tag{17}$$

$$\sum_{j \in \mathcal{S} \cup C} x_{ijt}^k = \sum_{j \in \mathcal{S} \cup C} x_{jit}^k, \quad i \in \mathcal{S} \cup C, k \in \mathcal{K}, t \in \mathcal{T} \tag{18}$$

$$\sum_{j \in \mathcal{S} \cup C} \sum_{k \in \mathcal{K}_s} x_{sjt}^k + \sum_{j \in \mathcal{S} \cup C} \sum_{k \in \mathcal{K}_s} x_{jst}^k = 2z_{st}, \quad s \in \mathcal{S}, t \in \mathcal{T} \tag{19}$$

$$\sum_{j \in \mathcal{S} \cup C} \sum_{k \in \mathcal{K}} x_{ijt}^k + \sum_{j \in \mathcal{S} \cup C} \sum_{k \in \mathcal{K}} x_{jit}^k = 2z_{it}, \quad i \in C, t \in \mathcal{T} \tag{20}$$

$$\sum_{i,j \in \mathcal{S} \cup C} w_{ij} x_{ijt}^k + \sum_{i \in C} \sum_{j \in \mathcal{S} \cup C} e_i x_{ijt}^k \leq W, \quad k \in \mathcal{K}, t \in \mathcal{T} \tag{21}$$

$$\sum_{i,j \in \alpha} x_{ijt}^k \leq |\alpha| - 1, \quad k \in \mathcal{K}_s, s \in \mathcal{S}, \quad \alpha \in \mathcal{S} \cup C \setminus \{s\}, |\alpha| \geq 2, t \in \mathcal{T} \tag{22}$$

$$x_{ijt}^k \in \{0, 1\}, \quad i, j \in \mathcal{S} \cup C, i \neq j, k \in \mathcal{K}, t \in \mathcal{T} \tag{23}$$

$$x_{ijt}^k = 0, \quad i, j \in \mathcal{S} \cup C, i = j, k \in \mathcal{K}, t \in \mathcal{T} \tag{24}$$

$$x_{ijt}^k = 0, \quad i, j \in \mathcal{S}, k \in \mathcal{K}, t \in \mathcal{T} \tag{25}$$

$$z_{st} \in \{0, \ldots, K_s\}, \quad s \in \mathcal{S}, t \in \mathcal{T} \tag{26}$$

$$z_{it} \in \{0, 1\}, \quad i \in C, t \in \mathcal{T} \tag{27}$$

$$y_{mt} \in \{0, 1\}, \quad m \in \mathcal{M}, t \in \mathcal{T} \tag{28}$$

$$p_{mt} \geq 0, \quad m \in \mathcal{M}, t \in \mathcal{T} \tag{29}$$

$$q_{ijmt} \geq 0, \quad i \in \mathcal{N} \setminus C, j \in \mathcal{N} \setminus \mathcal{P}, m \in \mathcal{M}, t \in \mathcal{T} \tag{30}$$

$$q_{ijmt} = 0, \quad i \in \mathcal{P}, j \in C, m \in \mathcal{M}, t \in \mathcal{T} \tag{31}$$

$$I_{imt} \geq 0, \quad i \in \mathcal{N} \setminus C, m \in \mathcal{M}, t \in \mathcal{T} \tag{32}$$

where $A = \sum_{i \in C^{\Delta}} \max(l - \beta_i, \beta_i - 1) + \left|C^{\Gamma} \cup C^{\Pi}\right| (l - 2)$.

The objective function (1) minimizes the production, inventory, shipping and routing costs, while (2) minimizes the so-called customer dissatisfaction. The latter is calculated as the percentage between how far from the target period the tasks are actually served (numerator), and how far they could have been served (denominator). Constraints (3) and (4) establish respectively the inventory balance at the production plant and at the SCs. Constraints (5) guarantee that the inventory limit of the production facility is respected, while (6) assure that the quantity of products at the SCs cannot exceed their inventory capacity after they receive the deliveries from their sourcing nodes. Constraints (7) are responsible for forcing a setup if there is production in period t and for limiting the production quantity to the plant capacity. Each task must be served only once in the time horizon (8), and the total quantity of products delivered in the time horizon to serve it needs to be equal to its demand (9). If a task is scheduled to be executed in period t, the quantity of products sent to its site in the same period needs to be at least equal to its demand (10). Constraints (11) guarantee the precedence between tasks and that both tasks from an emergency inquiry are served by the same SC. Tasks with successors and predecessors cannot be served in the last (12) and first period

(13), respectively. Each task can only be served by a technician with the minimum skills to do so (14), and it is visited by a single employee (15). Each technician can perform a maximum of one tour per period (16), and if a tour starts at a given SC, at least one task needs to be served by it (17). The flow conservation is ensured by constraint (18), while Eqs. (19) and (20) are the degree constraints at the SCs and task nodes, respectively. The time limit on tours is imposed by (21), while subtours are prevented by constraints (22). The non-negativity and integrality restrictions on the decision variables are defined by constraints (23)–(32).

5 Solution Procedure

In bi-objective optimization problems, some techniques are commonly used to account for the conflicting objective functions. One of the most common, which has also been successfully used in PRP applications (e.g., Kumar et al., 2016; Fattahi et al., 2017; Li et al., 2019b; Peng et al., 2019; Aazami et al., 2021), is the construction of a Pareto front/curve. The latter is formed by a set of solutions that can be chosen according to the trade-offs the decision maker is willing to make. An optimal solution to a bi-objective problem is called Pareto-optimal, or non-dominated, if improvements in one objective cannot be achieved without the deterioration of the other one. A set of Pareto-optimal solutions forms a Pareto front.

Generating the Pareto solutions, however, can be a cumbersome and time-consuming process. In the weighted-sum method, different linear weights are assigned to each objective function, which are then added to produce a new single goal that needs to be optimized. However, different combinations of weights can be associated with the same optimal solution, and the objective functions need to be converted to the same scale (Ehrgott, 2008; Li et al., 2019b), which in our case is complicated due to the nature of goals. The ϵ-constraint method, on the other hand, converts the bi-objective problem into a single-objective one, with the other goal being transformed into a constraint bounded by the parameter ϵ. Nonetheless, it is difficult to efficiently cover all the values of ϵ to be able to generate the exact Pareto front (Bérubé et al., 2009); moreover, a large number of problems would need to be solved, which is not practical for decision makers (Li et al., 2019b).

In order to solve our bi-objective model, we decided to solve two problems sequentially, namely, one aiming at minimizing the objective function (1) and one minimizing (2). The objective value of the first one is used to limit the admissible cost in the second problem. We also assumed the existence of cost tolerances (0%, 5% and 10%), in order to analyze the trade-offs between customer satisfaction and cost. This procedure requires substantially less computational time than producing an exact Pareto front, and still enables the decision maker to have useful managerial insights. Thus, the problems to be solved are given by

$$[PITRSP_C]: \min_{\substack{x,y,z\\p,q,I}} \sum_{t\in\mathcal{T}}\left(\sum_{m\in\mathcal{M}}\left(f_m y_{mt} + u_m p_{mt} + \sum_{i\in\mathcal{N}\setminus C} h_{im} I_{imt} + \right.\right.$$

$$\left.\left.\sum_{i\in\mathcal{N}\setminus C}\sum_{j\in\mathcal{N}\setminus\mathcal{P}} c_{ijm} q_{ijmt}\right) + \sum_{i,j\in\mathcal{S}\cup C}\sum_{k\in\mathcal{K}} r_{ij} x^k_{ijt}\right) \tag{1}$$

$$\text{s.t.} \quad (3)\text{–}(32)$$

$$[PITRSP_{CD}]: \min_{\substack{x,y,z\\p,q,I}} \frac{100}{A}\sum_{t\in\mathcal{T}}\sum_{i\in C}|tz_{it} - \beta_i| \tag{2}$$

$$\text{s.t.} \quad (3)\text{–}(32)$$

$$\sum_{t\in\mathcal{T}}\left(\sum_{m\in\mathcal{M}}\left(f_m y_{mt} + u_m p_{mt} + \sum_{i\in\mathcal{N}\setminus C} h_{im} I_{imt} + \right.\right.$$

$$\left.\left.\sum_{i\in\mathcal{N}\setminus C}\sum_{j\in\mathcal{N}\setminus\mathcal{P}} c_{ijm} q_{ijmt}\right) + \sum_{i,j\in\mathcal{S}\cup C}\sum_{k\in\mathcal{K}} r_{ij} x^k_{ijt}\right)$$

$$\leq Z_C(1 + tol) \tag{33}$$

where Z_C is the best objective value of $[PITRSP_C]$ (minimum cost) and tol is a given parameter that describes the allowed cost tolerance. In the problem $[PITRSP_{CD}]$, constraint (33) establishes that the production, inventory, shipping and routing costs need to be smaller or equal to the best objective value obtained in $[PITRSP_C]$ plus a given tolerance.

The $[PITRSP_C]$ and the $[PITRSP_{CD}]$ are NP-hard mixed-integer linear problems (MILP). One of the most used algorithms to solve this type of problem is the B&C (Karamanov, 2006), which solves the relaxed version of the original problem, and adds valid constraints to improve the solution. However, in the case of problems with complicating variables, such as the one in this study, decomposition methods can be adopted. In the latter, the resolution of the problem is decentralized, and the decision-making process occurs in several stages, which makes the solution procedure easier (Conejo et al., 2006). In order to incorporate the benefits of both approaches, we use the BD algorithm in our study.

BD is a decomposition method that uses B&C to handle the great number of constraints. Its idea is to take advantage of the structure of large problems for which the variables can be partitioned in two subsets. The original model is decomposed into a MP that depends only on the complicating variables (e.g. integer variables), and SPs that are described in relation to the remaining ones. The MP is initially solved in its reduced form, without any cuts. The complicating variables are temporarily fixed to the values obtained, generating a more tractable problem (Geoffrion, 1972; Rahmaniani et al., 2017). The SPs are then solved. If all of them are feasible, optimality cuts are added to the MP; otherwise, feasibility cuts are produced. The MP is solved again with the new incorporated cuts. This procedure is iterated until no improvements can be achieved in the complicating variables, and all the SPs are feasible (Benders, 1962).

In order to perform the decomposition of both $[PITRSP_C]$ and $[PITRSP_{CD}]$ models, we keep the integer variables $\boldsymbol{x}, z$ and $\boldsymbol{y}$ in their MPs, i.e., the decisions related to the maintenance visits, assignment of technicians to serve a task, routing, and production setups. Then, we assign to the SPs the decisions related to inventory, delivery and production quantities, i.e., the variables $\boldsymbol{p}, \boldsymbol{q}$ and $\boldsymbol{I}$. This decomposition is in accordance with real-world practice, in which visits are scheduled in advance, to allow some time to organize the necessary staff and products to execute a maintenance task. Moreover, production plans are also usually made in advance to avoid significant disruptions in the supply chain (Adulyasak et al., 2015a).

It is argued that in the BD algorithm, most of the computational time is allocated to solve the MP (Magnanti & Wong, 1981; Rahmaniani et al., 2017; Porumbel, 2018), thus, it is essential to be able to manage the size of the problem in order to improve the performance of the solution procedure. Due to the decomposition herein proposed, the subtour elimination constraints—SECs—(22) remain in the MP. As there is a large quantity of them, they are removed from the problem, and a relaxed version of the MP is solved. We then check for the existence of subtours by implementing the Lysgaard's separation procedure (Lysgaard et al., 2004). The latter is a well known and used package that uses multiple heuristics to detect violated inequalities, allowing for bound improvement and speed up of the solution process. If subtours are identified, we exclude them from the current solution by adding the corresponding inequalities (22). This process is repeated iteratively until we reach the optimal solution, or the time limit, whichever occurs first.

6 Computational Experiments

In this section, we describe the computational experiments performed, and the considered instances. The BD algorithm was coded in C++ on Visual Studio 2017, using Cplex 20.1.0. All the experiments are conducted using a machine with a 11th Gen Intel Core i5 (2.4 GHz) processor and 8 Gb of RAM. Our goal is to evaluate the potential improvements in customer satisfaction, given the cost minimization results, perform a trade-off analysis between costs and customer satisfaction and investigate the role of the inventory costs in some of the results obtained.

To solve each problem individually, $[PITRSP_C]$ and $[PITRSP_{CD}]$, a time limit of one hour is set, totalizing a maximum of two hours of running time for each instance. The problem herein studied does not have any benchmark instances that can be used to perform the experiments. Thus, we modify the first class of instances proposed by Archetti et al. (2011) and Adulyasak et al. (2014a), which are used to solve variants of the PRP. Our instances are generated as described in Table 1, and are named according to the notation Ta_Pb_Mc, where a, b, and c are the number of tasks, periods, and products, respectively.

Initially, 27 instances are used to solve the $[PITRSP_C]$ model. For each one, we evaluated the cost objective function (Z_C), the gap percentage with respect to the lower bound (LB) given by Cplex, and the running time in seconds spent both in the MP (RT_{MP}) and the SP (RT_{SP}). Additionally, we report the corresponding customer dissatisfaction percentage (Z_{CD}) and the number of feasibility (FC) and optimality cuts (OC) generated. The results obtained are described in Table 2.

It is possible to observe that the problem under study generally becomes more difficult to solve as the number of tasks and periods increase, with a corresponding growth in the number of Benders cuts generated. The same behavior is present for cases where $m = 4$, which are characterized by significantly higher running times and gaps. In comparison to the time spent solving the MP, the running times of the SP are generally negligible. Moreover, production decisions are directly related to the feasibility and complexity of the SP, as feasibility cuts are only generated when production takes place and larger values of RT_{SP} are also associated with these instances. The customer dissatisfaction percentage varies significantly among the analyzed instances, ranging from 16.4% to 38.5% for cases solved to optimality. This demonstrates that pursuing only a goal of cost minimization might compromise significantly how the clients perceive the services provided by a company. Moreover, the customer satisfaction level does not seem to have any clear dependency with the quantity of tasks and periods. In fact, its value depends on how the target times are scattered throughout the time horizon, and not directly on how many periods are considered.

In order to perform the subsequent experiments, the values in the column Z_C were used as an input in constraint (33). We disconsidered the cases with gaps larger than 5%, as we believe it is difficult to draw precise conclusions from them. The problem $[PITRSP_{CD}]$ was then solved for the remaining 23 instances, considering three different tolerances for the admissible cost, namely 0%, 5% and 10%. The results for a tolerance of 0% are provided in Table 3, where we report the best customer dissatisfaction percentage obtained for the given tolerance $(Z_{CD_{tol}})$, the running time (RT) in seconds, and the customer dissatisfaction reduction for the given tolerance $(\Delta Z_{CD_{tol}})$, which is calculated with respect to the values of Table 2. The results obtained while considering tolerances equal to 5% and 10% are reported respectively in Tables 4 and 5, which, in addition to the information of Table 3, also includes the cost $(Z_{C_{tol}})$ and the cost increase percentage for each case $(\Delta Z_{C_{tol}})$, both for the considered tolerance.

Table 1 Parameters used to generate instances

Parameters	Values	Details
C	6, 9, 12	The total number of tasks is equally distributed between the three existing types (scheduled, first emergency task and second emergency task).
M	1, 2, 4	–
l	4, 6, 8	Each period is a week.
K	10	Each SC has the same number of technicians.
C_m	304[a]	The production capacity is the same for all the products.
L_i	152[a]	$i \in \{0, 3\}$.
L_i	76	$i \in S \setminus \{3\}$.
I_{im0}	$\lceil 0.05L_i/M \rceil$, $\lceil 0.1L_i/M \rceil$	$I_{im0} = \lceil 0.05L_i/M \rceil, i \in \mathcal{N} \setminus C, m = 1$, and $I_{im0} = \lceil 0.1L_i/M \rceil, i \in \mathcal{N} \setminus C, m = 2, 4$.
h_{0m}	3, 8[a]	–
h_{im}	[1,10][a]	Random value in $U(1, 10), i \in S$.
u_m	$10h_{0m}$ [a]	–
f_m	$100u_m$ [a]	–
d_{im}	[1, 10]	Random value in $U(1, 10), i \in C^{\Delta} \cup C^{\Pi}$.
r_{ij}	$\lfloor \sqrt{(X_i - X_j)^2 + (Y_i - Y_j)^2} + 0.5 \rfloor$[a]	Nodes i and j are the real locations of the SCs and of ski stations in North America. Each inquiry is randomly allocated to one of the possible locations.
c_{ijm}	$50 + \beta_m r_{ij}$	Flat rate plus a factor that depends on the distance. $\beta_m \in \{1, \ldots, 3\}$ is a multiplier that characterizes the weight of product m.
w_{ij}	–	Travel times were obtained using Google Maps by calculating the average travel time of the shortest path connecting nodes i and j. We considered every day of the week and trips starting at 8 AM.
e_i	$N(5, 0.5)$	Random value in the considered distribution, $i \in C^{\Delta}$.
e_i	$N(2, 0.5)$	Random value in the considered distribution, $i \in C^{\Gamma}$.
e_i	$N(5, 1)$	Random value in the considered distribution, $i \in C^{\Pi}$.
W	40	–
β_i	1, 2, [1, l]	$\beta_i = 1, i \in C^{\Gamma}$, $\beta_i = 2, i \in C^{\Pi}$, and β_i is a random value in $U(1, l), i \in C^{\Delta}$.

(continued)

Table 1 (continued)

Parameters	Values	Details
a_i^k	0, 1	We considered 4 skills and 3 levels of expertise, and built requirements/skills matrices for each task/technician, as Cordeau et al. (2010) and Kovacs et al. (2012). For each task, the requirement level of each skill was generated with a discrete distribution, and levels 1, 2, and 3 have a probability of occurrence of 0.6, 0.3 and 0.1, respectively. The expertise of each technician was generated similarly, considering probabilities of occurrence of 0.3, 0.5, and 0.2. Due to the location of the SCs, a fully skilled technician was added to SCs 2 and 3 to guarantee the feasibility of the problem.
tol	0%, 5%, 10%	–

[a] Data imported from or calculated as in the works of Archetti et al. (2011) and Adulyasak et al. (2014a)

The results in Table 3 show that for cases in which tasks are served by the existing inventory, it is difficult to improve the satisfaction of customers, as inventory would need to be held for longer, and higher costs would be incurred. This behaviour is exemplified by the trade-off curve in Fig. 2a, which shows an instance in which improvements in customer satisfaction cannot be obtained with a tolerance of 0%. For those instances in which production takes place, which usually are those that consider more types of products, some tasks are served only by items that have been just produced and were not held in inventory. This allows for more flexibility in scheduling a task while maintaining the same costs, resulting in an improvement in customer satisfaction. This can be illustrated by the trade-off curve (b) of Fig. 2. For cases in which the corresponding $[PITRSP_{CD}]$ was not solved to optimality, the cost Z_C used in constraint (33) is not minimum. In these situations, more alternative solutions are available for the same or smaller cost and higher customer satisfaction, and can be produced, for example, by changing routes.

We can observe in Table 4 that, with an average cost increase of 3.4%, it is possible to increase the customer satisfaction by 86.6%, in comparison with a problem aiming solely at minimizing costs. If we increase the cost tolerance to 10%, it is possible to serve almost all the tasks in their target period, with the average customer satisfaction improving by 97.5%, with a corresponding average cost increase of 5.2%. In general, the larger the cost tolerance, the higher the improvement of the customer satisfaction, and the higher the cost increase. Moreover, problems with a tolerance of 10% have a smaller average running time, indicating that they are easier to solve. In these cases, the feasible region is larger, and more solutions are available with the same objective function value.

The results also show that for higher values of the right-hand side of inequality (33), it is easier to reschedule tasks to satisfy the customer expectations. This occurs because the allowed absolute cost margin is larger, which makes it easier to adjust

Table 2 Results of the $\lfloor PITRSP_C \rfloor$ problem

Instance	Z_C	Gap	Z_{CD}	RT_{MP}	RT_{SP}	FC	OC
$T6\_P4\_M1$	1278	0.0	38.5	1.81	0.05	0	45
$T6\_P4\_M2$	3002	0.0	38.5	1.28	0.05	0	31
$T6\_P4\_M4$	31,632	0.0	38.5	29.20	0.83	57	527
$T6\_P6\_M1$	1350	0.0	29.2	1.95	0.02	0	23
$T6\_P6\_M2$	3166	0.0	29.2	4.12	0.05	0	49
$T6\_P6\_M4$	31,632	0.0	29.2	609.83	5.89	325	2079
$T6\_P8\_M1$	1422	0.0	25.0	9.89	0.09	0	62
$T6\_P8\_M2$	3330	0.0	25.0	11.60	0.03	0	81
$T6\_P8\_M4$	31,737	2.3	33.3	3580.87	19.13	926	4274
$T9\_P4\_M1$	1923	0.0	25.0	7.71	0.06	0	67
$T9\_P4\_M2$	3700	0.0	25.0	3.39	0.05	0	33
$T9\_P4\_M4$	43,696	0.0	20.0	256.47	3.05	133	1467
$T9\_P6\_M1$	1971	0.0	30.6	53.53	0.14	0	120
$T9\_P6\_M2$	4010	0.0	30.6	25.17	0.16	0	102
$T9\_P6\_M4$	43,826	3.9	25.0	3564.93	35.07	618	7488
$T9\_P8\_M1$	2019	0.0	16.4	148.79	0.27	0	233
$T9\_P8\_M2$	4102	0.0	16.4	164.76	0.44	0	229
$T9\_P8\_M4$	44,045	7.2	40.0	3586.39	13.61	1022	2290
$T12\_P4\_M1$	5809	0.0	33.3	218.41	0.14	9	111
$T12\_P4\_M2$	8654	0.0	33.3	208.57	0.32	40	180
$T12\_P4\_M4$	50,708	6.0	44.4	3593.76	6.24	511	1318
$T12\_P6\_M1$	5814	0.5	27.1	3598.74	1.26	137	490
$T12\_P6\_M2$	8665	1.3	29.2	3598.62	1.38	200	520
$T12\_P6\_M4$	51,165	6.9	33.3	3590.84	9.16	480	1707
$T12\_P8\_M1$	5849	4.4	45.9	3598.52	1.48	332	551
$T12\_P8\_M2$	8699	2.9	28.4	3596.79	3.21	726	551
$T12\_P8\_M4$	54,304	16.3	56.8	3586.40	13.60	69	2894

the tasks throughout the time horizon without compromising significantly on the relative costs. On the other hand, for instances in which the total costs are lower and the inventory costs represent a significant share of them (usually smaller instances with 1 or 2 types of products), it is difficult to reschedule tasks. In these situations, the absolute tolerance is smaller, not giving too much flexibility to change the service periods; the situation is worsened if the demand of products and the unit inventory holding costs are high.

Taking the aforementioned into consideration, we decided to perform a sensitivity analysis on the unit inventory holding costs for the cases in which we could not reach 100% of customer satisfaction, i.e., instances $T6\_P4\_M1$, $T6\_P6\_M1$, and $T6\_P8\_M1$. In order to do that, the original values of the parameters h_{0m} and h_{im} were respectively substituted by $\lfloor h_{0m}/2 \rfloor$ and $\lfloor h_{im}/2 \rfloor, i \in \mathcal{N} \setminus C, m \in \mathcal{M}$. We

Table 3 Results of the $[PITRSP_{CD}]$ problem for $tol = 0\%$

Instance	Z_{CD_0}	ΔZ_{CD_0}	RT
$T6\_P4\_M1$	38.5	0.0	3
$T6\_P4\_M2$	38.5	0.0	1
$T6\_P4\_M4$	38.5	0.0	25
$T6\_P6\_M1$	29.2	0.0	4
$T6\_P6\_M2$	29.2	0.0	2
$T6\_P6\_M4$	29.2	0.0	831
$T6\_P8\_M1$	25.0	0.0	6
$T6\_P8\_M2$	25.0	0.0	5
$T6\_P8\_M4$	19.4	−41.7	2098
$T9\_P4\_M1$	25.0	0.0	4
$T9\_P4\_M2$	25.0	0.0	3
$T9\_P4\_M4$	15.0	−25.0	83
$T9\_P6\_M1$	30.6	0.0	38
$T9\_P6\_M2$	30.6	0.0	22
$T9\_P6\_M4$	11.1	−55.6	3245
$T9\_P8\_M1$	16.4	0.0	92
$T9\_P8\_M2$	16.4	0.0	28
$T12\_P4\_M1$	33.3	0.0	135
$T12\_P4\_M2$	33.3	0.0	151
$T12\_P6\_M1$	27.1	0.0	3600
$T12\_P6\_M2$	22.9	−21.4	3600
$T12\_P8\_M1$	14.9	−67.6	3600
$T12\_P8\_M2$	20.3	−28.6	3600
Average	25.8	−10.4	921

performed the same experiments as before, for the tolerances of 0%, 5%, and 10%, and the obtained results are reported in Table 6.

It is possible to see that the reduction in the unit inventory holding costs allows for an improvement in the customer satisfaction. In comparison to the experiments on the same instances using the original parameters, we are able to increase the customer satisfaction from an average of 53.8% to 96.3%, for a tolerance of 5%, and from 81.2% to 100% in the case of 10% of tolerance.

Moreover, in the original results of the considered instances, the inventory costs accounted for an approximate average of 18.6% of the total costs. With the reduction in the unit holding costs, this value decreases to 9.3%. This corroborates the fact that adjustments in the schedule, are easier to be obtained when the inventory costs represent a smaller fraction of the total costs. This is an essential factor to be taken into account when one needs to guarantee higher levels of customer satisfaction.

Table 4 Results of the $[PITRSP_{CD}]$ problem for $tol = 5\%$

Instance	Z_{CD_5}	ΔZ_{CD_5}	Z_{C_5}	ΔZ_{C_5}	RT
$T6\_P4\_M1$	15.4	−60.0	1342	5.0	2
$T6\_P4\_M2$	15.4	−60.0	3145	4.8	1
$T6\_P4\_M4$	0.0	−100.0	32,026	1.2	1
$T6\_P6\_M1$	12.5	−57.1	1417	5.0	2
$T6\_P6\_M2$	8.3	−71.4	3287	3.8	4
$T6\_P6\_M4$	0.0	−100.0	32,109	1.5	4
$T6\_P8\_M1$	13.9	−44.4	1493	5.0	10
$T6\_P8\_M2$	5.6	−77.8	3492	4.9	4
$T6\_P8\_M4$	0.0	−100.0	32,210	1.5	207
$T9\_P4\_M1$	5.0	−80.0	2005	4.3	5
$T9\_P4\_M2$	0.0	−100.0	3882	4.9	5
$T9\_P4\_M4$	0.0	−100.0	43,978	0.6	10
$T9\_P6\_M1$	5.6	−81.8	2039	3.5	11
$T9\_P6\_M2$	5.6	−81.8	4162	3.8	6
$T9\_P6\_M4$	0.0	−100.0	44,290	1.1	80
$T9\_P8\_M1$	1.8	−88.9	2117	4.9	60
$T9\_P8\_M2$	1.8	−88.9	4287	4.5	34
$T12\_P4\_M1$	0.0	−100.0	5979	2.8	24
$T12\_P4\_M2$	0.0	−100.0	8928	3.2	26
$T12\_P6\_M1$	0.0	−100.0	5977	2.8	175
$T12\_P6\_M2$	0.0	−100.0	8932	3.1	27
$T12\_P8\_M1$	0.0	−100.0	6015	2.8	344
$T12\_P8\_M2$	0.0	−100.0	9030	3.8	97
Average	3.9	−86.6		3.4	49

7 Conclusions

We have addressed a deterministic PITRSP, which is inspired by a real case. The initial model, $[PITRSP]$, aims at minimizing the operational costs and the customer dissatisfaction simultaneously, and it is divided into two models that are sequentially solved using BD. The first model, $[PITRSP_C]$, aims at minimizing costs and the second one, $[PITRSP_{CD}]$, minimizes the customer dissatisfaction. A constraint establishing a cost tolerance was imposed in $[PITRSP_{CD}]$ in order to evaluate how much we can improve its objective function.

Results considering three different cost tolerances show that a value of 5% allows for significant improvements in customer satisfaction levels without large compromises in terms of cost. Compared to the problem that solely minimizes costs, it is able to increase the average customer satisfaction by 86.6%, with a corresponding average cost growth of 3.4%. Setting the cost tolerance to 10% leads to higher improvements in customer satisfaction, on average 97.5%, which means that most tasks can be served in the expected period. However, this behaviour is usually accompanied by an increase in cost of, on average 5.2%. An important factor

Table 5 Results of the $[PITRSP_{CD}]$ problem for $tol = 10\%$

Instance	$Z_{CD_{10}}$	$\Delta Z_{CD_{10}}$	$Z_{C_{10}}$	$\Delta Z_{C_{10}}$	RT
$T6\_P4\_M1$	7.7	−80.0	1405	9.9	1
$T6\_P4\_M2$	0.0	−100.0	3244	8.1	0
$T6\_P4\_M4$	0.0	−100.0	32,018	1.2	2
$T6\_P6\_M1$	4.2	−85.7	1485	10.0	3
$T6\_P6\_M2$	0.0	−100.0	3444	8.8	2
$T6\_P6\_M4$	0.0	−100.0	32,108	1.5	4
$T6\_P8\_M1$	5.6	−77.8	1564	10.0	7
$T6\_P8\_M2$	0.0	−100.0	3640	9.3	8
$T6\_P8\_M4$	0.0	−100.0	32,232	1.6	326
$T9\_P4\_M1$	0.0	−100.0	2025	5.3	5
$T9\_P4\_M2$	0.0	−100.0	3894	5.2	7
$T9\_P4\_M4$	0.0	−100.0	44,516	1.9	10
$T9\_P6\_M1$	0.0	−100.0	2111	7.1	8
$T9\_P6\_M2$	0.0	−100.0	4304	7.3	8
$T9\_P6\_M4$	0.0	−100.0	44,266	1.0	93
$T9\_P8\_M1$	0.0	−100.0	2146	6.3	15
$T9\_P8\_M2$	0.0	−100.0	4334	5.7	14
$T12\_P4\_M1$	0.0	−100.0	5982	3.0	14
$T12\_P4\_M2$	0.0	−100.0	8936	3.3	8
$T12\_P6\_M1$	0.0	−100.0	5958	2.5	46
$T12\_P6\_M2$	0.0	−100.0	8935	3.1	14
$T12\_P8\_M1$	0.0	−100.0	6016	2.9	142
$T12\_P8\_M2$	0.0	−100.0	9028	3.8	57
Average	0.8	−97.5		5.2	34

that can interfere in the results is the inventory cost. If it represents a significant share of the total costs, it might be difficult for the decision maker to reschedule tasks, and consequently guarantee higher levels of customer satisfaction.

Although integrated problems have received more attention in recent years, this is the first study on the PITRSP. Future research should focus on the development of faster algorithms to solve the problem, allowing for the solution of larger instances, which can include the consideration of more tasks, SCs, technicians and products. The use of BD enhancements and extensions are encouraged, as methods like the partial BD (Crainic et al., 2021) and the Benders dual decomposition (Rahmaniani et al., 2020) have showed potential to speed up the solution process. Moreover, in order to improve the practical aspects of the problem, interesting new variants of the problem can consider data uncertainty, such as demand, travel times and execution times, or the possibility of technicians teaming up. For those decision makers interested in a full understanding of the potential trade-offs between the competing objectives, the construction of the exact Pareto front is advisable.

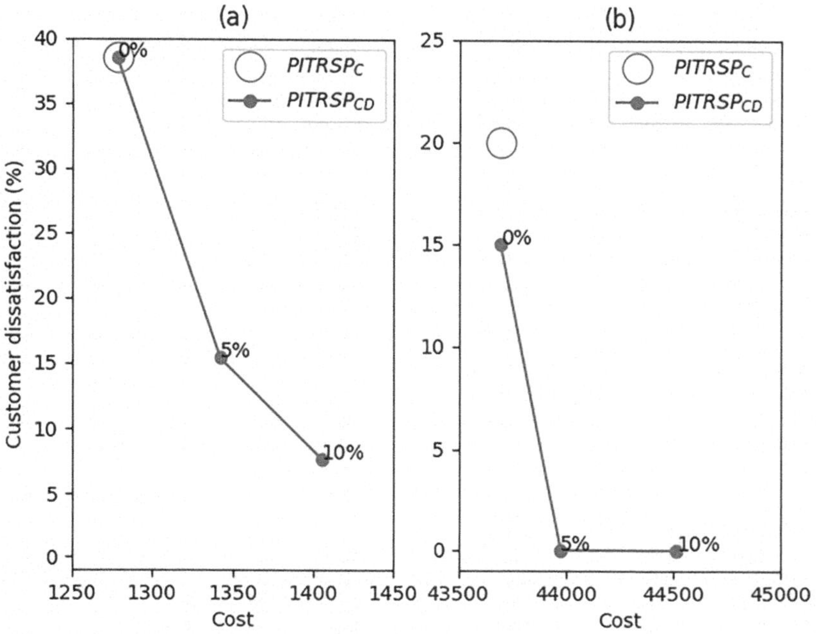

Fig. 2 Trade-off curves for instances (**a**) $T6\_P4\_M1$ and (**b**) $T9\_P4\_M4$

Table 6 Sensitivity analysis on the inventory cost

Instance	Z_C	Z_{CD}	ΔZ_{C_0}	ΔZ_{CD_0}	ΔZ_{C_5}	ΔZ_{CD_5}	$\Delta Z_{C_{10}}$	$\Delta Z_{CD_{10}}$
$T6\_P4\_M1$	1141	38.5	0.0	0.0	5.0	−100.0	10.0	−100.0
$T6\_P6\_M1$	1181	29.2	0.0	0.0	5.0	−100.0	10.0	−100.0
$T6\_P8\_M1$	1209	25.0	0.0	0.0	5.0	−88.9	9.9	−100.0

Acknowledgments The first author has been supported by grants from the Canadian Natural Sciences and Engineering Research Council, MITACS, and Conseil 2.0. We also acknowledge Jean-François Cordeau and Roberto Baldacci for providing insightful comments and suggestions on the chapter.

References

Aazami, A., Saidi-Mehrabad, M., & Seyedhosseini, S. M. (2021). A bi-objective robust optimization model for an integrated production-distribution problem of perishable goods with demand improvement strategies: A case study. *International Journal of Engineering, Transactions A: Basics, 34*(7), 1766–1777.

Absi, N., Archetti, C., Dauzère-Pérès, S., Feillet, D., & Speranza, M. G. (2018). Comparing sequential and integrated approaches for the production routing problem. *European Journal of Operational Research, 269*(2), 633–646.

Adulyasak, Y., Cordeau, J. F., & Jans, R. (2014a). Formulations and branch-and-cut algorithms for multivehicle production and inventory routing problems. *INFORMS Journal on Computing, 26*(1), 103–120.
Adulyasak, Y., Cordeau, J. F., & Jans, R. (2014b). Optimization-based adaptive large neighborhood search for the production routing problem. *Transportation Science, 48*(1), 20–45.
Adulyasak, Y., Cordeau, J. F., & Jans, R. (2015a). Benders decomposition for production routing under demand uncertainty. *Operations Research, 63*(4), 851–867.
Adulyasak, Y., Cordeau, J. F., & Jans, R. (2015b). The production routing problem: A review of formulations and solution algorithms. *Computers and Operations Research, 55*, 141–152.
Alkaabneh, F., Diabat, A., & Gao, H. O. (2020). Benders decomposition for the inventory vehicle routing problem with perishable products and environmental costs. *Computers and Operations Research, 113*, 1–13.
Archetti, C., Bertazzi, L., Paletta, G., & Speranza, M. G. (2011). Analysis of the maximum level policy in a production-distribution system. *Computers and Operations Research, 38*(12), 1731–1746.
Avci, M., & Yildiz, S. T. (2019). A matheuristic solution approach for the production routing problem with visit spacing policy. *European Journal of Operational Research, 279*(2), 572–588.
Benders, J. F. (1962). Partitioning procedures for solving mixed-variables programming problems. *Numerische Mathematik, 4*, 238–252.
Bérubé, J. F., Gendreau, M., & Potvin, J. Y. (2009). An exact ϵlunate-constraint method for bi-objective combinatorial optimization problems: Application to the traveling salesman problem with profits. *European Journal of Operational Research, 194*(1), 39–50.
Brahimi, N., & Aouam, T. (2016). Multi-item production routing problem with backordering: A MILP approach. *International Journal of Production Research, 54*(4), 1076–1093.
Castillo-Salazar, J. A., Landa-Silva, D., & Qu, R. (2016). Workforce scheduling and routing problems: Literature survey and computational study. *Annals of Operations Research, 239*(1), 39–67.
Çetinkaya, S., Üster, H., Easwaran, G., & Keskin, B. B. (2009). An integrated outbound logistics model for Frito-Lay: Coordinating aggregate-level production and distribution decisions. *Interfaces, 39*(5), 460–475.
Chen, X., Thomas, B. W., & Hewitt, M. (2016). The technician routing problem with experience-based service times. *Omega, 61*, 49–61.
Choudhary, D. (2015). An optimization model for scheduling tour of service personnel in aftersale service process with additional side conditions of responsiveness and FCFS service policy. *Industrial Engineering and Management, 04*, 1–5.
Conejo, A. J., Castillo, E., & García-bertrand, R. M. R. (2006). *Decomposition techniques in mathematical programming*. Berlin: Springer.
Cordeau, J. F., Stojković, G., Soumis, F., & Desrosiers, J. (2001). Benders decomposition for simultaneous aircraft routing and crew scheduling. *Transportation Science, 35*(4), 375–388.
Cordeau, J. F., Laporte, G., Pasin, F., & Ropke, S. (2010). Scheduling technicians and tasks in a telecommunications company. *Journal of Scheduling, 13*(4), 393–409.
Cortés, C. E., Gendreau, M., Rousseau, L. M., Souyris, S., & Weintraub, A. (2014) Branch-and-price and constraint programming for solving a real-life technician dispatching problem. *European Journal of Operational Research, 238*(1), 300–312.
Crainic, T. G., Hewitt, M., Maggioni, F., & Rei, W. (2021). Partial benders decomposition: General methodology and application to stochastic network design. *Transportation Science, 55*(2), 414–435.
de Souza, R., Tan, A. W. K., Othman, H., & Garg, M. (2011). A proposed framework for managing service parts in automotive and aerospace industries. *Benchmarking, 18*(6), 769–782.
Díaz-Madroñero, M., Peidro, D., & Mula, J. (2015) Integrated production and routing planning decisions under uncertainty: A case study. In *Proceedings of the 2015 Conference of the International Fuzzy Systems Association and the European Society for Fuzzy Logic and Technology*, December 2016 (pp. 1565–1570).

Du, M., Kong, N., & Hu, X. (2019). A new heuristic scheduling method for the make-pack-route problem in make-to-order supply chains. *INFOR: Information Systems and Operational Research, 57*(2), 296–313.

Ehrgott, M. (2008). Multiobjective optimization. *AI Magazine, 29*(4), 47–57.

Erengüç, S. S., Simpson, N. C., & Vakharia, A. J. (1999). Integrated production/distribution planning in supply chains: An invited review. *European Journal of Operational Research, 115*(2), 219–236.

Farghadani-Chaharsooghi, P., Kamranfar, P., Mirzapour Al-e-Hashem, M. S., & Rekik, Y. (2022). A joint production-workforce-delivery stochastic planning problem for perishable items. *International Journal of Production Research, 60*, 1–25.

Fattahi, P., Tanhatalab, M., & Bashiri, M. (2017). Bi-objectives approach for a multi-period two echelons perishable product inventory-routing problem with production and lateral transshipment. *International Journal of Engineering, 30*(6), 876–886.

Geoffrion, A. M. (1972). Generalized benders decomposition. *Journal of Optimization Theory and Applications, 10*(4), 237–260.

Grötschel, M., Krumke, S. O., Rambau, J., & Torres, L. M. (2003). Online-dispatching of automobile service units. In *Operations Research Proceedings 2002* (pp. 168–173).

Hadjiconstantinou, E., & Roberts, D. (2002). Routing under uncertainty: An application in the scheduling of field service engineers. In P. Toth & D. Vigo (Eds.), *The vehicle routing problem* (Chap. 13, pp. 331–352). Philadelphia: Society for Industrial and Applied Mathematics.

Hallikas, J., Immonen, M., Pynnönen, M., & Mikkonen, K. (2014). Service purchasing and value creation: Towards systemic purchases. *International Journal of Production Economics, 147*(Part A), 53–61.

Hooker, J. N. (2019). Logic-based benders decomposition for large-scale optimization. In J. M. Velásquez-Bermúdez, M. Khakifirooz, & M. Fathi (Eds.), *Large scale optimization in supply chains and smart manufacturing* (Vol. 149, pp. 1–26). Cham: Springer Nature Switzerland AG.

Ji, S. F., Peng, X. S., & Luo, R. J. (2019). An integrated model for the production-inventory-distribution problem in the Physical Internet. *International Journal of Production Research, 57*(4), 1000–1017.

Jia, Y. J., Wang, C. J., & Wang, B. (2008). Multi-objective vehicle scheduling problem based on customer satisfaction and hybrid genetic algorithm. In *Proceedings of the IEEE International Conference on Automation and Logistics, ICAL 2008* September 1934–1937.

Karamanov, M. (2006). Branch and cut: An empirical study. PhD thesis, Carnegie Mellon University.

Kovacs, A. A., Parragh, S. N., Doerner, K. F., & Hartl, R. F. (2012). Adaptive large neighborhood search for service technician routing and scheduling problems. *Journal of Scheduling, 15*(5), 579–600.

Kumar, R. S., Kondapaneni, K., Dixit, V., Goswami, A., Thakur, L. S., & Tiwari, M. K. (2016). Multi-objective modeling of production and pollution routing problem with time window: A self-learning particle swarm optimization approach. *Computers and Industrial Engineering, 99*, 29–40.

Lei, L., Liu, S., Ruszczynski, A., & Park, S. (2006). On the integrated production, inventory, and distribution routing problem. *IIE Transactions, 38*(11), 955–970.

Li, Y., Chu, F., Chu, C., Zhou, W., & Zhu, Z. (2016). Integrated production inventory routing planning with time windows for perishable food. In *IEEE Conference on Intelligent Transportation Systems, Proceedings, ITSC* (pp. 2651–2656).

Li, Y., Chu, F., Chu, C., & Zhu, Z. (2019a). An efficient three-level heuristic for the large-scaled multi-product production routing problem with outsourcing. *European Journal of Operational Research, 272*(3), 914–927.

Li, Y., Chu, F., Feng, C., Chu, C., & Zhou, M. C. (2019b). Integrated production inventory routing planning for intelligent food logistics systems. *IEEE Transactions on Intelligent Transportation Systems, 20*(3), 867–878.

Liu, Y., Gao, B., Liu, H., & Wei, M. (2019). Optimal production, delivery and inventory policies for perishable products with replenishment lead time in two-echelon distribution network. In

2019 IEEE 6th International Conference on Industrial Engineering and Applications, ICIEA 2019 (pp. 262–266).
Liu, Z., Zhou, C., Chen, H., & Zhao, R. (2021). Impact of cost uncertainty on supply chain competition under different confidence levels. *International Transactions in Operational Research, 28*(3), 1465–1504.
Lysgaard, J., Letchford, A. N., & Eglese, R. W. (2004). A new branch-and-cut algorithm for the capacitated vehicle routing problem. *Mathematical Programming, 100*(2), 423–445.
Magnanti, T. L., & Wong, R. T. (1981). Accelerating benders decomposition: Algorithmic enhancement and model selection criteria. *Operations Research, 29*(3), 464–484.
Mostafa, N. A., & Eltawil, A. B. (2015). The production-inventory-distribution-routing problem: An integrated formulation and solution framework. In *IEOM 2015 - 5th International Conference on Industrial Engineering and Operations Management, Proceeding* (pp. 1–10).
Murthy, D. N., Solem, O., & Roren, T. (2004). Product warranty logistics: Issues and challenges. *European Journal of Operational Research, 156*(1), 110–126.
Neves-Moreira, F., Almada-Lobo, B., Cordeau, J. F., Guimarães, L., & Jans, R. (2019). Solving a large multi-product production-routing problem with delivery time windows. *Omega, 86*, 154–172.
Parasuraman, A., Zeithaml, V. A., & Berry, L. L. (1985). A conceptual model of service quality and its implications for future research. *Journal of Marketing, 49*(4), 41.
Peng, X. s., Feng Ji, S., & Ting Ji, T. (2019). Promoting sustainability of the integrated production-inventory-distribution system through the physical Internet. *International Journal of Production Research, 58*, 6985–7004.
Pillac, V., Guéret, C., & Medaglia, A. L. (2013). A parallel matheuristic for the technician routing and scheduling problem. *Optimization Letters, 7*(7), 1525–1535.
Porumbel, D. (2018). From the separation to the intersection sub-problem in Benders decomposition models with prohibitively-many constraints. *Discrete Optimization, 29*, 148–173.
Qiu, Y., Wang, L., Fang, X., Pardalos, P. M., & Goldengorin, B. (2018a). Formulations and branch-and-cut algorithms for production routing problems with time windows. *Transportmetrica A: Transport Science, 14*(8), 669–690.
Qiu, Y., Wang, L., Xu, X., Fang, X., & Pardalos, P. M. (2018b). Formulations and branch-and-cut algorithms for multi-product multi-vehicle production routing problems with startup cost. *Expert Systems with Applications, 98*, 1–10.
Qiu, Y., Qiao, J., & Pardalos, P. M. (2019). Optimal production, replenishment, delivery, routing and inventory management policies for products with perishable inventory. *Omega, 82*, 193–204.
Rahmaniani, R., Crainic, T. G., Gendreau, M., & Rei, W. (2017). The Benders decomposition algorithm: A literature review. *European Journal of Operational Research, 259*(3), 801–817.
Rahmaniani, R., Ahmed, S., Crainic, T. G., Gendreau, M., & Rei, W. (2020). The benders dual decomposition method. *Operations Research, 68*(3), 878–895.
Restrepo, M. I., Gendron, B., & Rousseau, L. M. (2018). Combining benders decomposition and column generation for multi-activity tour scheduling. *Computers and Operations Research, 93*, 151–165.
Ruokokoski, M., Solyali, O., Cordeau, J., Jans, R., & Sural, H. (2010). Efficient formulations and a branch-and-cut algorithm for a production-routing problem. Accessed April 25, 2023. https://numerique.banq.qc.ca/patrimoine/details/52327/2772132
Shiguemoto, A. L., & Armentano, V. A. (2010). A tabu search procedure for coordinating production, inventory and distribution routing problems. *International Transactions in Operational Research, 17*(2), 179–195.
Syahrial, E., Suzuki, H., & Schvaneveldt, S. J. (2019). The impact of serviceability-oriented dimensions on after-sales service cost and customer satisfaction. *Total Quality Management and Business Excellence, 30*(11–12), 1257–1281.
Tang, H., Miller-Hooks, E., & Tomastik, R. (2007). Scheduling technicians for planned maintenance of geographically distributed equipment. *Transportation Research Part E: Logistics and Transportation Review, 43*(5), 591–609.

Tversky, A., & Kahneman, D. (2019). Advances in prospect theory: Cumulative representation of uncertainty. *Choices, Values, and Frames, 323*, 44–66.

Vahdani, B., Niaki, S. T., & Aslanzade, S. (2017). Production-inventory-routing coordination with capacity and time window constraints for perishable products: Heuristic and meta-heuristic algorithms. *Journal of Cleaner Production, 161*, 598–618.

Wang, S., & Chu, F. (2022). A decomposition-based heuristic method for inventory routing problem. *IEEE Transactions on Intelligent Transportation Systems, 23*(10), 18352–18360.

Weintraub, A., Aboud, J., Fernandez, C., Laporte, G., & Ramirez, E. (1999). An emergency vehicle dispatching system for an electric utility in Chile. *Journal of the Operational Research Society, 50*(7), 690–696.

Wu, B., Jiang, H. J., Wang, C., & Dong, M. (2021). Knowledge and behavior-driven fruit fly optimization algorithm for field service scheduling problem with customer satisfaction. *Complexity, 2021*

Xu, J., & Chiu, S. Y. (2001). Effective heuristic procedures for a field technician scheduling problem. *Journal of Heuristics, 7*(5), 495–509.

Zhang, Q., Sundaramoorthy, A., Grossmann, I. E., & Pinto, J. M. (2017). Multiscale production routing in multicommodity supply chains with complex production facilities. *Computers and Operations Research, 79*(April 2016), 207–222.

Zhang, Y., Alshraideh, H., & Diabat, A. (2018). A stochastic reverse logistics production routing model with environmental considerations. *Annals of Operations Research, 271*(2), 1023–1044.

Zhang, Z., Luo, Z., Baldacci, R., & Lim, A. (2021). A benders decomposition approach for the multivehicle production routing problem with order-up-to-level policy. *Transportation Science, 55*(1), 160–178.

Express Package Delivery Optimization Using Walkers, Cargo Tricycles and Delivery Trucks

Patrick Meredith-Karam, Jia Hui Jiang, Sina Bahrami, and Matthew J. Roorda

1 Introduction

Sustainable last-mile logistics and efficient vehicle routing are ongoing city logistics challenges faced by any company that moves goods in large urban areas. In dense, often-congested centers, trucks or vans do not always present the most time-efficient, economical or sustainable delivery option. Therefore, many freight and courier companies opt to serve demand using a heterogeneous fleet of vehicle types, often including walkers and cyclists. One courier company exemplifies this fleet model in its downtown Toronto operations. The company hosts a diverse fleet of delivery vehicles and on-foot delivery personnel, and is seeking to use cargo tricycles in the future. Downtown Toronto poses challenges for express package delivery companies due to congestion, lack of parking, and tight delivery time windows.

This paper investigates the use of various vehicle types for delivery in urban centers. It does this by modifying a heuristic algorithm for the vehicle routing problem with multiple vehicle types and time windows (VRPMVTTW) to be more applicable to urban centers. This revised heuristic algorithm is applied in a case study of downtown Toronto operations for a courier company, to investigate the feasibility of cargo tricycles for deliveries.

P. Meredith-Karam
Department of Urban Studies and Planning, Massachusetts Institute of Technology, Cambridge, MA, USA

J. H. Jiang
Division of Engineering Science, University of Toronto, Toronto, ON, Canada

S. Bahrami · M. J. Roorda (✉)
Department of Civil and Mineral Engineering, University of Toronto, Toronto, ON, Canada
e-mail: matt.roorda@utoronto.ca

T. G. Crainic et al. (eds.), *Combinatorial Optimization and Applications*,
International Series in Operations Research & Management Science 358,
https://doi.org/10.1007/978-3-031-57603-4_17

The heuristic builds upon previous operational research/city logistics research that addresses the vehicle routing problem, including the Vehicle Routing Problem with Multiple Vehicle Types (VRPMVT) and VRPMVTTW (Ferland & Michelon 1988; Liu & Shen 1999). This previous research provides a preliminary formulation of VRPMVTTW solution heuristics, but leaves an opportunity to bring the problem into operational practice. This paper makes three key contributions to the VRP-MVTTW problem by: (1) revising the insertion-based heuristic proposed by Liu and Shen (1999) to be more applicable to operations in urban centers, (2) identifying operational characteristics that would make cargo tricycles preferable to truck/van deliveries and vice-versa, and (3) proposing future revisions to routing algorithms within urban centers.

The remainder of this paper is organized as follows. The literature review examines city logistics research that assesses the addition of cargo bicycles to delivery fleets in downtown cores, and previous research on the VRPMVTTW. Then, the methods section discusses the rationale for the selection of the VRP-MVTTW heuristic developed by Liu and Shen (1999), a detailed formulation of the VRPMVTTW problem, various adaptations to the heuristic to improve its performance, and the application of the solver to a downtown Toronto case study. Next, the results section provides sample results and an interpretation of the optimal parameter values found by the solver. The discussion and conclusion sections present a sensitivity analysis, discuss the viability of cargo tricycles, and highlight the significance and limitations of the study.

2 Literature Review

The literature review is composed of two parts: (1) an examination of cargo bicycle technologies, and viability for implementation in urban environments; and (2) a review of the VRPMVTTW problem and approaches used to solve it.

2.1 Evaluation of Cargo Bicycle Applications

Cargo bicycles are a promising mode that can address city logistics objectives of efficiency, sustainability, congestion and safety. Several studies have been conducted to evaluate cargo bicycles towards these objectives. Naumov and Starczewski (2019) analyze the effectiveness of cargo bicycles for deliveries in urban areas via simulation. Their simulation system integrates three subsystem models, including a model of the transport network (on which deliveries are conducted), a model for goods delivery demand, and a model for the servicing system (including cargo bicycles and loading locations). Using this framework, the system determines optimal cargo bicycle location points within a network and calculates the related transportation costs of a cargo bicycle delivery system, and cargo bicycles are demonstrated to be an applicable solution for last-mile deliveries in urban areas.

Sheth et al. (2019) compare the cost trade-offs between delivery trucks and electrically assisted cargo bicycles in urban areas, given the same set of route demand points. Four delivery scenarios are evaluated, with varying distance from distribution center to demand point, number of stops, distance between stops, and number of parcels per stop. Cargo bicycles are found to be more cost effective than delivery trucks in cases where the distribution center is within two miles of demand points.

Arnold et al. (2018) consider the adoption of cargo bicycle deliveries for e-commerce applications by examining their real-world cost structure. A real-world dataset from Antwerp, Belgium is used to predict demand volumes, and compare cost structures for bicycle deliveries and customer self-pick-up programs using a two-step simulation model for route computation. The simulation approach first generates customer demand using a spatial distribution from historic demand patterns and assigning demand points accordingly to 400m^2 cells in a grid-mapping of the city, then computes delivery routes using the Clark-Wright savings algorithm. The study finds that operational costs may be reduced by implementing self-pick-up, but that cargo bicycle deliveries may decrease negative system externalities (Arnold et al. 2018).

Lebeau et al. (2023) compare the external costs of transport (climate change, air pollution, noise pollution, well-to-tank (WTT) costs, accidents and congestion), for cargo bikes and light commercial vehicles. They found, via a simple travelling salesman problem analysis with simplifying assumptions, that the greatest external cost savings occurred in environments with smaller and more fragmented freight flows, where cargo bikes could replace vehicles one-to-one. Schliwa et al. (2015) provide a review of e-cargo cycle use in urban freight transportation, and discuss the role of local authorities to create conditions to incentivize e-cargo cycle adoption by large logistics service providers. Conway et al. (2012) consider opportunities for e-cargo cycle operations in New York City, and find that cargo tricycles can accommodate urban freight at lower social and environmental impact without increasing overall cost. Gruber et al. (2014) investigate the market for e-cargo bikes and find that they have an overlapping market with trucks and bicycles, that there is a general acceptance amongst bike and vehicle couriers towards their use, and that the key factors towards implementation include range, purchase price and availability of public information.

Overall, there is significant evidence from the city logistics literature that e-cargo bicycles are a promising means of transporting goods in urban areas.

2.2 Vehicle Routing Problem with Multiple Vehicle Types and Time Windows (VRPMVTTW)

Part of the mandate of city logistics (Marcucci et al. 2023) is to address questions of efficiency in vehicle routing. The challenge addressed by the vehicle routing problem (VRP) is the minimization of transportation costs for the distribution of goods or services from a depot to a set of demand points (or customers). The VRP was formulated in 1959 by Dantzig and Ramser (1959), who examined the truck dispatching problem for a fleet of gasoline trucks which provided fuel to service stations. Review papers on the VRP, including Laporte (2009), and books by Golden et al. (2008) and Toth and Vigo (2014) describe the many variants, and solution algorithms.

Several extensions to the basic vehicle routing problem have been made throughout its history (Mutar et al. 2017; Kim et al. 2015). Early approaches include the vehicle routing problem with time window constraints (Solomon 1984, 1987), the multi-depot vehicle routing problem (Sumichras & Markham 1995; Renaud et al. 1996), and the dynamic vehicle routing problem (Wilson & Colvin 1977; Larsen & Madsen 2000; Van Woensel et al. 2008; Pillac et al. 2013). More recently, eco-friendly vehicle routing problems have been researched, which seek to minimize air pollution (Quak & de Koster 2009; Bektaş & Laporte 2011; Franceschetti et al. 2013).

This section focuses on the extension of the vehicle routing problem to consider multiple vehicle types, or modes of delivery (VRPMVT). As described in review papers by Baldacci et al. (2008), Irnich and Schneider (2014), Koç et al. (2016) and SteadieSeifi et al. (2014), freight systems are increasingly multimodal throughout their corresponding supply chains, and particularly so in multimodal city logistics for last-mile delivery. The VRPMVT is alternatively described as the heterogeneous fleet vehicle routing problem HFVRP, first introduced by Golden et al. (1984). Some examples of more recent research on the heterogeneous vehicle routing problem include (Perboli et al. 2018), who develop a simulation-optimization framework to operationalize a multimodal last-mile delivery fleet of trucks and cargo bicycles in Turin, Italy. Sun et al. (2019) also examine a heterogeneous fleet of vehicles in application to pickup and dropoff of passengers with the objective of minimizing carbon emissions. The study applies an exact solution algorithm to a case study of three vehicle types with variable capacities but constant travel speeds in China, finding that a heterogeneous vehicle fleet can always achieve emissions reductions. Scherr et al. (2019) approach the problem of mixed vehicle fleets from the perspective of autonomous parcel delivery, seeking to design fleet size, fleet mix, and routing.

Although various studies examine VRPMVT, apply exact (Baldacci & Mingozzi 2009; Baldacci et al. 2010) or heuristic (Li et al. 2007; Liu et al. 2009; Penna et al. 2013; Máximo et al. 2022) solutions, and differentiate vehicle types along different parameters, these studies are generally limited to smaller scale problems (less than 100 demand points) with the exception of Máximo et al. (2022) (360 demand points). Jamil et al. (2021) apply a bio-inspired firefly algorithm to a case study of 14 demand points in Balikpapan, Indonesia, while Gan et al. (2014) use a particle swarm optimization approach for a case study of 14 demand points. Huai et al. (2019) and Wang et al. (2015) use the VRPMVT to differentiate refrigerated and non-refrigerated trucks with the same operating speed, seeking to minimize both cargo damage and delivery cost through case studies of 37 customers and 82 customers, respectively. For example, Máximo et al. (2022) used an adaptive iterative local search heuristic to solve benchmark instances of up to 360 customers, building upon earlier applications of iterative local search by Penna et al. (2013). An opportunity for advancing the state of the VRPMVTTW is by operationalizing solution heuristics for city-wide use by logistics companies. This study achieves this by scaling up to a case study of over 1000 demand points, and considering vehicle types with different travel speeds and costs.

Ferland and Michelon (1988) describe a formulation of the formal VRPMVTTW problem, and offer alternative methods to solve it. Liu and Shen (1999) assess solution algorithms for the VRPMVTTW, in conjunction with proven methods for the VRPMVT and VRPTW, and offer a means of assessing heuristic effectiveness. Liu and Shen (1999) also develop an insertion-based heuristic method for solving the VRPMVTTW, which is used as a starting point for the heuristic developed in this study.

Ferland and Michelon (1988) formulate the VRPMVTTW using linear programming to minimize the total cost of operations. The problem definition includes time windows for delivery, and allows an unlimited number of vehicle types. Their objective function includes both operations and capital costs for additional vehicles. Ferland and Michelon (1988) propose three heuristic methods for solving the VRPMVTTW, using: (1) discrete approximation of the time windows, (2) iterative generation and improvement upon feasible solutions, and (3) division of the problem into subsets of tasks, and iterative consideration of the tasks which must be performed last.

Liu and Shen (1999) expand on the VRPMVTTW, by comparing the performance of several heuristic methods for route construction. By testing several insertion-based savings heuristics on 24 instances with 100 customers, Liu and Shen (1999) demonstrate that heuristics using sequential route construction performed significantly better than others at the time. Liu and Shen (1999) present a heuristic method developed in consideration of this analysis, which constructs a route through the sequential insertion of demand points into trip chains. It is selected as the starting point from which the VRPMVTTW solver used for this study is adapted.

3 Methods

3.1 Problem Formulation

The formulation of the VRPMVTTW problem is adapted from Munari et al. (2016). The problem is represented by the graph $G(\mathcal{N}, \mathcal{E})$ in which $\mathcal{N} = \{0, \ldots, n\}$ is the set of nodes associated to customers including the depot, 0 (Munari et al. 2016). $\mathcal{E}$ contains the arcs (i, j) for each pair of nodes $i, j \in \mathcal{N}$. The cost of traversing an arc for vehicle type k is denoted by C_{ij}^k and the time required to traverse that arc is represented by t_{ij}^k.

Each demand point is enumerated i (with a range of $0 \leq i \leq n$) with $i = 0$ being the depot. Each demand point must be serviced beginning at a time within some time interval $[a_i, b_i]$ and has vehicle capacity utilization (demand), q_i.

Each vehicle type is enumerated k (with a range of $1 \leq k \leq K$) and has a capacity Q^k, initialization cost F^k, and speed v^k. To service a demand point, each vehicle expends service time, s^k, and service cost S^k. Because there is a chance that a vehicle may arrive at a demand point prior to the start of its service time window, a_i, the time a vehicle spends waiting for the service time window to open is w_i^k and the associated cost per unit time is W^k. Finally, the number of nodes serviced by vehicle type k is represented by c^k and the number of vehicles of type k is represented by u^k. There is no upper constraint on u^k, in other words, there is no maximum fleet size for each vehicle type k. The decision variables are x_{ij}^k which is a binary variable that equals 1 if a vehicle of type k traverses arc $(i, j) \in \mathcal{E}$ and is 0 otherwise.

The objective function seeks to minimize the costs associated with travel, wait, service, and initial vehicle procurement. This is represented by:

$$\min \sum_{k=1}^{K} \left(\sum_{(i,j)\in\mathcal{E}} \left(C_{ij}^k x_{ij}^k\right) + W^k \sum_{i=1}^{n} w_i^k + S^k c^k + F^k u^k \right) \tag{1}$$

The decision variables x_{ij}^k are subject to a number of constraints. Constraint (2) ensures that each customer is only serviced once. Constraint (3) ensures that the flow of vehicles is internally consistent, in other words, a vehicle that arrives at node h must also depart from node h. Constraints (4) and (5) together guarantee that capacity constraints are followed, where y_j^k is the cumulated demand on the route of vehicle type k that visits node j up to that visit. Constraints (6) and (7) define variables u^k, and c^k:

$$\sum_{k=1}^{K} \sum_{j\in\mathcal{N}, j\neq i} x_{ij}^k = 1 \qquad i = 1, \ldots, n \tag{2}$$

$$\sum_{i\in\mathcal{N}, i\neq h} x_{ih}^k = \sum_{j\in\mathcal{N}, j\neq h} x_{hj}^k \qquad h = 1, \ldots, n\,,\; k = 1, \ldots, K \tag{3}$$

$$y_j^k = \sum_{i \in N, i \neq j} x_{ij}^k \cdot (q_i + y_i^k) \qquad j = 1, \ldots, n, \; k = 1, \ldots, K \tag{4}$$

$$y_j^k \leq Q^k \qquad k = 1, \ldots, K \tag{5}$$

$$u^k = \sum_{j=1}^{n} x_{0j}^k \qquad k = 1, \ldots, K \tag{6}$$

$$c^k = \sum_{i,j \in N} x_{ij}^k - u^k \qquad k = 1, \ldots, K \tag{7}$$

Finally, each node, h, has an arrival time A_h, and a departure time D_h, that are calculated by Eqs. (8) and (9), respectively. These yield the time window constraint (10) and a definition of wait times (11):

$$A_h = \sum_{k=1}^{K} \sum_{i \in N, i \neq h} \left(x_{ih}^k (t_{ih}^k + D_i \right) \qquad h = 1, \ldots, n \tag{8}$$

$$D_h = \max(a_h, A_h) + \sum_{k=1}^{K} \sum_{i \in N, i \neq h} \left(x_{ih}^k s^k \right) \qquad h = 1, \ldots, n \tag{9}$$

$$A_h \leq b_h \qquad h = 1, \ldots, n \tag{10}$$

$$w_h^k = \sum_{i \in N, i \neq h} \left(x_{ih}^k \right) \max(0, a_h - A_h) \qquad h = 1, \ldots, n, \; k = 1, \ldots, K \tag{11}$$

3.2 *Selection of VRPMVTTW Heuristic Method*

The VRPMVTTW heuristic outlined by Liu and Shen (1999) is selected for this project. The method proposed by Liu and Shen (1999) identifies several modifications to the traditional Clarke Wright algorithm that improves performance by approximately 20% in a wide array of demand contexts (Liu & Shen 1999).

Liu and Shen (1999)'s heuristic features several parameters that could be varied to alter the final results. Given that heuristics do not generate exact solutions, a parameterized heuristic allows for a grid-search on the parameters to find a better solution in the objective function space. Moreover, analyzing the optimal values of the parameters also enables interesting qualitative deductions.

Liu and Shen (1999)'s algorithm is modified for the application in this paper to better address the challenge of urban freight operations. The modifications include accounting for costs associated with time spent waiting to begin servicing a customer (due to time window constraints) as well as time spent parking the vehicle and servicing the customer. Secondly, Liu and Shen (1999)'s algorithm only considers upgrading to a higher-cost vehicle type when capacity demands it. However, there is the potential that a higher-cost vehicle would make feasible a more cost-efficient route. Hence, all vehicle types are considered at every stage of the algorithm. Additionally, our proposed savings function considers the fixed cost savings by upgrading to a larger vehicle as a proportion of remaining capacity, whereas Liu and Shen (1999) simplify to consider the entire fixed cost of the smallest vehicle that would service the excess capacity of the new vehicle. Finally, Liu and Shen (1999)'s algorithmic approach is revised to be more computationally efficient. By storing a list of feasible insertions and corresponding savings and updating the list when an insertion is made (rather than computing the list for each iteration), the complexity of the heuristic is reduced from $O(n^3)$ to $O(n^2)$.

3.3 Implementation of VRPMVTTW Solver

The heuristic implementation begins by initiating all routes. This module generates a separate route for each destination, i, that goes from the depot to the destination and then back to the depot, $[0, i, 0]$. The vehicle type is initialized as the vehicle with the smallest capacity that is capable of servicing i.

The algorithm then proceeds to insert routes into other routes until it terminates. For example, the algorithm may choose to insert a route $[0, i, j, 0]$ into link $l - m$ of route $[0, k, l, m, 0]$ to produce the final route of $[0, k, l, i, j, m, 0]$.

It chooses which insertions to make by iterating over all possible insertion combinations with each vehicle type to determine the insertion that yields the largest savings. This insertion is then applied to the route and all parameters for the route are updated including the vehicle type, route schedule, route capacity, departure times, and other computational elements. Note that for each possible insertion, the savings are evaluated with each vehicle type. The algorithm continues iterating over all possibilities until there are no non-zero savings remaining.

Prior to computing the savings, the prospective insertion is assessed for feasibility. The required capacity is assessed against the vehicle capacity. If the vehicle has sufficient capacity then the arrival and departure times for each node are computed and compared against the demand time constraints to determine if the vehicle can service all of the demand points within the demand window.

Each feasible insertion is assessed via a savings score that is correlated with the cost-savings associated with the insertion. Liu and Shen (1999)'s "Modified

Combined Savings" function, for an insertion of route R^{II} into link i, j of route R^I to produce route R^{III}, is defined as:

$$MCS\left(R^I, R^{II}, R^{III}\right) = \\ \alpha\left(T^I + T^{II} - T^{III}\right) - (1-\alpha)\left(\frac{D_j^{III} - D_j^I}{cus^{II}}\right) + V^I + V^{II} - V^{III}, \tag{12}$$

where T^R denotes the travel cost of route R, cus^{II} represents the number of customers on route II, and V^R denotes the initial cost of the vehicle used for route R. α is a parameter between 0 and 1 to be varied in order to optimize the cost function. This cost function accounts for the travel cost savings, the cost savings of reducing the number of vehicles. Finally, the function penalizes insertions that push back the departure time of the original route to discourage insertions that harm route feasibility for future insertions.

The Modified Optimistic Opportunity Savings (MOOS) function is adjusted from the one provided by Liu and Shen (1999), and is defined as follows:

$$MOOS\left(R^I, R^{II}, R^{III}\right) = MCS\left(R^I, R^{II}, R^{III}\right) + V^{III}\left(\frac{K^{III} - K^{II} - K^I}{K^{III}}\right), \tag{13}$$

where K^R denotes the capacity of the vehicle used for route R. This addition to the savings function accounts for the excess capacity left in the vehicle that can potentially be used to service more locations in the future. While Liu and Shen (1999) consider only the fixed cost of the smallest vehicle which could serve the excess capacity, the proposed savings formula considers this cost as a proportion of the remaining capacity. This term can generally be interpreted as a method of balancing out the cost of upgrading a vehicle type across the routes which the upgrade may render feasible, by awarding extra savings to that upgrade (Liu & Shen 1999).

This implementation also differs from Liu and Shen (1999) in that it keeps track of the time the vehicle spends waiting for the location's service time window to begin. Moreover, the service time of each vehicle is accounted for and a cost is attached to it. This is to account for the scenario when one mode of transportation has some cost attached to stopping (e.g. parking tickets). Hence, the final savings function is

$$SAVINGS = MOOS\left(R^I, R^{II}, R^{III}\right) + \left(W^I w^I + W^{II} w^{II} - W^{III} w^{III}\right) \\ + \left(S^I N^I + S^{II} N^{II} - S^{III} N^{III}\right), \tag{14}$$

where W^R denotes the cost ($/min) of waiting on route R, w^R is the time in minutes spent waiting on route R, S^R is the service cost ($) for the vehicle on route R, and N^R represents the number of locations serviced on route R.

Finally, the cost-savings involving a route that only services one location are scaled by a factor of β in order to either incentivize or disincentivize the algorithm to construct routes sequentially (by adding one demand stop at a time).

3.4 Case Study

A case study is undertaken using the described VRPMVTTW heuristic, to investigate the benefits of using cargo tricycles within a heterogenous vehicle fleet for last mile delivery operations in downtown Toronto. The study is formulated as described in the following sections, informed by both operating data and discussions with a partner express courier delivery company in Toronto, Canada. Various assumptions are made regarding the vehicle fleet operating characteristics and the demand profile of deliveries, as discussed.

The case study is implemented following several steps. The vehicle data (discussed in Sect. 3.4.1) is first initialized, establishing the fleet of available vehicles for the given service area. Operational demand data provided by the partner company is then processed and geocoded. Following this, demand points are clustered according to the process described in Sect. 3.4.2, and travel time matrices are calculated between all demand points for each vehicle type. The developed heuristic is applied to this processed demand data, to provide a set of proposed routes which serve all of the demand points given the vehicle operating parameters provided.

3.4.1 Vehicle Operating Characteristics

Vehicle characteristics for each vehicle type are established using specification documents, and reference studies (Chevrolet 2016; Sweet et al. 2015; Salbach et al. 2015; Deng et al. 2017; Chander 2019), as well as through discussion with the partner company. These characteristics include vehicle capacities, average downtown operating speeds, amortized initial vehicle costs, per-minute operating and idling costs, and parking/service times. The resultant values are summarized in Table 1.

3.4.2 Delivery Demand Data

The parcel demand data used in the case study is extracted directly from historic delivery records provided by the partner express courier company. These records reflect the single-day demand for a given service area, and describe the delivery time

Table 1 Vehicle characteristics used for application to case study

Vehicle type	Avg. speed (km/h)	Volume capacity (m^3)	Amortized vehicle cost ($/day)	Operating cost ($/min)	Waiting/ service cost ($/min)	Service time (min)
Delivery van	25	6.79	13.17	1.13	1.02	4.16
Cargo tricycle	10	2.00	23.95	0.72	0.72	3.33
Walking	5	0.50	0	0.72	0.72	2.75

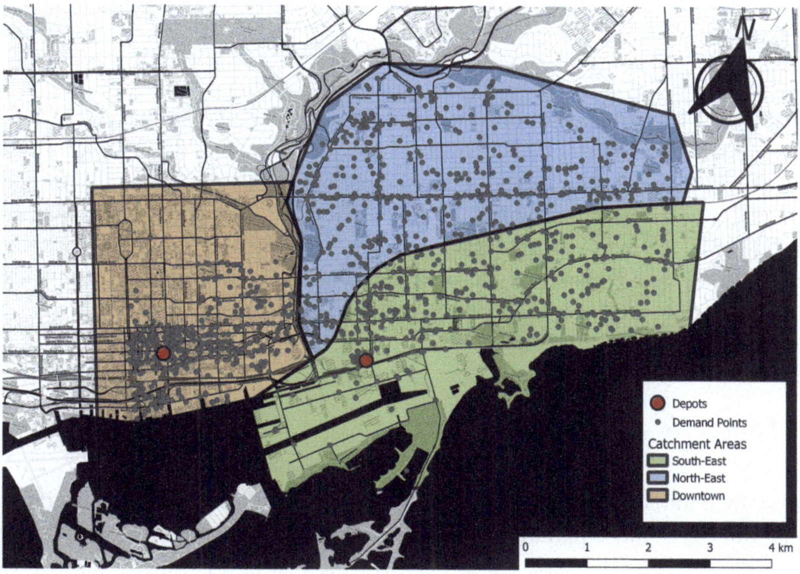

Fig. 1 Service areas, depots, and distribution of delivery demand points

window, delivery location (geocoded from a street address), and package volume (based on either recorded dimensions or parcel type) for each parcel. The geographic distribution of demand points are shown in Fig. 1. This is grouped into three sub-areas according to the feasible delivery vehicle types and appropriate depot locations in each area, dividing along natural boundaries in the built environment including the Don River and the Lakeshore East rail corridor. Subsequently, three 'service areas' are developed which are considered independently. The west 'downtown' service area uses a downtown depot and considers walking and cargo tricycle vehicle types for this dense urban setting, while the eastern two service areas consider delivery van and cargo tricycle vehicle types, using a depot located in Toronto's port lands to serve a more sparsely distributed demand profile.

A total of 5092 demand points is considered in this analysis. Deliveries that share both a delivery address and time window are combined into a single demand point, with an added marginal service time of 2 min per delivery for movement within the delivery location. The maximum combined delivery demand is also capped at the capacity of a walking delivery-person to ensure that it may still be served by all vehicle types. This yields a total of 1217 demand points for analysis by the vehicle routing heuristic. For the implementation of the case study, an origin-destination (OD) matrix of travel times between all delivery points and the depot within each service area is calculated. These matrices are used in the calculation of cost savings by the heuristic algorithm.

4 Results

The VRPMVTTW solver is applied to the case study demand points to assess cargo tricycles as a prospective vehicle type for a courier company in downtown Toronto. The heuristic model is first tuned using a grid search approach to set appropriate parameter values, following which the tuned heuristic is applied to the case study demand profile. A sensitivity analysis is also performed to determine the consistency of results relative to changes in the operating characteristics of cargo tricycles, as they are a novel mode for which limited historic operating data is available.

4.1 Heuristic Parameter Tuning

Because the parameters α and β alter the results of the algorithm, various combinations of values for those parameters are tested and the best result is selected. A grid search approach is used to tune the model using the case study vehicle characteristics and service area, with a limited set of demand points. Through this method, a wide variety of parameter combinations are tested with the intention of minimizing the heuristic objective function.

The parameter α is bounded between 0.25 and 1.00 because a value less than 0.25 indicates that saving travel cost is less significant than pushing back departure times whereas a value of 1.00 causes the algorithm to disregard the impact of pushing back departure times. β is bounded between 0.5 and 2.0 to provide a range of values both greater than and less than 1.0, because a value less than 1.0 indicates no preference towards or against sequential construction (inserting nodes one at a time) whereas a value greater than 1.0 indicates a preference towards sequential construction. The results of the objective function for different combinations of these parameters are illustrated in Table 2.

The objective function is minimized for (α, β) equals (1.0, 1.5–2.0). This implies that saving travel cost is much more important than pushing back departure times. However, the α parameter appears to be substantially more important to the cost than the β value. A β value of 2.0 implies that sequential construction is preferred over combining two routes which each serve multiple locations.

While the parameter values obtained in Table 2 are useful for determining appropriate heuristic parameters for the given case study, they may not necessarily

Table 2 Objective function values along grid search points for α and β

$\alpha \backslash \beta$	0.5	1.0	1.5	2.0
0.25	1344	1344	1479	1345
0.50	589	589	589	589
0.75	480	480	478	478
1.00	464	461	456	456

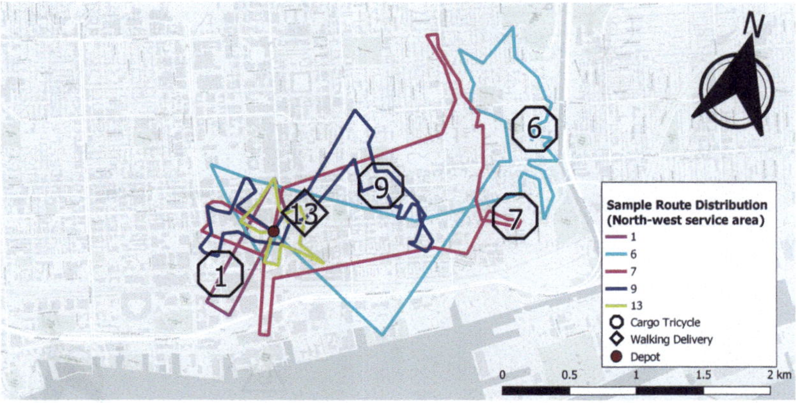

Fig. 2 Sample of calculated routes for the north-west service area

be useful for making generalized claims about the nature of the VRPMVTTW problem. Given a different route configuration, it may be advantageous to use a different configuration of α and β and thus new applications of this heuristic should be tuned independently.

4.2 *Case Study: Proposed Vehicle Routing*

Using the solver and heuristic developed, a strategy is generated which could be adopted to dispatch a fleet of vehicles to serve the desired demand points. It identifies the sequence of points served, vehicle type used for delivery, and cost per route. Cargo tricycles are the recommended vehicle type for the vast majority of routes under default vehicle operating characteristics. This is likely due to their low-cost as well as their relatively high capacity and travel speed. However, given the uncertainty around operating characteristics for this emerging vehicle type, the robustness of this result is tested through sensitivity analysis in Sect. 4.3.

A sample plot of the routes generated by the algorithm is shown in Fig. 2 for the downtown service area, Fig. 3 for the south-east service area, and Fig. 4 for the north-east service area. For visual clarity, a subset of the routes generated is provided. These routes generally cover distinct geographic areas, although there is some level of overlap when necessary to accommodate delivery time windows or

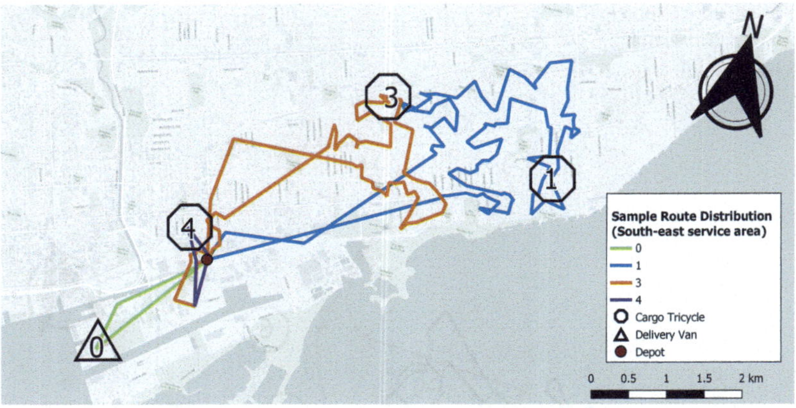

Fig. 3 Sample of calculated routes for the south-east service area

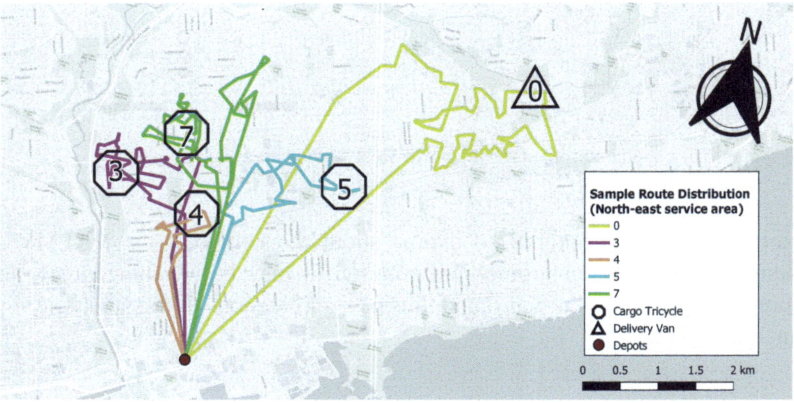

Fig. 4 Sample of calculated routes for the north-east service area

vehicle capacity constraints. In particular, several overlapping routes are found in the downtown area, which has a very dense demand profile along with low costs of travel between demand points.

4.3 Sensitivity Analysis

Sensitivity analysis is performed to demonstrate the robustness of the solution when uncertainty in the vehicle operating data is considered. In particular, cargo tricycles are an emerging vehicle type for commercial operations, for which limited historical operating data are available to frame the case study. It is thus important to explore scenarios to test how their efficacy changes under various sets of operating parameters.

Sensitivity analysis is performed by altering four operating characteristics for cargo tricycles within a reasonable range, including their operating cost per minute, service time, operating speed, and amortized vehicle cost. The dependent variable is the percentage of demand points serviced by cargo tricycles. The total cost of operations is plotted to indicate the variation in operating cost for the freight carrier under input variable uncertainty. The results are illustrated in Figs. 5, 6, 7, and 8, with initial estimated values (see Table 1) indicated with a red dashed outline.

Figure 5 indicates that the usage of cargo tricycles and related total cost impact varies significantly by service area for variations in the cargo tricycle operating cost. In the downtown area, cargo tricycles remain the dominant vehicle type despite significant increases in operating cost (up to \$1.17/min), resulting in significant increases in the total cost to serve the area. By contrast, both the south-east and north-east service areas begin to be dominated by van deliveries once cargo tricycle operating costs surpass \$0.95, above which total delivery cost increases are minimal.

Increases in the cargo tricycle parking and service time substantially change their modal share for all service areas (Fig. 6), with marginal increases of around 2 min resulting in a shift toward van deliveries for north-east and south-east service areas, while increases of 4–7 min cause a shift toward walking deliveries in the downtown area. This parameter is a significant cause for consideration in cargo tricycle implementation, as parking times may be impacted by municipal regulations of where the vehicles are allowed to park (e.g. on sidewalks or bicycle racks, or using curbside parking).

Decreases in cargo tricycle speed generally cause limited decreases in total delivery cost across all service areas, as shown in Fig. 7. Speeds of 7.5 km/h cause significant shifts away from cargo tricycles in the south-east and north-east areas, however, in the downtown area cargo tricycles remain dominant until their speed is

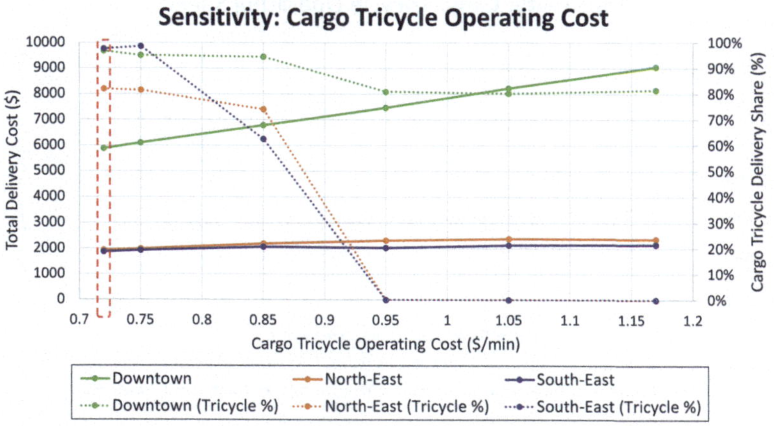

Fig. 5 Sensitivity analysis for cargo tricycle operating cost

decreased to that of walking deliveries. This shift in delivery share is accompanied by marginal increases in the total service cost as the primary vehicle type changes.

Finally, increases in cargo tricycle amortized vehicle cost do not significantly impact the results of this analysis as illustrated in Fig. 8. The amortized vehicle cost represents a small fraction (approximately 10%) of the overall operating cost, and as a result, we have found that the results are insensitive to this cost. This result demonstrates the robustness of preference for cargo tricycle deliveries across a wide range of potential upfront vehicle cost values, which is relevant as cargo tricycle capital costs may vary considerably as the technology is relatively new and may change over time.

Overall, the results show that the robustness of efficacy for cargo tricycles varies considerably by implementation area. Where delivery vans provide a means

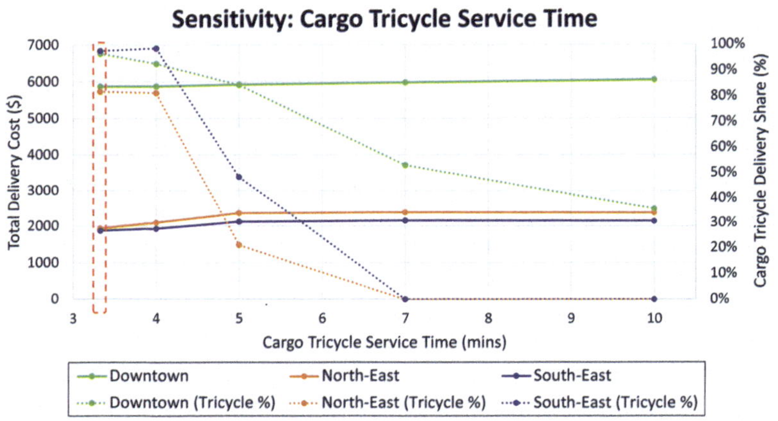

Fig. 6 Sensitivity analysis for cargo tricycle service time

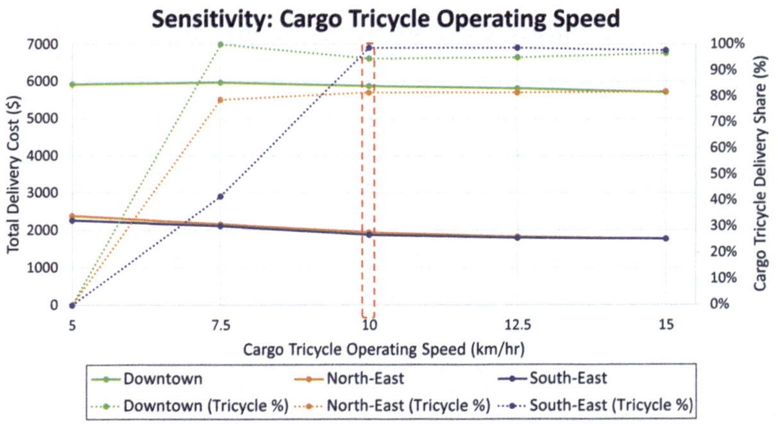

Fig. 7 Sensitivity analysis for cargo tricycle speed

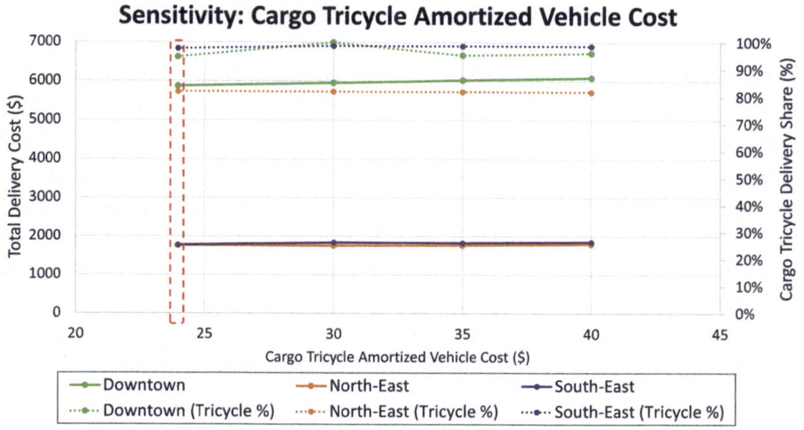

Fig. 8 Sensitivity analysis for cargo tricycle amortized vehicle cost

of quick travel across longer distances (such as in the south-east and north-east service areas), the effectiveness of cargo tricycles is highly susceptible to their operating parameters. However, in the downtown service area, the increase in speed with minimal changes to operating cost or service time reflects a significant improvement over walking deliveries, and reliance on walking delivery-people over cargo tricycles would reflect significant increases in cost. Furthermore, these savings are especially apparent for longer-distance downtown deliveries, as shown by the balanced route sample (obtained through sensitivity analysis with increased cargo tricycle service time) between walkers and cargo tricycles in Fig. 9. This sample

Fig. 9 Sample routes for balanced fleet of walking deliveries and cargo tricycles for downtown service area

routing was obtained through the sensitivity analysis, and selected as it represents a case where there is a relatively similar volume of cargo tricycle and walking deliveries, based on the percentage of total demand points serviced by each mode. The sample taken is from the case with cargo tricycle service time increased to 7 min, while other operating parameters remain consistent with the baseline assumptions. It was found that in this mixed-fleet circumstance where both modes provide some utility in the lowest-cost delivery routing, the selection of stops to be serviced is not randomly distributed across delivery modes, instead favouring cargo bicycles for longer distances and walking deliveries for shorter trips. This effectively demonstrates how a mixed fleet of vehicles could be used in certain circumstances to minimize overall delivery costs, in consideration of advantages (relative operating speeds, service speeds, and costs) between the various modes available.

5 Discussion

The results above demonstrate the ability of the developed heuristic to output practical results across different operating conditions. They make a strong argument in favor of the applicability of cargo tricycles for urban deliveries, particularly in dense urban centers. This section discusses the effectiveness of cargo tricycles and their robustness to various operating conditions, as well as some of the current limitations of the VRPMVTTW heuristic and solver developed.

5.1 Effectiveness of Cargo Tricycles

The results generally indicate that cargo tricycles are a promising, cost-effective vehicle type for urban last-mile delivery operations. Cargo tricycles serve the majority of demand points under default operating assumptions, and are a particularly robust option under changing assumptions for downtown operations. This result supports the potential for cargo tricycles to provide an effective delivery option for freight companies in dense urban areas.

Sensitivity analysis assesses the stability of this result. For the downtown area, results remain robust to significant changes in cargo tricycle operating parameters, demonstrating particular value of cargo tricycles over walking deliveries in dense downtown cores. The suitability of each vehicle type by context is also shown to be robust to changes in operating characteristics, with delivery vans serving less densely populated (though still dense relative to suburban) areas, and mixed vehicle fleets providing a suitable option even under conservative assumptions of cargo tricycle operating characteristics (lower operating speed, higher operating cost, and higher service times). Most notably, cargo tricycle suitability was sensitive to changes in the required parking time, identifying a potentially significant impact of municipal regulations for vehicle parking or stopping locations.

Variability in cost effectiveness by delivery type is also discovered, shown by Fig. 9. Given the greater suitability for cargo tricycles to make medium-range deliveries in dense areas, they provide a promising option for mixed-fleet implementation with another vehicle type, particularly walking deliveries in downtown Toronto.

Ultimately, this study demonstrates that cargo tricycles are a promising option for the courier company's delivery operations in downtown Toronto. Despite potentially significant upfront costs, cargo tricycles increase the available capacity and average speed when compared to walking, while only marginally increasing operating costs. The sensitivity analysis corroborates this finding in the face of uncertain operating characteristics. Cargo tricycles therefore present a legitimate opportunity to decrease operating costs for courier companies operating in downtown Toronto, and should be explored further.

5.2 Limitations of the VRPMVTTW Solver

There are several key limitations of the current VRPMVTTW solver implementation. This section describes these limitations, in addition to potential next steps for improving the algorithm.

The largest limitation of the algorithm is that it runs in $O(n^2)$ time. While this is an improvement upon previous approaches which run in $O(n^3)$ time, it does limit the number of demand points that the algorithm can be executed on. This challenge imposed a constraint on the granularity of values explored by the parameter grid search, as well as a need to consider catchment areas separately as opposed to in a single, combined solver run.

Additionally, the use of deterministic vehicle operating characteristics may not be entirely accurate, especially pertaining to the van and cargo tricycle vehicle types. Although these values were determined based on historical data, the use of deterministic values presents an inherent limitation. In particular, speeds and travel times are likely to vary considerably with trip distance and by time of day (as well as due to stochastic fluctuations), particularly for traffic-prone vehicles such as delivery vans and (to some extent) cargo tricycles. Hence, the algorithm runs the risk of undervaluing more reliable vehicle types and is insensitive to missed delivery penalties due to stochastic travel times under congestion. A more accurate operational implementation could be achieved by expanding the heuristic proposed in this paper with a stochastic consideration of travel time.

The algorithm could also be extended by adding constraints related to human resources. We have not imposed constraints on tour duration, meaning that short shifts are not penalized despite potentially requiring employees to be paid beyond the completion of their deliveries, to adhere to labor policy. The algorithm could be adjusted to combine multiple shorter routes to be served by a single vehicle, shifting delivery times and removing initialization costs accordingly. The use of amortized initialization costs for a single service day also ignores the long-term staffing commitments and investments in a fleet of vans and cargo tricycles. A longer-term dataset could be used to determine the consistency of fleet requirements over several days, and better-calculate longer term vehicle requirements. Adjusting the heuristic to consider a fixed fleet size would further expand its potential for operational use.

6 Conclusion

This study outlines a revised implementation of a VRPMVTTW solver and applies it to a heterogenous fleet in urban service areas allowing for walking deliveries, cargo tricycles, and delivery vans. The VRPMVTTW solver implementation builds upon prior studies with several key revisions to adapt to challenges specific to urban areas. Operational characteristics for a case study application, including demand distribution and volume, as well as vehicle characteristics such as speed and capacity are determined through collaboration with an industry partner.

Results from the case study show cargo tricycles provide a cost-effective option within the service areas examined, and are highly competitive with existing vehicle types. Sensitivity analysis is conducted to assess the robustness of these findings under uncertain operating conditions. The results indicate that cargo tricycles are highly suitable in the downtown core even under conservative assumptions for operating cost, service time and travel speed, while being more closely matched with delivery vans for operation in less densely populated urban areas. Furthermore, assessments of mixed fleet operations identify that cargo tricycles can efficiently operate alongside existing vehicle types, by covering medium-distance deliveries downtown and shorter-distance deliveries in other areas of the city.

Acknowledgments The authors would like to thank several University of Toronto researchers who laid the groundwork for this study. This includes Onkar Chander, Mehdi Nourinejad, Mahyar Jahangiriesmaili, and Puyuan Deng. We would also like to graciously thank Purolator Inc. for technical collaboration, data access, and funding support provided.

References

Arnold, F., Cardenas, I., Sörensen, K., & Dewulf, W. (2018). Simulation of b2c e-commerce distribution in antwerp using cargo bikes and delivery points. *European Transport Research Review, 10*(1), 1–13.

Baldacci, R., Bartolini, E., Mingozzi, A., & Roberti, R. (2010). An exact solution framework for a broad class of vehicle routing problems. *Computational Management Science, 7*, 229–268.

Baldacci, R., Battarra, M., Vigo, D. (2008). Routing a Heterogeneous Fleet of Vehicles. In: B. Golden, S. Raghavan & E. Wasil (Eds.), The vehicle routing problem: latest advances and new challenges. *Operations Research/Computer Science Interfaces, 43*, Springer, Boston, MA. https://doi.org/10.1007/978-0-387-77778-8_1

Baldacci, R., & Mingozzi, A. (2009). A unified exact method for solving different classes of vehicle routing problems. *Mathematical Programming, 120*, 347–380.

Bektaş, T., & Laporte, G. (2011). The pollution-routing problem. *Transportation Research Part B: Methodological, 45*(8), 1232–1250.

Chander, O. (2019). Cargo tricycle parcel delivery pilot study at the university of toronto. Master's Thesis, University of Toronto.

Chevrolet. (2016). 2016 chevrolet express cargo van. https://media.gm.ca/media/ca/en/chevrolet/vehicles/express_cargo_van/2016.tab1.html

Conway, A., Fatisson, P. E., Eickemeyer, P., Cheng, J., & Peters, D. (2012). Urban micro-consolidation and last mile goods delivery by freight-tricycle in Manhattan: Opportunities and challenges. In: *Conference Proceedings, Transportation Research Board 91st Annual Meeting* (pp 1–17).

Dantzig, G. B., & Ramser, J. H. (1959). The truck dispatching problem. *Management Science, 6*(1), 80–91.

Deng, P., Nourinejad, M., Jahangiriesmaili, M., & Roorda, M. (2017). Design and analysis of city logistics systems in toronto for express courier companies. In *Canadian Transportation Research Forum 52nd Annual Conference-Canadian Transportation: 150 Years of Progress//Les transports au Canada: 150 ans de progrès-Winnipeg, Manitoba, May 28–31, 2017*

Ferland, J. A., & Michelon, P. (1988). The vehicle scheduling problem with multiple vehicle types. *Journal of the Operational Research Society, 39*(6), 577–583.

Franceschetti, A., Honhon, D., Van Woensel, T., Bektaş, T, & Laporte, G. (2013). The time-dependent pollution-routing problem. *Transportation Research Part B: Methodological, 56*, 265–293.

Gan, X., Kuang, J., & Niu, B. (2014). Particle swarm optimizations for multi-type vehicle routing problem with time windows. In *International Conference on Intelligent Computing* (pp. 808–815). Springer.

Golden, B., Assad, A., Levy, L., & Gheysens, F. (1984). The fleet size and mix vehicle routing problem. *Computers & Operations Research, 11*(1), 49–66.

Golden, B. L., Raghavan, S., & Wasil, E. A., et al. (2008). *The vehicle routing problem: Latest advances and new challenges* (vol. 43). Springer.

Gruber, J., Kihm, A., & Lenz, B. (2014). A new vehicle for urban freight? An ex-ante evaluation of electric cargo bikes in courier services. *Research in Transportation Business & Management, 11*, 53–62.

Huai, C. X., Sun, G. H., Qu, R. R., Gao, Z., & Zhang, Z. H. (2019). Vehicle routing problem with multi-type vehicles in the cold chain logistics system. In *16th International Conference on Service Systems and Service Management (ICSSSM)* (pp. 1–4). IEEE.

Irnich, S., & Schneider, M. (2014). Four variants of the vehicle routing problem. In P. Toth & D. Vigo (Eds.), *Vehicle routing: Problems, methods, and applications* (pp. 241–271)

Jamil, A., Abdallah, B. N., & Leksono, V. A. (2021). Firefly algorithm for multi-type vehicle routing problem. In *Journal of Physics: Conference Series* (vol. 1726, p. 012006). IOP Publishing

Kim, G., Ong, Y. S., Heng, C. K., Tan, P. S., & Zhang, N. A. (2015). City vehicle routing problem (city vrp): A review. *IEEE Transactions on Intelligent Transportation Systems, 16*(4), 1654–1666.

Koç, Ç., Bektaş, T., Jabali, O., & Laporte, G. (2016). Thirty years of heterogeneous vehicle routing. *European Journal of Operational Research, 249*(1), 1–21.

Laporte, G. (2009). Fifty years of vehicle routing. *Transportation Science, 43*(4), 408–416.

Larsen, A., & Madsen, O. (2000). The dynamic vehicle routing problem. Ph.D. Thesis, Institute of Mathematical Modelling, Technical University of Denmark.

Lebeau, P., Cok, B., Kees, C., & Macharis, C. (2023). Towards more sustainable vehicles for the last mile? Cycle logistics as a part of the solution. In *Handbook on city logistics and urban freight* (p. 178). Edward Elgar Publishing.

Li, F., Golden, B., & Wasil, E. (2007). A record-to-record travel algorithm for solving the heterogeneous fleet vehicle routing problem. *Computers & Operations Research, 34*(9), 2734–2742.

Liu, F. H., & Shen, S. Y. (1999). A method for vehicle routing problem with multiple vehicle types and time windows. *Proceedings-National Science Council Republic of China Part A Physical Science And Engineering, 23*, 526–536.
Liu, S., Huang, W., & Ma, H. (2009). An effective genetic algorithm for the fleet size and mix vehicle routing problems. *Transportation Research Part E: Logistics and Transportation Review, 45*(3), 434–445.
Marcucci, E., Gatta, V., & Le Pira, M. (2023). *Handbook on city logistics and urban freight: 0.* Edward Elgar Publishing.
Máximo, V. R., Cordeau, J. F., & Nascimento, M. C. (2022). An adaptive iterated local search heuristic for the heterogeneous fleet vehicle routing problem. *Computers & Operations Research, 148*, 105954.
Munari, P., Dollevoet, T., & Spliet, R. (2016). A generalized formulation for vehicle routing problems. arXiv:160601935.
Mutar, M., Aboobaider, B., & Hameed, A. (2017). Rev vehicle routing problem and future research trend. *International Journal of Applied Engineering Research, 12*(22), 12279–12283.
Naumov, V., & Starczewski, J. (2019). Approach to simulations of goods deliveries with the use of cargo bicycles. In *AIP Conference Proceedings* (vol. 2078, p. 020070). AIP Publishing LLC.
Penna, P. H. V., Subramanian, A., & Ochi, L. S. (2013). An iterated local search heuristic for the heterogeneous fleet vehicle routing problem. *Journal of Heuristics, 19*(2), 201–232.
Perboli, G., Rosano, M., Saint-Guillain, M., & Rizzo, P. (2018). Simulation–optimisation framework for city logistics: An application on multimodal last-mile delivery. *IET Intelligent Transport Systems, 12*(4), 262–269.
Pillac, V., Gendreau, M., Guéret, C., & Medaglia, A. L. (2013). A review of dynamic vehicle routing problems. *European Journal of Operational Research, 225*(1), 1–11.
Quak, H., & de Koster, M. B. (2009). Delivering goods in urban areas: How to deal with urban policy restrictions and the environment. *Transportation Science, 43*(2), 211–227.
Renaud, J., Laporte, G., & Boctor, F. F. (1996). A tabu search heuristic for the multi-depot vehicle routing problem. *Computers & Operations Research, 23*(3), 229–235.
Salbach, N. M., O'Brien, K. K., Brooks, D., Irvin, E., Martino, R., Takhar, P., Chan, S., & Howe, J. A. (2015). Reference values for standardized tests of walking speed and distance: A systematic review. *Gait & Posture, 41*(2), 341–360.
Scherr, Y. O., Saavedra, B. A. N., Hewitt, M., & Mattfeld, D. C. (2019). Service network design with mixed autonomous fleets. *Transportation Research Part E: Logistics and Transportation Review, 124*, 40–55.
Schliwa, G., Armitage, R., Aziz, S., Evans, J., & Rhoades, J. (2015). Sustainable city logistics—making cargo cycles viable for urban freight transport. *Research in Transportation Business & Management, 15*, 50–57.
Sheth, M., Butrina, P., Goodchild, A., & McCormack, E. (2019). Measuring delivery route cost trade-offs between electric-assist cargo bicycles and delivery trucks in dense urban areas. *European Transport Research Review, 11*(1), 1–12.
Solomon, M. M. (1984). Vehicle routing and scheduling with time window constraints: Models and algorithms (Heuristics). Ph.D. Thesis, University of Pennsylvania.
Solomon, M. M. (1987). Algorithms for the vehicle routing and scheduling problems with time window constraints. *Operations Research, 35*(2), 254–265.
SteadieSeifi, M., Dellaert, N. P., Nuijten, W., Van Woensel, T., & Raoufi, R. (2014). Multimodal freight transportation planning: A literature review. *European Journal of Operational Research, 233*(1), 1–15.
Sumichras, R. T., & Markham, I. S. (1995). A heuristic and lower bound for a multi-depot routing problem. *Computers & Operations Research, 22*(10), 1047–1056.
Sun, W., Yu, Y., & Wang, J. (2019). Heterogeneous vehicle pickup and delivery problems: Formulation and exact solution. *Transportation Research Part E: Logistics and Transportation Review, 125*, 181–202.
Sweet, M., Harrison, C., & Kanaroglou, P. (2015). Congestion trends in the city of Toronto (2011–2014). McMaster Institute for Transportation and Logistics, Hamilton, Ontario, Canada.

Toth, P., & Vigo, D. (2014). *Vehicle routing: Problems, methods, and applications*. SIAM.
Van Woensel, T., Kerbache, L., Peremans, H., & Vandaele, N. (2008). Vehicle routing with dynamic travel times: A queueing approach. *European Journal of Operational Research, 186*(3), 990–1007.
Wang, Z., Li, Y., & Hu, X. (2015). A heuristic approach and a tabu search for the heterogeneous multi-type fleet vehicle routing problem with time windows and an incompatible loading constraint. *Computers & Industrial Engineering, 89*, 162–176.
Wilson, N. H. M., & Colvin, N. J. (1977). *Computer control of the Rochester dial-a-ride system* (vol. 77). Massachusetts Institute of Technology, Center for Transportation Studies.

Integrated Location, Sizing, and Pricing for EV Charging Stations

Miguel F. Anjos, Ikram Bouras, Luce Brotcorne, Alemseged G. Weldeyesus, Clémence Alasseur, and Riadh Zorgati

1 Introduction

With the growing adoption of electric vehicles (EVs), the demand for EV charging stations (CSs) is increasing, especially in urban areas. Research in this area has considered various aspects of the problem as well as different types of mathematical models. Table 1 summarizes recent research on location and pricing for EV charging stations, including the different characteristics of the model (type of the models, objective functions, decision variables, constraints and the proposed algorithms). To the best of our knowledge, most of the research on EV charging infrastructure has focused on the placement of charging stations. These optimization models differ

M. F. Anjos (✉)
University of Edinburgh, Edinburgh, UK
GERAD, Polytechnique Montréal, Montréal, QC, Canada
e-mail: anjos@stanfordalumni.org

I. Bouras
Centre Inria de l'Université de Lille, Lille, France
Université Ferhat Abbas, Sétif, Algeria
e-mail: ikram.bouras@univ-setif.dz

L. Brotcorne
Centre Inria de l'Université de Lille, Lille, France
e-mail: luce.brotcorne@inria.fr

A. G. Weldeyesus
University of Edinburgh, Edinburgh, UK
e-mail: a.weldeyesus@ed.ac.uk

C. Alasseur · R. Zorgati
EDF R&D, Paris, France
e-mail: clemence.alasseur@edf.fr; riadh.zorgati@edf.fr

T. G. Crainic et al. (eds.), *Combinatorial Optimization and Applications*,
International Series in Operations Research & Management Science 358,
https://doi.org/10.1007/978-3-031-57603-4_18

with respect to objective functions, decision types, station types, application sizes, and also the type of the optimization models. In terms of the objective function, maximizing the number of EV users, maximizing profit, and minimizing the total costs (energy and social costs) are very common. The demand in these works is defined in different ways: traffic flow, grid specifications, queuing models, etc. Many studies have also covered specific case studies of optimal deployment of EV charging stations. Furthermore, different types of optimization models have been applied in this area. They include bilevel optimization models, multi-objective models, and stochastic models. Most of the models are solved using heuristic algorithms, as is the case in this work as well, due to the complex structure and the large size of the models.

From Table 1, we remark that the studies on the provision of EV charging mostly consider two main questions. The first question, and the most commonly addressed, is to optimize the location and sizing of charging stations for an EV charging network, given objective(s) and demands. Various objective functions and demand definitions have been considered. Domínguez-Navarro et al. (2019) proposed a design problem of EV fast charging stations where the demand is modeled using Monte Carlo method and considering a variety of energy sources, the problem is solved using genetic algorithm. In Faridimehr et al. (2018), two-stage stochastic programming model is proposed to determine the optimal network of charging stations, considering uncertainties in the arrival and dwell times of EVs. Studies in Huang and Kockelman (2020) and Xiong et al. (2017) present bilevel optimization models to find the optimal locations for charging stations so as to maximize the total profit (respectively, minimizing the social cost) in the upper level, and considering the traffic flow of the users in the lower level problem. The second question considered in the literature is to optimize the pricing of charging at each station of the network. In Xiong et al. (2016), a pricing model is presented using EV charging behavior, operating expense of charging stations, and the traffic congestion. The problem is formulated as a mixed integer non-convex optimization problem and a scalable algorithm is proposed to solve the problem.

Our main goal in this chapter is to propose a strategic model that combines the two questions of location and sizing of stations, and of pricing the charging. The motivation is to have a model that informs the strategic design decisions of locating and sizing by including information about the users' choices, and hence about how much revenue the charging will bring in the medium- to long-term, given possible location and sizing decisions.

Our proposed model simultaneously locates charging stations, sets their size (number of charging points), and determines the prices for charging dynamically over the course of a day. This motivates the use of a bilevel model as it is the privileged means to model the reaction of one or more agents (the EV users) to the decisions of another agent (the charging service provider). The key to achieve this is to integrate a more detailed description of the user's behaviour, which we model by taking into account their geographical location at different times during the time duration of interest (typically a 24-h day), and by integrating the concept of preference lists. To the best of our knowledge, it is the first time that preference

Table 1 Summary of the literature on the provision of public EV charging station (EVCS) services

Paper	Problem	1-level	Bilevel	Multi-objec.	Stochastic	Design	Pricing	Objectives	Variables	Constraints	Algorithms	Other aspects
Domínguez-Navarro et al. (2019)	Design of EV fast CS	×	–	–	–	×	–	Max profit	Energy amount	Charging process	Genetic	Renewable energy
Faridimehr et al. (2018)	EVC network design	× 2 stages	–	–	×	×	–	Opt network of CS	CS installation + uncertain demand	Budget, capacity, demand	Heuristic	
Faridpak et al. (2019)	Opt placement & operation of EVCS	× 2 steps charging interval	–	–	×	×	–	Min supply cost + opt operation of CS	Charging start time+	Different charging modes	LP	
Dait et al. (2018)	FCS placement	–	×	–	–	×	–(fixed)	Max profit + Opt capacity	Design var+ assignment	M/M/r queuing system	Heuristic	Using zones
Aljaidi et al. (2019)	Opt placement	×	–	–	–	×	–	Opt capacity& Placement	Design var.	–	–	Zones, urban areas, survey
Canizes et al. (2019)	Opt distribution	×	–	–	–	–	×	Dynamic energy pricing	–	–	–	Smart cities
Xiong et al. (2016)	Opt pricing	-	×	–	–	–	×	Minimum social cost	Congestion	Strategy variables	Rules alg.	–
Celli et al. (2018)	Opt allocation FCS	–	–	×	–	×	–	Min cost + waiting time, Max traffic flow		M/M/s	Genetic+ MILP	Traffic flow
Anjos et al. (2020)	Opt planning	×	–	–	–	×	–	Max number of EV	Number of EV	Budget, EV number	Greedy	
Zhang et al. (2015)	EVCS layout research	–	×	–	×	×	–	Min cost(operator+ users), Max traffic flow	Path variables	1-cost, capacity, energy 2-Traffic flow constraints	ICA & ILP	Traffic flow
Chen et al. (2019)	Optimal EVCND	×	–	–	–	×	–	Min system cost	Assignment variables	M/M/c system	B&B	Non-linear
Xiong et al. (2017)	EVFCS	–	×	–	–	×	–	Min social cost, charging cost of users	Design var	Budget, equilibrium strategy	Heuristic	Traffic flow
Gan et al. (2020)	EVFCS, elastic demand	× (NL)	–	–	×	×	–	Max profit	Design variables	Distance penalty, elastic demand, waiting time	Genetic	Fixed price
Xiang et al. (2016)	Economic planning	×	–	–	–	×	–	Min cost	Design	Traffic, capacity	B&B	Flow
Lee and Han (2017)	EVCS location	× (NL)	–	–	×	×	–	Max expected demand	Design+Flow variables	Flow constraints	Benders	
Huang and Kockelman (2020)	EVCS location with elastic demand	–	×	–	–	×	–(fixed)	Max profit, Assignment	Design + path variables	Flow constraints	Genetic	
This paper	EVCS pricing and design	–	×	–	–	×	×	Max profit	Design + price + assignment	Preference lists, capacities, price threshold	ILP	Strategic model

lists are used to model the choice of the decision agents within a bilevel model. The idea of the predefined preference list is to allow each user to select a set of charging station candidates based on the distance and the expected cost. The users also define a price threshold beyond which they are unwilling to pay for the charging, and this threshold can change from one time period to another. One of the advantages of modelling users' behaviour using this combination of preference lists and price thresholds is that we explicitly allow the possibility for users to decide to not charge at any of the provider's stations, which makes the model applicable in the presence of competitors in the market for charging services.

More specifically, we propose a strategic bilevel model that involves two agents, namely the charging service provider (represented by the upper level problem) and the users (represented by the lower level problem). The model combines the location, sizing, and pricing decisions of the charging provider with the preferences of the users. We assume that the users want to charge precisely once during the given day, and that the energy demand is the same for every user. We model the users' preferences using a preference-list or rank-based approach (Farias et al. 2013) that characterizes users by a set of distinct ordered sets of predefined preference lists or products, which is a collection of pairs of charging station and corresponding time period. Moreover, we assume users set their own price thresholds which will give them the option to opt-in or -out of using the charging stations. On the upper level of the bilevel model, the charging service provider is in charge of selecting the location of charging stations and how many charging points in each charging stations, as well as defining the charging price in order to maximize his profit. In response, on the lower level, each user chooses the first available charging station from its predefined preference list. It is possible that the number of users wanting to charge at a given station and time is greater than the number of charging points of that station. The model addresses this by including each user's distance from the station in the objective function of the lower level, so that priority is given to the users geographically closer to the station at that time.

In general, solving bilevel optimization problems is a challenging problem, see, e.g., Colson et al. (2005) and Dempe and Zemkoho (2012). We apply the commonly used solution method for bilevel optimization in which we first derive the optimality conditions (KKT) of the lower level problem and then include them in the upper level problem to find a single-level reformulation of the bilevel model. In this case, the new problem formulation has complementarity constraints, which are hard to tackle for optimization solvers. To overcome the challenge, we use a common approach called big-M relaxation to linearize the complementarity constraints.

We summarize the novelties and contributions of this chapter as follows:

- We study the problem of integrated locating, sizing and pricing EV charging on a charging provider's network.
- We formulate the problem as a bilevel optimization model where, for the first time, the behaviour of users is modelled using preference lists and a time-dependent price threshold.

- We provide managerial insights based on computational experiments on a large collections of instances

The chapter is organized as follows. We present the proposed bilevel optimization model in Sect. 2. In Sect. 3, we describe the single-level reformulation of the model used to solve the problem. In Sect. 4, we present and analyze computational results on a set of randomly generated instances. Section 5 concludes the chapter.

2 Bilevel Optimization Model

We consider a charging service provider installing and operating charging infrastructure to serve the needs of EV users in an urban region. Let K be the set of candidate station locations. For each location $k \in K$, δ_k is the maximum number of charging points that can be installed at that station, and γ_k is the maximum number of users who can charge at a charging point of station k. Moreover, for each $k \in K$, let c_k be the installation cost of a charging spot, and let e_k be the cost to the provider to make available at station k the energy required for the recharge for any user. (Recall that we assume that the energy demand is the same for every user.) Let β denote the installation budget of the provider; if $\beta < \sum_{k \in K} c_k \delta_k$ then the budget constraint is binding, otherwise it is omitted.

Let I be the set of users and T be the set of time periods in the given day. We integrate the behaviour of users in the model using a preference-list (or rank-based) approach in which each user is characterized by an ordered set of s "products", where a product is a combination of a charging location, a time period and a charging price threshold. In other words, for every user $i \in I$, the preference list is an ordered set of s distinct pairs is defined by

$$R_i = \{(k_i^1, t_i^1), ..., (k_i^{l-1}, t_i^{l-1}), (k_i^l, t_i^l), (k_i^{l+1}, t_i^{l+1}), \ldots, (k_i^s, t_i^s)$$
$$|k_i^l \in K, t_i^l \in T\}.$$

The tuple (k_i^l, t_i^l) implies that the lth choice of the user i is station k^l at time t^l. These are employed in the model using a binary parameter v_{kt}^{il}, such that $v_{kt}^{il} = 1$ if $(k_i^l, t_i^l) \in R_i$ for each $i \in I$. Additionally, we assume every user $i \in I$ will have a charging price threshold $P_i^t \in \mathbb{R}^+$ at time t that will be used as a criterion whether the user will opt-in or -out of using any of the stations.

As the elements in the preference list are distinct for every user $i \in I$, it follows that $\sum_l v_{kt}^{il} \leq 1$ for each $k \in K$ and $t \in T$. Finally, the parameter $\lambda_{ik}^t \in [0, 1]$ represents the normalized distance between the location of user i and station k at time t.

The optimization variables are defined as follows. The integer variables y_k are the number of charging points installed at station k. The continuous variables p_k^t are the charging price at station k at time period t. The auxiliary binary variables w_{kt}^{il} are used to check the availability of the lth choice of user i at time t. Specifically, if $v_{kt}^{il} = 0$, then $w_{kt}^{il} = 0$, and if $v_{kt}^{il} = 1$, then $w_{kt}^{il} = 1$ unless $p_k^t > P_i^t$, in which case $w_{kt}^{il} = 0$ (see also (2)). The binary variables x_{ik}^t indicate if station k is selected by user i at time t. Finally, the binary variables z_i track the use of the provider's stations, so that $z_i = 1$ if user i uses one of the stations, and 0 otherwise.

There are three scenarios in which a station included in a user's preference list with a corresponding time may not be available for that user at that time:

1. The charging station has no charging points installed, i.e., $y_k = 0$.
2. The capacity of the charging station at the given time is fully taken up by other users.
3. The charging price at the station at the given time exceeds the threshold price of the user at that time.

If any of the above scenarios happens, then the value of the corresponding auxiliary binary variable w_{kt}^{il} will be zero. Note that the values of w_{kt}^{il} cannot be precomputed because the prices p_i^t are variables representing the pricing decisions of the provider.

The optimization variables and parameters defined above are summarized in Tables 2 and 3.

Table 2 List of variables in the bilevel model (1a)–(1p)

x_{ik}^t	:	1 if user i uses charging station k at time t, 0 otherwise.
z_i	:	1 if user i uses any of the charging stations, 0 otherwise.
y_k	:	number of charging points installed at station k.
w_{kt}^{il}	:	1 if the lth choice of user i, which is station k at time t, is available, 0 otherwise.
p_k^t	:	charging price of charging station k at time t.

Table 3 List of parameters in the bilevel model (1a)–(1p)

δ_k	:	maximum number of charging points that can be installed at station k
γ_k	:	maximum number of users who can charge at a charging point of station k
c_k	:	amortized (per day) installation cost of charging point at station k (\$)
e_k	:	energy unit cost for charging one user at station k (\$)
P_i^t	:	charging price threshold at time t set by user i (\$)
λ_{ik}^t	:	normalized distance between user i and charging station k at time t ([0, 1])
v_{kt}^{il}	:	1 if the lth preference of user i is station k at time t, 0 otherwise
β	:	daily budget of the provider for installing charging points (\$)

Next, we propose the bilevel optimization model (1a)–(1p), where the constant $M_0 \gg \max\{P_i^t, i \in I, t \in T\}$ and the parameter ϵ is a given small number.

$$\max_{p,y,z,w,x} \quad \sum_{i\in I,k\in K,t\in T} p_k^t x_{ik}^t - \sum_{k\in K} c_k y_k - \sum_{i\in I,k\in K,t\in T} e_k x_{ik}^t \tag{1a}$$

$$\text{s.t.} \quad 0 \le y_k \le \delta_k, k \in K, \tag{1b}$$

$$\sum_{k\in K} c_k y_k \le \beta, \tag{1c}$$

$$w_{kt}^{il} \ge \frac{v_{kt}^{il}(P_i^t - p_k^t)}{M_0}, i \in I, t \in T, k \in K, l \in L, \tag{1d}$$

$$w_{kt}^{il} \le v_{kt}^{il} + \frac{v_{kt}^{il}(P_i^t - p_k^t)}{M_0}, i \in I, t \in T, k \in K, l \in L, \tag{1e}$$

$$z_i \le \sum_{t\in T}\sum_{k\in K}\sum_{l\in L} w_{il}^{kt}, i \in I, \tag{1f}$$

$$z_i \ge \sum_{t\in T}\sum_{k\in K}\sum_{l\in L} \frac{w_{kt}^{il}}{s}, i \in I, \tag{1g}$$

$$z_i \in \{0, 1\}, i \in I, \tag{1h}$$

$$w_{kt}^{il} \in \{0, 1\}, i \in I, t \in T, k \in K, l \in L, \tag{1i}$$

$$y_k \in \mathbb{Z}, k \in K, \tag{1j}$$

$$p_k^t \ge 0, k \in K, t \in T, \tag{1k}$$

$$\min_x \quad \sum_{t\in T}\sum_{i\in I}\sum_{k\in K}(\lambda_{ik}^t + \sum_{l=1}^{s} l w_{il}^{kt}) x_{ik}^t \tag{1l}$$

$$\text{s.t.} \quad \sum_{i\in I} x_{ik}^t \le \gamma_k y_k, lk \in K, t \in T, \tag{1m}$$

$$\sum_{t\in T}\sum_{k\in K} x_{ik}^t = z_i, i \in I, \tag{1n}$$

$$x_{ik}^t \le \sum_{l=1}^{s} w_{kt}^{il}, i \in I, t \in T, k \in K, \tag{1o}$$

$$x_{ik}^t \in \{0, 1\}, i \in I, k \in K, t \in T. \tag{1p}$$

Equations (1a)–(1k) represent the upper level problem, and the Eqs. (1l)–(1p) correspond to that of the lower level problem. The objective function (1a) of the upper level represents the profit to be maximized . The constraints (1b) and (1c) are the limits on the number of charging points in each charging station, and the daily (amortized) budget, respectively. The constraints (1d)–(1e) are used for updating the predefined preference lists by comparing the charging price thresholds set by the users and the charging prices decided by the charging service provider. In particular, constraints (1d) and (1e) are the linearization of Eq. (2):

$$w_{il}^{kt} = \lceil \frac{v_{il}^{kt}(P_i^t - p_k^t)}{M_0} \rceil, \; i \in I, t \in T, k \in K, l \in L, \tag{2}$$

which means if $v_{kt}^{il} = 1$, then the corresponding element in the preference list will still be removed if $p_k^t > P_i^t$. Then, the constraints (1f) and (1g) determine whether user will opt-in or -out to use one of the charging stations. This is done by setting the variable z_i to 0 if no w_{il}^{kt} equals 1 (constraint (1f)) and by setting z_i to 1 if at least one w_{il}^{kt} equals 1 (constraint (1g)).

The first term in the objective function (1l) of the lower level problem gives priority to the closest user when there are more than one users wishing to select the same station at the same time. The second term of the objective function ensures that users select their first available charging station from their preference lists. Equation (1m) describes capacity constraints of the stations. The constraint (1n) enforces every user who opted-in to use the charging stations will be assigned to exactly one charging station at a certain time. Finally, the constraint (1o) implies that a user can only charge if at least one of the auxiliary binary variables makes an option from the preference list available.

3 Single-Level Reformulation

In this section, we present the single-level reformulation of the bilevel problem (1a)–(1p).

We first derive the KKT optimality conditions of the lower level problem and then include them in the upper level problem in order to reformulate the bilevel problem as a single-level problem. To do this, we derive the Lagrangian of the lower level problem, which is

$$\begin{aligned} \mathcal{L} = & \sum_{t \in T} \sum_{i \in I} \sum_{k \in K} (\lambda_{ik}^t + \sum_{l=1}^{s} l w_{il}^{kt}) x_{ik}^t \\ & + \sum_{t \in T, k \in K} \pi_k^t (\sum_{i \in I} x_{ik}^t - \gamma_k y_k) + \sum_{i \in I} \rho_i (\sum_{t \in T} \sum_{k \in K} x_{ik}^t - z_i) \\ & \sum_{t \in T, i \in I, k \in K} + \phi_{ik}^t (x_{ik}^t - \sum_{l=1}^{s} w_{il}^{kt}) - \sum_{t \in T, i \in I, k \in K} \eta_{ik}^t x_{ik}^t, \end{aligned} \tag{3}$$

where $\pi_k^t \geq 0$, $\phi_{ik}^t \geq 0$, $\eta_{ik}^t \geq 0$, and ρ_i free. We observe that the coefficient matrix of the constraints (1m) and (1o) is totally unimodular (TU), using, e.g., (Wolsey & Nemhauser 1999, Theorem 2.7). However, this is no longer the case with the addition of (1n). Nevertheless, we can write the optimality conditions of the lower level problem using the KKT conditions of its linear relaxation:

$$\lambda_{ik}^t + \sum_{l=1}^{s} l w_{kt}^{il} + \pi_k^t + \rho_i + \phi_{ik}^t \geq 0, i \in I, k \in K, t \in T, \tag{4a}$$

$$(\lambda_{ik}^t + \sum_{l=1}^{s} l w_{kt}^{il} + \pi_k^t + \rho_i + \phi_{ik}^t) x_{ik}^t = 0, i \in I, k \in K, t \in T, \tag{4b}$$

$$(\gamma_k y_k - \sum_{i \in I} x_{ik}^t) \pi_k^t = 0, k \in K, t \in T, \tag{4c}$$

$$(\sum_{l=1}^{s} w_{kt}^{il} - x_{ik}^t) \phi_{ik}^t = 0, i \in I, t \in T, k \in K, \tag{4d}$$

$$\rho_i \in \mathbb{R}, i \in I, \tag{4e}$$

$$\rho_i \in \mathbb{R}, \;\; \pi_k^t, \;\phi_{ik}^t \geq 0, i \in I, k \in K, t \in T. \tag{4f}$$

We proceed with replacing the lower level problem with these optimality conditions, plus the binary constraints (1p). It is important to note that this is a heuristic approach that this approach does not guarantee neither optimality nor bilevel feasibility. The resulting single-level reformulation of the bilevel model (1a)–(1p) is the problem (6a)–(6u).

We follow the standard big-M approach (Wolsey 1998) to linearize the complementarity conditions (6i), (6k), and (6n). With the introduction of the binary variable ξ_k^t for each $k \in K$ and for each $t \in T$, and the big-M's M_1, M_2, M_3, and M_4, the complementarity conditions are replaced by (5a)–(5e):

$$\lambda_{ik}^t + \sum_{l=1}^{m} l w_{kt}^{il} + \pi_k^t + \rho_i + \phi_{ik^t} \leq M_1(1 - x_{ik}^t) \quad i \in I, k \in K, t \in T, \tag{5a}$$

$$(\gamma_k y_k - \sum_{i \in I} x_{ik}^t) \leq M_2(1 - \xi_k^t), \quad k \in K, t \in T, \tag{5b}$$

$$\pi_k^t \leq M_3 \xi_k^t, \quad k \in K, t \in T, \tag{5c}$$

$$\phi_{ik}^t \leq M_4(1 + x_{ik}^t - \sum_{l=1}^{s} w_{kt}^{il}), \quad i \in I, k \in K, t \in T, \tag{5d}$$

$$\xi_k^t \in \{0, 1\}, \quad k \in K, t \in T. \tag{5e}$$

To verify the validity of the linearization, one needs to take into account the associated inequalities (6h), (6j), and (6m). There remains the potential difficulty in determining valid and effective values for the big-Ms which is a limitation of this approach.

$$\max_{p,x,z,w,y,\pi,\rho,\phi} \sum_{i\in I,k\in K,t\in T} p_k^t x_{ik}^t - \sum_{k\in K} c_k y_k - \sum_{i\in I,k\in K,t\in T} e_k x_{ik}^t \tag{6a}$$

$$\text{s.t.} \quad 0 \leq y_k \leq \delta_k, k \in K, \tag{6b}$$

$$\sum_{k\in K} c_k y_k \leq \beta, \tag{6c}$$

$$w_{kt}^{il} \geq \frac{v_{kt}^{il}(P_i^t - p_k^t)}{M_0}, i \in I, t \in T, k \in K, l \in L, \tag{6d}$$

$$w_{kt}^{il} \leq v_{kt}^{il} + \frac{v_{kt}^{il}(P_i^t - p_k^t)}{M_0}, i \in I, t \in T, k \in K, l \in L, \tag{6e}$$

$$z_i \leq \sum_k \sum_l \sum_t w_{kt}^{il}, i \in I, \tag{6f}$$

$$z_i \geq \sum_k \sum_l \sum_t \frac{w_{kt}^{il}}{s}, i \in I, \tag{6g}$$

$$\lambda_{ik}^t + \sum_{l=1}^{s} l w_{kt}^{il} + \pi_k^t + \rho_i + \phi_{ik}^t \geq 0, i \in I, k \in K, t \in T, \tag{6h}$$

$$(\lambda_{ik}^t + \sum_{l=1}^{s} l w_{kt}^{il} + \pi_k^t + \rho_i + \phi_{ik}^t) x_{ik}^t = 0, i \in I, k \in K, t \in T, \tag{6i}$$

$$\sum_{i\in I} x_{ik}^t \leq \gamma_k y_k, k \in K, t \in T, \tag{6j}$$

$$(\gamma_k y_k - \sum_{i\in I} x_{ik}^t)\pi_k^t = 0, k \in K, t \in T, \tag{6k}$$

$$\sum_{t\in T} \sum_{k\in K} x_{ik}^t = z_i, i \in I, \tag{6l}$$

$$x_{ik}^t \leq \sum_{l=1}^{s} w_{kt}^{il}, i \in I, t \in T, k \in K, \tag{6m}$$

$$(\sum_{l=1}^{s} w_{kt}^{il} - x_{ik}^t)\phi_{ik}^t = 0, i \in I, \forall t \in T, k \in K, \tag{6n}$$

$$z_i \in \{0, 1\}, i \in I, \tag{6o}$$

$$w_{kt}^{il} \in \{0, 1\}, i \in I, t \in T, k \in K, l \in L, \tag{6p}$$

$$y_k \in \mathbb{Z}, k \in K, \tag{6q}$$

$$p_k^t \geq 0, k \in K, t \in T, \tag{6r}$$

$$x_{ik}^t \in \{0, 1\}, i \in I, k \in K, t \in T, \tag{6s}$$

$$\rho_i \in \mathbb{R}, i \in I, \tag{6t}$$

$$\pi_k^t \geq 0, \ \phi_{ik}^t \geq 0, i \in I, k \in K, t \in T \tag{6u}$$

We point out that formulation (6a)–(6u) has bilinear terms of the form $p_k^t x_{ik}^t$ in the objective function. We leave these terms unchanged because CPLEX is able to handle such bilinear terms in the objective automatically, so we use CPLEX directly to solve (6a)–(6u).

4 Computational Experiments

In this section, we solve the problem (6a)–(6u) on several randomly generated instances using various combinations of the numbers of users $n = |I|$ and charging stations $m = |K|$, which we set to $0.1n$ and $0.2n$, see column 1 of Tables 4, 5, 6, and 7. In all cases, we assume three time periods ($|T| = 3$). The first four parameters in Table 3 are set as $\delta_k = 2$, $\gamma_k = 3$, $c_k = 4$, and $e_k = 0.1$ for each $k \in K$. The charging price threshold P_i^t is randomly generated between 3 and 5 for each $i \in I$ and $t \in T$. Below, we describe how the values of λ_{ik}^t are generated, and how the preference lists are predefined.

The position of the users and charging stations are randomly generated on a 10×10 square region in which every user has 3 time-dependent locations at t_1, t_2 and t_3, see e.g. Fig. 1 when $n = 20$ and $m = 4$. Additionally, we enforce 50% of the locations at t_1 and t_3 to be the same. This is to imply that most users will be at the same locations in the mornings and late evenings, let us say around residential areas. The parameter λ_{ik}^t is computed as a normalized Euclidean distance between

Table 4 Results for $s = 2$ with the budget constraints (1c)

(n, m)	Active charging Stations	Active charging Points	EV users served: (#), %	% of EV users served who got 1st, 2nd choice: (#), %	Profit	CPU (sec)
(20, 2)	2	2	(17), 85%	(4, 13), 23%, 76%	66.3	< 1
(20, 4)	4	4	(20), 100%	(13, 7), 65%, 35%	62.3	< 1
(40, 4)	4	4	(35), 87%	(20, 15), 57%, 42%	116.4	< 2
(40, 8)	7	7	(40), 100%	(16, 24), 40%, 60%	119.0	< 3
(80, 8)	7	8	(69), 86%	(30, 39), 43%, 56%	223.1	46
(80, 16)	15	15	(80), 100%	(40, 40), 50%, 50%	231.0	1427

Table 5 Results for $s = 3$ with the budget constraints (1c)

(n, m)	Active charging Stations	Points	EV users served: (#), %	% of EV users served who got 1st, 2nd, 3rd choice: (#), %	Profit	CPU (sec)
(20, 2)	2	2	(18), 90%	(4, 10, 4), 22%, 55%, 22%	71.2	< 1
(20, 4)	3	3	(20), 100%	(7, 5, 8), 35%, 25%, 40%	73.0	< 1
(40, 4)	4	4	(35), 87%	(10, 7, 18), 28%, 20%, 51%	134.5	12
(40, 8)	6	6	(40), 100%	(10, 11, 19), 25%, 27%, 47%	140.0	30
(80, 8)	7	8	(70), 87%	(14, 26, 30), 20%, 37%, 42%	268.6	103
(80, 16)	14	14	(80), 100%	(21, 31, 28), 26%, 38%, 35%	254.0	290794

Table 6 Results for $s = 2$ without the budget constraint (1c)

(n, m)	Active charging Stations	Points	EV users served: (#), %	% of EV users served who got 1st, 2nd choice: (#), %	Profit	CPU (sec)
(20, 2)	2	3	(19), 95 %	(11, 8), 57%, 42%	69.1	< 1
(20, 4)	4	4	(20), 100%	(11, 9), 55%, 40 %	63.0	< 1
(40, 4)	4	5	(39), 97%	(17, 22), 43%, 56%	123.1	2
(40, 8)	7	7	(40), 100 %	(17, 23), 42%, 57%	119.0	3
(80, 8)	8	11	(80), 100 %	(39, 41), 48%, 51%	245.0	41
(80, 16)	16	16	(80), 100%	(40, 40), 50%, 50%	231.0	487

Table 7 Results for $s = 3$ without the budget constraint (1c)

(n, m)	Active charging Stations	Points	EV users served: (#), %	% of EV users served who got 1st, 2nd, 3rd choice: (#), %	Profit	CPU (sec)
(20, 2)	2	3	(20), 100%	(3, 7, 10), 15%, 35%, 50%	77.0	< 1
(20, 4)	3	3	(20), 100%	(7, 5, 8), 35%, 25%, 40 %	73.0	< 1
(40, 4)	4	5	(40), 100%	(11, 11, 18), 27%, 27%, 45%	147.0	10
(40, 8)	6	6	(40), 100%	(10, 11, 19), 25%, 27%, 47 %	140.0	9
(80, 8)	8	10	(78), 97%	(21, 29, 28), 26%, 37%, 35%	282.2	712
(80, 16)	14	14	(80), 100%	(19, 30, 31), 23%, 37%, 38%	254.0	46071

EV user i and charging station k at time t. To generate the predefined preference lists, we suppose that there are three types of users S_1, S_2, and S_3, defined by

S_1: the set of users who prefer to minimize the price,

S_3: the set of users who prefer to minimize the distance,

S_3: the set of users who prefer to minimise the sum of the price and the distance,

where $I = S_1 \cup S_2 \cup S_3$. Hence, to accommodate these diverse groups, we first define randomly the parameter $\alpha_i \in [0, 1]$ for every user $i \in I$, and then compute convex combinations of the normalized distances d_{ik}^t between users and stations, and the proposed price threshold P_i^t as

$$\alpha_i d_{ik}^t + (1 - \alpha_i) P_i^t / P_{\max}, \qquad i \in I, k \in K, t \in T, \tag{7}$$

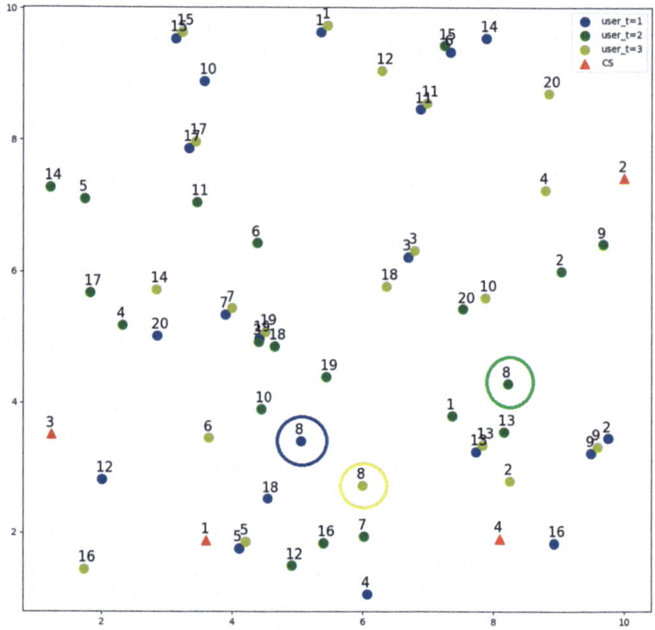

(a)

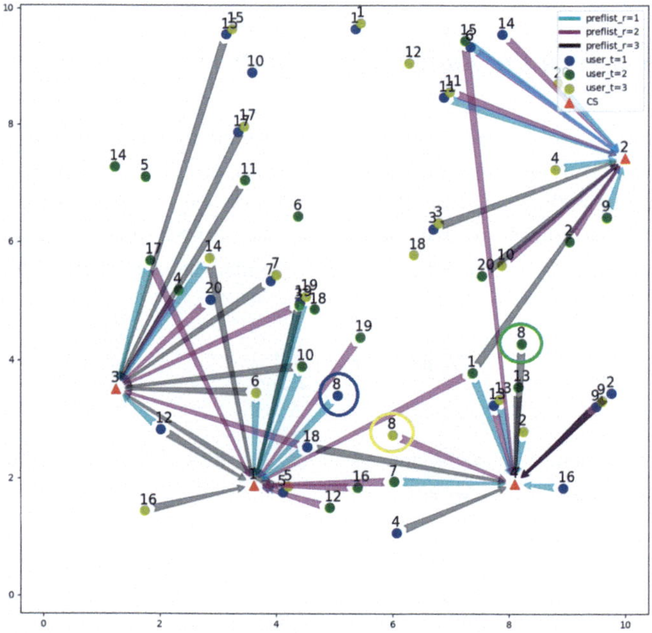

(b)

Fig. 1 Inputs for the case $n = 20$, $m = 4$, $s = 3$, $r = 3$. (**a**) User and station locations (blue $t = 1$, green $t = 2$, yellow $t = 3$). (**b**) User preference lists (cyan 1st choice, magenta 2nd choice, black 3rd choice)

where $P_{\max} = \max\{P_i^t, i \in I, t \in T\}$. Then, the preference list will be the set of combinations of the charging stations and the time periods producing the first s smallest values of (7). This is demonstrated in Fig. 1b when $n = 20$ and $m = 4$ and a preference list of length $s = 3$. In this case, most of the cyan arrows that represent the first choices are shorter and many of the black arrows that show last choices are longer. This shows the contribution of the proximity of users to charging station to determine the predefined preference lists.

We set the big-M's as: $M_0 = M_1 = M_3 = M_4 = 1000$, $M_2 = \delta_k \gamma_k$ for each $k \in K$, and $M_5 = (3 \cdot n \cdot m \cdot s + \sum_{\in I, k \in K, t \in t} \lambda_{ik}^t)/(3 \cdot m)$. These choices were made based on computational experimentation.

All problem instances are solved for two lengths of the preference lists, specifically $s = 2$ and $s = 3$, and with and without the budget constraint (1c). When the budget constraint is considered, we set the budget to $\beta = \frac{1}{2} \sum_{k \in K} c_k \delta_k$, that would mean at most 50% of the charging points can be installed. This will have an effect on the global capacity of the charging service for some choice of the parameters described above to set up the problem instances. This is described in Sect. 4.1.

The computational tests were carried out using CPLEX on a PC with Intel(R) Core(TM) i5-8350U CPU running at 1.90 GHz with 8 GB RAM. The results are reported in Tables 4, 5, 6, and 7. Note that the percentages sometimes do not add up to 100% due to rounding.

4.1 Results With the Budget Constraints (1c)

Here, we discuss the computational results reported in Tables 4 and 5 when the length of the preference lists is 2 and 3, respectively. The number of users and charging stations is the same in both cases as can be seen in column 1 of the tables. If the budget constraint (1c) is activated to limit the number of active charging points at a maximum of 50% of the total candidate points, one can easily verify that the global capacity GE is given by

$$GE = 0.5|T| \sum_{k \in K} \gamma_k \delta_k. \quad (8)$$

Setting the parameters $\delta_k = 2$ and $\gamma_k = 3$, the global capacity of the stations described in Tables 4 and 5 is

$$(18, 36, 36, 72, 72, 114),$$

where the order of the elements follow the order of the rows of the tables. This shows that the global capacity of the instances in rows 1, 3, and 5 is below the total number of users, whereas that of the instance in rows 2, 4, and 6 is beyond the number of users. The former will be probably more realistic since it is unlikely that every user will use one of the stations on daily basis.

From the results in Tables 4 and 5, we see that for most of the computational results, the numbers of active stations/points and users served are similar in both tables. However, looking at column 6 of both tables, we see that the profit of the provider increases as the length of the preference list increases. This is because increasing the length preference list means users are more flexible, and this creates more options for the provider to meet their demand, thus increasing the profit.

Another observation is when we compare the results of the instances where the global capacity of candidate charging stations is set below and above the number of candidate users. In this case, when the global capacity is slightly smaller than the total number of EV users, we do not see a significant decrease in the profit of the provider. This is because meeting the additional demand would require increasing the number of charging points and reducing the prices. This follows directly from the consideration of the users' behaviour in the model. Similar results on the effect of the users' reaction to having more than a sufficient number of candidate stations and consequently a lower profit have been also been observed in Sect. 4.2.

Figure 1 is a visualisation of the inputs for the case $n = 20$, $m = 4$, $s = 3$. Figure 1a gives the locations of the 4 stations (red triangles) as well as the location of each user at each time period (blue for $t = 1$, green for $t = 2$, yellow for $t = 3$). Figure 1b depicts the preference lists of each user, indicating each pair (station, time) by an arrow from the user's location at that time to the station, and with the colour of the arrow corresponding to the order of the pair within the preference list.

Figure 2 is a visualisation of the optimal solution for the same case. Figure 2a indicates the chosen option for each user by picking one of the arrows from Fig.,2b, and with the colour of the arrow indicating the time period when the charging occurs. Figure 2b reports, for the chosen option for each user in the optimal solution, the position of that option in the user's preference list.

4.2 Results Without the Budget Constraints (1c)

We discuss the computational results reported in Tables 6 and 7 when the length of the preference lists is 2 and 3, respectively, and without the budget constraint (1c). In this case, it is clear that we do not have the global capacity issue (unlike in Sect. 4.1).

We observe from column 3 of Tables 6 and 7 that more stations are activated that in the results in Sect. 4.1. This is because with no budget constraint, more stations and points can be installed, and this makes sense for the provider because it allows the demand of a greater proportion of users (in fact 100% in most instances) to be met, resulting in a higher profit for the provider than for the same instances with the budget constraint enforced. Similar to the results in Sect. 4.1, we also observed an increase in the profit of the provider when the length of the preference list is increased (compare columns 6 of Tables 6 and 7).

Another interesting observation from Tables 6 and 7 is that for a fixed number of users, the model produces solutions in which the provider's profit decreases even

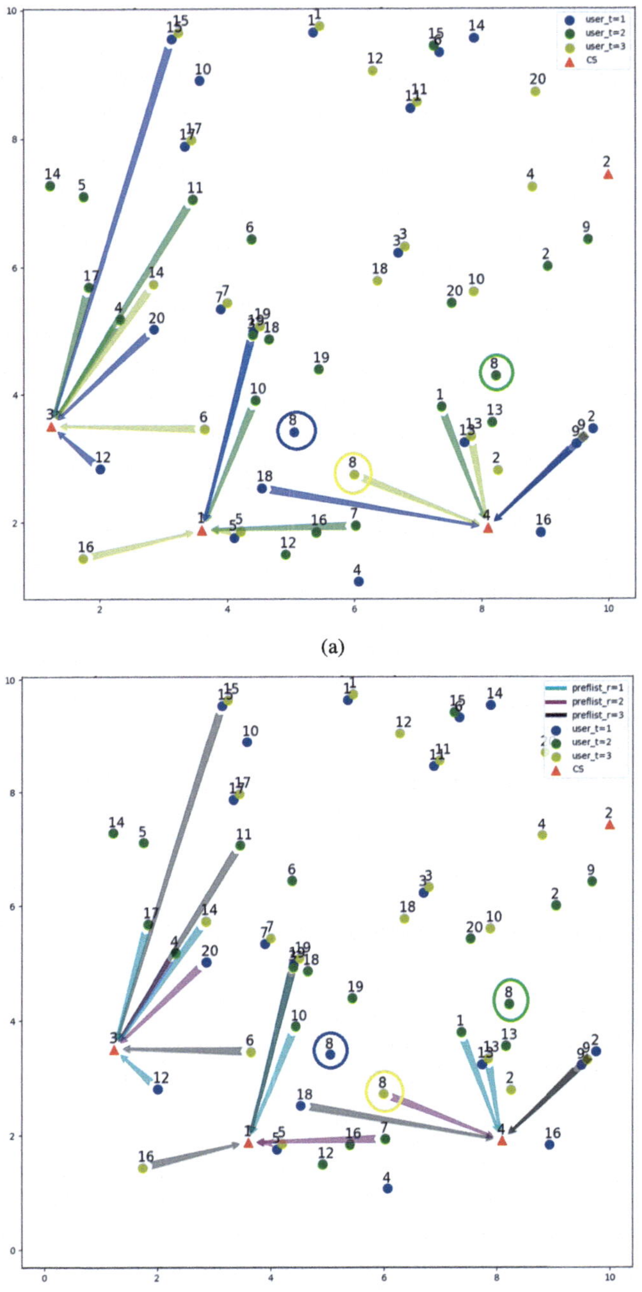

(a)

(b)

Fig. 2 Optimal solution for the case $n = 20$, $m = 4$, $s = 3$, $r = 3$. (**a**) Arrow colours indicate charging time periods (blue $t = 1$, green $t = 2$, yellow $t = 3$). (**b**) Selected options from the preference lists (cyan 1st choice, magenta 2nd choice, black 3rd choice)

though the number of stations increases. This again shows that it is not necessarily always in the interest of the provider to open more stations, unless doing so is profitable.

5 Conclusions and Future Research

We have proposed a bilevel optimization model to support decision-making about the price for charging, the number of charging points, and the location of charging stations. The model integrates the charging behaviour of EV users in the formulation by characterizing the users using a set of distinct ordered set of preference. At the upper-level, the model maximizes the profit of the charging service provider by deciding the charging price, the location of the charging stations, and the number of charging points in each station. At the lower level, EV users select the first available charging station (and corresponding time period) from their predefined preference lists. The model is solved using a KKT-based heuristic method. Our computational results show the capabilities of our approach.

In future research, we plan to extend this model to several private stations managers leading to a case of multiple charging providers. Moreover, the current model uses predefined preference lists, which are somehow updated during the solution process. Hence, one can look into the possibility of embedding the preference list as optimization variables in the model. Finally, it would be very interesting to test the model on real-life instances.

Acknowledgments This research was partially supported by the Gaspard Monge Program for Optimization, Operations Research, and their interactions with data science via grant P-2019-0037.

References

Aljaidi, M., Aslam, N., & Kaiwartya, O. (2019). Optimal placement and capacity of electric vehicle charging stations in urban areas: Survey and open challenges. In *2019 IEEE Jordan International Joint Conference on Electrical Engineering and Information Technology (JEEIT)* (pp. 238–243). IEEE.

Anjos, M. F., Gendron, B., & Joyce-Moniz, M. (2020). Increasing electric vehicle adoption through the optimal deployment of fast-charging stations for local and long-distance travel. *European Journal of Operational Research, 285*(1), 263–278.

Canizes, B., Soares, J., Vale, Z., & Corchado, J. M. (2019). Optimal distribution grid operation using DLMP-based pricing for electric vehicle charging infrastructure in a smart city. *Energies, 12*(4), 686.

Celli, G., Pilo, F., Monni, G., & Soma, G. G. (2018). Optimal multi-objective allocation of fast charging stations. In *2018 IEEE International Conference on Environment and Electrical Engineering and 2018 IEEE Industrial and Commercial Power Systems Europe (EEEIC/I&CPS Europe)* (pp. 1–6). IEEE.

Chen, Y., Qin, Z., & Gan, X. (2019). Optimal ev charging network design: When users have choices. In *2019 11th International Conference on Wireless Communications and Signal Processing (WCSP)* (pp 1–6). IEEE.

Colson, B., Marcotte, P, & Savard, G. (2005). Bilevel programming: A survey. *4OR, 3*, 87–107.

Dait, W., Lit, Y., Gan, X., & Xie, G. (2018). Fast charging station placement with elastic demand. In *2018 IEEE Global Communications Conference (GLOBECOM)* (pp. 1–7). IEEE.

Dempe, S., & Zemkoho, A. (2012). On the Karush–Kuhn–Tucker reformulation of the bilevel optimization problem. *Nonlinear Analysis: Theory, Methods & Applications, 75*(3), 1202–1218. https://doi.org/10.1016/j.na.2011.05.097

Domínguez-Navarro, J., Dufo-López, R., Yusta-Loyo, J., Artal-Sevil, J., & Bernal-Agustín, J. (2019). Design of an electric vehicle fast-charging station with integration of renewable energy and storage systems. *International Journal of Electrical Power & Energy Systems, 105*, 46–58.

Farias, V. F., Jagabathula, S., & Shah, D. (2013). A nonparametric approach to modeling choice with limited data. *Management Science, 59*(2), 305–322.

Faridimehr, S., Venkatachalam, S., & Chinnam, R. B. (2018). A stochastic programming approach for electric vehicle charging network design. *IEEE Transactions on Intelligent Transportation Systems, 20*(5), 1870–1882.

Faridpak, B., Gharibeh, H. F., Farrokhifar, M., & Pozo, D. (2019). Two-step lp approach for optimal placement and operation of EV charging stations. In *2019 IEEE PES Innovative Smart Grid Technologies Europe (ISGT-Europe)* (pp. 1–5). IEEE.

Gan, X., Zhang, H., Hang, G., Qin, Z., & Jin, H. (2020). Fast charging station deployment considering elastic demand. *IEEE Transactions on Transportation Electrification, 6*(1), 158–169.

Huang, Y., & Kockelman, K. M. (2020). Electric vehicle charging station locations: Elastic demand, station congestion, and network equilibrium. *Transportation Research Part D: Transport and Environment, 78*, 102–179.

Lee, C., & Han, J. (2017). Benders-and-price approach for electric vehicle charging station location problem under probabilistic travel range. *Transportation Research Part B: Methodological, 106*, 130–152.

Wolsey, L. A. (1998). *Integer programming* (vol. 42). Wiley.

Wolsey, L. A., & Nemhauser, G. L. (1999). *Integer and combinatorial optimization* (vol. 55). Wiley.

Xiang, Y., Liu, J., Li, R., Li, F., Gu, C., & Tang, S. (2016). Economic planning of electric vehicle charging stations considering traffic constraints and load profile templates. *Applied Energy, 178*, 647–659.

Xiong, Y., Gan, J., An, B., Miao, C., & Soh, Y. C. (2016). Optimal pricing for efficient electric vehicle charging station management. In *Proceedings of the 2016 International Conference on Autonomous Agents & Multiagent Systems* (pp. 749–757).

Xiong, Y., Gan, J., An, B., Miao, C., & Bazzan, A. L. (2017). Optimal electric vehicle fast charging station placement based on game theoretical framework. *IEEE Transactions on Intelligent Transportation Systems, 19*(8), 2493–2504.

Zhang, G., Yang, H., & Dong, J. (2015). Electric vehicle charging stations layout research based on bi-level programming. In *2015 5th International Conference on Electric Utility Deregulation and Restructuring and Power Technologies (DRPT)* (pp. 609–614). IEEE.

Improving Service Quality of an Urban Rail Transit Line by Integrating Passenger and Freight Train Transportation

Krissada Tundulyasaree, Layla Martin, Rolf van Lieshout, and Tom Van Woensel

1 Introduction

Integrating freight into public transport is a promising, sustainable last-mile solution. For example, the Cargo and E-Tram project is a bulky waste collection service in Zurich that allows residents to dispose of trash, ranging from bottles and cartons to electrical appliances, at tram or bus stops (Stadt Zürich 2023). These wastes are loaded on a tram with one or two cargo carriages. The integration of freight into public transport also decreases vehicle, and driver cost (Savelsbergh & Van Woensel 2016), reduces transportation cost (Ghilas et al. 2016), and improves the utilization of trains (Cochrane et al. 2017). Those cost savings are due to using the unfilled public transport capacity for long-distance freight delivery, which decreases vehicle delivery distance. Thus, integrating freight into urban railways also decreases road congestion, road accidents, and noise pollution (Arvidsson & Browne 2013; Cochrane et al. 2017; Villa & Monzón 2021). The scheme also has the potential to increase the quality of public transport service in peripheral areas as a result of additional services (Bruzzone et al. 2021).

Despite the mentioned benefits, pilot projects in several European cities had mixed success. CityCargo in Amsterdam used trams to deliver freight to business locations, such as restaurants or stores. Although this project seemed auspicious

K. Tundulyasaree (✉) · R. van Lieshout
Operations, Planning, Accounting and Control, Eindhoven University of Technology, Eindhoven, The Netherlands
e-mail: k.tundulyasaree@tue.nl; r.n.v.lieshout@tue.nl

L. Martin · T. Van Woensel
Operations, Planning, Accounting and Control, Eindhoven AI Systems Institute, Eindhoven University of Technology, Eindhoven, The Netherlands
e-mail: l.martin@tue.nl; t.v.woensel@tue.nl

T. G. Crainic et al. (eds.), *Combinatorial Optimization and Applications*,
International Series in Operations Research & Management Science 358,
https://doi.org/10.1007/978-3-031-57603-4_19

with societal and economic benefits, it went bankrupt after a short period due to interference with personal transport resulting from a lack of dedicated infrastructure (Arvidsson & Browne 2013). Cochrane et al. (2017) partially attribute the success of the Dresden-based CarGoTram project to minimal interaction with passenger transport. Deutsche Post DHL recently launched a pilot project in Schwerin, using custom trams for parcel distribution (van Leijen 2022). Any attempts to integrate freight into public transport require minimal impact on passengers, which demands improved system operations.

Capacity and service quality are two main planning issues for freight integration in public transport. According to 34 transportation experts, the main operational challenge is finding slack capacity in crowded public transport networks and vehicles, especially in metropolises (Cochrane et al. 2017). However, with the availability of passenger demand data, operators can tailor their services to demand. Moreover, maintaining the service quality for passengers is essential to success in implementation, as in the CarGoTram case, because freight integration can deteriorate the service quality. For example, passengers on board may wait longer at the station due to the (un)loading time of freight or may need to wait for the next train as the current train has reached capacity with cargo.

To address service quality and capacity concerns, we formulate a joint optimization model that determines the train schedule, the rolling stock schedule, and freight allocation to maximize profits from passengers and freight. Freight and passengers travel in dedicated carriages. Our model results in a tactical plan that will be published and active for months. In contrast to previous integrating passenger and freight studies, we also consider the required train resources, increasing the practical feasibility of the resulting schedule. Moreover, we explicitly incorporate passenger travel behavior. From the published schedule, passengers choose not to travel on the train and use another mode of transport if waiting and travel times are too long. Consequently, our model represents served demand more realistically. We combine this mode choice with time-dependent passenger demands. This allows for utilizing slack capacities in off-peak periods. Previous studies either assume given slack capacities from exogenous timetables or ignore passengers' preferences and alternatives when generating timetables. From a managerial point of view, we identify under which circumstances the integrated freight and passenger demand planning yields benefits.

The contributions of this chapter are twofold:

1. The proposed model is the first in the integrated passenger and freight transportation literature to incorporate mode choices with integrated timetabling and rolling stock planning. This allows us to realistically test the service quality changes due to freight integration.
2. To enable successful freight integration in public transport, we identify circumstances under which operators and passengers benefit. Freight integration benefits both operators and passengers in time-varying arrival demand. Operator profit increases, more passengers board, and their average trip duration is reduced. The benefits are higher for passengers in the off-peak than in the peak

period. Furthermore, we can observe that freight integration allows the reduction of passenger revenue while operators maintain the same service quality.

The remainder of this chapter is structured as follows: Sect. 2 reviews related research. In Sect. 3, we describe the problem and define how to measure the passenger service quality. Section 4 introduces the proposed joint optimization model and describes the benchmark models for experiments. Section 5 outlines the experimental setup and managerial contributions. Section 6 concludes this chapter.

2 Literature Review

This section reviews the relevant studies to our paper related to four different aspects of our proposed model. First, we identify the gap in the tactical scheduling in integrated passenger and freight transportation. Second, we investigate the timetabling literature, which explicitly incorporates passenger demands. Third, we look at the rolling stock literature and study the integrated planning of both literature streams.

2.1 Integrated Passenger and Freight Transportation

Historically, the transportation of people and goods developed into two separate systems. Recently, practitioners and academics began to integrate both systems. Road-based transport easily permits sharing infrastructure and vehicles (Li et al. 2014, 2016a,b; Lu et al. 2022; Zhan et al. 2023; Vincent et al. 2023), including crowdsourced delivery (Arslan et al. 2019; Sampaio et al. 2019). Recently, passengers and freight-sharing railway transportation also became of interest (Elbert & Rentschler 2022).

The most prominent decision in the literature on integrated passenger and freight railway transportation is freight allocation, as visible from the overview Table 1. Behiri et al. (2018), Sahli et al. (2022), and Zuo et al. (2023) fit freight delivery into existing trains running on pre-determined schedules to minimize freight delivery time (Behiri et al. 2018; Sahli et al. 2022) and to maximize the modal share of rail (Zuo et al. 2023). Li et al. (2021) include additional freight trains if the inter-arrival time of passenger trains allows doing so. Hörsting and Cleophas (2022) and Li et al. (2022) further adapt the schedules of the passenger trains to minimize travel times. In addition to allocating freight, Di et al. (2022) and Xu et al. (2022) allow operators to decide the number of freight and passenger carriages, i.e., capacity allocation. In this chapter, we combine all three decisions with passenger mode choice.

Table 1 Overview of related literature (TD Demand: Time-dependent passenger demand, TS: Train Scheduling, FA: Freight Allocation, RS: Rolling stock Scheduling)

	Mode choice	TD demand	Decision		
			TS	FA	RS
Behiri et al. (2018) and Sahli et al. (2022)				x	
Li et al. (2021)			x	x	
Hörsting and Cleophas (2022)		x	x	x	
Li et al. (2022)		x	x	x	
Di et al. (2022)		x		x	x
Xu et al. (2022)		x		x	x
Zuo et al. (2023)				x	
This paper	x	x	x	x	x

2.2 Demand-Oriented Railway Scheduling

While the literature on passenger and freight transportation integration does not yet consider the demand-side implications of their decisions, such as an increased modal share due to a higher train frequency, the literature on traditional railway operations already shows the benefits of aperiodic schedules on service quality. Niu and Zhou (2013), Sun et al. (2014), and Barrena et al. (2014a) incorporate time-dependent origin-destination demands and optimize the arrival and departure times of trains. Niu and Zhou (2013) and Barrena et al. (2014a) shows that an aperiodic timetable can reduce the passenger waiting time, while Sun et al. (2014) found that an aperiodic timetable is more resilient to demand changes. Further, varying speed, as made possible by aperiodic timetables, reduces operating costs (Barrena et al. 2014b; Yin et al. 2017). They show that aperiodic timetables also reduce train energy consumption. Li et al. (2019) further explore the trade-off between operational cost efficiency and maximum waiting time using a multi-objective approach.

This chapter incorporates freight demands with delivery deadlines and time-dependent passenger demand. Moreover, in contrast to minimizing operating costs or passenger waiting time, we maximize the profit from serving passengers and freight. Further, we model passenger behavior more accurately by considering mode choice.

2.3 Rolling Stock Scheduling

Rolling stock scheduling assigns rolling stock compositions to each scheduled trip in the timetable and schedules train units to match these compositions. 'Composition' refers to the combination of train units consisting of passenger carriages and propelled units. Previous studies propose different model formulations depending on the operational context (Hoogervorst 2021). For example, several studies (Fioole et al. 2006; Kroon et al. 2015; Wagenaar et al. 2017) base their model on the Dutch railway setting in which trains have pre-defined successors and short turn-around times due to the dense timetable. Borndörfer et al. (2016) use the German railway setting in which the number and orientation of the train units are essential.

Cacchiani et al. (2010) propose a train unit assignment problem in which trains are composed of train units or carriages and are assigned to the timetable. In this setting, they consider the number of available rolling stock and allow the flexible successor train trip. The objective is then to minimize the assignment cost. Cadarso and Marín (2011) incorporate the robustness into the train unit assignment problem by introducing the penalty cost of deadheading and shunting cost, as these are usually tricky operations, especially in the crowded network. Lin and Kwan (2016) propose a branch and price approach to solve this problem.

This chapter then adopts the train unit assignment as a part of our model.

2.4 Integrated Demand-Oriented Timetabling and Rolling Stock Scheduling

The studies reviewed in the last two sections focus on timetabling or rolling stock scheduling. This aligns with the sequential railway planning process, i.e., scheduling trains and assigning available rolling stocks. Often, the complexity of the underlying problem justifies such a sequential approach. However, the sequential process may result in an unfeasible solution (Pan et al. 2023). For instance, the designed timetable may lead to an infeasible rolling stock plan due to the lack of rolling stock units or insufficient time for coupling or decoupling. In addition, integrated planning theoretically gives a better solution than sequential planning. Schiewe and Schöbel (2022) show that even partial integration leads to a better solution than the sequential approach in public transport planning. Partial integration refers to combining some parts of the sequential decision process. We briefly review recent studies using integrated planning with time-dependent passenger demand here.

Most studies formulate the integrated problem on a bidirectional corridor with a depot for composition changes at each end. However, each model incorporates different operating strategies or constraints. Wang et al. (2018) formulate the integrated timetabling and rolling stock circulation with time-dependent demand as multi-objective mixed-integer nonlinear programming. Their objectives are minimizing load factor, the train's headway variation, and the number of train composition changes. In addition to the scheduling decision, they also incorporate train speed.

Mo et al. (2021) propose an asymmetric operating strategy for the integrated problem. In other words, the decisions on upstream and downstream directions are determined separately. They found that this strategy suits well for asymmetric demands. Zhao et al. (2021) incorporate fairness for passengers and trains into their integrated model. Fairness for passengers is defined as the maximum waiting time for individual passengers, while the maximum operating time is used for the rolling stock. Yuan et al. (2022) allow short turning and confirm that such a strategy can provide more efficient operations with higher train frequency. Pan et al. (2023) investigate the impact of the flexible composition of integrated planning. Flexible composition means the train can change its capacity after finishing a round trip. This strategy matches the demand better and saves the rolling stock cost.

In contrast to previous studies, this chapter considers freight demand, which does not require fast delivery as long as it can meet the deadline. With the additional demands, we can study how it affects service quality.

3 Problem Statement

This study considers a joint optimization model for three decisions: train schedule, freight carriage assignment, and freight allocation for trains on a single corridor. These trains have dedicated carriages for passengers and freight. The operator schedules trains to provide service to passengers and satisfy freight commodities to maximize profits.

3.1 Network

We consider a single line with a set of stations S $(i, j \in S)$. Since freight handling between different lines is very costly, freight does not switch trains, making the single-line case a reasonable approximation. The line is operated in both directions, with train trips traveling from the first station to the last station and making the return trip back to the first station. Along the journey, the trip stops at all intermediate stations to load and unload passengers and freight. The network is double-tracked with one track per direction. Consequently, trains cannot overtake each other. The running time between stations i and j is denoted as T_{ij}, and the minimum headway time between departure, arrival, and trips are $\mathrm{H}^{\mathrm{dep}}$, $\mathrm{H}^{\mathrm{arr}}$, and $\mathrm{H}^{\mathrm{com}}$, respectively.

Passenger Behavior Passengers choose to travel by train or other modes based on the travel time. They determine the train travel time to their destination from the published train schedule. We assume that they use the train only if the service quality is sufficient: if the travel time, which is the sum of waiting and in-vehicle time, exceeds a certain threshold T_{ij}^{wait}, passengers seek an alternative mode of transport. Specifically, we discretize the time horizon into periods $p \in P$ and let λ_{ijp} denote the arrival rate at station i in period p for passengers traveling to station j.

Freight Demand Shippers book capacity for freight requests at the train operator, allowing them to ship the same amount daily for a fixed remuneration, which must be paid even if the shipper does not utilize the capacity on a given day. For simplicity, we aggregate all requests with identical ready time, pickup station, delivery deadline, and delivery station, regardless of the shipper, to one commodity. A commodity $r \in R$ is thus characterized by a ready time (T_r^{ready}), a pickup station (s_r^{start}), a delivery deadline (T_r^{due}), a delivery station (s_r^{end}) and a total amount (Q_r). Since requests belonging to the same commodity share identical attributes, the operator can thus split the same commodity across multiple trains (or trips). Moreover, we measure the quantity of a commodity in volume. In one unit of time, e.g., a minute, the operator can load or unload β^{f} units.

3.2 *Operator's Decisions*

The operator determines the number of trains that are used, the number of trips that are operated, the composition for each trip, the schedule for each trip, which freight commodities to satisfy, and how to allocate these to the trips.

Train Composition Every train has a fixed total number of carriages, denoted by N^{car}, and may contain both passenger, and freight carriages, which can vary per trip. Each passenger (freight) carriage can contain up to C^{pas} passengers (C^{fre} freight units). Consequently, a trip can carry freight, passengers or both as long as that trip consists of the corresponding carriages. A train can perform multiple consecutive trips if the time between arrival and departure is sufficient. The train composition can change between trips.

Scheduling The operator must determine departure from and arrival times to all stations for all trips. These must respect station running times and minimum headway time between trains. In addition, dwell times should be sufficient to board and un-board passengers and freight and lie within an acceptable range from minimum dwell time, H^{min}, to maximum dwell time, H^{max}. We assume a passenger (un-)boarding rate of β^{p} per passenger carriage.

Freight Assignment For each freight commodity $r \in R$, the operator decides which fraction of the freight demand to transport on which train. Operators can also reject the transportation of some freight to increase profits or due to insufficient capacity.

3.3 Service Quality

Integration of freight into passenger rail systems affects operators, passengers, and freight owners. For operators, the main interest is profit maximization, i.e., the objective function. For simplicity, we assume that operators have a fixed freight rate and passenger ticket price. Freight owners are predominantly interested in transporting as much of their freight within the time limits. However, the public acceptance of freight using passenger rail systems depends on the service quality offered to passengers. We outline the passengers' interests and service quality measures and show how these measures can change due to freight integration.

Passengers value the time they spend traveling and the ability to board a train. In a "passenger-optimal" public transport system, passengers board the train immediately after arriving at the station and reach their destination directly. Due to the service schedules and physical realities, this is impossible, resulting in **waiting time** and **in-vehicle time**. Waiting and in-vehicle times can decrease due to higher frequencies and decrease due to freight loading times. Boarding a train depends on the train's capacity and outside options: If it takes too long for the next train to arrive, passengers will choose another mode of transport. We refer to this metric as **served demand**. Served demand may increase due to more frequent trips but may decrease due to substitution by freight.

We measure the waiting time as the average time traveling passengers spend at the station between arrival and boarding a train. The travel time is the average time all passengers spend from arriving at the origin station to reaching their destination. Passengers who do not board any train do not account for waiting time but account for travel time since they choose an alternative mode of transport, which we assume takes their maximum acceptable travel time. We measure served demand S^{D} as the fraction of passengers who board a train relative to the entire demand.

To sum up, considering passengers' interests and freight integration's effect on service quality, the operator maximizes the profits from the served demand by considering that passengers will only board the train if the service quality based on travel time is sufficient.

4 Mathematical Models

This section presents the passenger and freight railway scheduling model. The model uses a Mixed-Integer Linear Programming formulation to optimize the railway schedule, freight carriage assignment, and freight commodity allocation decisions, which we call the Passengers-and-Freight model.

4.1 Passengers-and-Freight Model

Since the number of trips is variable in our model, we introduce the sufficiently large, ordered set K to represent all trips that can be operated. Next, for every $k \in K$, we introduce the decision variables o_k, a binary variable indicating whether trip k is operated and d_{ik} and a_{ik}, continuous variables representing the departure time and arrival time of trip k from and to station i, respectively.

4.1.1 Objective Function

The operator maximizes profit. Their revenues are from the number of boarding passenger tickets and the transported commodities, while the costs are from the operated number of trains and trips. We introduce decision variables b_{ijk}, which represents the number of passengers who board train trip k and travel from station i to station j. For freight, the continuous variable q_{rk} refers to the amount of freight commodity r that is loaded onto trip k. Moreover, the integer variable m^{train} indicates the number of required trains. Then, the objective of our model is given by

$$\max \quad \sum_{k\in K}\left(\sum_{(i,j)\in W_{od}} R^{\text{pas}}\cdot b_{ijk}+\sum_{r\in R} R_r^{\text{fre}}\cdot q_{rk}-C^{\text{trip}}\cdot o_k\right)-C^{\text{train}}\cdot m^{\text{train}} \tag{1}$$

In the objective (1), we sum the revenue from passengers and freight for each trip and subtract the costs per trip and per required train.

4.1.2 Train Scheduling Constraints

Due to track restrictions and for safe operation, the operator incorporates headway, running time, and dwell time constraints.

We introduce decision variables h_{ik}, which represent the dwell time of trip k at the station i, and binary variables y_{ikp}, which equals 1 if the departure time of trip k from station i is in period p and 0 otherwise. The integer variable n_k^{fre} specifies the number of freight carriages on trip k. The notation pred(i) refers to the predecessor station of $i \in S$. M is a sufficiently large constant. Moreover, s^1 and s^2 denote the first and last station on the train line, respectively. k^1 denotes the first departed train trip. Parameter δ_p and $T^{\max}$ denotes the length of time interval p and the considered time horizon, respectively.

$$d_{ik} - d_{i\,\mathrm{pred}(k)} \geq H^{\mathrm{dep}} \cdot o_k \qquad i \in S, k \in K \setminus \{k^1\} \tag{2}$$

$$d_{ik} \geq \sum_{p \in P} y_{ikp} \cdot t_p \qquad i \in S, k \in K \tag{3}$$

$$d_{ik} \leq \sum_{p \in P} y_{ikp} \cdot \left(t_p + \delta_p\right) \qquad i \in S, k \in K \tag{4}$$

$$o_{\mathrm{pred}(k)} \geq o_k \qquad k \in K \setminus \{k^1\} \tag{5}$$

$$a_{ik} \geq d_{\mathrm{pred}(i)k} + T_{\mathrm{pred}(i)i} \cdot o_k \qquad i \in S \setminus \{s^1\}, k \in K \tag{6}$$

$$d_{ik} \geq a_{ik} + h_{ik} - M \cdot (1 - o_k) \qquad i \in S, k \in K \tag{7}$$

$$a_{s^2k} \leq T^{\max} \qquad k \in K \tag{8}$$

$$a_{ik} \geq d_{i\,\mathrm{pred}(k)} + H^{\mathrm{arr}} \cdot o_{\mathrm{pred}(k)} \qquad i \in S, k \in K \setminus \{k^1\} \tag{9}$$

$$h_{s^1k} \geq \beta^{\mathrm{pas}} \cdot \sum_{j \in \mathrm{Ori}_i} b_{ijk} \qquad k \in K \tag{10}$$

$$h_{ik} \geq \beta^{\mathrm{pas}} \cdot \left(\sum_{j \in \mathrm{Ori}_i} b_{ijk} + \sum_{j \in \mathrm{Des}_i} b_{jik} \right) \qquad i \in S \setminus \{s^1, s^2\}, k \in K \tag{11}$$

$$h_{s^2k} \geq \beta^{\mathrm{pas}} \cdot \sum_{j \in \mathrm{Des}_i} b_{jik} \qquad k \in K \tag{12}$$

$$n_k^{\mathrm{fre}} \cdot h_{ik} \geq \beta^{\mathrm{fre}} \cdot \left(\sum_{r \in O_i} x_{rk} + \sum_{r \in D_i} x_{rk} \right) \qquad i \in S, k \in K \tag{13}$$

$$H^{\min} \leq h_{ik} \leq H^{\max} \qquad i \in S \setminus \{s^2\}, k \in K \tag{14}$$

Constraints (2) ensure that two consecutive trips departing from a station are at least a headway apart. Constraints (3) and (4) track the time period p of departure time d_{ik}. To break symmetries, constraints (5) ensure that if m trips are scheduled, these are the first m trips of set K. Although set K is predefined, we denote the decisions to operate trip k as binary variables, o_k. Constraints (6) define the arrival time of the trip k at station i from the departure time from the previous station and the running time between the stations. Constraints (7) define that the train departs after dwelling at the station. The arrival and departure times are absolute values for the starting time of the system. Furthermore, constraints (8) ensure that each trip returns to the depot before the end of the planning horizon. Constraints (9) ensure that a train occupies a station at a time. Constraints (10), (11) and (12) describe the required dwell time for boarding and alighting passengers. Constraints (13) specify the dwell time due to the (un)loading of the freight process. Constraints (14) define the dwell time range for all scheduled trips.

4.1.3 Mode Choice Constraints

Passengers may travel by train or other modes depending on whether they can reach the destination in the desired time. We introduce decision variables, t^{arr}_{ijk}, for the earliest time a passenger who travels in trip k can arrive at station i to travel to station j. A binary variable, z_{ijkp}, is 1 if the earliest arrival time t^{arr}_{ijk} is in time period p or 0 otherwise. W_{od} denotes the set of origin-destination pairs.

$$t^{\text{arr}}_{ijk} \geq a_{jk} - T_{ij} \qquad i, j \in W_{od}, k \in K \qquad (15)$$

$$t^{\text{arr}}_{ijk} \geq d_{i\text{pred}(k)} \qquad i, j \in W_{od}, k \in K \qquad (16)$$

$$t^{\text{arr}}_{ijk} \leq d_{ik} \qquad i, j \in W_{od}, k \in K \qquad (17)$$

$$t^{\text{arr}}_{ijk} \geq z_{ijkp} \cdot t_p \qquad i, j \in W_{od}, k \in K, p \in P \qquad (18)$$

$$t^{\text{arr}}_{ijk} \leq M \cdot (1 - z_{ijkp}) + t_p + \delta_p \qquad i, j \in W_{od}, k \in K, p \in P \qquad (19)$$

$$\sum_{p \in P} z_{ijkp} = 1 \qquad i, j \in W_{od}, k \in K \qquad (20)$$

Constraints (15) and (16) define the lower bound for the latest earliest arrival time to board trip k at the originating station by the time to reach the destination and departure time of the previous train. Constraints (17) give the upper bound for the latest earliest arrival time to board trip k as trip k departure time. Constraints (18) and (19) track the time period p for the t^{arr} using z_{ijkp}. Constraints (20) ensure that the departure time of a train trip from a station, t^{arr}, is only in a single period.

4.1.4 Passenger (Un)Boarding Constraints

The number of boarding passengers for each trip depends on the train schedule.

$$b_{ijk} \leq \sum_{p \in P} \lambda_{ijp} \cdot z_{ijkp} \cdot \left(t_p + \delta_p - t_{ijk}^{\text{arr}}\right)$$

$$+ \sum_{p \in P} \delta_p \cdot \lambda_{ijp} \cdot \left(\sum_{p'<p} \left[z_{ijkp'} - y_{ikp'}\right] - y_{ikp} \right)$$

$$+ \sum_{p \in P} \lambda_{ijp} \cdot \left(d_{ik} - t_p\right) \cdot y_{ikp} \qquad i, j \in W_{od}, k \in K \qquad (21)$$

Constraints (21) track the number of boarding passengers on trip k. To illustrate, Fig. 1 shows the passenger arrival rate in three time periods at station i. The number of passengers boarding the departed train trip k is the shaded area under the graph, which includes all arrival passengers who arrive after the earliest arrival time to board trip k, t_{ijk}^{arr}, until the departure time d_{ik}. The first term in (21) calculates the number of passengers boarding the train in the first period, i.e., between t_{ijk}^{arr} and t_p. The second term calculates the number of passengers boarding the train in intermediate periods, i.e., between t_p and t_{p+1}. The third term calculates the number of passengers boarding between t_{p+1} and d_{ik}, i.e., in the period in which the train departs.

4.1.5 Freight (Un)Loading Constraints

We introduce binary variables x_{rk} to model the freight loading and unloading constraints to track if shipment r is loaded onto trip k.

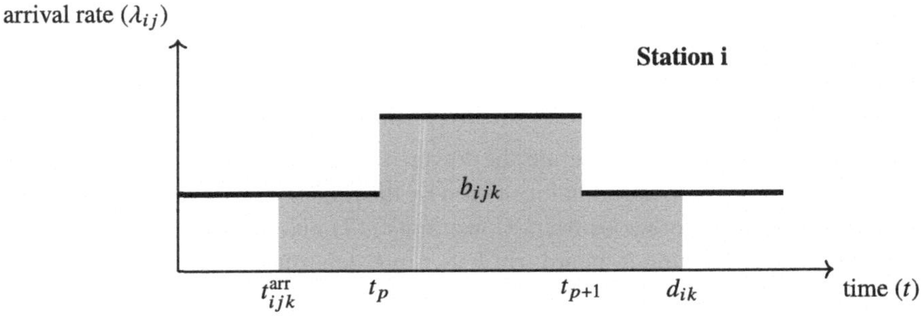

Fig. 1 Arrival rate of passengers at station i traveling to station j (λ_{ij}) and time (t) (shaded area refers to passengers who board train trip k (b_{ijk}))

$$d_{s_r^{\text{start}}k} + t_r^{\text{start}} \cdot (1 - x_{rk}) \geq T_r^{\text{start}} \qquad k \in K, r \in R \tag{22}$$

$$a_{s_r^{\text{end}}k} - (T^{\max} - t_r^{\text{due}}) \cdot (1 - x_{rk}) \leq T_r^{\text{due}} \qquad k \in K, r \in R \tag{23}$$

$$x_{rk} \leq n_k^{\text{fre}} \qquad k \in K, r \in R \tag{24}$$

$$x_{rk} \leq o_k \qquad k \in K, r \in R \tag{25}$$

$$q_{rk} \leq Q_r \cdot x_{rk} \qquad k \in K, r \in R \tag{26}$$

$$\sum_{k \in K} q_{rk} \leq Q_r \qquad k \in K, r \in R \tag{27}$$

Constraints (22) and (23) ensure that trip k can deliver a shipment r if it departs after the ready-time of shipments and arrives at the destination before the delivery deadline. In addition, constraints (24) and (25) restrict that only scheduled trains with freight carriages deliver shipments. Constraints (26) and (27) ensure that the transported amount is at most the available amount while allowing a commodity to be split across multiple trips.

4.1.6 Rolling Stock Scheduling Constraints

The continuous variables π_{ik} track the number of passengers onboard train trip k after departing from station i. The binary variables w_{kl} denote if trip k and trip l use the same train. Moreover, Des_i is a set of originating stations for passengers whose destination is station i and vice versa for Ori_i. For freight, L_i is a freight set whose intermediate station includes station i.

$$\pi_{s^1k} = \sum_{j \in \text{Des}_i} b_{ijk} \qquad k \in K \tag{28}$$

$$\pi_{ik} - \pi_{\text{pred}(i)k} = \sum_{j \in \text{Des}_i} b_{ijk} - \sum_{j \in \text{Ori}_i} b_{jik} \qquad i \in S \setminus \{s^1, s^2\}, k \in K \tag{29}$$

$$\pi_{s^2k} = \pi_{\text{pred}(s^2)k} - \sum_{j \in \text{Ori}_{s^2}} b_{js^2k} \qquad k \in K \tag{30}$$

$$\pi_{ik} \leq n_k^{\text{pas}} \cdot A^{\text{pas}} \qquad i \in S, k \in K \tag{31}$$

$$\sum_{r \in L_i} q_{rk} \leq n_k^{\text{fre}} \cdot A^{\text{fre}} \qquad i \in S, k \in K, r \in R \tag{32}$$

$$n_k^{\text{pas}} + n_k^{\text{fre}} \leq N^{\text{car}} \cdot o_k \qquad k \in K \tag{33}$$

$$d_{s^1l} + M \cdot (1 - w_{kl}) \geq a_{s^2k} + H^{\text{com}} \qquad k, l \in K, l > k \tag{34}$$

$$\sum_{l \in K, k < l} w_{kl} \leq o_k \qquad k \in K \tag{35}$$

$$\sum_{k \in K, k<l} w_{kl} \leq o_l \qquad l \in K \qquad (36)$$

$$m^{\text{train}} \geq \sum_{k \in K} (o_k - \sum_{l \in K, k<l} w_{kl}) \qquad (37)$$

Constraints (28), (29) and (30) track the number of passengers on board after departing from a station. Constraints (31) and (32) limit the capacity for passengers and freight. Constraints (33) specify the number of carriages that can be attached. Constraints (34), (35), (36) and (37) track the minimum number of trains needed for the schedule. Constraints (34) check if two consecutive train trips can use the same train only if that train returns before the departure of the subsequent trip. Constraints (35) and (36) match the scheduled trip to the subsequent train trip. Constraints (37) determine the number of trains required for the scheduled trips.

4.1.7 Model Tightening

After some preliminary testing, we found that the mode choice constraints contribute to a long solution time. We propose boarding-approximation constraints in place of constraints (15)–(20) to simplify the model and reduce the computation time. Instead of precisely calculating the number of passengers who board, we upper-bound it utilizing the travel time threshold, T_{ij}^{wait}, and relax variables t_{ijk}^{arr}. Let $\lambda_{ijp}^{\max} = \max(\lambda_{ijp}, \lambda_{ij\text{pred}(p)})$. Then,

$$b_{ijk} \leq T_{ij}^{\text{wait}} \cdot \sum_{p \in P} o_{ik} \cdot y_{ikp} \cdot \lambda_{ijp}^{\max} \qquad i, j \in W_{od}, k \in K \qquad (38)$$

$$\sum_{k \in K_0} b_{ijk} \leq T_{ij}^{\text{wait}} \cdot \sum_{p \in P} \sum_{k \in K} \left(o_{ik} \cdot y_{ikp} - 1\right) \cdot \lambda_{ijp} + T_{ij}^{\text{wait}} \sum_{p \in P} \lambda_{ijp}^{\max} \qquad i, j \in W_{od} \qquad (39)$$

$$\sum_{p \in P} y_{ikp} = 1 \qquad i \in S, k \in K \qquad (40)$$

Constraints (38) limit the number of passengers boarding by the travel time threshold and the maximum arrival rate from the current and previous period. Constraints (39) limit the number of passengers boarding for an OD pair. This upper bound is the sum of the number of boarding for departed trains except the first and the first train in the period. Constraints (40) specify that every trip can only depart once.

Since decision variables, t_{ijk}^{arr}, are relaxed, constraints (21) can be simplified into following constraints.

$$b_{ijk^1} \leq \sum_{p \in P} y_{i1p} \left(\lambda_{ijp} \cdot (d_{i1} - t_p) + \sum_{p' < p} \lambda_{ijp_1} \cdot \delta_p \right) \qquad i, j \in S \tag{41}$$

$$\begin{aligned} b_{ijk} \leq & \sum_{p \in P} \lambda_{ijp} \cdot y_{i\text{pred}(k)p} \cdot \left(t_p + \delta_p - d_{i\text{pred}(k)}\right) \\ & + \sum_{p \in P} \delta_p \cdot \lambda_{ijp} \cdot \left(\sum_{p' < p} \left[y_{i\text{pred}(k)p'} - y_{ikp'} \right] - y_{ikp} \right) \\ & + \sum_{p \in P} \lambda_{ijp} \cdot \left(d_{ik} - t_p\right) \cdot y_{ikp} \qquad i, j \in W_{od}, k \in K \end{aligned} \tag{42}$$

These constraints use the departure time of the previous train as a reference for boarding calculation. Constraints (41) and (42) track the number of passengers boarding the first and the rest of the scheduled train, respectively.

Moreover, the model above contains several nonlinear constraints due to the product of variables. Constraints (13), (38), (39), (41), and (42) contain nonlinear terms, $n_k^{\text{fre}} \cdot h_{ik}$, and $o_k \cdot y_{ikp}$, and $d_{ik} \cdot y_{ikp}$,respectively. Appendix 1 contains the linearization of these terms.

4.1.8 Domain

The following constraints are the domain for all decision variables in the model. Appendix 2 contains all the notation, sets and indices (Table 8), variables (Table 9) and parameters (Table 10).

$$a_{ik} \geq 0 \qquad i \in S, k \in K \tag{43}$$

$$b_{ijk} \geq 0 \qquad i, j \in W_{od}, k \in K \tag{44}$$

$$d_{i,k} \geq 0 \qquad i \in S, k \in K \tag{45}$$

$$h_{i,k} \geq 0 \qquad i \in S, k \in K \tag{46}$$

$$n_k^{\text{pas}} \in \mathbb{N} \qquad k \in K \tag{47}$$

$$n_k^{\text{fre}} \in \mathbb{N} \qquad k \in K \tag{48}$$

$$m^{\text{train}} \in \mathbb{N} \tag{49}$$

$$w_{kl} \in \{0, 1\} \qquad k, l \in K \tag{50}$$

$$x_{rk} \in \{0, 1\} \qquad k \in K, r \in R \tag{51}$$

$$q_{rk} \in \{0, 1\} \qquad k \in K, r \in R \tag{52}$$

$$z_{ijkp} \in \{0, 1\} \qquad i, j \in W_{od}, k \in K, p \in P \tag{53}$$

$$t_{ijk}^{\text{arr}} \geq 0 \qquad i, j \in W_{od}, k \in K, p \in P \tag{54}$$

$$\pi_{ik} \geq 0 \qquad i \in S, k \in K \tag{55}$$

$$o_k \in \{0, 1\} \qquad k \in K \tag{56}$$

5 Numerical Experiments

We analyze the effect of freight integration on profit and passenger service quality. First, we measure the impact of freight integration on operator profit. Second, we observe how the served demand changes due to integrating freight. Third, we study the impact of freight on passengers' travel time.

5.1 *Experimental Design*

We perform numerical experiments on artificial instances with five stations per direction. The instances are characterized by the number of arrivals (#arr), arrival demand pattern (DT), passenger (PR), and freight revenue (FR). At each station, 1500 or 2500 passengers arrive throughout the 2-h interval. Passenger revenue is the unit revenue for a passenger, which is equal for all origin-destinations. Similarly, freight revenue only depends on the shipment quantities, and we exclude the cost of delays due to deterministic freight availability. We differentiate between a "flat" demand pattern in which the demand does not change during this interval (demand pattern '1') and a time-varying demand pattern in which the demand in the first hour is twice as high as in the second hour (demand pattern '2'). We randomly generate ten origin-destination pairs for freight commodities with different time windows in each instance. The number of requests for each origin-destination pair differs and adds up to 300 requests in terms of volume. Passenger and freight revenue varies between 1 and 2, and 10 and 20, respectively. Since one train car can hold 180 passengers or five freight units, passenger revenues slightly exceed freight revenues. Overall, 16 instances are generated. Table 2 lists all parameters for instance generation. See Table 11 in Appendix 3 for the rest of the parameter settings.

Table 3 shows the computation time required to solve each instance optimally. All instances are solved within 2 h. In the following section, we compare the results of the Passenger model with the Passenger-and-Freight model to show how freight can affect operator profits, served demand, and the average travel time. We use the mathematical model without freight commodities to obtain metrics for the

Table 2 Experimental design

Parameters	Values
# arrivals	Low (1500), High (2500)
Demand type	One (1:1), Two (2:1)
Passenger revenue	Low (1), High (2)
Freight revenue	Low (10), High (20)

Table 3 Solution time without (P) and with freight (PAF). Time is shown in minutes

				Time	
#arr	DT	PR	FR	P	PAF
1500	1	1	10	2	7
			20	2	11
		2	10	3	16
			20	2	17
	2	1	10	2	8
			20	2	30
		2	10	9	22
			20	9	38
2500	1	1	10	3	24
			20	3	44
		2	10	3	24
			20	3	29
	2	1	10	12	15
			20	13	56
		2	10	6	32
			20	6	70

passenger-only setting. Moreover, to compute the average travel time and served demand, we assume deterministic passenger arrivals, i.e., in every time interval, the number of arriving passengers equals the expected value.

5.2 Impact of Freight Integration on Operator Profits

In the first experiment, we examine the impacts of freight integration on operator profits. Table 4 lists the profit, total passenger revenue, and operating cost, separately with and without freight for all instances. We observe that the service is unprofitable in several instances.

Result 1 *If the train service without freight is unprofitable due to low arrival rates and low passenger revenues, freight integration can allow operators to offer their service.*

Hereafter, we refer to these cases as unprofitable instances.

Table 4 Profit, ticket revenue and operation costs without (P) and with freight (PAF). All numbers shown are in thousands

				Profit			Total passenger revenue			Operating cost		
#arr	DT	PR	FR	P	PAF	Diff	P	PAF	%	P	PAF	%
1500	1	1	10	–	1.1	–	–	6.3	–	–	6.6	–
			20	–	4.5	–	–	9.3	–	–	9.6	–
		2	10	9.3	9.2	0.0	21.9	19.6	−10.4	12.6	12.6	0.0
			20	9.3	13.1	3.8	21.9	20.8	−5.0	12.6	12.6	0.0
	2	1	10	–	1.2	–	–	5.8	–	–	6.0	–
			20	–	3.9	–	–	8.0	–	–	9.0	–
		2	10	5.1	8.9	3.7	16.5	18.6	12.7	11.4	11.4	0.0
			20	5.1	10.8	5.6	16.5	17.5	5.9	11.4	11.4	0.0
2500	1	1	10	6.3	5.2	−1.2	16.5	16.3	−1.3	10.2	12.6	23.5
			20	6.3	10.1	3.8	16.5	18.0	9.1	10.2	12.6	23.5
		2	10	24.6	24.5	−0.1	37.8	36.1	−4.5	13.2	13.2	0.0
			20	24.6	30.0	5.4	37.8	39.1	3.6	13.2	13.8	4.6
	2	1	10	3.8	5.8	2.0	13.4	14.8	10.4	9.6	10.2	6.3
			20	3.8	8.7	4.9	13.4	15.5	15.9	9.6	10.2	6.3
		2	10	17.1	20.0	3.0	29.1	31.0	6.7	12.0	12.0	0.0
			20	17.1	24.6	7.5	29.1	33.4	14.9	12.0	12.6	5.0

Result 2 *Operators benefit most from integrating freight if passenger arrival demand is time-varying, due to the served freight or additional passengers. The total passenger revenue increases more if passenger revenue is low.*

Freight integration increases revenues from freight shipments and passenger revenues. As Table 4 shows, the profit gain is higher under time-varying demand (DT: 2) than under flat demand (DT: 1) (unless the instance is unprofitable without freight). Without freight integration, the flat demand generates higher profits than the time-varying demand as more passengers board trains (higher total passenger revenue). With freight integration, the total passenger revenue decreases in most instances with flat demand unless demand and passenger revenue are high. This implies that freight commodities replace some passengers. On the other hand, all time-varying instances generate higher total passenger revenues when integrating freight, and the increase is higher if freight integration allows for a profitable schedule of additional train trips, even though this results in higher operating costs. This means that operators can maintain or increase the number of passengers. Thus, the operator profits increase more if demand varies.

5.3 Impact of Freight Integration on Served Demand

In the second experiment, we study the impact of freight integration on served demand.

Result 3 *Freight integration increases served demand when the arrival pattern is time-varying. The served demand increases more in the off-peak period than in the peak period and decreases with the passenger revenue.*

Table 5 shows average travel time, served demand, and the number of trains and trips with and without freight for all instances. For flat demand, freight integration decreases the served demand unless revenue from freight is high because adding trips or trains is unprofitable. Table 6 shows served demand and the number of departures in the peak and off-peak periods with and without freight under the time-varying arrival pattern. Time-varying demand instances mostly have higher served demand with freight integration because of added departures in the off-peak period.

Figure 2 illustrates the served demand with and without freight for different passenger revenues for flat and time-varying arrival patterns. As the served demand without freight increases in passenger revenue, the additionally gained served demand decreases in passenger revenue. Compared to instances with time-varying demand, instances with flat demand reach a high service level for lower passenger revenues. Thus, only very low passenger revenues allow for increasing served demand by integrating freight. This implies that public authorities may mandate

Table 5 Average travel time, served demand (SD), number of trains and trips from without (P) and with freight (PAF)

				Travel time			SD		#trips		#train	
#arr	DT	PR	FR	P	PAF	%	P	PAF	P	PAF	P	PAF
1500	1	1	10	–	13.6	–	–	52	–	7	–	2
			20	–	12.8	–	–	77	–	10	–	3
		2	10	11.5	12.2	6.0	91	82	13	13	4	4
			20	11.5	12.4	8.3	91	87	13	13	4	4
	2	1	10	–	13.4	–	–	48	–	6	–	2
			20	–	12.2	–	–	67	–	9	–	3
		2	10	12.7	12.2	−3.5	69	78	11	11	4	4
			20	12.7	11.9	−6.6	69	73	11	11	4	4
2500	1	1	10	11.9	12.3	3.8	83	82	11	13	3	4
			20	11.9	12.2	2.4	83	90	11	13	3	4
		2	10	11.3	12.2	7.8	94	90	14	14	4	4
			20	11.3	11.9	4.8	94	98	14	15	4	4
	2	1	10	12.7	12.3	−3.6	67	74	10	11	3	3
			20	12.7	11.4	−10.3	67	77	10	11	3	3
		2	10	12.5	11.7	−6.2	73	78	12	12	4	4
			20	12.5	11.0	−11.8	73	83	12	13	4	4

Table 6 Served Demand in the peak and off-peak period, the average number of departures in the peak and off-peak period, and overall served demand obtained with the passenger model (P) and the passenger-and-freight model (PAF) for the time-varying demand pattern

			Peak SD			Off-peak SD			Peak average # trips			Off-peak average # trips			SD	
# arr	PR	FR	P	PAF	%	P	PAF	%	P	PAF	%	P	PAF	%	P	PAF
1500	1	10	–	56	–	–	32	–	–	3.2	–	–	2.6	–	–	48
		20	–	78	–	–	43	–	–	4.6	–	–	3.8	–	–	67
	2	10	86	90	5	34	52	53	5.7	5.6	−2	3.6	3.8	6	69	78
		20	86	89	3	34	41	21	5.7	5.7	0	3.6	3.8	6	69	73
2500	1	10	81	86	6	38	50	32	5.1	5.1	0	3.3	4.2	27	67	74
		20	81	83	2	38	66	74	5.1	4.9	−4	3.3	4.4	33	67	77
	2	10	89	92	3	40	49	23	6.0	6.0	0	3.4	3.8	12	73	78
		20	89	97	9	40	57	43	6	6.2	3	3.4	4.6	35	73	83

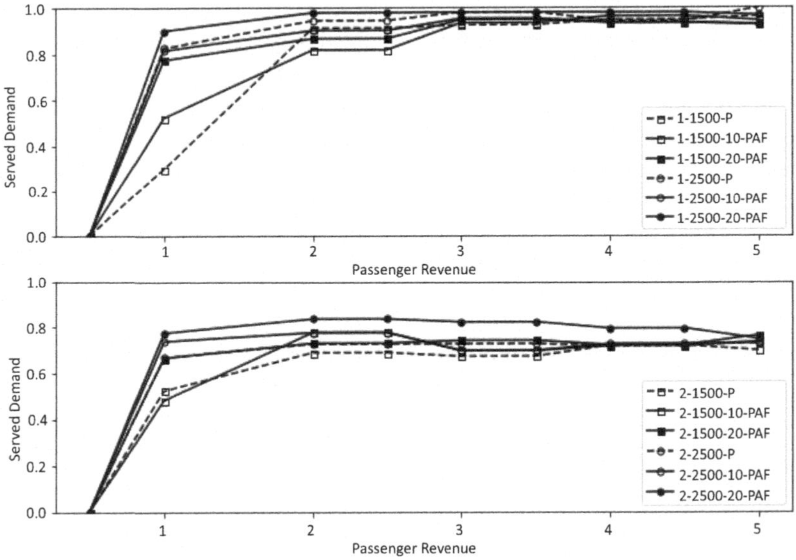

Fig. 2 Sensitivity of served demand to passenger revenue for flat (upper) and time-varying demand (lower) (label: demand type—# arrivals—freight Revenue—type)

reduced fares charged to passengers when integrating freight without operators drastically reducing the service level. Moreover, the served demand from the same timetable subject to randomly generated passengers are as shown in Table 12 in Appendix 4. They show that the served demand from the simulation is roughly the same as our calculation. This implies that our results related to served demands are robust to the randomness of passenger arrivals.

Table 7 Overall Average travel time, Average travel time for peak and off-peak periods, for time-varying demand instances from without (P) and with freight (PAF)

			Travel time			Peak Travel time			Off-peak Travel time		
# arr	PR	FR	P	PAF	%	P	PAF	%	P	PAF	%
1500	1	10	–	13.32	–	–	13.35	–	–	13.61	–
		20	–	13.32	–	–	12.31	–	–	12.18	–
	2	10	12.69	12.24	−3.53	11.89	12.44	4.60	14.27	12.26	−14.10
		20	12.69	11.85	−6.57	11.89	12.13	2.02	14.27	11.79	−17.44
2500	1	10	12.72	12.26	−3.63	12.11	12.34	1.90	14.09	12.34	−12.44
		20	12.72	11.41	−10.28	12.11	11.54	−4.68	14.09	11.42	−18.95
	2	10	12.46	11.68	−6.24	11.67	11.81	1.15	14.16	11.75	−16.99
		20	12.46	10.99	−11.77	11.67	11.24	−3.70	14.16	11.00	−22.33

5.4 *Impact of Freight Integration on Travel Time*

In a third experiment, we measure the impact of freight integration on travel time.

Result 4 *Freight integration decreases the average travel time, particularly in the off-peak period if demand is time-varying. The reduced average travel time rises with increasing demand, especially if passenger revenues are high.*

Table 7 compares the average travel time and the travel time in the peak and the off-peak periods with and without freight. The average travel times are shorter if demand does not vary since more trains and trips run. The decrease in average travel times is caused by (i) additional trains and trips and (ii) increased served demand. Instances with flat demand result in shorter average travel time than time-varying demand as more trains or trips are operated (higher operating cost in Table 4). Moreover, as previously mentioned, the served demand for the time-varying cases also increases with freight integration. Thus, the average travel time gain is higher in the time-varying demand cases with freight integration. Moreover, average travel times decrease mostly in the off-peak period due to the shorter waiting time as more trips are scheduled (see Table 6).

Figure 3 illustrates the sensitivity of average travel time to increasing demand. The average reduction in travel time from the freight integration decreases as the demand increases. With high passenger revenue, the gain seems to increase, which may be because scheduling more trips is profitable. This implies that freight integration allows the operator to improve their services with less demand.

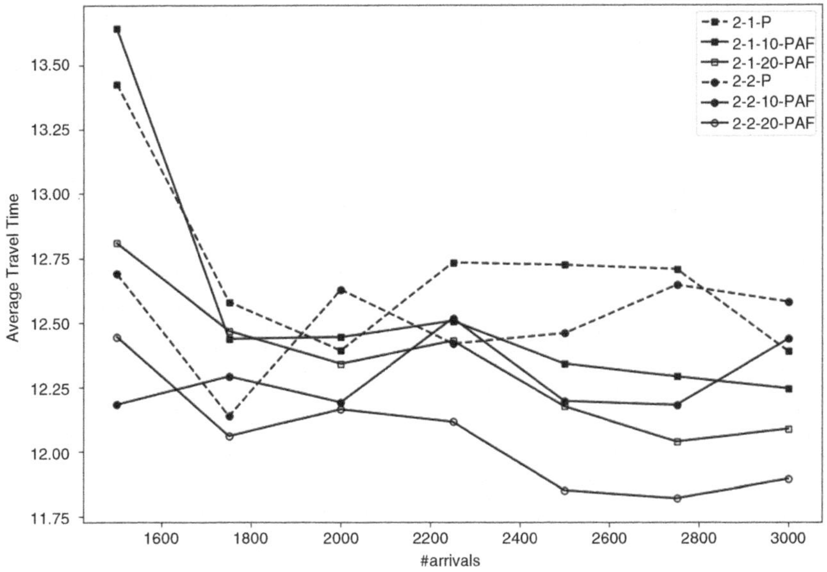

Fig. 3 Sensitivity of average travel time to the demand for time-varying demand (label: demand type—passenger revenue—freight revenue—type)

6 Conclusions

Projects aiming to integrate freight into public transport have had mixed success due to freight transport interfering with passenger transport or lacking slack capacity. To facilitate the integration, this study determines the circumstances in which freight integration with public transport benefits both passengers and the operator. While operators always increase their profits, we find that passengers can experience delays due to freight integration. If demand remains constant over time, served demand decreases, and travel times increase. In addition, low freight revenues do not allow for scheduling additional trains and trips, which can also result in longer travel times. On the other hand, freight integration during the off-peak period substantially improves passenger service through additional trains and trips. This results in lower travel times and thus makes public transport more competitive, resulting in higher served demand.

The possible negative effects of freight integration on passenger service quality limit its practicability. More research is required to ensure that travelers do not switch to less sustainable modes of transport. For example, operators could counteract discomfort by charging passengers lower or higher prices for transporting freight in peak periods. Optimizing the ticket price or freight rate provides an interesting avenue for future research. We may consider the cost of delay packages to evaluate how it would affect passenger, operator, and shipper decisions. Moreover, it is interesting to consider stochasticity in freight commodities since they directly affect the operator's decisions.

Acknowledgments Krissada Tundulyasaree would like to thank the Royal Thai Government for the Ph.D. Scholarship.

Appendix 1: Linearization

We formulate the linearized constraints as follows:

Let: $z^1_{ik} = h_{ik} \cdot n^{\text{fre}}_k$, $z^2_{cik} = z^3_{ck} \cdot h_{ik}$, $z^4_{ikp} = d_{ik} \cdot y_{ikp}$, $o_{ikp} = o_k \cdot y_{ikp}$

$$n^{\text{fre}}_k = \sum_{c \in N_t} c \cdot z^3_{ck} \qquad k \in K \tag{57}$$

$$\sum_{c \in N_t} z^3_{ck} = 1 \qquad k \in K \tag{58}$$

$$z^1_{ik} = \sum_{c \in N_t} c \cdot z^2_{ick} \qquad i \in S, k \in K \tag{59}$$

$$z^2_{cik} \leq h_{ik} \qquad i \in S, c \in N_t, k \in K \tag{60}$$

$$z^2_{cik} z^3_{ck} \cdot T^{\max} \cdot N^{\text{car}} \qquad i \in S, k \in K, c \in N_t \tag{61}$$

$$z^2_{cik} \geq h_{ik} + T^{\max} \cdot N^{\text{car}} \cdot (z^3_{ck} - 1) \qquad i \in S, k \in K, c \in N_t \tag{62}$$

$$z^4_{ikp} \leq d_{ik} \qquad i \in S, k \in K, p \in P \tag{63}$$

$$z^4_{ikp} \leq T^{\max} \cdot y_{ikp} \qquad i \in S, k \in K, p \in P \tag{64}$$

$$z^4_{ikp} \geq d_{ik} - T^{\max} \cdot (1 - y_{ikp}) \qquad i \in S, k \in K, p \in P \tag{65}$$

$$\sum_{p \in P} o_{ikp} = o_k \qquad i \in S, k \in K \tag{66}$$

$$o_{ikp} \leq o_k \qquad i \in S, k \in K, p \in P \tag{67}$$

$$o_{ikp} \leq y_{ikp} \qquad i \in S, k \in K, p \in P \tag{68}$$

$$o_{ikp} \geq y_{ikp} + o_k - 1 \qquad i \in S, k \in K, p \in P \tag{69}$$

Appendix 2: Lists of Indices, Set, Parameters, and Decision Variables

The notations in this paper are summarized in Tables 8, 9, and 10.

Table 8 List of sets and indices

Notation	Description
S	Set of stations on a train line ($i, j \in S$)
K	Set of all train trips ($k \in K$)
P	Set of all time period ($p \in P$)
R	Set of all freight commodities ($r \in R$)
O_i	Set of freight commodities whose origin is at station i
D_i	Set of freight commodities whose destination is station i
L_i	Set of freight commodities which travel via station i $\{r \mid s_r^{\text{start}} \leq i < s_r^{\text{end}}, r \in R\}$
W_{od}	Set of passenger OD pairs
s^1	The first station on a train line
s^2	The last station on a train line
k^1	The first train on the corridor
Ori_i	The set of origin station for station i
Des_i	The set of destination stations for station i

Table 9 Variables

Notation	Description
a_{ik}	Arrival time of a train trip k at station i
d_{ik}	Departure time of a train trip k from station i
h_{ik}	Dwell time of a train trip k at station i
n_k^{pas}	Number of passenger carriages for train k
n_k^{fre}	Number of freight carriages for train k
m^{train}	Number of trains required
w_{kl}	1 if train trip k and train trip l use the same train
x_{rk}	1 if shipment commodities r is delivered by train trip k
q_{rk}	The amount of shipment commodities r loaded on train trip k
y_{ikp}	1 if departure time of a train trip k from station i is in the interval p
z_{ijkp}	1 if the earliest passenger arrival time who can board train trip k at station i to go to station j is in the interval p
t_{ijk}^{arr}	Earliest passenger arrival time who board train trip k at station i and travel to station j
b_{ijk}	Number of passengers board train k at station i and travel to station j
π_{ik}	Number of passengers on-board train k after departure from station i
o_k	1 if train run k operates

Table 10 Parameters

Notation	Description
δ_p	Length of a time interval p
T_{ij}^{wait}	Travel time threshold from station i to j
λ_{ijp}	Passenger arrival rate in the interval p at station i go to station j
T_{ij}	Link running time between station i and station j
H^{dep}	Minimum headway time between consecutive departures
H^{arr}	Minimum headway time between consecutive arrivals
H^{com}	Minimum time for altering train composition
H^{min}	Minimum dwell time
H^{max}	Maximum dwell time
β^{pas}	Boarding and alighting time unit per number of passenger carriages
β^{fre}	Loading and unloading time unit per number of freight carriages
s_r^{start}	An originating station for freight commodities r
s_r^{end}	A destination station for freight commodities r
T_r^{ready}	Ready time for freight commodities r
T_r^{due}	Due time for freight commodities r
N^{car}	Total number of carriages for a train trip
A^{pas}	Capacity for a passenger carriage
A^{fre}	Capacity for a freight carriage
M	A big value number
R^{pas}	Ticket revenue for a passenger
R_r^{fre}	Revenue from delivery a unit of shipment commodities r
C^{trip}	Operating cost for a train trip per time unit
C^{train}	Fixed cost for a train per time unit
t_p	Starting time of interval p
T^{max}	Length of the planning horizon

Appendix 3: Parameter Values for Instances

The parameter values for instances in the numerical experiments are shown in Table 11.

Table 11 Other parameter values for all scenarios

Notation	Values
Timetabling	
H^{dep}	2 min
H^{arr}	2 min
H^{com}	2 min
H^{min}	1 min
H^{max}	1 min
Train carriage	
N_{car}	4 cars
Passenger-related	
β^{pas}	0.12 s / person
β^{fre}	15 s / unit
T_{ij}	600 s
Cost-revenue	
C^{trip}	$25/\text{trip} \cdot$ running time
C^{train}	$50/\text{train} \cdot$ running time

Appendix 4: Passenger Simulation

We conduct a simulation to verify the performance of the optimized schedule for random Poisson passenger arrivals. In the simulation, we impose that passengers can only board the first train after arriving at the station. They will use an alternative mode of transport if this train is full or the travel time exceeds the threshold travel time.

Since we aim to test our schedule solution against the passenger arrival time randomness, we adopt the same parameter setting from the numerical experiment described in Sect. 5. The mean arrival rate to the exponential distribution is derived from the total arrivals per unit of time. In other words, we assume that inter-arrival time is exponentially distributed with the average arrival rate adopted from the numerical experiments. To address the varying arrival rate between periods, we adopt the thinning method, where we generate the number of passengers according to the highest arrival rate but accept or deny arrivals based on the ratio between the current arrival rate and the maximum rate. Next, we evaluate the passenger list with our scheduling solution previously obtained from Gurobi and calculate all measures. For each instance, we generate 100 different sets of arrival passengers to ensure that our evaluation is reliable.

Table 12 compares the model presented in Sect. 4 and the simulation results. We report the average value from the simulation where the confident interval width

Table 12 Average travel time and served demand(SD) obtained with the Passenger model (P), the Passenger-and-Freight model (PAF), simulated Passenger model (SP) and simulated Passenger-and-Freight model (SPAF)

				Travel time				SD			
#arr	DT	PR	FR	P	SP	PAF	SPAF	P	SP	PAF	SPAF
1500	1	1	10	–	–	13.6	13.4	–	–	52.0	52.4
			20	–	–	12.8	12.2	–	–	77.0	77.5
		2	10	11.5	11.5	12.2	12.2	91.0	91.1	82.0	81.8
			20	11.5	11.5	12.5	11.8	91.0	91.2	87.0	86.5
	2	1	10	–	–	13.4	13.6	–	–	48.0	48.0
			20	–	–	12.2	12.8	–	–	67.0	66.2
		2	10	12.7	12.6	12.2	12.1	69.0	68.8	78.0	77.7
			20	12.7	12.6	11.9	12.4	69.0	68.9	73.0	73.0
2500	1	1	10	11.9	11.9	12.3	12.2	83.0	82.8	82.0	81.5
			20	11.9	11.9	12.2	11.4	83.0	82.8	90.0	89.7
		2	10	11.3	11.3	12.2	11.7	94.0	94.5	90.0	90.3
			20	11.3	11.3	11.9	11.0	94.0	94.5	98.0	97.6
	2	1	10	12.7	12.7	12.3	12.3	67.0	66.9	74.0	73.7
			20	12.7	12.7	11.4	12.2	67.0	66.9	77.0	77.4
		2	10	12.5	12.4	11.7	12.1	73.0	72.7	78.0	77.5
			20	12.5	12.4	11.0	11.8	73.0	72.6	83.0	83.2

Table 13 Served Demand in the peak and off-peak periods obtained with the Passenger model (P), simulated Passenger model (SP), Passenger-and-Freight model (PAF) and the simulated Passenger-and-Freight model (SPAF) for the time-varying demand pattern.

			Peak SD				Off-peak SD			
# arr	PR	FR	P	SP	PAF	SPAF	P	SP	PAF	SPAF
1500	1	10	–	–	56.0	55.9	–	–	32.0	32.1
		20	–	–	78.0	77.9	–	–	43.0	42.9
	2	10	86.0	86.2	90.0	90.4	34.0	34.0	52.0	52.3
		20	86.0	86.2	89.0	88.8	34.0	34.4	41.0	41.2
2500	1	10	81.0	81.1	86.0	85.7	38.0	38.4	50.0	49.8
		20	81.0	81.0	83.0	83.3	38.0	38.6	66.0	65.8
	2	10	89.0	89.2	92.0	91.8	40.0	39.8	49.0	48.9
		20	89.0	89.1	97.0	96.2	40.0	39.7	57.0	57.2

can be differentiated with one and two decimal points, respectively. We can see that both served demand and average travel time from the simulation are close to the calculated results from the model. Also, the served demand in the peak and off-period are shown in Table 13. They are also close to the calculated ones from the optimization model. We may conclude that the average travel time and served demand are robust to the random arrivals.

References

Arslan, A. M., Agatz, N., Kroon, L., & Zuidwijk, R. (2019). Crowdsourced delivery–a dynamic pickup and delivery problem with ad hoc drivers. *Transportation Science, 53*(1), 222–235.

Arvidsson, N., & Browne, M. (2013). A review of the success and failure of tram systems to carry urban freight: The implications for a low emission intermodal solution using electric vehicles on trams. *European Transport\Trasporti Europei, 54*, 18.

Barrena, E., Canca, D., Coelho, L. C., & Laporte, G. (2014a). Exact formulations and algorithm for the train timetabling problem with dynamic demand. *Computers & Operations Research, 44*, 66–74.

Barrena, E., Canca, D., Coelho, L. C., & Laporte, G. (2014b). Single-line rail rapid transit timetabling under dynamic passenger demand. *Transportation Research Part B: Methodological, 70*, 134–150.

Behiri, W., Belmokhtar-Berraf, S., & Chu, C. (2018). Urban freight transport using passenger rail network: Scientific issues and quantitative analysis. *Transportation Research Part E: Logistics and Transportation Review, 115*, 227–245.

Borndörfer, R., Reuther, M., Schlechte, T., Waas, K., & Weider, S. (2016). Integrated optimization of rolling stock rotations for intercity railways. *Transportation Science, 50*(3), 863–877.

Bruzzone, F., Cavallaro, F., & Nocera, S. (2021). The integration of passenger and freight transport for first-last mile operations. *Transport Policy 100*, 31–48.

Cacchiani, V., Caprara, A., & Toth, P. (2010). Solving a real-world train-unit assignment problem. *Mathematical Programming, 124*, 207–231.

Cadarso, L., & Marín, Á. (2011). Robust rolling stock in rapid transit networks. *Computers & Operations Research, 38*(8), 1131–1142.

Cochrane, K., Saxe, S., Roorda, M. J., & Shalaby, A. (2017). Moving freight on public transit: Best practices, challenges, and opportunities. *International Journal of Sustainable Transportation, 11*(2), 120–132.

Di, Z., Yang, L., Shi, J., Zhou, H., Yang, K., & Gao, Z. (2022). Joint optimization of carriage arrangement and flow control in a metro-based underground logistics system. *Transportation Research Part B: Methodological, 159*, 1–23.

Elbert, R., & Rentschler, J. (2022). Freight on urban public transportation: A systematic literature review. *Research in Transportation Business & Management, 45*, 100679.

Fioole, P. J., Kroon, L., Maróti, G., & Schrijver, A. (2006). A rolling stock circulation model for combining and splitting of passenger trains. *European Journal of Operational Research, 174*(2), 1281–1297.

Ghilas, V., Demir, E., & Van Woensel, T. (2016). The pickup and delivery problem with time windows and scheduled lines. *INFOR: Information Systems and Operational Research, 54*(2), 147–167.

Hoogervorst, R. (2021). Improving the scheduling and rescheduling of rolling stock: Solution methods and extensions. EPS-2021-534-LIS.

Hörsting, L. L., & Cleophas, C. (2022). Scheduling shared passenger and freight transport on a fixed infrastructure. *European Journal of Operational Research, 36*(3), 1158–1169.

Kroon, L., Maróti, G., & Nielsen, L. (2015). Rescheduling of railway rolling stock with dynamic passenger flows. *Transportation Science, 49*(2), 165–184.

Li, B., Krushinsky, D., Reijers, H. A., & Van Woensel, T. (2014). The share-a-ride problem: People and parcels sharing taxis. *European Journal of Operational Research, 238*(1), 31–40.

Li, B., Krushinsky, D., Van Woensel, T., & Reijers, H. A. (2016a). An adaptive large neighborhood search heuristic for the share-a-ride problem. *Computers & Operations Research, 66*, 170–180.

Li, B., Krushinsky, D., Van Woensel, T., & Reijers, H. A. (2016b). The share-a-ride problem with stochastic travel times and stochastic delivery locations. *Transportation Research Part C: Emerging Technologies, 67*, 95–108.

Li, D., Zhang, T., Dong, X., Yin, Y., & Cao, J. (2019) Trade-off between efficiency and fairness in timetabling on a single urban rail transit line under time-dependent demand condition. *Transportmetrica B: Transport Dynamics, 7*(1), 1203–1231.

Li, F., Guo, X., Zhou, L., Wu, J., & Li, T. (2022). A capacity matching model in a collaborative urban public transport system: Integrating passenger and freight transportation. *International Journal of Production Research, 60*(20), 6303–6328.

Li, Z., Shalaby, A., Roorda, M. J., & Mao, B. (2021). Urban rail service design for collaborative passenger and freight transport. *Transportation Research Part E: Logistics and Transportation Review, 147*, 102205.

Lin, Z., & Kwan, R. S. (2016). A branch-and-price approach for solving the train unit scheduling problem. *Transportation Research Part B: Methodological, 94*, 97–120.

Lu, C. C., Diabat, A., Li, Y. T., & Yang, Y. M. (2022). Combined passenger and parcel transportation using a mixed fleet of electric and gasoline vehicles. *Transportation Research Part E: Logistics and Transportation Review, 157*, 102546.

Mo, P., D'Ariano, A., Yang, L., Veelenturf, L. P., & Gao, Z. (2021). An exact method for the integrated optimization of subway lines operation strategies with asymmetric passenger demand and operating costs. *Transportation Research Part B: Methodological, 149*, 283–321.

Niu, H., & Zhou, X. (2013). Optimizing urban rail timetable under time-dependent demand and oversaturated conditions. *Transportation Research Part C: Emerging Technologies, 36*:212–230

Pan, H., Yang, L., & Liang, Z. (2023). Demand-oriented integration optimization of train timetabling and rolling stock circulation planning with flexible train compositions: A column-generation-based approach. *European Journal of Operational Research, 305*(1):184–206

Sahli, A., Behiri, W., Belmokhtar-Berraf, S., & Chu, C. (2022). An effective and robust genetic algorithm for urban freight transport scheduling using passenger rail network. *Computers & Industrial Engineering, 173*, 108645.

Sampaio, A., Savelsbergh, M., Veelenturf, L., & Van Woensel, T. (2019). Crowd-based city logistics. In *Sustainable transportation and smart logistics* (pp. 381–400). Elsevier.

Savelsbergh, M., & Van Woensel, T. (2016). City logistics: Challenges and opportunities. *Transportation Science, 50*(2), 579–590.

Schiewe, P., & Schöbel, A. (2022). Integrated optimization of sequential processes: General analysis and application to public transport. *EURO Journal on Transportation and Logistics, 11*, 100073.

Stadt Zürich. (2023). Cargo-Tram und E-Tram. https://www.stadt-zuerich.ch/ted/de/index/entsorgung_recycling/entsorgen/wo-wann-entsorgen/cargo-tram_und_e-tram.html#

Sun, L., Jin, J. G., Lee, D. H., Axhausen, K. W., & Erath, A. (2014). Demand-driven timetable design for metro services. *Transportation Research Part C: Emerging Technologies, 46*, 284–299.

van Leijen, M. (2022). Another cargo tram in germany to change the way mail is delivered. https://www.railfreight.com/rolling-stock/2022/11/02/another-cargo-tram-in-germany-to-change-the-way-mail-is-delivered/

Villa, R., & Monzón, A. (2021). A metro-based system as sustainable alternative for urban logistics in the era of e-commerce. *Sustainability, 13*(8), 4479.

Vincent, F. Y., Zegeye, M. M., Gebeyehu, S. G., Indrakarna, P. A., & Jodiawan, P. (2023). The multi-depot general share-a-ride problem. *Expert Systems with Applications, 213*, 119044.

Wagenaar, J. C., Kroon, L. G., & Schmidt, M. (2017). Maintenance appointments in railway rolling stock rescheduling. *Transportation Science, 51*(4), 1138–1160.

Wang, Y., D'Ariano, A., Yin, J., Meng, L., Tang, T., & Ning, B. (2018). Passenger demand oriented train scheduling and rolling stock circulation planning for an urban rail transit line. *Transportation Research Part B: Methodological 118*, 193–227.

Xu, G., Zhong, L., Wu, R., Hu, X., & Guo, J. (2022). Optimize train capacity allocation for the high-speed railway mixed transportation of passenger and freight. *Computers & Industrial Engineering, 174*, 108788.

Yin, J., Yang, L., Tang, T., Gao, Z., & Ran, B. (2017). Dynamic passenger demand oriented metro train scheduling with energy-efficiency and waiting time minimization: Mixed-integer linear programming approaches. *Transportation Research Part B: Methodological, 97*, 182–213.

Yuan, J., Gao, Y., Li, S., Liu, P., & Yang, L. (2022). Integrated optimization of train timetable, rolling stock assignment and short-turning strategy for a metro line. *European Journal of Operational Research, 301*(3), 855–874.

Zhan, X., Szeto, W., & Wang, Y. (2023). The ride-hailing sharing problem with parcel transportation. *Transportation Research Part E: Logistics and Transportation Review, 172*, 103073.

Zhao, S., Yang, H., & Wu, Y. (2021). An integrated approach of train scheduling and rolling stock circulation with skip-stopping pattern for urban rail transit lines. *Transportation Research Part C: Emerging Technologies 128*, 103170.

Zuo, T., Li, B., Zhang, F., & Yin, Y. (2023). Analysis of a collaborative transport model mixing passengers with freights in metro system. *Journal of Rail Transport Planning & Management, 25*, 100358.

Two-Stage Stochastic Optimization for the Extended Aircraft Arrival Management Problem Under Uncertainty

Fabian Bastin, Sonia Cafieri, Ahmed Khassiba, and Marcel Mongeau

1 Introduction

We address an operations research problem, the aircraft arrival scheduling problem, which is crucial in air traffic management, as one of its major bottlenecks is the airport landing runways (Kistan et al. 2017). We take account of the arriving aircraft a long time before landing, so as to have more degrees of freedom to schedule landings. However, increasing thereby the considered time horizon brings more uncertainty that needs to be handled.

The aircraft arrival scheduling problem consists in determining a sequence of a set of aircraft and their schedule on reference air-traffic route points in the *terminal maneuvering area* (air space around the destination airport), as well as a landing schedule satisfying time-window constraints and safety separation constraints for any pair of aircraft. The separation constraints are of particular importance in air traffic control. When aircraft are airborne, a fixed minimal inter-aircraft distance (typically 5 NM, where NM stands for Nautical Mile, which corresponds to approximately 1.85 km) constraint must be satisfied, whereas at the runway this safety distance depends on the types of aircraft involved, due to so-called *wake-vortex turbulence* (which affects more severely a light aircraft landing immediately

F. Bastin (✉)
CIRRELT and DIRO, Université de Montréal, Montréal, QC, Canada
e-mail: bastin@iro.umontreal.ca

S. Cafieri · M. Mongeau
ENAC, Université de Toulouse, Toulouse, France
e-mail: sonia.cafieri@enac.fr; marcel.mongeau@enac.fr

A. Khassiba
Capgemini Engineering, Toulouse, France
e-mail: ahmed.khassiba@capgemini.com

T. G. Crainic et al. (eds.), *Combinatorial Optimization and Applications*,
International Series in Operations Research & Management Science 358,
https://doi.org/10.1007/978-3-031-57603-4_20

after a large aircraft) (Breitsamter 2011). Several other operational features can be taken into account, allowing thereby one to assess better the quality of the aircraft arrival schedule.

We assume that each aircraft considered in the problem comes with an initial (default) reference route from its origin airport to its destination airport. We also assume that the following information is known for each aircraft: speed, initial (planned) departure time, and wake-vortex turbulence category (light, medium, or heavy), as introduced by the International Civil Aviation Organization (ICAO) (Antolovic & Franjkovic 2020).

The framework of the aircraft arrival scheduling problem addressed in this chapter is related to the AMAN (Arrival MANager) decision-aid tool (Hasevoets & Conroy 2010) currently in use by air traffic controllers. Air traffic controllers can issue various possible instructions to pilots in order to impose delays (positive or negative) to flights and thereby satisfy the constraints and improve the schedule. Among these instructions used to implement the solution proposed by AMAN, let us mention aircraft speed reduction, path stretching or shortening, or *stack holding* (airborne aircraft "waiting room", as depicted in Fig. 1). Typically, scheduling starts around 30 min before landing. The problem addressed here is in the more general *extended-AMAN* context: the scheduling horizon is extended to aircraft that are within 200–500 NM from runway threshold (Itoh et al. 2017). This corresponds to 2–3 h before landing (Tielrooij et al. 2015), yielding more uncertainty on predicted approach and landing times.

We first consider the even more general operational setup that involves a set of aircraft flying over multiple *route points*, which are the vertices of the network of air routes. In particular, before landing, each aircraft must fly through a route point called *Initial Approach Fix* (IAF) where aircraft are sequenced by air-traffic controllers, using current decision-aid tools such as AMAN. The terminal maneuvering area of an airport can have a single IAF, gathering all arriving aircraft prior to landing, or several IAFs. Typically, aircraft sharing neighboring routes go through a same IAF, as illustrated on Fig. 1 for the case involving two IAFs. In the case involving several IAFs, aircraft are often simply assigned to the IAF closest to their arrival route. However, in practice, air traffic controllers can change an aircraft's IAF assignment to improve the arrival sequence, as explored in Khassiba et al. (2022). Additionally, the flight *status* of each considered aircraft can be taken into account, distinguishing between an aircraft that is still at its departure airport gate from one that is already airborne. More precisely, when making decisions on arrival times to IAFs, the former can be delayed on ground if needed.

The sequencing and scheduling aircraft landing problem has been studied since the 1970s (see for instance Dear 1976; Psaraftis 1978; Dear and Sherif 1991). The deterministic case has received the most attention (Balakrishnan & Chandran 2010; Beasley et al. 2000) until recently. We refer the reader to Bennell et al. (2011) and Ikli et al. (2021) for recent surveys. However, the importance of taking uncertainty into account in air traffic management is now well recognized (Shone et al. 2021).

Specific optimization paradigms such as two-stage stochastic programming (Birge & Louveaux 2011), have been applied to the aircraft scheduling

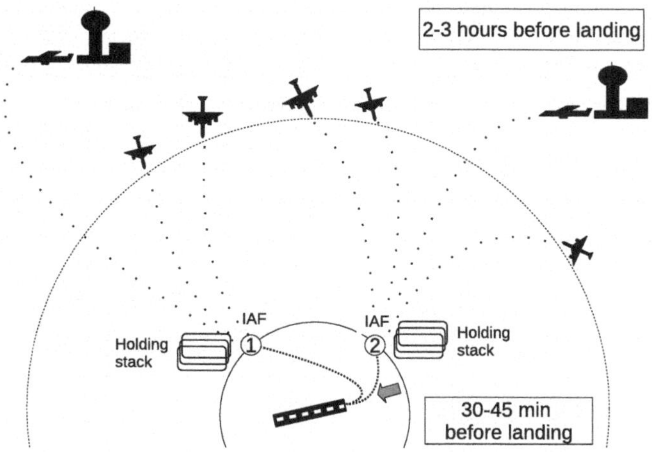

Fig. 1 Terminal maneuvering area and airport approach

problem under uncertain arrival times with short operational horizon (Liu et al. 2018; Sölveling & Clarke 2014; Sölveling et al. 2011). Extensions to various types of disturbances and uncertainties have also been addressed in a number of papers related to aircraft scheduling in the terminal maneuvering area (Huo et al. 2021; Samà et al. 2014; Scala et al. 2021; Vié et al. 2022). Khassiba et al. (2019, 2020, 2022) introduce two-stage stochastic mixed-integer programming models to address the problem of arrival management under uncertainty with an extended operational time horizon. In the first-stage problem, target times of arrival of aircraft at the IAF, and a target IAF aircraft sequence are determined. Actual arrival times at the IAF are assumed to deviate randomly from these target times, following known probability distributions. They are assumed to be revealed at the second stage, so that the second-stage problem determines landing times (keeping the same IAF aircraft sequence) in view of minimizing a cost function. The numerical experiments presented in Khassiba et al. (2019, 2020) show that the obtained stochastic solutions are more robust to the uncertainty occurring within the 2–3 h before landing than their deterministic counterparts.

In Khassiba et al. (2019, 2020), one single IAF is considered. In Khassiba et al. (2019, 2020), all aircraft are considered airborne, while on-ground vs. airborne flight status are taken into account in Khassiba et al. (2022).

The present chapter first introduces, in Sect. 2, our main contribution: a detailed high-level multi-stage stochastic optimization formulation of the extended aircraft arrival management problem under uncertainty that can consider *several* air network points of interest (previous literature commonly considers at most two such points: the IAF and the landing runway). It takes into account other practical operational features such as the airborne/on-ground status of aircraft, and the possibility to delay the departure of on-ground aircraft, as this was also the case for the two-stage study of Khassiba et al. (2022). The focus of this chapter is then made on the two-stage special case which corresponds to recent studies on the aircraft arrival management problem (Khassiba et al. 2019, 2020, 2022): a landing order is decided at one particular air network point, the IAF (first stage), and a recourse cost is proposed so as to ensure that aircraft separation constraints are satisfied at the landing runway (second stage). Section 3 presents the sample average approximation (SAA) method (Shapiro 2021) that is applied to perform numerical tests. Section 4 reports new numerical experiments on realistic instances based on Paris-Charles de Gaulle arrival data, with discussions on the appropriate number of scenarios, the value of the stochastic solutions, and the effect of using chance constraints. Section 5 presents conclusions and propositions of avenues for future research.

2 Optimization Models

This section first introduces the high-level multi-stage stochastic optimization model we are proposing for the extended aircraft arrival management problem under uncertainty. Details on the two-stage special case are then presented, along with the full two-stage stochastic optimization model on which we shall rely for our numerical experiments. It is a concise form (without the detailed description of all the operational constraints, and keeping the most generic setting) of the two-stage model presented in Khassiba et al. (2022). A summary of the notations used is provided in Appendix.

2.1 High-Level Multi-Stage Stochastic Optimization Model

The last decades have experienced a fast increase of air traffic, and while the aviation industry had to reduce significantly its activities during the Covid-19 pandemic, the demand has returned to high levels, raising various challenges in air traffic management (Kistan et al. 2017; Bolić and Ravenhill 2021). As underlined by Itoh et al. (2021), a current research avenue to improve aircraft arrival scheduling is the use of a multi-stage decision sequence model, while accounting for uncertainty. They also suggest to consider time management rather than flow management.

In line with these considerations, we introduce a new model, as general as possible, using a stochastic dynamic programming formulation over $K (\geq 2)$ stages. Each stage is indexed by $k = 1, 2, \ldots, K$. We assume that for each stage $k = 1, 2, \ldots, K - 1$, we are given the number, $n^{(k)}$, of considered aircraft, and $\mathcal{A}^{(k)}$, the associated index set of aircraft. The latter is partitioned further into two subsets: $\mathcal{A}_G^{(k)}$ and $\mathcal{A}_A^{(k)}$, the index set corresponding to the aircraft that are still on the ground at stage k, and the index set related to the aircraft that are airborne at stage k, respectively. We also assume that we are given an index set, $\mathcal{I}^{(k)}$, of possible routes for the aircraft in $\mathcal{A}^{(k)}$. We suppose here that $n^{(k)}$ can change over time as with extended aircraft arrival management systems, airborne aircraft can appear on the radar at a stage $k > 1$. Similarly, additional on-ground aircraft may have to be considered from a stage $k > 1$, for instance if their planned departure time have been delayed.

At each stage k, the following decisions must be made. First, one decides for each aircraft $a \in \mathcal{A}^{(k)}$, whether its *route* should be modified (the route is represented by a sequence of *waypoints*, which are air network vertices) chosen within $\mathcal{I}^{(k)}$. Second, one decides an aircraft sequence at a particular future air route point of interest (that we shall call *target point*) associated to stage k. Third, one decides a target time, within a given time window, for each aircraft, $a \in \mathcal{A}^{(k)}$, to reach this (next-stage) target point. Finally, one decides a target take-off time for each on-ground aircraft $a \in \mathcal{A}_G^{(k)}$.

More precisely, for $k = 1, 2, \ldots K-1$, the kth-stage problem **decision vector**, noted $x^{(k)}$, reads:

$$x^{(k)} = (z^{(k)}, t^{(k)}, \delta^{(k)}, \zeta^{(k)}),$$

where $z^{(k)} \in \mathbb{R}^{n^{(k)}}$ is the vector of target-time variables (associated to the next-stage target point); $t^{(k)} \in \mathbb{R}^{|\mathcal{A}_G^{(k)}|}$is the vector of target take-off time variables for on-ground aircraft; $\delta^{(k)}$ is a vector of discrete variables (*sequencing variables*: one binary variable for each possible pair of aircraft) used to decide a sequence of aircraft at the next-stage target point, and $\zeta^{(k)}$ is the vector of the aircraft route assignment variables (each of its $n^{(k)}$ components is a route index from $\mathcal{I}^{(k)}$). The last-stage decision vector, $x^{(K)}$, is particular: it is simply the vector of target landing time variables, $z^{(K)} \in \mathbb{R}^{n^{(K-1)}}$.

In this general multi-stage context, the last stage, K, is indeed particular: it focuses on the decision related to a special target point: the (landing) runway threshold. The penultimate stage, $K-1$, is concerned with another special target point: the IAF. Stage $K-1$ thereby concentrates on the decisions related to the aircraft passing through their assigned IAF. For example, deciding a route assignment $\zeta_a^{(K-1)}$ for aircraft $a \in \mathcal{A}^{(K-1)}$, boils down to deciding whether the initial IAF assignment for aircraft a should be modified. In a context involving more than two stages, the previous stage, $K-2$, could involve deciding whether yet another specialized air-traffic-control waypoint, the merge point, planned on the initial route of a, should be modified or not.

At each stage $k = 1, 2, \ldots, K$, we define a state vector, $s^{(k)}$, whose ath component,

$$s_a^{(k)} = (p_a^{(k)}, r_a^{(k)}, v_a^{(k)}, \kappa_a^{(k)}, \chi_a^{(k)}),$$

gives the following five features related to aircraft $a \in \mathcal{A}^{(k)}$: $p_a^{(k)}$ is the position of aircraft a at stage k, $r_a^{(k)}$ is the *current* (at stage k) reference route of aircraft a, $v_a^{(k)}$ is its speed at stage k, $\kappa_a^{(k)}$ gives the status (on-ground or airborne) of a

at stage k, and $\chi_a^{(k)}$ stores other useful characteristics of aircraft a (that does not change through the stages in the current study) such as its wake-vortex turbulence category (light, medium, or heavy). Remark that at stage $k = 1$ the state vector, $s_a^{(1)}$, is known for every aircraft $a \in \mathcal{A}^{(1)}$, and similarly for any new aircraft appearing at some subsequent stage (*pop-up flight*).

Let us introduce, for each stage $k = 1, 2, \ldots, K$, the following notation: the objective function $f^{(k)}$, capturing the operating costs at stage k (for instance, penalties on the deviations from the target times), the separation-constraint mapping $S^{(k)}$ to be satisfied by the aircraft pairs at the next-stage target points, the time-window constraint mapping $T^{(k)}$ representing the feasible passing times at next-stage target points, and a feasible set $\mathcal{X}^{(k)}$ to account for the domain definition of the vector of decision variables, $x^{(k)}$.

The stochastic dynamic optimization model therefore reads as follows, for $k = 1, 2, \ldots, K$:

$$V^{(k)}(s^{(k)}) = \min_{x^{(k)}} f^{(k)}(x^{(k)}, s^{(k)}) + \mathbb{E}_{\boldsymbol{\xi}^{(k+1)}}[V^{(k+1)}(s^{(k+1)}) \mid s^{(k)}] \tag{1}$$

$$\text{s.t.} \quad T^{(k)}(x^{(k)}, s^{(k)}) \le 0 \quad \text{time windows at next-stage target points} \tag{2}$$

$$S^{(k)}(x^{(k)}, s^{(k)}) \le 0 \quad \text{separation at next-stage target points} \tag{3}$$

$$x^{(k)} \in \mathcal{X}^{(k)}, \tag{4}$$

and $V^{(K+1)}(\cdot) = 0$. For $k = 1, \ldots, K - 1$, the state transition function is

$$s^{(k+1)} = g^{(k)}(x^{(k)}, s^{(k)}, \boldsymbol{\xi}^{(k+1)}),$$

where $\boldsymbol{\xi}^{(k+1)}$ is some random vector capturing the uncertainty affecting the problem at the next stage (stage $k + 1$). In line with other studies (see for instance Itoh et al. 2021), in our context, $\boldsymbol{\xi}^{(k+1)}$ captures the random time deviations (advances or delays) affecting the aircraft with respect to their target times at their next target points, due to various operational and environmental factors, such as the wind, and limited operational adjustments in order to extend engine lifetime. Other sources of uncertainty could also be taken into account, at the price of additional modeling and computational complexity.

2.2 Two-Stage Special Case

Practitioners commonly concentrate on decisions related to the last two target points: the IAF and the runway threshold. Two-stage stochastic optimization is suitable for this operational context. The exposition can be generalized to more than two stages as introduced above but specific operational constraints (including the smooth management of aircraft engine to optimize its lifetime) renders the two-stage framework best fit to this application. For this reason, and in order to simplify the presentation, in the sequel, we set $K = 2$ (two-stage optimization): the first target point ($k = 1$) is the IAF, and the second ($k = 2$) is the runway threshold. Also, for notational simplicity, we shall use $x \in \mathcal{X}$ instead of our general multi-stage notation $x^{(1)} \in \mathcal{X}^{(1)}$, and $y \in \mathcal{Y}$ instead of $x^{(2)} \in \mathcal{X}^{(2)}$. Regarding uncertainty, we shall simply use $\boldsymbol{\xi}$ instead of $\boldsymbol{\xi}^{(2)}$ (and there is no random variable $\boldsymbol{\xi}^{(1)}$ in the special case we are considering here).

The **first-stage** problem considers aircraft that are 2–3 h away from the IAFs (see Fig. 1), when each aircraft time to cross the IAF is still uncertain. The first-stage problem aims thereby at deciding a target aircraft sequence, together with an IAF affectation and an IAF target time for each aircraft, under the constraints of IAF time windows and safe separation of aircraft at IAF. Depending on the modeling assumptions, for each aircraft the IAF may be selected before solving the optimization model (typically, the closest IAF), or explicitly be a first-stage decision variable. The **second-stage** problem deals with the same set of aircraft, now close to the IAF, when the actual time to cross the IAF is known for each aircraft. The second-stage problem aims at deciding a target landing time for each aircraft, under the constraints of landing time windows and safe aircraft separations at the runway.

This leads to the following two-stage stochastic optimization model:

$$\min_{x=(z,t,\delta,\zeta)} f^{(1)}(x) + \mathbb{E}_{\boldsymbol{\xi}}\left[Q(x, \boldsymbol{\xi})\right] \tag{5}$$

$$\text{s.t.} \quad T^{(1)}(z, t, s^{(1)}) \leq 0 \qquad \text{time windows at IAF} \tag{6}$$

$$S^{(1)}(x, s^{(1)}) \leq 0 \qquad \text{separation at IAF} \tag{7}$$

$$x \in \mathcal{X}, \tag{8}$$

with

$$Q(x, \xi) = \min_{y} f^{(2)}(y, \xi) \tag{9}$$

$$\text{s.t.} \quad T^{(2)}(y, s^{(2)}) \leq 0 \qquad \text{time windows at runway} \tag{10}$$

$$S^{(2)}(y, s^{(2)}) \leq 0 \qquad \text{separation at runway} \tag{11}$$

$$y \in \mathcal{Y}, \tag{12}$$

where $s^{(2)} = g(x, s^{(1)}, \xi)$, g is the transition function that updates the state vector from stage 1 to stage 2, and the first-stage vector of decision variables, $x = (z, t, \delta, \zeta)$, and the second-stage vector of decision variables, y, are defined in the following.

Let $\mathcal{A}$ be the index set of all aircraft, which is partitioned into the index sets $\mathcal{A}_A$ and $\mathcal{A}_G$, respectively the subset of airborne and the subset of on-ground aircraft, and let $\mathcal{I}$ be the index set of IAFs. At the first stage, one decides:

- a target arrival time for each aircraft $a \in \mathcal{A}$ at its assigned IAF, which is denoted by the optimization variable $z \in \mathbb{R}^{|\mathcal{A}|}$;
- a target take-off time for each on-ground aircraft, $a \in \mathcal{A}_G$, represented by the optimization variable $t \in \mathbb{R}^{|\mathcal{A}_G|}$;
- a target sequence of all aircraft $a \in \mathcal{A}$ which is coded in the binary optimization variable $\delta \in \{0, 1\}^{|\mathcal{A}| \times |\mathcal{A}|}$;
- the assignment of an IAF $i \in \mathcal{I}$ for each aircraft $a \in \mathcal{A}$, encoded via the binary optimization variable $\zeta \in \{0, 1\}^{|\mathcal{A}| \times |\mathcal{I}|}$.

At the second stage, one decides:

- a target landing time for each aircraft $a \in \mathcal{A}$ which is denoted by $y \in \mathbb{R}^{|\mathcal{A}|}$.

Remark that for each IAF, δ specifies the sequence order, and z specifies each aircraft time at its IAF, while y encodes each aircraft landing time. The landing sequence must be *coherent* with the IAF sequences, since aircraft coming from a same IAF cannot overtake each other.

The objective and constraint functions of the general formulation (5)–(12) are made explicit in the next subsections, following the model introduced in Khassiba et al. (2022). Note that in Khassiba et al. (2022) some variants of the problem (depending on the considered operational setup) are considered and associated models are proposed. Here we focus only on the two-stage model that was identified as the closest to the air traffic operating environment, i.e., the one incorporating decisions on multiple IAF points and taking into account multiple aircraft flight status, and we present it in a concise way which is moreover easier to generalize.

2.2.1 Objective Functions

The objective function in (5) is the sum of the first-stage function that captures the time-deviation costs at take-off for the on-ground aircraft, and the recourse cost, expressed as the expectation of the optimal value, $Q(x, \boldsymbol{\xi})$, of the second-stage objective function. Note that we could weigh the recourse cost by multiplying it by a parameter $\lambda \in \mathbb{R}^+$, as done in Khassiba et al. (2020), but for simplicity, we keep λ equal to one in the following.

Given a scenario ξ (a realization of the random vector $\boldsymbol{\xi}$), actual times of arrival at the IAFs are assumed to be known, and for each IAF, the corresponding aircraft sequence computed at the first stage is enforced and only the target landing time, y_a, is to be decided for each aircraft $a \in \mathcal{A}$. The complete landing sequence, gathering the aircraft coming from any IAF, is thus implicitly determined by the landing times. The second-stage objective function corresponds to the total time-deviation (delay and advance) cost incurred during the *en-route* (relative here to the part of the flight that takes place before the IAF) and the *approach* (after the IAF) phases. As in Khassiba et al. (2022), the time-deviation cost depends on the deviation duration, and the delay cost is determined according to the reference values proposed in the report authored by Cook and Tanner (2015), while the approach cost is minimized with respect to the landing times y_a for each $a \in \mathcal{A}$. The second-stage objective function is typically a sum of convex piecewise linear functions, and can be easily replaced by an equivalent linear objective function (Bradley et al. 1977, Section 9.2).

Alternative formulations, and related interpretations are possible (see for instance Khassiba et al. 2019, 2020), but are not considered here.

2.2.2 Time-Window Constraints at IAF

Let us consider an aircraft $a \in \mathcal{A}$. The time-window constraint (6) expresses that its *target IAF time*, z_a, must lie within a time window $\left[E_a^i, L_a^i\right]$, where E_a^i and L_a^i denote respectively the *earliest time* and the *latest time* of a at its assigned IAF $i \in \mathcal{I}$:

$$z_a \in \left[E_a^i, L_a^i\right].$$

Recall that during the en-route flight phase, with 2–3 h look-ahead time, it is possible to expedite or delay aircraft a by a given amount of time through its speed change. These earliest and latest times can be computed by taking into account the *maximal possible time saving*, and the *maximal possible delay* for aircraft a during its en-route phase, denoted as $\underline{d}_a^R$ and $\overline{d}_a^R$ respectively, as follows:

$$E_a^i = P_a^i - \underline{d}_a^R$$
$$L_a^i = P_a^i + \overline{d}_a^R,$$

where P_a^i is the *planned IAF time* of aircraft a at IAF i.

Remark that P_a^i can be determined, and hence E_a^i and L_a^i can be further made explicit, according to the flight status (airborne or on-ground) of aircraft a. Let us first observe that the initially-assigned IAF of aircraft a is assumed to be the IAF $i^* \in \mathcal{I}$ that is closest to the reference (initially-planned) route of a. As a consequence, deciding to assign aircraft a to an IAF j that is different from i^* ($j \in \mathcal{I} \setminus \{i^*\}$) incurs a positive rerouting delay, noted $r^{i^*j} > 0$ (with $r^{i^*i^*} = 0$). Recalling that variables ζ are used to decide the assignment of each aircraft a to an IAF i, $\sum_{j \in \mathcal{I}} r^{i^*j} \zeta_a^j$ denotes the actual rerouting delay for aircraft a.

For an airborne aircraft, $a \in \mathcal{A}_A$, and an IAF $i \in \mathcal{I}$, the planned IAF time is then straightforwardly computed from input data (the planned time to the initial IAF, denoted $P_a^{i^*}$), adding r^{i^*j} in the case where the aircraft is rerouted to an IAF different from its initial one. The time-window constraint (6) at the IAF for an airborne aircraft then reads as

$$-\underline{d}_a^R \leq z_a - P_a^{i^*} - \sum_{j \in \mathcal{I}} r^{i^*j} \zeta_a^j \leq \overline{d}_a^R.$$

For an on-ground aircraft, $a \in \mathcal{A}_G$ (still at its departure gate), the planned IAF time depends on further input data (the flight time from the origin airport of a to its initial IAF $i^* \in \mathcal{I}$, denoted $\hat{V}_a^O$) as well as on the decision variable t_a representing the target take-off time. This variable t_a is in turn bounded by taking into account the planned take-off time, P_a^{TOT}, and the maximal possible delay $\overline{d}^G > 0$ to take off: $0 \leq t_a \leq P_a^{\text{TOT}} + \overline{d}^G$. The time-window constraint at the IAF for an on-ground aircraft then reads:

$$-\underline{d}_a^R \leq z_a - t_a - \hat{V}_a^O - \sum_{j \in \mathcal{I}} r^{i^* j} \zeta_a^j \leq \overline{d}_a^R.$$

The way E_a^i and L_a^i are computed for aircraft with different flight status is detailed in Khassiba et al. (2022).

2.2.3 Separation Constraints at IAF

In the first-stage problem, a target sequence of aircraft at each IAF and target IAF times of each aircraft are determined in such a way that, for safety reasons, successive aircraft crossing the same IAF are pairwise separated by a minimal time separation, $\underline{S}$. The assignment of each aircraft $a \in \mathcal{A}$ to an IAF $i \in \mathcal{I}$ is decided through the binary decision variable ζ_a^i (whose value is equal to 1 if aircraft a is assigned to IAF i, 0 otherwise), while binary variables δ decide the landing sequence of all aircraft during the second stage. This in turn determines the sequence of arrivals at each IAF since the corresponding relative order is kept unchanged for landing at the runway. More precisely, a binary decision variable δ_{ab} for each ordered pair of aircraft $(a, b) \in \mathcal{A} \times \mathcal{A}$, $a \neq b$ is defined as follows:

$$\delta_{ab} = \begin{cases} 1 & \text{if aircraft } a \textbf{ lands before } \text{aircraft } b, \\ 0 & \text{otherwise.} \end{cases}$$

Remark that these variables are defined, like in Khassiba et al. (2022), as in scheduling theory, where *direct* precedence is *not* required, as opposed to what is done in Khassiba et al. (2020), where the sequencing variables are defined so as to indicate *direct precedence* between a pair of aircraft (like in the classical model for the traveling salesman problem).

In order to formulate the separation constraints at each IAF, we further introduce the following auxiliary binary decision variables for each pair of aircraft $(a, b) \in \mathcal{A} \times \mathcal{A}$, $a \neq b$:

$$\phi_{ab} = \begin{cases} 1 & \text{if aircraft } a \text{ and } b \text{ are assigned to a same IAF,} \\ 0 & \text{otherwise,} \end{cases}$$

together with defining constraints:

$$\sum_{i \in \mathcal{I}} \zeta_a^i = 1 \qquad a \in \mathcal{A}$$

$$\phi_{ab} = \phi_{ba} \quad (a,b) \in \mathcal{A} \times \mathcal{A}, \qquad a < b$$

$$\phi_{ab} \geq \zeta_a^i + \zeta_b^i - 1 \qquad i \in \mathcal{I}\,, \quad (a,b) \in \mathcal{A} \times \mathcal{A}, \quad a < b$$

$$\delta_{ab} + \delta_{ba} = 1 \qquad (a,b) \in \mathcal{A} \times \mathcal{A}, \quad a < b.$$

Finally, the first-stage constraints expressing pairwise aircraft separation at each IAF take the form:

$$z_b \geq z_a + \underline{S} - M_{ab}(2 - \phi_{ab} - \delta_{ab}) \qquad (a,b) \in \mathcal{A} \times \mathcal{A}, \quad a \neq b,$$

where M_{ab} is a sufficiently large constant defined for each pair $(a,b) \in \mathcal{A} \times \mathcal{A}$ such that $a \neq b$.

2.2.4 Time-Window Constraints at Runway

Let us consider an aircraft $a \in \mathcal{A}$. The target landing time, y_a, must lie within a time window:

$$y_a \in [E_a,\ L_a]\,,$$

where E_a and L_a are its earliest and latest landing times, computed as the actual time of a at its assigned IAF i when uncertainty is revealed: $z_a + \xi$, to which we add the minimal and maximal flight times from the IAF i to the runway threshold. These minimal and maximal flight times are respectively defined as

$$\underline{V_a}^i = \hat{V_a}^i + \underline{d}_a^T, \qquad \overline{V_a}^i = \hat{V_a}^i + \overline{d}_a^T,$$

where, $\hat{V_a}^i$ is the unconstrained flight time of the aircraft a from i to the runway threshold, $\overline{d}_a^T$ is the maximal possible delay during the approach phase, and $\underline{d}_a^T$ is the maximal possible time saving during the approach phase. Recall that the assigned IAF i of each aircraft a is decided through binary variables ζ_a^i, for all i. Hence, minimal and maximal flight times from the IAF to the runway threshold are to be selected according to the value of such variables ζ. Finally, time windows constraints at runway read:

$$\sum_{i \in \mathcal{I}} \underline{V_a}^i \zeta_a^i \leq y_a - (z_a + \xi_a) \leq \sum_{i \in \mathcal{I}} \overline{V_a}^i \zeta_a^i.$$

Table 1 Final-approach separations (in seconds) at runway according to wake-vortex turbulence categories of the leader aircraft, l, and the follower aircraft, f

	H_f	M_f	L_f
H_l	96	157	207
M_l	60	69	123
L_l	60	69	82

2.2.5 Separation Constraints at Runway

In the second-stage problem, when uncertainty on target times at the IAFs is revealed, for each aircraft, we decide on a target time, y_a, at the runway threshold. A minimal time separation, S_{ab}, must then be satisfied for each pair of aircraft landing successively:

$$y_b \geq y_a + S_{ab} - M^L_{ab}(1 - \delta_{ab}) \quad (a, b) \in \mathcal{A} \times \mathcal{A}, \quad a \neq b,$$

where M^L_{ab} is a sufficiently large constant defined for each $(a, b) \in \mathcal{A} \times \mathcal{A}$ $(a \neq b)$, and where S_{ab} is determined from minimal separation distances (Gerz et al. 2002; Breitsamter 2011). These separation distances are defined according to the wake-vortex turbulence category of aircraft a and that of aircraft b as introduced by the International Civil Aviation Organization: heavy (H), medium (M), and light (L) (Antolovic & Franjkovic 2020). They are converted into minimal *time* separations, and reported in Table 1.

2.3 *The Two-stage Stochastic Optimization Model*

The two-stage stochastic optimization model of the extended aircraft arrival management under uncertainty presented above (and introduced in Khassiba et al. 2022) can be summarized as follows:

$$\min_{\substack{z,t,\delta,\\ \zeta,\phi}} \sum_{a \in \mathcal{A}_G} f^{\mathrm{G}}_a \left(t_a - P^{\mathrm{TOT}}_a\right) + \mathbb{E}_{\boldsymbol{\xi}}[Q(z, t, \zeta, \delta, \boldsymbol{\xi})] \tag{13}$$

$$\text{s.t.} \sum_{i \in \mathcal{I}} \zeta^i_a = 1, \qquad a \in \mathcal{A} \tag{14}$$

$$\phi_{ab} = \phi_{ba}, \qquad (a, b) \in \mathcal{A} \times \mathcal{A}, a < b \tag{15}$$

$$\phi_{ab} \geq \zeta^i_a + \zeta^i_b - 1, \qquad i \in \mathcal{I}, (a, b) \in \mathcal{A} \times \mathcal{A}, a < b \tag{16}$$

$$\delta_{ab} + \delta_{ba} = 1, \qquad (a, b) \in \mathcal{A} \times \mathcal{A}, a < b \tag{17}$$

$$z_b \geq z_a + \underline{S} - M_{ab}(2 - \phi_{ab} - \delta_{ab}), \qquad (a,b) \in \mathcal{A} \times \mathcal{A}, a \neq b \tag{18}$$

$$-\underline{d}_a^R \leq z_a - t_a - \hat{V}_a^O - \sum_{j \in \mathcal{I}} r^{i^*j} \zeta_a^j \leq \overline{d}_a^R, \; i^* \in \mathcal{I}, a \in \mathcal{A}_G^{i^*} \tag{19}$$

$$-\underline{d}_a^R \leq z_a - P_a^{i^*} - \sum_{j \in \mathcal{I}} r^{i^*j} \zeta_a^j \leq \overline{d}_a^R, \qquad i^* \in \mathcal{I}, a \in \mathcal{A}_A^{i^*} \tag{20}$$

$$0 \leq t_a - P_a^{\text{TOT}} \leq \overline{d}_a^G, \qquad a \in \mathcal{A}_G \tag{21}$$

$$\zeta_a^i \in \{0,1\}, \qquad i \in \mathcal{I}, a \in \mathcal{A} \tag{22}$$

$$\phi_{ab} \in \{0,1\}, \qquad (a,b) \in \mathcal{A} \times \mathcal{A}, a \neq b \tag{23}$$

$$\delta_{ab} \in \{0,1\}, \qquad (a,b) \in \mathcal{A} \times \mathcal{A}, a \neq b, \tag{24}$$

where

$$Q(z,t,\zeta,\delta,\xi) =$$

$$\min_y \sum_{a \in \mathcal{A}_G} f_a^{\text{R}} \left(z_a + \xi_a - \left(t_a + \hat{V}_a^O \right) \right) +$$

$$\sum_{a \in \mathcal{A}_A} f_a^{\text{R}} \left(z_a + \xi_a - P_a^{i^*} \right) +$$

$$\sum_{a \in \mathcal{A}} f_a^{\text{T}} \left(y_a - \left(z_a + \xi_a + \sum_{i \in \mathcal{I}} \hat{V}_a^i \zeta_a^i \right) \right) \tag{25}$$

$$\text{s.t. } y_b \geq y_a + S_{ab} - M_{ab}^L(1 - \delta_{ab}), \qquad (a,b) \in \mathcal{A} \times \mathcal{A}, a \neq b \tag{26}$$

$$\sum_{i \in \mathcal{I}} \underline{V_a}^i \zeta_a^i \leq y_a - (z_a + \xi_a) \leq \sum_{i \in \mathcal{I}} \overline{V_a}^i \zeta_a^i, \qquad a \in \mathcal{A}, \tag{27}$$

and where M_{ab} and M_{ab}^L are sufficiently large (*big-M*) constants, defined for each $(a,b) \in \mathcal{A} \times \mathcal{A}$, $a \neq b$, and whose value can be estimated from input data (see Khassiba et al. 2022).

The first-stage objective function (13), to be minimized, is the sum of the total at-gate time deviation and the expectation of the second-stage objective function (see Sect. 2.2.1). Constraints (14)–(17) are defining constraints on variables ζ, ϕ and δ (see Sect. 2.2.3). Constraints (18) ensure separation (in terms of target IAF times) for any pair of successive aircraft assigned to a same IAF (Sect. 2.2.3). Constraints (19) and (20) are time-window constraints for on-ground and airborne aircraft, respectively (see Sect. 2.2.2). Constraints (21) ensure that the take-off time of an on-ground aircraft is chosen in the appropriate time window. Constraints (22), (23), and (24) simply stipulate that the decision variables ζ_a^i, ϕ_{ab} and δ_{ab} are binary.

The second-stage objective function (25), to be minimized, is the total time-deviation (delay and advance) cost during the en-route and the landing phases. Constraints (26) are separation constraints at runway between any pair of landing aircraft (see Sect. 2.2.5). Constraints (27) ensure, for a given aircraft, that its landing time between its assigned IAF and the runway, lies within an appropriate time window (see Sect. 2.2.4).

Remark The first-stage problem can be enriched by chance constraints expressing a protection level against any separation loss over each IAF.

Letting $0 \le \alpha \le 1$ be a given confidence-level parameter, and incorporating the random time deviations at the IAFs into (18), we can introduce the chance constraints

$$\mathcal{P}_{\xi}\left(z_b + \xi_b \ge z_a + \xi_a + \underline{S} - M_{ab}(2 - \phi_{ab} - \delta_{ab})\right) \ge \alpha \quad (a, b) \in \mathcal{A} \times \mathcal{A}, a \ne b. \tag{28}$$

Chance constraints guarantee that any pair of aircraft are separated at the IAF after uncertainty is revealed (actual IAF time deviations with respect to target IAF times) with a probability α, at least. Thus, $1 - \alpha$ represents the risk level at which a separation constraint is violated. Equivalently, we can rewrite (28) as the linear constraints

$$z_b \ge z_a + \underline{S} + F_{\psi_{ab}}^{-1}(\alpha) - M_{ab}(2 - \phi_{ab} - \delta_{ab}) \quad (a, b) \in \mathcal{A} \times \mathcal{A}, a \ne b, \tag{29}$$

where $F_{\psi_{ab}}(\alpha)$ is the distribution function of $\psi_{ab} = \xi_a - \xi_b$. It is noteworthy that given an aircraft pair (a, b), $a \ne b$, (29) dominates the original non-buffered IAF separation constraint (18) when $F_{\psi_{ab}}^{-1}(\alpha) > 0$.

We numerically explore the effect of such chance constraints in Sect. 4.4. In our setting, similar to Khassiba et al. (2020), we assume that the random deviations ξ_a's, affecting the aircraft during the en-route phase, are independent and identically distributed, whatever the aircraft status in the first phase (airborne or on-ground). Quantifying the precise distribution of uncertainty is a tricky task, especially since the random noise is not directly observable (Tielrooij et al. 2015). For the sake of simplicity, we assume that for any a in $\mathcal{A}$, ξ_a is normally distributed with variance σ^2, ψ_{ab} is normally distributed with zero mean and variance $2\sigma^2$. In this case, $F_{\psi_{ab}}^{(-1)}(\alpha)$ is positive if and only if $\alpha > 0.5$. In a more realistic setting, that we leave as a future avenue of investigation, it is possible to consider that an on-ground aircraft, $a \in \mathcal{A}_G$, is affected by two time deviations: its actual take-off time deviates randomly from its target take-off time, t_a, and its actual IAF time deviates as well from its target IAF time, z_a.

3 Solution Method

Objective function (5) typically has no closed-form expression due to the expectation operator, hence the problem (5)–(12), referred to as the *true problem*, cannot be solved directly. A classical remedy is to sample a finite set, $\mathcal{S}$, of scenarios over the random vector $\boldsymbol{\xi}$ to produce the *sample average approximation (SAA)* problem (Shapiro 2021). One practical difficulty of the SAA method is the determination of an appropriate size of the scenario set. Indeed, large scenario sets capture better uncertainty. However, computing times increase exponentially with the number of scenarios. An appropriate number of scenarios can be defined as the smallest scenario-set size that ensures an acceptable representation level of the uncertainty. After solving a two-stage stochastic problem, it is key to validate the benefit from taking into account uncertainty, against the situation where the decision-maker computes his decision assuming mean values of the uncertain parameters. One classical metric to measure the benefit from two-stage stochastic programming over a deterministic approach is the *value of the stochastic solution*. The SAA method applied to a two-stage stochastic optimization problem is presented in Sect. 3.1. In Sect. 3.2, we develop on the metrics used in our study to quantify the appropriateness of a given scenario-set size. Section 3.3 presents a formal definition of the value of the stochastic solution adapted to our problem, where SAA is applied as a solution method.

3.1 Sample Average Approximation

The SAA problem can be formulated as follows:

$$\begin{aligned} \min_{x} \quad & \hat{f}(x, \mathcal{S}) = f^{(1)}(x) + \frac{1}{n_{\mathcal{S}}} \sum_{\xi \in \mathcal{S}} Q(x, \xi) \\ \text{s.t.} \quad & (6)\text{–}(8) \\ \text{where} \quad & Q(x, \xi) \text{ is defined by } (9)\text{–}(12), \end{aligned} \tag{SAA}$$

and where $n_{\mathcal{S}} = |\mathcal{S}|$ is the number of scenarios. In our case, the SAA problem turns out to be a mixed-integer linear program, which can be solved by state-of-the-art solvers.

Let $\hat{v}(\mathcal{S})$ be the optimal value of the problem (SAA) when considering the set, $\mathcal{S}$, of scenarios. The consistency of the SAA problems have been studied by several authors (Shapiro & Homem-de Mello 2000; Shapiro 2021; Bastin et al. 2006). In particular, under mild conditions, $\hat{v}(\mathcal{S})$ is an asymptotically unbiased estimator of v^*, the optimal value of the true problem (5), as $\hat{v}(\mathcal{S})$ converges towards v^*, with probability one as $n_{\mathcal{S}}$ tends towards infinity. In other words, with a probability rising to one, the SAA optimal value can be made arbitrarily close to the true optimal value for a sufficiently large number of scenarios. However, for any given number, $n_{\mathcal{S}}$, of scenarios, one has $\mathbb{E}\left[\hat{v}(\mathcal{S})\right] \leq v^*$, i.e., the optimal value of the SAA problem is negatively biased, and, in practice, the required computing time grows rapidly with $n_{\mathcal{S}}$.

3.2 Appropriate Number of Scenarios

The SAA method requires to decide whether a given number of scenarios, $n_{\mathcal{S}}$, is large enough to approximate correctly the original problem, i.e., to ensure that the solution obtained is a satisfying approximation of an optimal solution of the original problem. The optimal value of the SAA problem, $\hat{v}(\mathcal{S})$, is not necessarily a good-quality indicator with respect to the original problem, due to the SAA bias and variance of the SAA optimal value. Hence, a post-optimization validation step is needed to evaluate the quality of any SAA solution. In our study, we rely on the so-called *out-of-sample validation*, that consists in re-evaluating a first-stage SAA solution, $\hat{x}(\mathcal{S})$, with a *validation set* $\mathcal{S}_V$, containing many more scenarios than the training set $\mathcal{S}$ used to find the solution. The validation set is believed to represent the complete set of all possible scenarios, and can be used as a reference scenario set to compare the quality of different feasible solutions. The evaluation of a feasible solution x on the validation set $\mathcal{S}_V$, denoted $\hat{f}(x, \mathcal{S}_V)$, is called the *validation score*. When the gap between the SAA optimal value, $\hat{v}(\mathcal{S})$, and the validation score $\hat{f}(\hat{x}(\mathcal{S}), \mathcal{S}_V)$ is small (according to the analyst), the SAA problem can be considered as stable and the training set used large enough. As SAA solutions depend on the random scenario set $\mathcal{S}$, we build $n_{\mathcal{R}}$ replicated SAA problems with $n_{\mathcal{S}}$ scenarios each, independently generated, to illustrate better the average behavior of SAA problems with a given number of scenarios, $n_{\mathcal{S}}$. Let $\bar{v}\,(n_{\mathcal{S}}, n_{\mathcal{R}})$ be the *average optimal value* obtained over $n_{\mathcal{R}}$ such replications:

$$\bar{v}\,(n_{\mathcal{S}}, n_{\mathcal{R}}) = \frac{1}{n_{\mathcal{R}}} \sum_{r=1}^{n_{\mathcal{R}}} \hat{v}\,(\mathcal{S}_r)\,, \tag{30}$$

where $\mathcal{S}_r, r = 1, \ldots, n_{\mathcal{R}}$, are scenario sets of size $n_{\mathcal{S}}$, independently generated, and independent of the validation set $\mathcal{S}_V$. The corresponding *average validation score* over $n_{\mathcal{R}}$ replications, denoted $\overline{f}(n_{\mathcal{S}}, n_{\mathcal{R}}, \mathcal{S}_V)$, is

$$\overline{f}(n_{\mathcal{S}}, n_{\mathcal{R}}, \mathcal{S}_V) = \frac{1}{n_{\mathcal{R}}} \sum_{r=1}^{n_{\mathcal{R}}} \hat{f}(x^*(\mathcal{S}_r), \mathcal{S}_V), \tag{31}$$

where $x^*\,(\mathcal{S}_r)$ is the computed optimal solution of the problem (SAA), with the scenario set $\mathcal{S}_r$.

3.3 Value of the Stochastic Solution

Consider the two-stage stochastic problem (5)–(12). The incorporation of uncertainty, while making the model more realistic, comes with additional modeling questions and computational complexity, as we must decide on the random distributions to capture the uncertainty, and deal with more numerically intensive computations. It is therefore interesting to evaluate the relevance of the stochastic approach by comparing it to a simpler deterministic program. We first define the *expected value problem* (EVP) as the deterministic problem in which we replace the random vector, $\boldsymbol{\xi}$, by its expectation, $\mathbb{E}[\boldsymbol{\xi}]$:

$$\min_{x,y} f^{(1)}(x) + f^{(2)}(y, \mathbb{E}[\boldsymbol{\xi}]) \tag{32}$$

$$\text{s.t.} \quad T^{(1)}(y, t, s^{(1)}) \leq 0 \tag{33}$$

$$S^{(1)}(x, s^{(1)}) \leq 0 \tag{34}$$

$$T^{(2)}(y, s^{(2)}) \leq 0 \tag{35}$$

$$S^{(2)}(y, s^{(2)}) \leq 0 \tag{36}$$

$$x \in \mathcal{X}, \ \ y \in \mathcal{Y}, \tag{37}$$

where

$$s^{(2)} = g(x, s^{(1)}, \mathbb{E}[\boldsymbol{\xi}]).$$

Denote by x^*_{EVP} the first-stage optimal solution found when solving (32)–(37), and assume that the two-stage stochastic program (5)–(12) has a relatively complete recourse. The *value of the stochastic solution* (VSS) (Birge & Louveaux 2011) captures the improvement achieved when taking uncertainty into account:

$$\text{VSS} = f^{(1)}(x^*) + \mathbb{E}\left[Q(x^*, \boldsymbol{\xi}))\right] - f^{(1)}\left(x^*_{EVP}\right) - \mathbb{E}\left[Q\left(x^*_{EVP}, \boldsymbol{\xi}\right)\right]. \tag{38}$$

We again rely on the SAA approach to approximate the true VSS (38), defining the *SAA VSS* with respect to the set of scenarios $\mathcal{S}$ and the validation set $\mathcal{S}_V$ as

$$\widehat{\text{VSS}}(\mathcal{S}, \mathcal{S}_V) = \hat{f}\left(x^*(\mathcal{S}), \mathcal{S}_V\right) - \hat{f}\left(x^*_{EVP}, \mathcal{S}_V\right). \tag{39}$$

In other terms, the true solution x^* is replaced by the SAA solution $x^*(\mathcal{S})$, and the approximate VSS is computed as the difference of the validation scores computed at $x^*(\mathcal{S})$ and x^*_{EVP}. We can also compute the relative approximate VSS as

$$\widehat{\text{VSS}}_{\%}(\mathcal{S}, \mathcal{S}_V) = \widehat{VSS}(\mathcal{S}, \mathcal{S}_V)/\hat{f}\left(x^*_{EVP}, \mathcal{S}_V\right). \tag{40}$$

4 Numerical Study

We now illustrate our approach on various numerical experiments. Section 4.1 presents the realistic instances of the two-stage stochastic optimization problem we are addressing, together with the considered test data to conduct our numerical study. The aim of the numerical study is threefold. First, in Sect. 4.2, we seek the appropriate number of scenarios in order to apply the SAA method. Second, in Sect. 4.3, we compute the value of the stochastic solution for different uncertainty amplitudes in order to assess the benefit from taking account of the uncertainty via two-stage stochastic programming. Third, we study in Sect. 4.4 the effect of chance constraints on the value of the stochastic solution, and on the solution quality. All numerical tests are conducted on a Windows platform with an Intel i5-10310U and 16 GB RAM using the solver CPLEX 22.1.0.0.

4.1 Instances and Test Data

We focus on Paris Charles-de-Gaulle airport (CDG) and its northern landing runway (named 27R), depicted in Fig. 2, along with its surrounding IAFs. We use the five instances, of $n = 10$ aircraft each, described in Khassiba et al. (2022). They correspond to aircraft planning to land on the northern runway 27R, and to cross the two IAFs, MOPAR or LORNI, between 5:59 AM and 6:59 AM, on May 5, 2015. Instance characteristics are summarized in Table 2. We assume that time deviations of the IAF actual times, for all aircraft, are independent and normally distributed, with zero mean of variance σ^2. Additional test data are summarized in Table 3. Delay and advance unit costs, in euros per second, are the same as in Khassiba et al. (2022, Appendix C).

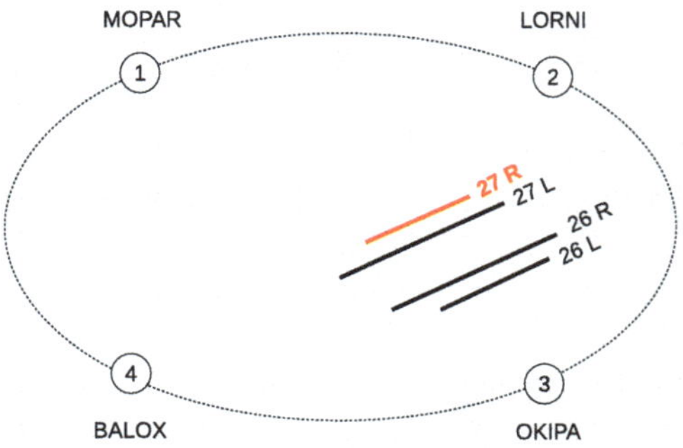

Fig. 2 Simplified scheme of the four IAFs surrounding the four runways of CDG airport

Table 2 Instance characteristics

Instance id	559_618	607_623	619_634	624_640	634_659
$\lvert\mathcal{A}\rvert$	10	10	10	10	10
Time span	10 min 30 s	8 min 42 s	8 min 30 s	8 min 4 s	12 min 29 s
$\lvert\mathcal{A}_A\rvert$	6	4	3	5	7
$\lvert\mathcal{A}_G\rvert$	4	6	7	5	3
# Medium (M)	4	6	7	7	7
# Heavy (H)	6	4	3	3	3
# assigned to IAF 1	6	5	5	4	3
# ass. to IAF 2	4	5	5	6	7

Table 3 Test data

Parameter	Definition	Test values
$\overline{d}^G$	Maximal delay at-gate, seconds	900
$\underline{d}_a^R$	Maximal advance en-route, seconds	60
$\overline{d}_a^R$	Maximal delay en-route, seconds	300
$\underline{d}_a^T$	Maximal advance approach, seconds	0
$\overline{d}_a^T$	Maximal delay approach, seconds	1200
r	Rerouting time to change IAF, seconds	300
$\underline{S}$	Minimal IAF time separation, seconds	72
$\hat{V}_a^1$	Unconstrained flight time from MOPAR to runway, seconds	780
$\hat{V}_a^2$	Unconstrained flight time from LORNI to runway, seconds	660
σ	Standard deviation of the normal random deviation ξ, seconds	60, 120
$n_{\mathcal{S}}$	Number of scenarios in the SAA problem	10, 50, 100, 200
$n_{\mathcal{S}_V}$	Number of scenarios in the validation set	1000
$n_{\mathcal{R}}$	Number of replications	20

4.2 *Appropriate Number of Scenarios*

We consider the problem (SAA), and explore an increasing number of scenarios $n_{\mathcal{S}} = 10, 50, 100, 200$, to determine the number of scenarios required to accurately enough approximate the true problem. For each value of $n_{\mathcal{S}}$, we solve $n_{\mathcal{R}} = 20$ independently generated replications of the SAA problem. In our experiments, the scenario sets are built incrementally, i.e., for a given replication $r = 1, 2, \ldots, n_{\mathcal{R}}$ and two scenario sets $\mathcal{S}_r'$ and $\mathcal{S}_r$, such that $n_{\mathcal{S}_r'} < n_{\mathcal{S}_r}$, we have $\mathcal{S}_r' \subset \mathcal{S}_r$. The optimal solution of each replication is then evaluated on the same validation set $\mathcal{S}_V$, with $n_{\mathcal{S}_V} = 1000$. Figure 3 focuses on instance 559_618 and plots the average optimal value (30) and the average validation score (31) for an increasing number of scenarios, using boxplots to exhibit the variability of the optimal value $\hat{v}(\mathcal{S})$ and the validation score $\hat{f}(\hat{x}(\mathcal{S}), \mathcal{S}_V)$ over the replications.

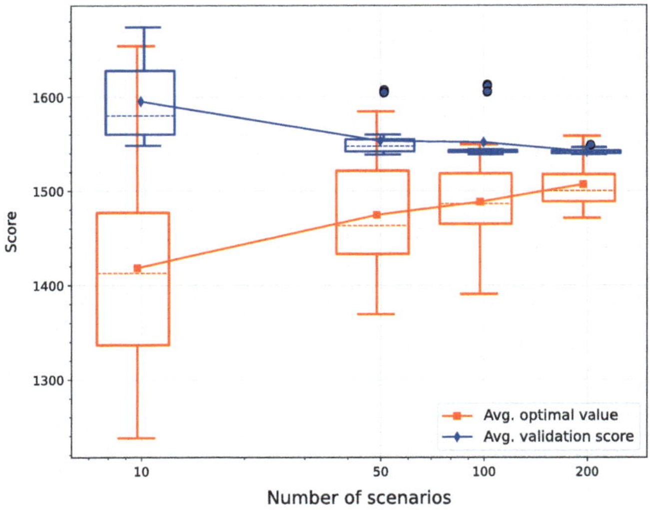

Fig. 3 Objective function's optimal value and validation score as functions of the number of scenarios, for the instance 559_618

As $n_{\mathcal{S}}$ grows, we observe that the average optimal value increases, whereas the average validation score decreases, while their variances decrease. This behavior can be expected as a SAA problem with a scenario set $\mathcal{S}' \subset \mathcal{S}$ can be viewed as a relaxation of the SAA problem with a scenario set $\mathcal{S}$ with respect to the first-stage decision x. However, due to the variability of the second-stage cost, the SAA optimal value is not constrained to be monotonically increasing with respect to the number of scenarios, as reflected by the whiskers of the boxplots, while this variability decreases as $n_{\mathcal{S}}$ increases. The evolution of the validation score, both in terms of average and variance, illustrates the convergence of the SAA optimal value towards the true optimal value when $n_{\mathcal{S}}$ rises, in line with consistency analysis results (Shapiro 2021). More specifically, we observe that, as the number of scenarios increases, the variances of the optimal value and validation score decrease, as well as the bias between the SAA and true optimal values, produced by the optimization over a finite sample size. For 200 scenarios, the variance of the validation score can be neglected, and the its average is close to the average SAA optimal value. Due to the numerical costs experienced with 200 scenarios (around 9 min on average to find the optimal solution of one replication), we decided not to increase the number of scenarios furthermore, and use $n_{\mathcal{S}} = 100$ as the number of scenarios for our subsequent tests.

4.3 Value of the Stochastic Solutions

Fixing the number of scenarios to $n_{\mathcal{S}} = 100$, we solve $n_{\mathcal{R}} = 20$ replications of the problem (SAA), for each of the five studied instances, and report the main results in Table 4. For each instance, we test two values of the standard error σ of the normally distributed random deviations, 60 and 120 s, corresponding to low and high uncertainty levels. We report in the column "CPU time" the computing time, in seconds, to find an optimal solution, averaged over the $n_{\mathcal{R}}$ replications, and the average validation score (31) in column "Validation score". The SAA VSS (39) and its relative counterpart (40), averaged over the $n_{\mathcal{R}}$ replications, are reported in columns "$\widehat{\text{VSS}}$" and "$\widehat{\text{VSS}}$ (%)", respectively. The ratio of aircraft that changed their IAF, averaged over all replications, is reported in column "Avg IAF changes". The number of different sequences at IAF 1, IAF 2, and runway, in the optimal solutions of the $n_{\mathcal{R}}$ replications, are listed in columns "# different sequences IAF 1", "IAF 2", and "RWY", respectively.

The increase in computational time alongside the uncertainty level exhibits that the problem then becomes more expensive to solve, even though the number of scenarios is fixed. Moreover, greater uncertainty leads to higher deviation costs, captured by the validation scores. Negative values of the stochastic solution, ranging from −2.03 to −21.61%, illustrate the benefit from taking into account uncertainty through two-stage stochastic programming. Moreover, the increase of the VSS in absolute value with the level of uncertainty suggests that the benefit of two-stage stochastic programming is greater for high uncertainty. Modifying the IAF assignment, by rerouting flights during the en-route phase, appears to be an efficient measure to hedge against uncertainty, as the average number of IAF changes increases as uncertainty grows, while deterministic expected value solutions do not propose any IAF change. Note that for σ equal to 60 s, most of the instances present two different sequences at each IAF and on the runway, across the $n_{\mathcal{R}}$ replications. This comforts the selection of $n_{\mathcal{S}} = 100$ as an appropriate number of scenarios. On

Table 4 Results summary of the stochastic solutions of all instances

	σ	CPU	Val. score	$\widehat{\text{VSS}}$		Avg IAF	# different sequences		
Instance	(s)	time (s)	(euros)	(euros)	(%)	changes	IAF 1	IAF 2	RWY
10_559_618	60	138.77	1552.38	−216.79	−12.25%	19%	2	2	2
	120	212.09	2782.65	−478.32	−14.67%	30%	1	1	1
10_607_623	60	78.78	1070.92	−295.27	−21.61%	1%	2	2	2
	120	106.62	1997.39	−448.40	−18.33%	10%	4	4	5
10_619_634	60	178.53	1062.28	−22.04	−2.03%	0%	2	2	3
	120	698.71	1899.01	−59.70	−3.05%	0%	2	2	6
10_624_640	60	575.54	944.40	−136.14	−12.60%	1%	2	2	2
	120	660.24	1673.17	−206.55	−10.99%	16%	4	3	7
10_634_659	60	37.63	670.34	−108.51	−13.93%	5%	1	2	2
	120	177.61	1445.17	−266.61	−15.58%	10%	1	2	2

the other hand, the higher uncertainty level, generated by fixing σ to 120 s, leads to a larger number of different sequences, in most cases, and more scenarios could be required to stabilize the solution.

4.4 *Effect of the Chance Constraints*

To study the effect of chance constraints, we run additional experiments with the same setup as in the previous Sect. 4.3, and the addition of the chance constraints (29), with $\alpha = 0.9$. Table 5 provides the values of the minimal IAF time separation, in the original case (without chance constraints) and when chance constraints are included, with $\alpha = 0.9$, and the uncertainty level, σ, equal to 60 or 120 s. A summary of the results is provided in Table 6.

A comparison with Table 4 reveals that the inclusion of chance constraints usually leads to higher computing times, suggesting that the problem with a buffered IAF separation is more difficult to solve. A possible explanation is that buffered IAF separations of 3 up to more than 4.5 min, as computed in Table 5, make difficult to find feasible sequences since in our experimental settings, described in Sect. 4.1, the maximal possible advance and delay at the IAF for a given aircraft are 1 and 5 min, respectively. This also contributes to produce more IAF changes, that offer additional en-route delays (recall that one IAF change represents 5 min of en-route delay), and larger validation scores, i.e., larger time deviation costs. This is consistent with the fact that buffered separation corresponds to

Table 5 Minimal IAF time separations (s) for the studied values of α and σ

	$\underline{S} + F^{-1}_{\psi_{ab}}(\alpha)$	
$\underline{S}$	$\alpha = 0.9, \sigma = 60$ s	$\alpha = 0.9, \sigma = 120$ s
72.0	180.7	289.5

Table 6 Results summary of the stochastic solutions of all instances with chance constraints ($\alpha = 0.9$)

	σ	CPU	Val. score	$\widehat{\text{VSS}}$		Avg IAF	# different sequences		
Instance	(s)	time (s)	(euros)	(euros)	(%)	changes	IAF 1	IAF 2	RWY
10_559_618	60	225.94	2361.37	−94.96	−3.87%	26%	2	2	2
	120	113.83	4855.63	−35.94	−0.73%	30%	2	2	4
10_607_623	60	324.40	1769.40	−246.86	−12.24%	12%	3	4	5
	120	398.99	3108.35	−118.49	−3.67%	20%	3	2	4
10_619_634	60	652.26	1146.99	−76.51	−6.25%	0%	1	1	3
	120	1256.40	2480.56	−36.36	−1.44%	20%	4	2	4
10_624_640	60	649.15	982.73	−37.51	−3.68%	11%	2	3	4
	120	646.20	1982.12	−39.69	−1.96%	12%	2	2	2
10_634_659	60	164.51	929.57	−47.60	−4.87%	10%	3	1	3
	120	433.82	2427.54	−15.61	−0.64%	23%	3	2	4

a conservative approach, hedging against uncertainty. However, for the case of instance 10_559_618 with σ equal to 120 s, the average computing time is the smallest over all cases involving this instance, with and without chance constraints, which might be explained by a set of feasible solutions being very small due to the large IAF separation requirement. The associated validation score is also the highest over all instances, highlighting that the solutions produced are averse to the risk of facing too small separations.

The chance-constrained variant of the problem presents smaller VSS in absolute value than the original formulation in the problem (SAA). This suggests that, when buffered IAF separations are used, there is a smaller room for the two-stage programming to improve on the deterministic EVP solutions, compared to the case with a basic IAF separation. This conclusion is consistent with the results obtained by Khassiba et al. (2020).

In summary, using buffered IAF separation seems inefficient to hedge against uncertainty for the problem of stochastic extended aircraft arrival scheduling with total deviation cost minimization, as validation scores are higher for both the stochastic and the deterministic EVP solutions.

5 Conclusions and Perspectives

This chapter introduced a high-level multi-stage stochastic optimization formulation to manage aircraft arrivals in the context of extended arrival manager systems, for which uncertainty is significant when predicting expected times to start the approach phase and landing times. It first considered a general setup involving several air network points of interest, and then presented a review of recent developments for the two-stage special case, which corresponds to recent studies on the aircraft arrival management problem taking account of practical operational constraints. Numerical experiments conducted on realistic instances based on Paris-Charles de Gaulle arrival data exhibited that the stochastic solutions are more robust than their deterministic counterparts, and can be obtained within a reasonable computing time budget using a limited number of randomly generated scenarios. Introducing chance constraints at the first stage to hedge against uncertainty led to an increase in required computing time and tended to yield overly conservative solutions. As a result, for this particular problem, this approach should not be prioritized.

Several avenues for future research were evoked, including the possibility to solve more realistic problems, relying for instance on a multi-stage formulation, and a better representation of the uncertainty affecting on-ground aircraft. More work on scenario generation and selection could be pursued to improve the solution quality while using a limited number of scenarios, allowing computing time reduction. Quantifying uncertainty in terms of random variables remains a challenging task due to the intricate interplay of random time deviations and air control orders in real delay data. To address this, adopting a distributionally robust optimization framework could provide a more accurate representation of the available information and lead to better results. Furthermore, extending the solution to a rolling horizon framework would be beneficial (Abba Rapaya et al. 2021; Furini et al. 2015). This advancement would facilitate the consideration of larger aircraft sequences and better reflect the dynamic nature of air traffic.

Acknowledgments We would like to express our deepest gratitude to Bernard Gendron, who played a major role in the development of this research. He will be deeply missed. We also would like to thank two anonymous referees for their valuable comments. The work of Fabian Bastin is supported by the Natural Sciences and Engineering Research Council of Canada [Discovery Grant 2022–04400].

Appendix: Summary of Notations Used

Sets	
Multi-stage formulation	
$\mathcal{A}^{(k)}$	Index set of aircraft at stage k
$\mathcal{A}_A^{(k)}$	Index set of airborne aircraft at stage k
$\mathcal{A}_G^{(k)}$	Index set of on-ground aircraft at stage k
$\mathcal{I}^{(k)}$	Index set of possible aircraft routes at stage k
$\mathcal{X}^{(k)}$	Feasible set for the decision vector $x^{(k)}$
Two-stage formulation	
$\mathcal{A}$	Index set of aircraft
$\mathcal{A}_G$	Index set of on-ground aircraft
$\mathcal{A}_A$	Index set of airborne aircraft
$\mathcal{I}$	Index set of IAFs
$\mathcal{A}^{i^*}$	Index set of aircraft whose initial IAF is $i^* \in \mathcal{I}$
$\mathcal{A}_G^{i^*}$	Index set of on-ground aircraft whose initial IAF is $i^* \in \mathcal{I}$
$\mathcal{A}_A^{i^*}$	Index set of airborne aircraft whose initial IAF is $i^* \in \mathcal{I}$
$\mathcal{X}$	Feasible set for the first-stage decision vector
$\mathcal{Y}$	Feasible set for the second-stage decision vector
$\mathcal{S}$	Set of uncertainty scenarios used for optimization
$\mathcal{S}_r$	rth-replication set of uncertainty scenarios used for optimization
$\mathcal{S}_V$	Set of uncertainty scenarios used for validation
Parameters	
Multi-stage formulation	
K	Number of stages
$n^{(k)}$	Number of aircraft at stage k
$s^{(k)}$	State vector at stage k
$s_a^{(k)}$	ath component of the state vector at stage k

(continued)

$p_a^{(k)}$	Position of aircraft a at stage k
$r_a^{(k)}$	Current (at stage k) reference route of aircraft a
$v_a^{(k)}$	Speed of aircraft a at stage k
$\kappa_a^{(k)}$	Status (on-ground or airborne) of a at stage k
$\chi_a^{(k)}$	Other characteristics of aircraft a such as its wake-vortex turbulence category
$\boldsymbol{\xi}^{(k)}$	Random vector capturing the uncertainty affecting the problem at stage k
Two-stage formulation	
P_a^i	Planned IAF time of aircraft a at an IAF i
P_a^{TOT}	Planned take-off time of aircraft $a \in \mathcal{A}_G$
E_a^i	Earliest time for aircraft a to reach IAF i
L_a^i	Latest time for aircraft a to reach IAF i
r^{ij}	Delay due to rerouting from IAF i to IAF j
$\hat{V}_a^O$	Unimpeded flight time from the origin airport of aircraft $a \in \mathcal{A}_G$
$\hat{V}_a^i$	Unimpeded time for aircraft a to fly from IAF i to the landing runway
$\underline{V}_a^i$	Minimal time for aircraft a to fly from IAF i to the landing runway
$\overline{V}_a^i$	Maximal time for aircraft a to fly from IAF i to the landing runway
$\overline{d}^G$	Maximum delay on-ground
$\underline{d}_a^R$	Maximal possible time saving for aircraft a during en-route phase
$\overline{d}_a^R$	Maximal possible delay for aircraft a during en-route phase
$\underline{d}_a^T$	Maximal possible time saving for aircraft a during approach phase
$\overline{d}_a^T$	Maximal possible delay for aircraft a during approach phase
$\underline{S}$	Minimal time separation between two consecutive aircraft at IAF
S_{ab}	Minimal final-approach time separation between leading aircraft a and following aircraft b
$\boldsymbol{\xi}_a$	Random variable of IAF time deviation of aircraft a
ξ_a	Realization of the random variable, IAF time deviation of aircraft a
M_{ab}	Big-M constant enabling/disabling the separation constraint at IAFs between aircraft a and b
M_{ab}^L	Big-M constant enabling/disabling the separation constraint at the runway between aircraft a and b
σ	Standard deviation of the normal random deviation $\boldsymbol{\xi}$
α	Desired probability of separation between any pair of aircraft over their IAF
$\boldsymbol{\psi}_{ab}$	Random variable defined as $\boldsymbol{\psi}_{ab} = \boldsymbol{\xi}_a - \boldsymbol{\xi}_b$
$n_{\mathcal{S}}$	Number of scenarios in uncertainty scenario set $\mathcal{S}$
$n_{\mathcal{S}_V}$	Number of scenarios in validation scenario set $\mathcal{S}_V$
$n_{\mathcal{R}}$	Number of replications of uncertainty scenario sets used for optimization

(continued)

Functions	
Multi-stage formulation	
$f^{(k)}$	Cost function at stage k
$S^{(k)}$	Separation-constraint mapping at stage-(k+1) target points
$T^{(k)}$	Time-window constraint mapping at stage-(k+1) target points
$g^{(k)}$	State transition function from stage k to stage $k+1$
$V^{(k)}$	Optimal value function at stage k
Two-stage formulation	
Q	Recourse cost
f_a^{G}	Cost function of the time deviation at gate of aircraft $a \in \mathcal{A}_G$
f_a^{R}	Cost function of the time deviation en route of aircraft a
f_a^{T}	Cost function of the time deviation in approach of aircraft a
$F_{\psi_{ab}}$	Probability distribution function of random variable ψ_{ab}
$\hat{f}$	Objective function of the SAA problem
$\bar{f}$	Average validation score over a number of replications
$\hat{v}$	Optimal value of the SAA problem
v^*	Optimal value of the true problem
$\bar{v}$	Average optimal value over a number of replications of the SAA problem
Decision vectors/variables	
Multi-stage formulation	
$x^{(k)}$	Decision vector at stage k
$z^{(k)}$	Vector of target-time variables (associated to stage-(k+1) target point)
$t^{(k)}$	Vector of target take-off time variables for on-ground aircraft at stage k
$\delta^{(k)}$	Vector of sequencing variables of aircraft at stage-(k+1) target point
$\zeta^{(k)}$	Vector of the aircraft route assignment variables at stage k
Two-stage formulation	
x	Decision vector in the first-stage problem
$\hat{x}(\mathcal{S})$	Feasible (first-stage) decision vector of the SAA problem with scenario set $\mathcal{S}$
$x^*(\mathcal{S})$	Optimal (first-stage) decision vector of the SAA problem with scenario set $\mathcal{S}$
x^*_{EVP}	Optimal (first-stage) decision vector of the expected value problem
z_a	Target IAF time of aircraft $a \in \mathcal{A}$
t_a	Target take-off time of aircraft $a \in \mathcal{A}_G G$
δ_{ab}	Sequencing variable of aircraft pair $(a, b) \in \mathcal{A} \times \mathcal{A}$
ζ_a^i	Assignment variable of aircraft a to IAF i
ϕ_{ab}	Same-IAF indicator variable for aircraft pair $(a, b) \in \mathcal{A} \times \mathcal{A}$
y	Decision vector in the second-stage problem
y_a	Target landing time of aircraft $a \in \mathcal{A}$

References

Abba Rapaya, S., Notry, P., & Delahaye, D. (2021). Coordinated sequencing of traffic on multiple en-route constraint points. In Electronic Navigation Research Institute (Ed.), *Air traffic management and systems IV* (pp. 41–57). Springer.

Antolovic, D., & Franjkovic, D. (2020). Calculation of airplane wake turbulence re-categorisation effects. *Transportation Research Procedia 51*, 179–189. https://doi.org/10.1016/j.trpro.2020.11.020

Balakrishnan, H., & Chandran, B. G. (2010). Algorithms for scheduling runway operations under constrained position shifting. *Operations Research, 58*(6), 1650–1665. https://doi.org/10.1287/opre.1100.0869

Bastin, F., Cirillo, C., Toint, Ph. L. (2006). Convergence theory for nonconvex stochastic programming with an application to mixed logit. *Mathematical Programming, 108*(2–3), 207–234. https://doi.org/10.1007/s10107-006-0708-6

Beasley, J. E., Krishnamoorthy, M., Sharaiha, Y. M., & Abramson, D. (2000). Scheduling aircraft landings: The static case. *Transportation Science, 34*(2), 180–197. https://doi.org/10.1287/trsc.34.2.180.12302

Bennell, J. A., Mesgarpour, M., & Potts, C. N. (2011). Airport runway scheduling. *4OR, 9*(2), 115–138. https://doi.org/10.1007/s10288-011-0172-x

Birge, J. R., & Louveaux, F. (2011). *Introduction to stochastic programming* (2nd edn.). Springer.

Bolić, T., & Ravenhill, P. (2021). SESAR: The past, present, and future of European air traffic management research. *Engineering, 7*(4), 448–451. https://doi.org/10.1016/j.eng.2020.08.023

Bradley, S., Hax, A., & Magnanti, T. L. (1977). *Applied mathematical programming*. Addison-Wesley. http://web.mit.edu/15.053/www/AMP.htm

Breitsamter, C. (2011). Wake vortex characteristics of transport aircraft. *Progress in Aerospace Sciences, 47*, 89–134. https://doi.org/10.1016/j.paerosci.2010.09.002

Cook, A. J., & Tanner, G. (2015). European airline delay cost reference values. Tech. Rep., Eurocontrol Performance Review Unit, University of Westminster, London.

Dear, R. G. (1976). The dynamic scheduling of aircraft in the near terminal area. Tech. Rep. FTL Report R76-9. Massachusetts Institute of Technology, Cambridge.

Dear, R. G., & Sherif, Y. S. (1991). An algorithm for computer assisted sequencing and scheduling of terminal area operations. *Transportation Research Part A, 25*(2), 129–139. https://doi.org/10.1016/0191-2607(91)90132-A

Furini, F., Kidd, M. P., Persiani, C. A., & Toth, P. (2015). Improved rolling horizon approaches to the aircraft sequencing problem. *Journal of Scheduling, 18*(5), 435–447. https://doi.org/10.1007/s10951-014-0415-8

Gerz, T., Holzäpfel, F., & Darracq, D. (2002). Commercial aircraft wake vortices. *Progress in Aerospace Sciences, 38*(3), 181–208. https://doi.org/10.1016/S0376-0421(02)00004-0

Hasevoets, N, & Conroy, P. (2010). Arrival manager - Implementation guidelines and lessons learned. Tech. Rep., EUROCONTROL, Brussels. https://www.skybrary.aero/bookshelf/books/2416.pdf

Huo, Y., Delahaye, D., & Sbihi, M. (2021). A probabilistic model based optimization for aircraft scheduling in terminal area under uncertainty. *Transportation Research Part C, 132*, 103374. https://doi.org/10.1016/j.trc.2021.103374

Ikli, S., Mancel, C., Mongeau, M., Olive, X., & Rachelson, E. (2021). The aircraft runway scheduling problem: A survey. *Computers & Operations Research, 132*, 105336. https://doi.org/10.1016/j.cor.2021.105336

Itoh, E., Brown, M., Senoguchi, A., Wickramasinghe, N. K., & Fukushima, S. (2017). Future arrival management collaborating with trajectory-based operations. In Electronic Navigation Research Institute (Ed.), *Air traffic management and systems II: Selected papers of the 4th enri international workshop, 2015* (pp. 137–156). Springer. https://doi.org/10.1007/978-4-431-56423-2_7

Itoh, E., Miyazawa, Y., Finke, M., & Rataj, J. (2021). Macroscopic analysis to identify stage boundaries in multi-stage arrival management. In: Electronic Navigation Research Institute (Ed.), *Air traffic management and systems IV* (pp. 59–76). Springer.

Khassiba, A., Bastin, F., Gendron, B., Cafieri, S., & Mongeau, M. (2019). Extended aircraft arrival management under uncertainty: A computational study. *Journal of Air Transportation, 27*(3), 131–143. https://doi.org/10.2514/1.D0135

Khassiba, A., Bastin, F., Cafieri, S., Gendron, B., & Mongeau, M. (2020). Two-stage stochastic mixed-integer programming with chance constraints for extended aircraft arrival management. *Transportation Science, 54*(4), 897–919. https://doi.org/10.1287/trsc.2020.0991

Khassiba, A., Cafieri, S., Bastin, F., Gendron, B., & Mongeau, M. (2022). Two-stage stochastic programming models for the extended aircraft arrival management problem with multiple prescheduling points. *Transportation Research Part C, 142*, 103769. https://doi.org/10.1016/j.trc.2022.103769

Kistan, T., Gardi, A., Sabatini, R., Ramasamy, S., & Batuwangala, E. (2017). An evolutionary outlook of air traffic flow management techniques. *Progress in Aerospace Sciences, 88*, 15–42. https://doi.org/10.1016/j.paerosci.2016.10.001

Liu, M., Liang, B., Zheng, F., Chu, C., & Chu, F. (2018). A two-stage stochastic programming approach for aircraft landing problem. In *15th International Conference on Service Systems and Service Management (ICSSSM)* (pp. 1–6). https://doi.org/10.1109/ICSSSM.2018.8465107

Psaraftis, H. N. (1978). A dynamic programming approach to the aircraft sequencing problem. Tech. Rep. R78-4, Massachusetts Institute of Technology, Cambridge.

Samà, M., D'Ariano, A., D'Ariano, P, & Pacciarelli, D. (2014). Optimal aircraft scheduling and routing at a terminal control area during disturbances. *Transportation Research Part C, 47*, 61–85. https://doi.org/10.1016/j.trc.2014.08.005

Scala, P., Mota, M. M., Wu, C. L., & Delahaye, D. (2021). An optimization–simulation closed-loop feedback framework for modeling the airport capacity management problem under uncertainty. *Transportation Research Part C, 124*, 102937. https://doi.org/10.1016/j.trc.2020.102937

Shapiro, A. (2021). Statistical inference. In A. Shapiro, D. Dentcheva, & A. Ruszczyński (Eds.), *Lectures on stochastic programming: Modeling and theory* (chap. 5, 3rd edn., pp. 151–221). Springer.

Shapiro, A., & Homem-de Mello, T. (2000). On the rate of convergence of optimal solutions of Monte Carlo approximations of stochastic programs. *SIAM Journal on Optimization, 11*(1), 70–86. https://doi.org/10.1137/S1052623498349541

Shone, R., Glazebrook, K., & Zografos, K. G. (2021). Applications of stochastic modeling in air traffic management: Methods, challenges and opportunities for solving air traffic problems under uncertainty. *European Journal of Operational Research, 292*(1), 1–26. https://doi.org/10.1016/j.ejor.2020.10.039

Sölveling, G., & Clarke, J. P. (2014). Scheduling of airport runway operations using stochastic branch and bound methods. *Transportation Research Part C, 45*, 119–137. https://doi.org/10.1016/j.trc.2014.02.021

Sölveling, G., Solak, S., Clarke, J. P., & Johnson, E. (2011). Runway operations optimization in the presence of uncertainties. *Journal of Guidance, Control, and Dynamics, 34*(5), 1373–1382. https://doi.org/10.2514/6.2010-9252

Tielrooij, M., Borst, C., Van Paassen, M. M., & Mulder, M. (2015). Predicting arrival time uncertainty from actual flight information. In *Proceedings of the 11th USA/Europe Air Traffic Management Research and Development Seminar, FAA/EUROCONTROL* (pp. 577–586).

Vié, M. S., Zufferey, N., & Leus, R. (2022). Aircraft landing planning under uncertain conditions. *Journal of Scheduling, 25*(2), 203–228. https://doi.org/10.1007/s10951-022-00730-0

SPRINGER NATURE

GPSR Compliance

The European Union's (EU) General Product Safety Regulation (GPSR) is a set of rules that requires consumer products to be safe and our obligations to ensure this.

If you have any concerns about our products, you can contact us on ProductSafety@springernature.com

In case Publisher is established outside the EU, the EU authorized representative is:

Springer Nature Customer Service Center GmbH
Europaplatz 3
69115 Heidelberg, Germany

The manufacturer's authorised representative in the EU is Springer Nature Customer Service Centre GmbH, Europaplatz 3, 69115 Heidelberg, Germany. If you have any concerns regarding our products, please contact ProductSafety@springernature.com

Printed and bound by CPI Group (UK) Ltd, Croydon, CR0 4YY
01/10/2026
02225862-0015